REVOLUTION IN MIND

REVOLUTION IN MIND

The Creation of Psychoanalysis

GEORGE MAKARI

HARPER

An Imprint of HarperCollinsPublishers
www.harpercollins.com

HarperCollins books may be purchased for educational, business, or sales promotional use. For information, please write: Special Markets Department, HarperCollins Publishers, 10 East 53rd Street, New York, NY 10022.

A continuation of this copyright page appears on page 577–578.

FIRST EDITION

Designed by Joseph Rutt

Library of Congress Cataloging-in-Publication Data is available upon request.

ISBN: 978-0-06-134661-3

08 09 10 11 12   NMSG/RRD   10 9 8 7 6 5 4 3 2 1

For Arabella, Gabrielle, and Jack
my kind of wonderful

Contents

REVOLUTION IN MIND

Prologue

When the twenty-nine-year-old doctor stepped off the train in the fall of 1885, he was a failure. Ambitious but poor, he had tried his hand at a number of sciences but still had nothing to secure his future. As he made his way onto the boulevards of Paris, he left behind him a growing storm of controversy regarding his claims for a new wonder drug called cocaine. With hopes of marriage to his fiancée pressing upon him, the doctor accepted what now seemed unavoidable: he would not become a university scientist and would have to open a medical practice to earn a living. He might be forced to emigrate to England or Australia or America. But first, he would try to make a living in his hometown of Vienna. Before that inevitable fate, in a last gasp of high-minded scientific aspiration, he had applied for and received a grant to study in Paris. What he would discover in that city would propel him forward on a long, winding journey that led to one of the great intellectual revolutions of the twentieth century.

Or, perhaps not.

Today, this young man's identity and legacy are hotly disputed. Sigmund Freud was a genius. Sigmund Freud was a fraud. Sigmund Freud was really a man of letters, or perhaps a philosopher, or a crypto-biologist. Sigmund Freud discovered psychoanalysis by delving deep into his own dreams and penetrating the mysteries of his patients. Sigmund Freud stole most of his good ideas from others and invented the rest out of his own odd imagination. Freud was the maker of a new science of the mind that dominated the West for much of the twentieth century. Freud was an unscientific conjurer who created a mass delusion. Who was Freud? Who are the Freudians, Freudian psychoanalysts, and psychoanalysts? And who are we, those of us in the West who have found the terms and concepts of psychoanalysis permeating our everyday language, changing on

the most intimate levels the ways in which we think about ourselves, surrounding us in what the poet W. H. Auden called "a whole climate of
opinion"?

For many years, these questions seemed to have been answered. The
history of psychoanalysis had been handed down by Freud's compatriots.
They portrayed the father of their field as a man of stunning originality,
great virtue, and nearly unfathomable genius. Freud discovered everlasting truths about the mind, it was said, and these truths had been preserved by his followers. In postwar America and in parts of the Western
world, this Freud became an essential coin of intellectual life. But over the
last thirty years, these standard accounts have been increasingly questioned. New documents, new sources, and new histories have made the
older, adoring portrait more improbable. As Freud's genius and virtue
were cast into doubt, contemporary psychoanalysts struggled with numerous forces that seemed to undermine their enterprise—ranging from
improved pharmaceuticals and the rise of cognitive neuroscience to the
exigencies of insurance companies. Soon, a new coin began to circulate. It
read: "Freud is dead." As the twenty-first century unfolds, it would seem
we have to choose: Freud as everlasting genius, or Freud as relic and
fraud.

This book offers a different choice and another kind of history. In all the
recent tumult over Freud, it has often gone unnoticed that these seemingly
antithetical accounts are flip sides of the same coin. The most devout admirers and fiercest detractors of Sigmund Freud both assume that the answers to the critical questions posed by psychoanalysis can be found in the
biography of the young man who stepped off that train in Paris in 1885.
Consequently, while hundreds of Freud studies and biographies have been
written pro and con, no broader account has yet been given of the rise of
psychoanalysis in its birthplace: western and central Europe. As a result, a
wide array of ideas, experiences, judgments, and debates have disappeared.
We have lost a good deal of the logic and illogic of what was a very human
undertaking, but more than that we have lost a world, a world not so
distant, but one made more remote by the European slaughters of the
twentieth century. It was a world that made Freud, the Freudians, and
the psychoanalysts, and it was a world in part made by them.

Psychoanalysis emerged between 1870 and 1945 in European communities that were ultimately decimated and dispersed. While psychoanaly-

sis survived on foreign shores, it was severed from its own past. Remnants of a great discussion on the nature of the mind and its troubles continued in these new lands without the contexts that had once given these debates broader definition. With the rich tapestry of Mittel Europa shredded and Germany in ruins, it became simpler to imagine that one immortal figure was responsible for this strange new mode of understanding, whether it was a science or a massive hoax.

In 1993, *Time* magazine captured this odd state of affairs when it ran a cover story bearing the ghoulish headline: "Is Freud Dead?" Not to be outdone, thirteen years later *Newsweek*'s cover declared: "Freud Is Not Dead." After leaving the earth one autumn day in 1939, a ghostly Freud, it would seem, still walked outside of time. And yet, Sigmund Freud was very much a man *in* time. As a large number of historians have now shown, many aspects of Freud's thinking were dependent on ideas put forth by others in medicine, politics, theology, literature, philosophy, and science, ranging from the ancients to his contemporaries. This revisionist work has been so rich, so plentiful, and at times so promiscuous in its conclusions that it has been difficult to synthesize. When we step back and take in all these attributions, they can appear to cancel one another out. If Sigmund Freud really derived psychoanalysis from Aristotle, Sophocles, and the Bible, as well as Shakespeare, Wordsworth, Goethe, and Nietzsche, not to mention Johann Herbart, Ernst Brücke, and Pierre Janet (to name but a few), it seems only fair to conclude that this strange amalgam was his alone.

But such is not the case. Psychoanalysis emerged at a time when Europeans were dramatically changing the ways they envisioned themselves. It shot forth from a mass of competing theories that had all been thrown up by seismic shifts in philosophy, science, and medicine. This book is an attempt to take in those grand shifts and locate the specific origins of psychoanalysis as a body of ideas and a movement. A broad canvas is required to locate the particular influences that defined psychoanalysis, for Sigmund Freud did not derive the field's central tenets from any single thinker or field. Rather, he pulled together new ideas and evidence from a number of domains to fashion a new discipline. The goal was to win for science the traditional object of humanist culture—the inner life of human beings.

Freed from religious doctrines of the soul, many late nineteenth-century Europeans struggled to reconcile their own inner experience with the demands of scientific positivism, the mechanistic universe of Isaac Newton, and the evolutionary biology of Charles Darwin. They tried to make sense of what it meant, alongside all that, to have an interior world, a mental life, to be conscious and psychologically human. Freud was one of many late nineteenth- and early twentieth-century intellectuals who responded to this confusion by trying to forge a science of inner life. The rules for this new hybrid science would not stem from evolutionary biology or Newtonian physics alone, for there was something peculiar and distinctly problematic about this endeavor. How could one make an objective science of subjectivity? For centuries, Western science made great strides by insisting that reliable knowledge was only possible if the object of study was observable or quantifiable. But what about mental life, a realm that seemed to be neither? Such a vexing domain might be simply dismissed as unreal, if everyone didn't already know that the psychic realm existed, if only for themselves in their own consciousness. This was a critical conundrum that would-be scientists of the mind faced. Sigmund Freud was one of a number of thinkers who tried to solve this riddle, and ultimately his solutions won him followers and a great future.

Throughout this book, Freud will play a large part, as he must. But this is less the story of one man than it is the history of a series of heated intellectual contests. In the course of these struggles, individuals banded together, formed alliances, and faced off. In the end, these pitched disputes defined a way of thought that came to be closely allied with Freud's name. Alongside the doctor from Vienna, we will meet the creative men and women who contributed greatly to this new way of thinking about the mind. Some were skeptics and naysayers; others were innovators who were later marginalized, defamed, or just forgotten. Over time, Freud became the name for a whole community of seekers. Consequently, it has been difficult to discern the essential considerations that went into the making of psychoanalysis. They have often seemed to be only a question of one man's biography.

By pulling back our focus from Freud, however, we find a new history emerging. The making of psychoanalysis can be divided into three closely intertwined, sequential phases. First, Sigmund Freud created a scientifically tenable theory of the mind and a model for psychical therapy out of his engagement with three preexisting nineteenth-century intellectual communities. Freud immersed himself in these different fields of study, taking a great deal lock, stock, and barrel from each, while renaming and reconceptualizing critical elements along the way. He proposed creative solutions to long-standing problems that split those older fields, and then, in 1905, he pulled together an overarching synthesis that consolidated his prior work into a new Freudian field. Over the next decades, men and women migrated from those other disciplines to Freud. In this way, it can be said that Sigmund Freud did not so much create a revolution in the way men and women understood their inner lives. Rather, he took command of revolutions that were already in progress.

The second phase commenced during the first years of the twentieth century when a growing band of Freudians formed and began to spread their ideas throughout Europe and America. After only a decade, this community fractured and fell apart amid accusations that it had become authoritarian and unscientific. The schisms that resulted in the departure of Eugen Bleuler, Carl Jung, and Alfred Adler, among others, exposed the highly tenuous nature of the knowledge claims that were supposed to hold the Freudians together.

The third and last phase of this history came in the wake of these splits. After the Great War, a newly constituted community emerged that was not so much Freudian as more generally psychoanalytic. During the 1920s and 1930s, this pluralistic community drew up different boundaries and central commitments in an effort to stabilize their field and better manage the ever-troubling question of how to know the darkest recesses of another's inner world. The answers they settled on would help shape psychoanalysis for the next half century.

As the twenty-first century begins, there are compelling reasons to return to the great debates that defined psychoanalysis. The field is now in turmoil. Its future is said to be in doubt. Some believe psychoanalysis is a hopeless pseudoscience. Others want to save it by shoring up its scientific claims. Still others believe salvation will come only when psychoanalysts recognize their endeavor is not scientific but akin to work in the humanities. And yet despite this confusion, despite all its extravagant flaws, psychoanalysis remains the most nuanced general account of interior life we possess. Read between the lines of biographies, novels, journalistic portraits, and screenplays and you will find explanations of human character that are deeply, inextricably indebted to this history. Talk to the record numbers of people in some form of therapy derived from psycho-analysis, and you will hear echoes of this past. When we speak about who we are, wittingly or not, we often use the language of psychoanalysis.

Revolution in Mind is a historical examination of the core questions at the heart of this most influential theory of human inner life. Many of those questions remain unresolved to this day, for this is an unfinished story of a complex, perhaps impossible endeavor. It is the story of a group of doctors, philosophers, scientists, and writers trying to grasp that most ephemeral and yet maddeningly obvious thing: the mind. It is the story too of a political world that for a short, fertile time allowed men and women the freedom to examine the potentially explosive questions of what makes us human. And it is the story of how in the process some failed, some fell into despair, while others tried to refine their methods, attempting again and again to map out that place we all hide in our heads.

PART ONE

Making Freudian Theory

ONE

A Mind for Science

It's wrong to say I think. Better to say: I am thought . . . *I* is an *other*.

—Arthur Rimbaud, 1871

I.

As the Enlightenment cast scientific rationalism up to celestial bodies and down to squirming microscopic life, there was one object that seemed impossible to penetrate: the mind. The French champion of science and rational skepticism, René Descartes, established this in his *Discourse on Method* when he declared the "I" was beyond rational inquiry, being nothing other than the immaterial soul described by Church fathers. Religious beliefs regarding inner life would prove durable and influential, but during the second half of the nineteenth century such notions began to lose some credence, and in that ceded ground a science of mental life took root.

When Sigmund Freud arrived in Paris in 1885, France had established itself as the center for cutting-edge research on psychological matters. Few scientists in Berlin or Vienna bothered to investigate the psyche, the "I," the soul, the self, or the mind—realms tainted by religion or speculative metaphysics. In Paris, however, scientists were drawn to the study of the inner world, thanks to a new method. That method, the *psychologie nouvelle,* transformed France into a hotbed of study for somnambulism, human automatisms, multiple personality, double consciousness, and second selves, as well as demonic possessions, fugue states, faith cures, and waking dreams. The marvelous and miraculous made their way from isolated villages and abbeys and carnival halls, from exorcists and charlatans and old mesmerists, into the great halls of French academic science.

The birth of this new psychology came as France itself was being reborn. Nearly a century after its revolution, the French suffered a humiliating defeat to the Prussians in 1870, resulting in the fall of Emperor Louis Napoleon III and the birth of the Third Republic. Many blamed this military debacle on French science and its failure to keep up with the advances made in German lands. French Republicanism combined anticlericalism with a commitment to revitalizing science. As the authority of the French Catholic Church to dictate thinking on the soul waned, a bold, new scientific psychology emerged.

At the time, psychology was considered a branch of philosophy, not science, but the champion of the *psychologie nouvelle*, Théodule Ribot, set out to change that. Born in 1839, the son of a provincial pharmacist, Théodule was forced by his father to become a civil servant. After three years of drudgery, he announced that he was off to Paris to try and gain entrance into the elite École Normale Supérieure. Two years later, Ribot won a spot at that university, where he quickly took a dislike to the reigning spiritualist philosophy championed by Victor Cousin. A strange brew of reason and faith, Cousin's psychology mixed notions of the soul and God along with naturalistic descriptions of the mind.

Ribot could not abide this. Despite being denounced by local clergy, he set out in search of a method that might make psychology fully amenable to scientific inquiry. Plunging into the writings of British thinkers, Ribot emerged in 1870 with *Contemporary English Psychology (The Experimental School)*. Despite the dry title, the book opened with a spirited manifesto that would define psychology in France for decades to come.

Conventional notions of philosophy and science both made objective study of the mind impossible, Ribot explained. He attacked philosophies like those of Descartes and Cousin, insisting that psychology must rid itself of metaphysics and religion. Psychologists could not comment on transcendental questions, nor honestly speak of the soul. And they could not rely on the armchair methods of philosophy, but needed to employ the methods of natural science.

For all this, Ribot had an eager audience. Many of his contemporaries were ready to jettison older philosophies of the soul for naturalistic study. But how was psychology to be remade into a science? To answer that question, Ribot took on a different set of critics, led by the fiery prophet of science, Auguste Comte. Despite leading a marginal erratic life, Auguste Comte achieved extraordinary influence over late nineteenth-century European intellectuals, politicians, and scientists. In 1855, the Frenchman laid out a history of all human knowledge, declaring that the most primitive stage was theology, myth, and fiction, which then progressed to a second stage of metaphysical abstraction. In the end, philosophical notions would be surpassed by the most perfect state of knowledge which was scientific and "positive." Hence Comte's program was dubbed positivism. With the rise of the Third Republic in 1870, Comte's vision of progress was embraced by the French political elite as a model for both science and social reform.

Comte's thinking posed a great dilemma for Ribot, for the founder of positivism believed an insoluble problem lay at the heart of psychological knowledge. Psychologists relied on self-observation to get at things like thought, feeling, and desire. Such interior observation—the knowledge that came from a mind looking in at itself—was exactly what constituted subjectivity. Therefore, Comte concluded psychology could never be objective, and his quick survey of prior efforts seemed to support this damning conclusion:

> After two thousand years of psychological pursuit, no one proposition is established to the satisfaction of its followers. They are divided, to this day, into a multitude of schools, still disputing about the very elements of their doctrine. This interior observation gives birth to almost as many theories as there are observers.

In the second half of the nineteenth century, anyone who sought to establish principles for a scientific psychology—including John Stuart Mill in England, Franz Brentano in Austria, and William James in the United States—would have to take on Auguste Comte's devastating indictment.

Comte pointed positivists down the only tenable path he saw for psychology: the field should restrict itself to observable signs such as physiognomy or behavior. To the embarrassment of his admirers, Comte thereby predicted that the future of psychology lay in phrenology. Initially conceived as the study of brain localization, phrenology had degenerated into quackery and the study of cranial lumps and bumps, based on the belief that these protuberances reflected mental capacities and deficits. By the time Ribot took up his pen, Comte's suggestion was ridiculous.

Furthermore, Ribot was unwilling to gut psychology of thought, emotion, and all other inner experiences. Instead, he proposed a different kind of science of the mind, in which lawful claims might be made about that dark and shifting domain. Psychology needed to carefully mix introspection and external observation. Introspection was critical to get at mental phenomena, but those subjective impressions needed to be stabilized and corroborated by a myriad of methods, including "the perception of signs and gestures, the interpretation of signs, induction from effects to causes, inference, reasoning by analogy." Arguments between subjective and objective methods were sterile: Ribot's scientific psychology required both.

That was Ribot's hybrid method, but he still needed to circumscribe his object of study. If not overt behavior or cranial bumps, what would define the psyche in his psychology? Instead of taking any one approach, Ribot proposed three related perspectives. Inner experience could be studied by a bare-bones assessment of how perceptions, ideas, and feelings were linked, synthesized, and brought before consciousness. Such an "associational psychology" had been pioneered in seventeenth-century England by John Locke and David Hume, the philosophers who also founded scientific empiricism. The two bodies of thought were related. Empiricism sought to explain how humans came to know the world around them, placing emphasis on observation and the causal, synthetic connections that could be forged through human experience (even staged human experiences or experiments). Attempts to explain how humans came to know the outer world inevitably led these philosophers to model our knowing machine, the mind, and in this way inaugurated associational psychology.

Later developed by David Hartley, James Mill, John Stuart Mill, and Alexander Bain, associationalism did away with assumed, inborn faculties like reason, imagination, or morality, instead seeking to show how such complex functions could emerge solely from the combination of basic psychic elements like ideas and sensory perceptions. They thought of the mind as a loom, weaving together sights, sounds, ideas, and feelings into a unified whole. Of course much could go wrong in this process; mis-associations accounted for human errors, illusions, and delusions. John Locke thought such false linkages as common as unreason, as common as childhood, as common as the everyday madness of "most men."

Associationalism held great advantages for a scientific psychology, for it did not speak of the soul or insist on hypothetical faculties that in the end often seemed arbitrary. Instead, this theoretically minimal tool allowed for a close analysis of the fleeting currents of inner experience. Furthermore, this theory of the mind cohered nicely with the (implied) mind at work in empirical science. To know another's inner world, it sufficed to explore and draw associations about another person's associations. Ribot predicted—rightly it turned out—that associationalism would provide a sturdy framework for psychological experimentation.

This British doctrine, however, also had limitations. Associationalists pressed forward only one simple precept regarding emotion: humans were pleasure seeking and pain avoidant. Pleasure and pain, they argued, could serve as the building blocks for complex human passions like love, hatred, hope, and sorrow. Despite this powerful notion, as Ribot pointed out, associationalism generally led to a focus on the inner play of ideas, more than "the sentiments, the emotions, affective phenomena in general." Secondly, most associational psychology assumed that experience solely furnished a mind that was otherwise bare. To offset this prejudice, Ribot suggested a second focus for psychology: heredity. In 1873, Ribot published *Heredity: A Psychological Study of Its Phenomena, Laws, Causes, and Consequences,* where he argued that evolution and biologic inheritance accounted for a good deal of psychological functioning.

With that, Ribot created a sturdy framework that organized French psychological inquiry for the next thirty years. Psychological content would be studied by associational tenets, while claims regarding psychic capacities and functions would be based on hereditary theories. In addition, he added a final leg to this research program. Since lab experiments were difficult to perform on the brain and mind, Ribot proposed that mental disease would act as the experimental arm of psychology: "(T)he morbid derangements of the organism that produces intellectual disorders; the anomalies, the monsters of psychological order, are for us like experiments prepared by Nature and all the more precious since experimentation is more rare."

Théodule Ribot's solutions were adopted by many, and before long he

sat at the center of a growing interdisciplinary community of psychological researchers. Burning with new ideas and surrounded by an array of
brilliant colleagues, he exclaimed: "What a cerebral orgy!" Appointed
editor of the *Revue philosophique de la France et de l'étranger* in 1876, Ribot
proceeded to spread *la psychologie nouvelle* along a network of alienists,
doctors, philosophers, and scientists in Europe and the United States. Between 1881 and 1885, he published *Diseases of Memory*, *The Diseases of the
Will*, and *The Diseases of Personality*. All were wildly popular, going
through twenty to thirty-six editions in France alone. In 1888, Ribot was
awarded a chair in experimental psychology at the prestigious Collège de
France. Fourteen years later when he retired, his successor, Pierre Janet,
lauded him as the man most responsible for defining French psychology
and giving it such a highly original, rich orientation.

Janet did not exaggerate. Between 1870 and 1900, Ribot forged a scientific psychology that made France famous. But his fame would be eclipsed
by a physician who for years seemed to have no respect for psychology. In
1884, Ribot innocently reported that he had found an easy way to get new
articles for the *Revue*: "Charcot and his students (the Salpêtrière School)
would very much like to make a foray into physiological psychology.
Since I see them constantly and am on very good terms with them, I have
a good foothold there."

THE FRENCHMAN JEAN-MARTIN CHARCOT was one of the most fabled physicians in Europe, but before 1884 he had shown little interest in Ribot's
line of work. A physician, neurologist, and strict positivist, he believed the
mind was simply an epiphenomenon of brain functioning, nothing more
than the froth stirred up by the sea. But as Ribot himself discovered, the
famed neurologist had been forced to reconsider this assumption, and in
the process he began to make extraordinary claims about psychic life that
would captivate medical circles throughout the Western world.

Born and educated in Paris, Charcot saw his career take off in 1862
when he was appointed physician to the Salpêtrière, a sprawling complex
housing some 5,000 women, many of whom were insane, demented, destitute, or deemed incurable. A follower of Comte, Charcot and his team
of doctors proceeded to study the chaotic mass of suffering they found.
While many physicians hoped lab study of diseased tissue would make

medicine more scientific, Charcot adopted positivist methods for clinical medicine and advocated close observation of patients as a way of newly classifying diseases. By 1870, Charcot and his coworkers had succeeded in giving classic descriptions of amyotrophic lateral sclerosis and multiple sclerosis and made important contributions to the study of rheumatism, gout, arthritis, and locomotor ataxia.

Charcot then entered the dubious terrain of the *névroses*, or "neuroses" as the English called them. Defined by what they were not, the *névroses* were nervous disorders that showed no brain or spinal lesions. A tangle of difficult-to-define symptom complexes and disorders, they included one of the oldest and most mysterious of them all: hysteria. According to his assistant, Pierre Marie, Charcot began to investigate this enigmatic disease for the most serendipitous of reasons. Hospital administrators needed to repair a decrepit facility, so they moved a ward of epileptics into one filled with mentally ill women. Suddenly, the female hysterics began having seizures. The doctors now faced the quandary of trying to distinguish hysterical seizures from real ones. With that, Charcot and his coworkers were forced to confront an even more vexing question: what was hysteria?

A diagnosis first made over 2,500 years ago, hysteria was long thought to be a woman's disease. As the etymology of the word denoted, this affliction was first considered a wandering of the womb, and in the first half of the nineteenth century, hysteria remained tied to female sexuality. That began to change when in 1859, the Parisian physician Paul Briquet published a landmark study. Examining over four hundred cases, he found that hysteria, while predominantly found in females, was not exclusively so; for every twenty female cases, Briquet found one male case. The doctor also reported a low incidence of the disease among nuns and a high incidence in prostitutes, refuting the old idea that sexual frustration caused this illness. Hysteria, he concluded, was a neurosis of the brain that disrupted emotional expression. Briquet further emphasized how poor heredity worked in combination with violent emotions to set the disease in motion. While many gynecologists still insisted hysteria was due to *une chose génitale*, Briquet allowed neurologists and psychiatrists to see this disorder in these newer terms.

Following Briquet and others, Charcot took up this Proteus of illnesses. A shifting kaleidoscope of bewildering symptoms that long frustrated attempts at classification, hysteria appeared to have no objective pattern.

Many thought it was not a disease at all but rather female subterfuge and fakery. Jean-Martin Charcot found order where others saw none. Hysterics suffered from attacks that had discrete pathophysiological stages, he concluded after much study. In its purest state, "*grande hystérie*" was marked by the "*grande attaque*," in which sufferers marched through an elaborate four-stage sequence. The symptoms were readily observable; the cause was poor heredity. Nothing needed to be said about the hysteric's thoughts or feelings, her psychology, her subjective world. Hysteria could be understood by objectively observable outward signs alone.

Word of Charcot's achievement spread. Astonished onlookers filed into the auditorium at the Salpêtrière, where hysterics writhed and shook and froze during their elaborate attacks. Charcot and his group began to photograph hysterics in different stages of their illness, in the hope that this would be scientific proof, their version of the pathologist's microscopic slide.

A hysteric in a state of "provoked somnambulism." Salpêtrière Hospital, Paris, circa 1879.

Charcot's study reached beyond medical circles. Close to positivists and reformers in the government, he shared the belief that progress would come when religion yielded to science. During the first years of the Third Republic when clerical forces still had a foothold in political circles, spies who attended Charcot's classes reported his frequent anticlerical jokes. No spy, however, was needed to recognize the political impact of studies that pathologized ecstatic and holy visions. It was only necessary to read Charcot's colleague, Désiré-Magloire Bourneville, who predicted that before long both the miraculous and the demonic would be exposed as simply hysterical.

A demystifying, anticlerical agenda may have also encouraged Charcot to take his next fateful turn. In 1878, the neurologist took up the study of hypnotism. A century earlier, a Viennese doctor named Franz Anton Mesmer had arrived in Paris, having fled his hometown amid charges of quackery and sexual impropriety. Mesmer became a sensation in Paris with dramatic cures attributed to the invisible force of animal magnetism, but the French Academy of Sciences convened a panel to judge the merits of his claims and condemned him as a seducer and a fraud, thus pushing the study of altered mental states into the backwoods of France for decades to come.

The distinguished French physiologist Charles Richet reignited mainstream interest in mesmeric states during the 1870s. Using the British doctor James Braid's term, Richet attributed "hypnosis" to a physiological dysfunction. In 1878, Charcot brought his reputation to the study of these bizarre states, and five years later he appeared before the same Academy of Sciences that condemned Mesmer, to demonstrate how his own study of hypnotism would be different. Hypnotism was a physiological and neuropathological disruption, not some spooky mesmeric power. Two of Charcot's allies, Alfred Binet and Charles Féré, explained that unlike prior experimenters, they would not even bother with "complex psychical phenomena," for these lacked the material characteristics that would place them beyond question. And so, a revived study of hypnosis became scientifically legitimate, thanks to this strict emphasis on bodily symptoms. Speaking to the academy, Charcot detailed the dramatic contractures and seizures of the *"grand hypnotisme,"* all of which proved hypnosis was neither miraculous nor quackery, but simply the sad result of an abnormal nervous state.

With remarkable speed, Charcot had conquered two monumental medical mysteries: hysteria and hypnotism. All the while, he studiously kept his distance from magical interpersonal forces or obscure psychological influences that might in any way hint of immaterial, invisible forces. These mental states were all the result of neurological disruption. Causality was a one-way street that ran from body to mind. Or so Charcot thought.

The transformation of Jean-Martin Charcot began rather simply. He and his coworkers discovered that if they suggested to a hypnotized hysteric that her arm was paralyzed, a paralysis would ensue. Incredibly, in this strange state, the idea of a paralysis seemed to create a paralysis. To explain how this could possibly be, one needed a model for how an idea could affect the body. That is to say, Charcot needed a psychology. And with that, the renowned positivist and his followers headed straight into Auguste Comte's forbidden garden.

SIGMUND FREUD ARRIVED at the Salpêtrière in 1885 as Charcot and his team had become engrossed in the study of how unconscious ideas and emotions might cause neurological symptoms. Adopting Ribot's model, the French neurologist employed associational psychology alongside hereditary explanations. A hypnotic suggestion, he concluded, allowed an idea to enter the mind in a disassociated, unconscious, quite isolated state. Suggestions fell into a space distinct from the interwoven collection of associations that normally made up consciousness. In that dark region, disassociated ideas seemed to act on the body freely and automatically.

Notions of unconscious physiological action were commonplace in the late nineteenth century. In fact some, like William Carpenter in England and William James in America, speculated that human beings might be automata wholly governed by unconscious physiology. But Charcot's explanation of hypnotic suggestion did not rely on physiology but rather psychology. Unconscious ideas could take hold of a body. Suggest to a hypnotized hysteric that her leg was paralyzed and *voilà*! Without her knowing what was happening, the leg went dead.

The Salpêtrière doctors grew particularly fascinated by the strange cases of two men they named Pin and Porez. These French laborers presented with paralyses that were, anatomically speaking, impossible. At the same time, Pin and Porez didn't seem to be faking their illnesses. Perhaps they were hysterics under the sway of unconscious ideas. But neither man was hypnotizable, and for Charcot that meant they could not be hysterics. He believed all hysterics were hypnotizable; it was one of their most salient characteristics.

Pin and Porez suffered blows to their arms, but these injuries were too minor to result in real nerve damage. Each man shook himself off and went about his life, only to suffer a paralysis days later. Fascinated, Charcot examined the men and concluded that their traumas had acted on their minds as well as their bodies. He set out to investigate and was stunned to find that a sharp blow to the arm of a hysteric under hypnosis could create the same symptoms that afflicted Pin and Porez. The blow by itself had acted as if it were a verbal suggestion.

These were all psychical paralyses or paralyses of the imagination, Charcot concluded. In the cases of Pin and Porez, he reasoned that the shock of the initial trauma sent their nervous systems spiraling into something like a hypnotic state, at which point each man entertained the idea: *I can't move my arm.* This panicky thought normally would be greeted by a host of associated ideas, including reassuring ones that might follow testing the arm and seeing that it seemed fine. But "the annihilation of the ego" produced by the traumatic shock left that frightening idea—*I can't move my arm*—isolated, unconscious. From there, it worked with all the impunity of a hypnotic command. His fear of becoming paralyzed acted as an autosuggestion, and the paralysis became real.

Imagination, it seemed, could make a man ill. But only in cases of trauma. Borrowed from the lexicon of surgery, trauma emerged in nineteenth-century psychiatry and neurology to account for nervous shocks like "railway spine" and "railway brain," which were thought to be brought on by the jarring rides in that new monster, the locomotive. It was accepted that a traumatic shock might disrupt associative processes in the brain. But Charcot's focus on self-suggestion was novel and created confusion. If autosuggestion had its origin in the patient's own mind, how did that idea end up outside the confines of consciousness? Hypnosis demonstrated how external suggestions could land in the unconscious, but how could this be with one's own ideas? Charcot reasoned that a traumatized mind was prone to dissociation, so that ideas peeled off from the stable matrix of conscious associations. Moreover he suggested that strong emotions like rage or terror could serve as traumas, resulting in dissociation and self-suggestion.

Charcot's growing psychological theory held fascinating therapeutic implications. If an idea could make a paralysis, then perhaps an idea could cure one. From 1885 to 1886, Charcot and his colleagues tried a talking treatment on Pin and Porez:

> In the first place we acted, and continue to act every day on their minds as much as possible, affirming in a positive manner a fact of which we are ourselves perfectly convinced—that their paralysis, in spite of its long duration, is not incurable, and that, on the contrary, it will certainly be cured by means of appropriate treatment . . . if they will only be so good as to aid us.

Therapeutic suggestion aimed to counter autosuggestion and alleviate symptoms, though this was no cure. Charcot never wavered from his belief that traumatic neurosis could only befall individuals tainted by degenerative heredity. That no talk could remedy.

When Freud arrived in Paris, a whole community of French psychologists and physicians were busy tracking inner life by investigating associations and dissociations, the role of heredity, and the light that psychopathology might throw on normal mental functioning. Having first conquered hysteria and hypnotism without entering the scientifically iffy zone of psychology, Jean-Martin Charcot and his coworkers found themselves discussing the role of unconscious psychic states in cases of psychic automatism, dual consciousness, multiple personality, and fugue states. Doctors from around Europe flocked to Paris to witness stunning cases of hypnosis, strange dances performed by hysterics, and bizarre ailments provoked by ideas. They came to learn of studies based on the scientific method of the *psychologie nouvelle*, studies underwritten by the authority of men like Ribot and Charcot, studies based on a great deal that was about to crumble, for something had gone terribly wrong.

Sigmund Freud in 1885, the year he traveled to Paris to study with Charcot.

II.

WHEN SIGMUND FREUD received a traveling grant issued by the University of Vienna Jubilee Fund, he was a man who had tried on a number of futures, and none had quite fit. Having aspired to a career in zoology, then physiology and neuroanatomy, he had turned to medicine, where he considered specialties like neurology and psychiatry. At twenty-nine, he was still impoverished, no longer so young, with no prospects for a university position. His fiancée had been waiting for him to be able to afford marriage. Desperately looking for a break, he had set his hopes on a new histological method for staining nerve cells and then put his faith in a new pharmacologic agent called cocaine. But the wondrous effects of cocaine started to show a dark side, and so, having heard of Charcot's researches on the neuroses, Freud came to Paris to try again.

Born to Jewish parents in Freiberg, Moravia, on May 6, 1856, Sigismund Freud was actually his name. When the boy was four, his family moved to the capital of the Austro-Hungarian Empire, Vienna, and there "Sigmund" attended the Leopoldstädter Gymnasium, where he proved an extraordinary student. Schooled in Latin and Greek and the classics such as Ovid, Horace, Cicero, Virgil, Sophocles, Homer, and Plato, he quickly made his way to the front of his class. As a Jew, he was a member of a mistreated, marginalized minority, but these were liberalizing years in the Habsburg Empire. Emperor Franz Josef had increased civil rights for Jews, and had even included a number of Jewish ministers in his cabinet. These men were heroes to Freud and his young Jewish friends. Drawn to historical figures like Brutus and Hannibal, the boy imagined himself a defender against tyranny and considered a future in the law. He declared himself an antiaristocratic, anticlerical republican, and a staunch materialist. After matriculating at the University of Vienna in the fall of 1873, the youth proved himself to be outspoken, even when that meant standing in the opposition. Though supported by many, he confronted anti-Semitism all around him and once faced down a small mob forming against the "dirty Jew."

By the time he entered university life, however, Sigmund was no longer primarily interested in politics and law. Captivated by Goethe's essay on nature, he shifted his plans to science and medicine. After enrolling in the medical curriculum, he signed up for anatomy, chemistry, "General Biology and Darwinism," botany, physiology, and physics. In the winter of 1874, he also began studies in philosophy, the only nonscientific discipline he pursued, working with a professor who had recently taken refuge in Vienna, Franz Brentano.

A Catholic priest and philosopher, Brentano became estranged from the church after its declaration of papal infallibility. His loud disdain for this doctrine made his academic position in Würzburg increasingly untenable. At the same time, Brentano discovered Comte and the work of British associational philosophers. Brentano resigned his professorship, left the church, and began planning a new life for himself. His ticket would be a work on scientific psychology.

Brentano, like Ribot, strove to separate psychology from philosophy without letting the whole enterprise collapse before positivist notions of science. To do so, he took up the problem of introspection. In his 1874 *Psychology from an Empirical Standpoint*, Brentano took pains to distinguish introspection from inner perception. The former was a kind of trained inner observation that some claimed approximated empirical observation of the outer world. Brentano pronounced all this impossible rubbish. We cannot stand outside our own minds to observe our minds with our minds. But inner perception was a completely different matter. That was as common as feeling joy, recalling a memory, or considering a thought. Inner perception might not be objective, but it remained a critical starting point for any psychology. Luckily, human memory allowed for the recollection and examination of these transitory moments. In addition to emphasizing the stabilizing power of memory, Brentano called for a close study of language and gesture as a way of aiding our knowledge of another's inner world. Psychologists should also pay special attention to children and animals, as well as diseased mental states and weird psychological occurrences, he advised.

On the strength of this work, Brentano won a professorship in Vienna in 1874; the same year Sigmund Freud became one of his students. Initially amused that Brentano was arguing for the existence of God, Freud soon wondered if he could defend his materialism before Brentano's sharp logic. After sending their professor formal criticisms of his positions, Freud and his friend Josef Paneth found themselves invited to Brentano's home for discussions. Soon, Freud fell under the philosopher's sway. His professor was "a believer, a teleologist, (!) and a Darwinian and a damned clever fellow, a genius in fact," wrote the young man.

Brentano encouraged his student to see the whole tradition of philosophy as a road leading to science. He attacked theoretically driven approaches to psychology, railed against those who never bothered to test their ideas in the world, and "declared himself unreservedly a follower of the empiricist school which applies the method of science to philosophy and to psychology." Advising his students to study Locke, Hume, Kant, and Comte, Brentano also warned against any premature attempt to marry physiology with psychology, arguing that the science of the mind was too undeveloped for any such union. It was a lesson Freud would accept only after years of struggle, but it was one he would later repeat to his own students.

Simultaneously, this admirer of Hannibal began to reshape his notions of what made a man radical. Freud declared himself not unsympathetic to socialism, educational reform, the redistribution of wealth, and other reforms that might ease the Darwinian struggle for existence. But he believed true radicals manifested their revolutionary spirit by rejecting religious dogma and accepting the dictates of materialism and empiricism. Many of Freud's generation shared the belief that science would reform political and social life. Scientists would contribute to the defeat of superstitions, religious fictions, and ideological illusions, providing valid knowledge that allowed for a clearer vision of reality by which political elites could more justly and rationally govern.

After two and a half years of classes, Freud embarked on his first attempt to discover new knowledge by doing research in zoology, the field that had provided evolutionary theory with so much of its evidence. Six months after studying the gonads of eels, Freud joined the physiological lab of Ernst Wilhelm von Brücke, the man who had brought laboratory science to Vienna. For the next six years, Freud toiled in Brücke's lab, happily examining nerve cells. He made some minor discoveries, developed a new stain, and by the age of twenty-six could boast of a number of publications from his work.

In the middle of these studies, Freud served a year of compulsory military service, during which time to keep himself occupied, he translated some essays by John Stuart Mill on subjects like the emancipation of woman. Returning to Vienna, he finally sat for his medical exams in 1881, seven and a half years after he began his medical education and two and a half years longer than the average student. Freud passed and later attributed his success to his extraordinary memory, since he had not bothered to thoroughly prepare himself.

The fact was that becoming a physician was less of a priority for the young Freud than making scientific discoveries and becoming a university professor. Freud dreamed of staying in Brücke's lab, but in 1882 when he became engaged to Martha Bernays of Hamburg, this dream died. Freud informed Brücke of his intentions to marry, and his mentor took him aside and urged him to be realistic. Brücke's two assistants were extraordinary scientists and nowhere near retirement. There were no other paying positions to offer Freud, who now had a fiancée waiting. Disheartened, Freud accepted Brücke's advice and set out to become a practicing doctor.

For the next three years, Freud disappeared into the wards and clinics of the Vienna General Hospital. Living on the grounds, he returned home only on weekends. While continuing some lab research, he struggled to find his way as a clinician. He approached Hermann Nothnagel, a professor of medicine, hoping to become an *Aspirant* at the hospital, by which young doctors could work toward the role of *Sekundararzt* or assistant physician. Once a neuropathologist himself, Nothnagel was appreciative of Freud's histological work. He took Freud on and over the next two decades proved an important ally.

Nothnagel received a recommendation from another lab-oriented physician, the psychiatrist Theodor Meynert, with whom Freud had studied in the winter of 1877. Meynert's fame grew out of his anatomical studies of the nervous system, but he had also gained notoriety thanks to asylum doctors who cast doubts on his clinical skills. In 1875, the director of the asylum that housed Meynert's department even demanded his resignation, but the dean of the Vienna medical school flew into action and created a second chair in psychiatry for his protégé. Thanks to this accident of history, Vienna would retain two university chairs in psychiatry, allowing for a diversity of opinion that would prove critical to mavericks like Freud.

Secure in his academic position, Meynert had begun working on a magnum opus that he hoped would define psychiatry and elaborate the relative roles of mind and brain. For Meynert, brain disease was the sole cause of mental disorders; psychological factors were irrelevant. As Meynert put the finishing touches on the first volume of this work, Freud joined his department. From May to September of 1883, Freud confronted cases of alcoholism, progressive paralysis, and patients vaguely diagnosed as mad. He also encountered a few female hysterics, but they do not seem to have left much of an impression.

While immersed in clinical medicine, Freud remained ambitious, now searching for new breakthrough treatments. He stumbled upon an article touting cocaine, a new drug that had been used to treat morphine withdrawal in America. "We need no more than one stroke of luck of this kind to consider setting up house," he wrote Martha, his fiancée. Freud ordered cocaine, tried it, and became convinced that this astonishing substance could cure heart disease, nervous exhaustion, and mild depression, not to mention the agonies of morphine withdrawal. Freud's friend and teacher from Brücke's lab, Ernst Fleischl von Marxow, had grown addicted to morphine after an amputation left him in chronic pain. Freud supplied Fleischl with the new drug, hoping it might help end his addiction.

Six weeks after trying cocaine for the first time, Freud wrote an exuberant paper on the drug for the *Centralblatt für die gesammte Therapie*. He was eager to attract notice, especially after witnessing the praise heaped on a colleague who, on Freud's advice, had successfully used the drug as a surgical anesthetic. Freud championed the possible medical and psychiatric uses of cocaine, and his appeal began to gain attention. His monograph on cocaine was picked up in the prestigious Viennese newspaper the *Neue Freie Presse*. Before long, Freud was inundated with requests for information. Presenting his findings to the Vienna Physiologic Society and Vienna Psychiatric Society, he heralded the drug as effective and harmless.

But cocaine was not harmless. By the spring of 1885, Freud knew Fleischl's so-called cocaine treatment had not freed him from his addiction to morphine but had instead created a dependence on both drugs. Furthermore, Fleischl's escalating cocaine use led to horrifying toxic psychoses. It was only a matter of time before others became aware of these dangers and attacked Freud for rashly advocating cocaine's use. Such public opprobrium could do lasting damage to a young doctor's reputation, but Freud still had powerful backers at the university. As the cocaine debacle was coming to a head, Freud marshaled the support of Brücke, Meynert, Nothnagel, and others and won the university's Jubilee Fund travel grant to go to Paris. It was a good time for him to get out of town.

Before leaving for France, Freud resigned from the General Hospital. His engagement to Martha Bernays had now dragged on for three and a half years. He had not been able to support himself and was deeply dependent on a number of benefactors who had loaned him money to survive. He prepared to leave Vienna, having convinced university authorities that he would study atrophic neuropathologies in children while at the Salpêtrière. But Freud confided his true plan to his fiancée: he would make a name for himself in the nervous disorders. This trip would transform him into a famed nervous specialist. Upon winning the grant, a giddy Freud wrote Martha:

> Oh how wonderful it is! I am coming with money and staying a
> long time and bringing something beautiful for you and then go on

to Paris and become a great scholar and then come back to Vienna with a huge, enormous halo, and then we will soon get married, and I will cure all the incurable nervous cases and through you I shall be healthy and I will go on kissing you.

On September 29, 1885, Freud arrived in Paris and took a room at the Hôtel de la Paix in the Latin Quarter. While feverishly writing papers on neuropathology, he began to visit the Salpêtrière's famed clinic. On Mondays, Charcot gave public lectures focused on his latest research, while on Tuesdays, he discussed a puzzling case brought from the outpatient clinic for diagnosis. Wednesdays were for opthamological lectures, and the rest of the week was filled with hospital rounds. While eschewing numerous other lecturers, Freud found time to attend forensic autopsies at the Paris Morgue.

Dr. Charcot announced that the days of great discovery in pathological anatomy were over. The future lay in those nervous disorders with no anatomical lesions—the neuroses. During Freud's months in Paris, Charcot's focus of interest was male hysteria caused by trauma, such as the cases of Pin and Porez. Traumatic hysteria had encountered resistance from German neurologists, especially Hermann Oppenheim of Berlin. After his stay in Paris, a dutiful Freud traveled to Berlin and met with Oppenheim, who viewed these illnesses in purely anatomical terms. Freud came home still convinced Charcot was right.

Freud also returned from Paris certain that the altered states exhibited in hypnosis were real. He told his sponsors that he had witnessed the incredible phenomena of hypnotism, which "had to be wrung on the one side from skepticism and on the other from fraud." He understood, however, the events at the Salpêtrière were so bizarre that they would elicit grave doubts unless they were witnessed firsthand. He himself had been dubious when six years earlier the traveling hypnotist Carl Hansen came to Vienna, warning a friend: "keep your mind skeptical and remember 'wonderful' is an exclamation of ignorance and not the acknowledgement of a miracle."

Yet what Freud saw at the Salpêtrière was overwhelming. A routine demonstration might be this: a woman sits on a chair, hypnotized. A doctor informs her that upon awakening, she will not be able to move her right arm. The patient comes out of the trance and cannot move her right arm. She does not know why and perhaps fabricates a story that seems to make sense of her debility. The doctor puts her under a trance again, now suggesting her arm is fine. She emerges from the trance, and her arm is fine. This was not only great theater, it was also shocking for scientists schooled in a brain-based approach to the mind. And these astonishing effects were not just a source of wonder but also phenomena analyzed by that haut positivist, Charcot. French psychopathologists had proved that bizarre unconscious psychological states existed.

Freud's world began to turn upside down:

> I am really very comfortable now and I think I am changing a great deal. I will tell you in detail what is affecting me. Charcot, who is one of the greatest of physicians and a man whose common sense borders on genius, is simply wrecking all of my aims and opinions. I sometimes come out of his lectures as from out of Notre Dame, with an entirely new idea about perfection.

Afterward, he wrote a report for the university on his trip with vivid descriptions of Charcot's work on hysteria and hypnotism, and halfhearted apologies for spending so little time on organic diseases. He was not really sorry. Wowed by Charcot and his cadre of bright colleagues like Joseph Babinski, Georges Gilles de la Tourette, and Paul Richer, Freud returned from Paris with a new goal. He would become Charcot's man in Vienna.

Before leaving France, Freud had aggressively worked his way into Charcot's inner circle. While complaining to Martha that his French was so bad he could barely order food at a café, the young man offered his services to Charcot as a German translator. Charcot accepted. "It is bound to make me known to doctors and patients in Germany," Freud gushed. The two men conducted a correspondence as Freud translated the third volume of Charcot's *Lectures on Diseases of the Nervous System*, much of which was concerned with hysteria, hypnosis, and the traumatic paralyses. Freud himself became especially intrigued by paralyses created by the imagination.

At home, Freud readied for war in Vienna, knowing his colleagues were skeptical of the psychologic, the ideogenic, the hypnotic, the hysterical, not to mention the French. Nevertheless, he began to lecture to physiological and psychiatric societies on Charcot's theories and agreed to write a report on his experiences for the Viennese Medical Association. In that report, Freud presented French thinking on male hysteria. Some doctors in the audience granted that hysteria in men was possible, but others sharply took issue with Charcot's appointed stages. Meynert pointedly pressed Freud to find a single case of traumatic paralysis in Vienna.

A month later, Freud presented such a case to the group. But his victory immediately turned sour. A furious Meynert would have none of it, suggesting that the French had ruined his former pupil. Freud later recalled: "with my hysteria in men and my production of hysterical paralyses by suggestion, I found myself forced into the Opposition." Meynert, Freud bitterly noted, believed that he had been taken in by "the wickedness of Paris."

The effect of this minor controversy was that Sigmund Freud became a prominent Viennese representative of French ideas about hysteria, hypnosis, psychology, and psychopathology. While these notions were strongly resisted in Austrian circles, Freud seemed unimpressed. He had seen hysterics go through Charcot's stages, seen paralyses created by the mere mention of an idea, seen these things with his very eyes. What his colleagues in Vienna read about and disdained, Freud had witnessed. As a Jew and an outsider, he knew something about the power of prejudice to blind. Unafraid of being in the minority, he tied himself to the great Charcot and his theories of hysteria, trauma, and hypnotism, embracing associational psychology and for a while even his emphasis on heredity. The future for Sigmund Freud was now clear. He married his fiancée, opened a private medical practice, and took up his role as the loyal Viennese representative of Jean-Martin Charcot's thinking, just as the Parisian neurologist's reputation began to plummet.

III.

In 1886, a French professor of medicine from the provincial city of Nancy announced that Charcot, that master decoder of hysteria, had succumbed to a kind of hysteria. Over but a few years, it became apparent that this was true, and as a consequence, much of Charcot's work on hysteria and hypnotism was wrong. For Freud, this looming disaster forced him to quickly mature from an acolyte into a more independent thinker, as he desperately scrambled to reformulate his own positions. While holding fast to the goals of scientific psychology and Charcot's notions of psychic trauma, Freud would, in the end, accept that the Parisian's greatest achievements in the understanding of neuroses were figments of his own imagination.

The David who slew this medical Goliath was Hippolyte Bernheim. Before 1882, this Nancy doctor had little to do with nervous diseases. That year, one of his patients was cured of sciatic pain by a slightly disreputable country doctor named Ambroise Auguste Liébeault. Liébeault was an old-time hypnotist who had doggedly continued employing this method during the inhospitable 1850s and 1860s. With little fanfare, he had written *On Sleep and Analogous States*, in which he argued that hypnotic states were forms of sleep brought on by suggestion. Bernheim sought out Liébeault and became his student. In 1886, Bernheim published his own landmark study, *On Suggestion and Its Therapeutic Applications*, in which he put forward a purely psychological explanation of hypnosis.

Charcot had conquered hysteria and hypnotism by conceptualizing these mysteries as nothing more than inherited neural dysfunctions that resulted in altered states of consciousness. Unconvinced, Bernheim began experimenting with hypnosis and decided that such states were not pathological at all. In fact, he found hypnotic trances were easy to elicit among the great majority of men and women of all temperaments. Hypnosis simply exaggerated a common property of psychological life and was not a physiological dysfunction, he concluded.

Bernheim went further. Hypnosis, he believed, wasn't even necessary for suggestions to take hold of another person. Ideas passed from one unconscious mind to another all the time. The mind's windows were open, taking in commands, suggestions, and ideas from others and then mistaking foreign notions for their own. All of human psychology was characterized by this gross "credulity." False impressions and ideas were readily accepted by the mind thanks to automatic unconscious cerebration, the frailty of reason, and the all-too-human need to believe. Religion, education, tradition, morality, allegiance to the state, and social conventionality; the work of lawyers, politicians, professors, orators, charlatans, and seducers, all these were evidence of a world dominated by suggestion and credulity. Credulity was not odd or unusual, but rather was essential to normal psychological life. While Charles Richet and others had noted the possibility of suggestion taking hold without hypnosis, no one with scientific standing had the audacity to make such sweeping claims before.

Coming from the margins of French medicine and standing in stark contrast to the hard-won advances of the *psychologie nouvelle* and the prestige of Charcot, Bernheim's theory seemed to stand little chance. Besides, Charcot's theories were precise, logical, and based on broadly shared scientific principles, while Bernheim's analysis was nebulous and bloated. But Bernheim held a powerful trump card. As a prime example of suggestion and credulity, he pointed to the research of Jean-Martin Charcot. Charcot's stages of hypnotism, he insisted, were wholly imaginary. According to the Nancy doctor, Charcot and his followers had unwittingly suggested their stages of hypnotism to their patients, who then complied. Instead of looking into another's mind, these scientists had been staring in a mirror. Bernheim called this dance of expectation and mimicry a "culture of hysteria" and gravely informed his readers that none of Charcot's supposedly universal stages for hypnotism could be found in Nancy.

Thus began a furious battle. The Salpêtrière doctors lambasted Hippolyte Bernheim, calling him a confused buffoon who operated outside science. At the 1889 International Congress on Hypnotism in Paris, one of Charcot's allies, Pierre Janet, declared that Bernheim's opinions were "not only anti-scientific and anti-physiologic" but also "anti-psychologic." The attacks on Bernheim's scientific credibility grew harsher, for in this Nancy doctor, the Salpêtrière school confronted a rival they feared might topple not just a theory of hypnotism, but also the whole project of scientific psychology. If Bernheim was right, if suggestions and credulity were really so common, how could one ever hope to empirically know anything about another's inner world? If everyone was infected by suggestion and blind belief, who could be an impartial observer or an uninfluenced subject? In Bernheim's view, observer and observed, suggester and suggested, scientist and hysteric, ultimately subject and object, were impossible to distinguish with any clarity. There was no way for psychological scientists to stand outside this swim, for they were being suggested to even as they were suggesting.

A doctor from the Salpêtrière pointed out the seemingly absurd conclusion that Bernheim's followers were forced to accept. By his theory "(E)very reasonable man would thus be constantly under the influence of suggestion." Bernheim would not have disagreed. But such shocking claims spread unease. By 1886, reports of psychic infections and mass hysteria hit the French press. Could a woman murder her lover due to the suggestions of another? Could a people rise up against a government due to suggestion? The land of the French Revolution and the Paris Commune confronted the fear that another mass uprising was one evil hypnotist away.

For Sigmund Freud, Bernheim's challenges could not be shrugged off. Freud's reputation and practice were closely associated with the prestige of the Salpêtrière school. He understood that Bernheim's claims would give succor to those in Germany—and they were many—who always thought of hypnosis as a gross charade. Freud took on the German translation of Bernheim's book, convinced that it would help him get out in front of a potentially devastating critique.

In 1888, when curious German readers purchased a translation of Hippolyte Bernheim's *On Suggestion and Its Therapeutic Applications*, they encountered an intrusive translator who begged to differ with the author. The translator railed against those who might use Bernheim's work to deny the reality of hypnosis and conclude that all these accounts were based on a mixture of naive belief and trickery. Defending the scientists of hysteria from the charge that they were themselves hysterically deluded, he attacked those who dismissed Charcot's studies as worthless "errors in observation," and retreated to the belief that hypnosis was "beyond scientific understanding."

Freud defended Charcot, saying the neurologist's careful work had proved hypnotism was lawful and therefore, by scientific standards, real. Unfortunately, it was getting harder to so insist. The debates between the Nancy and Salpêtrière schools generated an avalanche of research, and the results were devastating. Otto Wetterstrand reported having hypnotized 3,589 people; he never saw Charcot's stages. Study after study showed that a large percentage of normal people were in fact hypnotizable. In 1892, Albert von Schrenck-Notzing published a study of 8,705 subjects, and only 519 could not be hypnotized. Hypnotizability did not seem to be a specific sign of hysteria or even pathology; it seemed commonplace.

Freud faced a dilemma: on the one hand, there was a psychological theory that undermined the scientific legitimacy of psychopathologic research on hypnotism and hysteria. On the other, there was a lawful, psychopathologic scientific theory that increasingly appeared to be just plain wrong. After 1888, Freud distanced himself from his colleagues in Paris, and took care not to defend Charcot's stages of hysteria or hypnotism, even explicitly stating it was incumbent on the workers at Salpêtrière to prove these theories. They never could.

Like his colleagues at the Salpêtrière, however, Freud was eager to defend the reality of hypnotism by formulating theories that conformed to scientific standards of knowledge. If hysteria was real, he argued, it must be based on something other than random acts of suggestion and credulity. Hypnotism must also be rule bound and follow some inherent laws. Hippolyte Bernheim, his German translator charged, simply did not ask what those laws might be.

To pursue this further, Freud suggested Bernheim's readers stop thinking of the hypnotic encounter as some interpersonal drama between a wide-eyed hypnotist and a swooning subject. Instead, they should turn their attention to the intrapsychic conditions that made a man prone to another's suggestion. If known, those internal changes would explain both the interpersonal phenomenon described by Bernheim and the bodily changes Charcot cataloged.

This critical strategy allowed Freud to reduce the overwhelmingly complex problems of how two minds interacted, and limit his exploration to the workings of one mind, the patient's. Borrowing from Charcot's work on traumatic paralysis, Freud argued that all suggestions were the result of prior, internal self-suggestions. While Bernheim believed "suggestion pushes open the doors," in fact the doors were "slowly opening of themselves for auto-suggestion," Freud declared. Some internal dissociated idea set the stage for a suggestion to take hold; all the Nancy doctor's results were thus the result of a receptive state brought on by autosuggestion.

Freud translated Bernheim to undercut his most damaging claims and secure the scientific respectability of research on hysteria and hypnosis. He understood that if Bernheim's model of interpersonal persuasion could be rooted in an internal mental process, suggestibility would remain a lawful object of study that could not be attributed to the provocations of the physician. Freud argued that far from the everyday realm of consciousness, inner unconscious psychophysical changes took place that created a state of suggestibility; later suggestions took root in this fertile soil and caused the specific symptoms of hysteria. If in this the Viennese doctor was right, scientists of hysteria could safely ignore Bernheim's confusing, interpersonal dynamics and follow Charcot by devoting themselves to the study of the inner world.

Freud's argument was deft and his conclusions enduring. An insistent, intrapsychic focus would orient Freudian approaches to the mind for the next century. Furthermore, Freud had showcased an impressive capacity to reframe debates and turn them on their head. While conceding the reality of Bernheim's observations, Freud pulled the rug out from under the Nancy doctor's central explanations so as to reaffirm the boundaries between observer and observed, so critical to scientific knowledge. He stood behind Charcot's commitment to scientific method while jettisoning the master's stages of hypnosis and hysteria, and sided with Bernheim's contention that hypnosis was fundamentally based on unconscious psychology. Then as Freud was creatively reframing Bernheim's challenge, the young Viennese doctor's capacity to take command of critical debates would be tested again, as another pillar that buttressed Charcot and the *psychologie nouvelle* began to crack.

IN THE SECOND half of the nineteenth century, hereditary causes were extremely popular in French medicine, particularly psychiatry. Faced with the overwhelming complexity of nerve, brain, mind, person, behavior, and social environment, French doctors saw heredity as a one-size-fits-all answer to questions they could not truly fathom. After 1870, biologic inheritance was widely accepted as the cause of psychic functions and the central precondition that led to a mind breaking during accidental events. This explanation of mental illness was championed by the devout Catholic Bénédict-Augustin Morel, whose notions of hereditary degeneracy echoed the fall from grace that plagued men since Adam. His theories were adopted by others like Valentin Magnan, who led discussions at the Medico-Psychological Society on degeneracy while Freud was in Paris. By then heredity had won over most doubters. Few doctors in France believed mental disorders could be acquired any other way.

Jean-Martin Charcot pushed these hereditary assumptions further. By borrowing his student Charles Féré's notion of a "neuropathic family" and studying genealogies, Charcot linked a number of illnesses together, attributing all to the same inherited defect. Charcot mapped out family trees that bloomed with hysteria, alcoholism, suicide, progressive paralysis, apoplexy, rheumatic and arthritic disorders. When challenged as to the common inheritance of these illnesses, Charcot pointed to the neuropathic constellations that could be found among "Israelites." Ribot also argued that the purest example of psychological heredity came from the study of the Jews, who by "jealously guarding the purity of their race" became a distinct example of hereditary forces. In Ribot's case, he concluded that their endowment made Jews overly sentimental and imaginative, with an aptitude for poetry and music but not sculpture or painting, and minds sadly ill-suited for science.

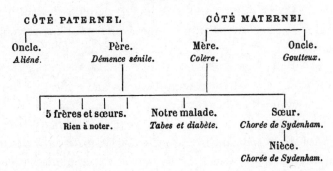

TABLEAU XXIII

FAMILLE ISRAÉLITE

Charcot's demonstration of a Jewish "neuropathic" family in which mental illness, dementia, gout, tabes dorsalis, Sydenham's chorea, and diabetes are all considered familial, and the result of degeneration.

During his stay in Paris, Freud came to believe he too was from a neuropathic Jewish family. The young doctor informed his fiancée that his uncle's family included a feeble-minded child and two children who succumbed to madness. Freud confessed that he had always thought his own family was free of hereditary taint, but while in Paris he saw things differently. To console Martha, he let her know: "(T)hese stories are very common in Jewish families."

Among the many diseases Charcot found in these familial clusters was tabes dorsalis, also known as locomotor ataxia. Charcot was an authority on this disease, for he was involved in its initial discovery. By the 1880s, however, a theory emerged that tied this disorder to syphilis, an infection that some believed also accounted for the epidemic of patients in asylums suffering from general paresis of the insane.

Charcot scoffed at all this. He had no doubt that these diseases were the result of bad heredity. But as was the case with hypnotism, the numbers began to pile up against him. By 1891, researchers presented data that showed between 90 percent and 91 percent of patients with tabes dorsalis had a prior infection of syphilis. Many younger doctors, even at the Salpêtrière, embraced the new germ theories. As it became clear that Charcot was wrong, his confident assertions about the role of heredity in other diseases—like hysteria—were also thrown into doubt.

While these clouds lowered over Charcot, Freud was busy translating the neurologist's *Tuesday Lessons* (*Leçons du Mardi*), which appeared in installments between 1892 and 1894. Again, in telling footnotes, the German translator begged to differ with the author, now over matters of heredity. Upon receiving page proofs with such amendments, Charcot replied to his Viennese acolyte:

> By the way! I am delighted with the notes and critical comments that I encountered at the bottom on the pages of "the Leçons." Go ahead—that's fine! Vive la liberté!! as we say here. After this declaration I shall ask the same from you, to tell you that I am astonished to see the extent to which the theory of the syphilitic nature of tabes, and P.G.P., wreaks havoc right now amongst the best minds. Really, the figure 90% (assuming it to be accurate?) can it have so much influence on a stable mind!—what do you do then with the other 10%?

Regarding the power of nervous heredity, Charcot encouraged Freud to check for himself and recommended he make the hunt easy by studying the genealogies of Jewish families.

With the scientific field tilting against Charcot, Freud also threw himself against degeneration theory. The theory of *famille névropathique* was in desperate need of reevaluation and could not be defended, he wrote. In footnotes, Freud made it clear that he also considered Charcot wrong about the hereditary nature of hysteria. In fact, he confessed that he and a colleague considered hysteria to be solely the result of trauma.

Before these remarks were written, Sigmund Freud could have been counted among a battalion of European thinkers who turned to French psychopathology, absorbing its methods, assumptions, and logic. Like others, he adopted a framework that included both associational psychology and hereditary explanation for the troubles of the mind. But as this theoretical bulwark began to fail, Freud fashioned a new one. His critiques of Bernheim and Charcot mark the outline of what would become distinctly Freudian ground. Unlike those in Paris who clung to claims regarding degeneration long after they had become difficult to defend, Freud, perhaps freed by his own marginality as a Jew and aided by anti-Semitic uses of this theory, let them go. It was a wise decision. Psychological heredity once gave French psychologists and psychopathologists an easy way to move mental phenomena from philosophy and mysticism to positivist science and biology, but it had become little more than a prejudice. While Freud readily acknowledged that "Charcot was the first to teach us that to explain the hysterical neurosis we must apply to psychology," he turned away from the biological buttressing that Charcot used to justify his psychology. Cut loose from that mooring, Freud floated forward. He would have to find an anchor of his own.

IV.

By 1892, Freud began to distinguish himself in a crowded field of psychopathologists and suggestive therapists. Bernheim's translator presented himself as a thinker—contra Bernheim—committed to studying intrapsychic processes. He would aggressively push this focus forward. Charcot's translator advertised himself as a man who did not believe degeneration caused hysteria or a number of other afflictions. Instead, he recognized the force of psychic trauma and unconscious autosuggestion as salient factors, another idea that he would carry with him.

As Freud developed the ideas that would form the basis of his theory, he confronted vehement opposition at home to French notions of hypnosis, suggestion, and psychic illness. The young doctor who had shown poor judgment about cocaine's safety now found himself defending hypnosis from those in Vienna who warned that this fantastic method could actually cause insanity. Freud would have been hard-pressed to take on these skeptics had he been alone, but he was not.

Hypnotism had found some allies in Vienna. In the 1870s, physiologists like Freud's friend Ernst Fleischl tried it on animals. And the method had been picked up by the maverick doctor Moritz Benedikt who, influenced by Charles Lasègue's work, tried hypnosis on hysterics in the late 1860s. Benedikt was confronted by his superior at the time, Dr. Josef Breuer, who told him to desist from such strange procedures. Benedikt agreed but, after meeting with Charcot in 1878, returned to hypnosis. In 1880, when the stage hypnotist Carl Hansen's performances resulted in the prohibition of any further such exhibitions in Vienna, Benedikt defended hypnotism before the Society of Physicians. It was Benedikt who supplied Freud with a letter of introduction to Charcot, and it was he more than anyone who created some credibility for hypnosis in Vienna prior to 1886.

When Freud returned home from Paris, he also discovered that two very prominent German-speaking psychiatrists had taken up hypnosis: Richard von Krafft-Ebing and Auguste Forel. Krafft-Ebing had been appointed to the second chair in psychiatry at the University of Vienna in 1889. Forel was a Swiss doctor with impeccable credentials as a brain anatomist, who had become director of the Zurich Burghölzli asylum and would later write a letter of introduction for Freud to Bernheim and Liébeault. Freud crowed over the arrival of the like-minded Krafft-Ebing and happily cited Forel as proof that "a man can be a brain anatomist and nevertheless see something in hypnosis other than a piece of absurdity."

But the man who provided the most support for Freud's work on hysteria and hypnosis was the internist and physiologist Josef Breuer. The son of a progressive Jewish scholar and a mother who had died when he was a child, Breuer graduated from the University of Vienna with a medical degree in 1864. Four years later while working with the physiologist Ewald Hering, Breuer became convinced that breathing was controlled by an automatic nervous process, which he proceeded to demonstrate as a fact. His fame as a physiologist was furthered when in 1873 he discovered the semicircular canals in the ear.

Breuer's research had a theme: he searched out the ways that reflexes regulated and stabilized human life. Somehow despite great success, Breuer's university career stalled. After numerous rejections and frustrations, in 1885 he resigned his academic position and became a private practitioner, though not just any private practitioner. Breuer developed into one of the most widely sought after doctors in Vienna, physician to aristocrats and members of Vienna's elite. He might have simply remained so had he not made the acquaintance of Sigmund Freud in the mid-1870s. By 1882, Breuer, along with Ernst Fleischl, Josef Paneth, and Samuel Hammerschlag, had become one of Freud's financial benefactors. The young Freud was often a guest at Breuer's home, and during one of those visits, his host told the story of a patient, Bertha Pappenheim, a woman whose case would be seen as foundational by legions of psychoanalysts who would come to know her as "Anna O."

Bertha came from an orthodox Jewish family, but unlike her parents she was far more entranced by literature and theater than religion. In the summer of 1880, her father fell ill, and soon after she began to suffer from violent, chaotic symptoms. Breuer was called in to cure Bertha; it would not be an easy task. Two years later when the young woman needed to be hospitalized, Breuer wrote up an extensive account of her illness, sharing his earliest attempts to make sense of this striking case.

According to her doctor, Bertha displayed classic signs of hysteria, caused by a hereditary taint mixed with exciting influences. When her beloved father became gravely ill, Bertha had held a vigil at his bedside. Suddenly she was beset by hallucinations of snakes. Afterward, she repeatedly suffered from "absences," accompanied by visions, paralyses, and physiological abnormalities. She complained that she had been split into "two selves, a real one and an evil one." Bertha could move her arm only to the right, and only see select parts of a face. She developed odd contractures, anesthesias, spasms, and periodically fell deaf. Even more bizarrely, Bertha was unable to speak in her mother tongue, though she communicated fluently in English.

These symptoms were quite dramatic, but in Breuer's view, not mysterious. Bertha suffered from a severe case of hysteria that had resulted in altered mental states and a maze of psychic and somatic ailments. In its treatment, however, the case was nothing short of astounding. The doctor who once opposed Benedikt's use of hypnosis tried this method on Bertha. Soon he discovered there was no need for hypnosis, for Bertha was often already in a similarly altered mental condition. With Bertha's guidance, Breuer came to understand that when in such a state, if she was allowed

to narrate her inner fantasies, her symptoms would abate. By simply talking, Bertha was relieved of some of her debilities.

After her father died in April of 1881, Bertha deteriorated and became
suicidal, at times refusing food from anyone except her beloved doctor.
Breuer began seeing Bertha more often, administering what she famously
called the "talking cure," in which she verbally cleansed her mind. Breuer
concluded that during the day, Bertha processed psychic events pathologically, but when they were narrated these psychic stimuli lost their power
to harm her.

Bertha's strange fantasies also held seeds of truth. For example, to her
caretakers' surprise, Bertha stubbornly insisted on wearing her stockings
to bed until one day she remembered that while her father was sick, she
would defy her doctor's orders and sneak into his room at night. She wore
stockings to bed to prepare for this nightly pilgrimage. After recalling
this, Bertha calmly removed her socks and went to bed. Remembering
had dissolved the symptom. On another occasion, Bertha refused to drink,
subsisting instead on fruits and melons. Upon recalling that she had seen
a dog drink from a glass and been disgusted, Bertha called for water and
drank. Instructed by these telling incidents, Breuer began to treat his patient by helping her recover her lost memories. Every evening, the doctor
arrived to sweep away the day's residues from Bertha's mind.

This was a new twist on medical attempts to use ideas and words to
relieve hysterical symptoms. Josef Breuer and Bertha Pappenheim jointly
constructed a method by which it was not the commanding suggestions of
a doctor, but rather the patient's narration and recollection that brought
relief. Unfortunately, the relief was often fleeting. Bertha's symptoms
took more and more of Breuer's time; he resorted to drugs like chloral
hydrate and morphine. In the end, he forcibly hospitalized Bertha. When
she was admitted to an asylum on July 21, 1882, she was addicted to morphine. The admitting physician tried to wean her of this drug and enlist
other cures—leeches, faradic electricity, and arsenic—all to no avail.
Meanwhile, Breuer took pains to assure the asylum director that Bertha's
illness was not faked.

When he first arrived in Paris to study with Charcot, Sigmund Freud
knew about the case of Bertha Pappenheim. In the French capital, he
learned more about autohypnotic states and traumatic neuroses and tried
to interest Charcot in Bertha's case, without success. When Freud re-

turned to Vienna filled with French ideas, he found a staunch ally and
steady source of referrals in Josef Breuer. After Bertha, Breuer had vowed
never to treat a case of hysteria again, and he eagerly referred new cases
to his junior colleague. The two men constantly discussed these patients,
and soon Breuer changed his thinking. If in 1882, Breuer conceptualized
Bertha's pathology by speaking of psychic stimuli and physiological excitations in a language common to psychophysics, under Freud's influence,
Breuer adopted Charcot's terms. Autohypnotic states came from trauma
and nervous shock; Bertha's symptoms were related to associations that
had become disassociated; her refusal to drink water was similar to
Porez's refusal to move his arm. It was not obstinacy, not even a refusal
so much as the result of an unconscious idea that had free rein over
her body.

In 1887, Freud began to experiment with suggestive treatments, including Josef Breuer's method. While others such as Alfred Binet and
Joseph Delboeuf had advised that in cases of traumatic paralysis one
should urge patients to recall their trauma so the doctor could suggest all
was well, no one thought that remembering alone would cure. In 1888,
Freud began to advertise this new twist. A year later, he argued that nervous functioning in hysteria could be altered by a pathogenic idea. If that
idea was gotten "rid of or its memory weakened," the disorder could be
cured.

Nervous shock, trauma, internal disassociation, unconscious ideas, a
cure by remembering—by 1892, Freud and Breuer began to entwine
these elements in a unique way. Soon, Freud would add a critical element
and synthesize these ideas by postulating the central role of an inner battle
of ideas, a mental conflict. The first hint of this novel integration came
when Freud wrote up the case of a patient cured by hypnosis. He had
been called in to see a woman who had become hysterical overnight.
Freud knew the family well and was convinced they had no hereditary
taint. This patient was a *hystérique d'occasion* who upon delivering a new
baby suddenly fell ill. Despite difficulties nursing in the past, the mother
was intent on breast-feeding. But she found herself unable to eat, unable
to nurse, at times unable to lift the baby to her breast. Freud began the
standard suggestive cure. She would eat, she would be a fine nurse, the
baby would thrive, he doggedly insisted. Later that day, the woman fed
herself and her baby, but a day later she lapsed into her prior state. Unable

to bring the child to her breast, the woman had made a mockery of Freud's treatment. He tried a new tack. Under hypnosis, he told her that once he left, she would demand food and ask her family how they could possibly starve her when they knew she needed nourishment to nurse her child. She did, and her troubles ended.

To explain this turnaround, Freud first insisted that the language of suggestion and countersuggestion be replaced by a division of mental life into intentions and expectations. Normally closely linked, the expectation—*I will fail, I will be unable to eat and nurse*—had become disassociated from the intention to nurse. It had existed as an unconscious idea, exerting a "counter-will" to the patient's conscious intention. Freud's suggestion subtly lifted this counter-will into the mother's consciousness, allowing it to return to the normal mass of associations, where it promptly lost its power.

Freud implied that dissociated counterforces might be common, asking his readers to recall Charcot's study of those from the Middle Ages who were demonically possessed. Wasn't it often the devout nun who began to blaspheme or indulge in outrageous erotic behavior? Wasn't it the well-behaved boy who during hysterical attacks became an unbridled rowdy? "It is the suppressed—the laboriously suppressed—groups of ideas that are brought into action in these cases," Freud declared. Hence, hysterical conditions might even be *produced* by "laborious" suppression. Freud quickly backpedaled from this last, staggering thought, but he would soon return to it.

Charcot had established that among susceptible people, trauma could cause neurosis. But Freud insisted this woman had no sign of degeneration. She had been traumatized by nothing more than her own thoughts. She had not been attacked; she had not fallen. She succumbed to illness due to an intrapsychic battle of ideas. Hypnotists long struggled with the interpersonal battle that took place between physician and subject. Freud witnessed such a test of wills during a visit to Nancy, when Bernheim berated a patient for failing to accept his suggestion. "You are counter-suggesting yourself!" the doctor furiously exclaimed. Afterward, Freud wondered if a man didn't have the right to defend himself with counter-suggestion when another tried to subdue him with suggestions. Freud internalized this battle, transforming conflict between a doctor's suggestion and a patient's defensive countersuggestion into one between an individual's intentions and his own desire to suppress those ideas.

Contemporary brain science may have aided Freud in this reconceptualization. Freud's old professor Theodor Meynert had postulated that the brain required inner controls over its primitive impulses; inhibition of this sort was critical to normal brain functioning. Of course, Meynert's model was neurological and never implied that control could be exerted psychologically. Freud intimated that the mind itself could control disruptive ideas, and in the process create illness. In this way, Freud proposed that the mind was self-regulating. It was a fascinating proposition that he and others would pursue for decades.

In 1892, Sigmund Freud and Josef Breuer wrote up their discoveries. A year later they rushed out "On the Psychical Mechanism of Hysterical Phenomena: Preliminary Communication" to protect their priority in a hot field. Defining their notion of "traumatic hysteria" as an extension of Charcot's traumatic paralyses, the Viennese acknowledged their model had been anticipated by the master and his followers such as Alfred Binet, Pierre Janet, and Joseph Delboeuf. They affirmed Charcot's contention that ideas could cause hysterical symptoms, noting that in this they were joined by few other German researchers, notably Paul Möbius. And they extended the notion of trauma to include emotions like fright, which they believed led to disassociated ideas and symbolically related symptoms. In cases of hysteria, they concluded, trauma resulted in a splitting of consciousness, which was the *double conscience* so commonly found in French case histories.

If in all that, Breuer and Freud were extending the work of the Salpêtrière school, each man also believed he had one major new contribution. For Freud it was that psychic conflict and the suppression of ideas was sufficient to create hysteria. Breuer's big idea was his memory therapy, in which the recollection of dissociated ideas could bring symptomatic relief. "Hysterics suffer mainly from reminiscences," the authors jointly declared. Separated from normal associations, certain recollections act like foreign bodies and are never dissipated. Treatment with hypnosis brought those memories to consciousness and gave a powerful feeling of relief—a "cathartic effect." An ancient term, catharsis had been employed by Aristotle to explain the emotional effects dramatic tragedy had on its audience. Martha Freud's uncle, Jacob Bernays, had written a scholarly treatise on this theory, which now served to name Breuer's innovation: it was the cathartic method.

Not long after Breuer and Freud published their "Preliminary Communication," Jean-Martin Charcot suddenly died, leaving his embattled legacy to others. Two years later, Breuer and Freud published *Studies on Hysteria*, a book that sought to extend one aspect of that legacy. The book was constructed around five case histories. First was Bertha Pappenheim, now reborn as "Anna O." Retelling this astonishing story, Breuer left the impression that Anna O. had been cured by her treatment. This deception is perplexing, for Breuer and Freud both knew that Bertha had bounced in and out of sanatoriums between 1883 and 1887. Moreover, Breuer's failure to cure this woman was hardly a condemnation of a treatment he touted only for symptomatic relief. Freud too had little stake in the fabrication, for he explicitly distanced himself from Breuer's handling of the case, implying the senior author's diagnosis was wrong and that other methods—for instance, Freud's—might have helped the patient more.

The other four cases came from Freud's practice. Frau Emmy von N. was a Viennese aristocrat Freud treated in 1888; she was prone to tics and sudden spasms of horror where she cried "Keep still! Don't say anything! Don't touch me!" Katharina was a sexually abused peasant girl Freud encountered while mountaineering in the Alps. In addition, there were two critical cases from 1892, Miss Lucy R. and Fräulein Elisabeth von R. In both cases, Freud had trouble getting the women to fall into a hypnotic trance, and so he resorted to Breuer's cathartic method but without hypnosis. At first, Freud experimented with a method in which he ordered his patients to lie down, shut their eyes, and concentrate. He recalled Bernheim once saying that hypnotic states could be recalled in waking states, if the physician gave a firm command and applied pressure on the patient's head, and Freud reasoned that the same might work for dissociated memories. He found it did.

After presenting these case histories, Josef Breuer composed a theoretical chapter on hysteria of his own. He made it clear that despite his training and prestige as a physiologist, he was now writing as a scientific psychologist: "In what follows little mention will be made of the brain and none whatever of molecules. Psychical processes will be dealt with in the language of psychology; and, indeed, it cannot be otherwise." No other lexicon could be used to discuss the central theme of this work: the power of unconscious ideas in hysteria. "We must recognize the fact that in reality, as has been shown by the valuable work carried out by French investigators, large complexes of ideas and involved psychical processes with important consequences remain completely unconscious in a number of patients and co-exist with conscious mental life."

Breuer's thinking had moved over to the French and Freud. But he did not accept his collaborator's central innovation. Breuer mocked the idea that all hysterias were caused by pathological ideas, a theory he diplomatically attributed to Paul Möbius, though he knew very well that his coauthor embraced this view too. Breuer found the theory ridiculous, rather like concluding that since an idea could cause an erection, ideas alone caused all erections. Perhaps some affectively charged ideas could be made unconscious by deliberate banishment, but this was due to nothing fancier than simple lack of attention. Far more important were ideas that could never be the objects of attention. They existed in an abnormal brain state, "hypnoid states," that only developed in those with pathological inheritances.

This theory would have sounded quite familiar in Paris, but Breuer's radical coauthor had other ideas. In 1894, Freud published a paper in which he developed his fledgling thoughts about psychological intentions and defenses, theorizing that the resulting internal warfare caused an acquired form of hysteria, obsessional neurosis, and hallucinatory psychosis. Freud tried to demonstrate that the splitting off of associations from consciousness was caused by the mind working against itself and was not a question of heredity.

Freud had given the mind the power to wound itself. Knowing and feeling too much could make you sick. Offending ideas disrupted the mind, and in response, the mind had developed the ability to guard itself. Suppression served the mind by robbing a threatening idea of its power and divesting it of affect. A terrifying thought could be banished, though in nonpsychotic illnesses, the *feeling* of terror remained, floating in consciousness and then attaching itself to some seemingly innocuous idea that then became strangely charged. This explained how irrational phobias or obsessions came into being. In other cases, the detached affect could be converted into a bodily change, such as in the hysterical paralyses. These neurotic pathologies developed from inner threat and defense, resulting in split-off ideas and erroneous links that Freud called "false connections."

In naming false connections, Freud broke some connections of his own. The man most responsible for reviving hypnotism in Austria and Germany, Auguste Forel, would have referred to this same process as autosuggestion. In fact, Forel read Freud's work as an extension of the theory of autosuggestion. However, in moving from autosuggestion to false connection, Freud began to sever himself from hypnotic discourse. Unlike Breuer, Freud used language more dependent on associational psychology, and for good reason. He had given up on hypnosis, which he found difficult to perform. In the *Studies*, he declared that his method now involved a conscious search for breaks in association and false connections, a process he called "psychical analysis."

Going further, Freud confessed to grave doubts regarding Breuer's theory of hypnoid states, daring to suggest that this splitting of the mind was not due to pathological brain function but rather psychic conflict. There was no inborn proclivity for psychoneuroses, only trauma, conflict, and ideas warring with inner defenses. Josef Breuer could not have been pleased to be undermined by his junior coauthor, and in fact Breuer and Freud would never write another work together again. As he had done with Charcot, Freud borrowed from Breuer and then, armed with his mentor's ideas, pivoted to face his teacher. Eagerly he followed others, only to stand against them, in open and often aggressive intellectual combat. It was a heroic stance, worthy of a Hannibal of the mind.

Having immersed himself in French psychological and psychopathological theory, Sigmund Freud was now eager to trumpet his own originality. Throughout *Studies on Hysteria*, he insisted that he was not just another follower of Charcot, because he rejected heredity as an explanation of mental disease, and he cautioned others to steer clear of the "theoretical prejudice that we are dealing with the abnormal brains of *dégénérés* and *déséquilibrés*." The benefit of Freud's contrarian position was potentially immense. He opened a door for doctors to do more that alleviate symptoms in biologically broken brains; they could now cure diseases that were the result of thoughts.

But lest there be confusion, Freud made it clear that he was not a child of Hippolyte Bernheim, either. Despite the fact that he too championed psychological causes, Freud did not support suggestion as a therapy, instead advocating psychical analysis. Analysts would dig into the strata of psychical life, forcing their way through resistances, tracking threads to nodal points, and chasing memories to the nucleus of some pathogenic organization. Freud's description conjures an adventurer in a foreign land. And while the patient's resistance must be "broken," he wrote, the adventurer need not worry about his objectivity:

> We learn with astonishment from this that we are not in a position to force anything on the patient about things of which he is ostensibly ignorant or to influence the products of the analysis by arousing an expectation. I have never once succeeded, by foretelling something in altering or falsifying the reproduction of memories or the connection of events.

Freud had worked hard to secure the doctor's scientific standing in a field that some described as riddled with credulity and suggestion. Now Freud went further, much further. He categorically stated that hysterics—those patients many thought were characterized by suggestibility—were immune to suggestion during an intimate analysis of their inner world. Freud wrote: "We need not be afraid, therefore, of telling the patient what we think his next connection of thought is going to be. It will do no harm." This was the initial theory of Freudian technique, a method intended to make manifest the patient's inner associations, one that focused on ideas, affects, memories, and gaps in inner experience, and a mode of inquiry in which suggestion was no problem. In order to stabilize the scientific foundation for his psychological work, Freud had pushed himself to the edge of credibility. Because his theory was founded on the recollection of memories, he felt compelled to assert—in opposition to a vast library of literature—that doctors could not possibly suggest false memories, even if they tried. It was a position he would live to regret.

Freud did acknowledge that some interpersonal troubles might complicate a psychical analysis, but unlike the Nancy school, he gave such troubles a small, subsidiary place. Personal estrangement was possible. Or, the patient (always referred to as "she") might be seized by a fear of sexual involvement with the doctor. More importantly, a "transference" might seize upon the figure of the physician. Freud wrote about a woman who once had an urge to kiss a man, a wish that horrified her, and which she had long ago banished. Now hysterical, she had come to Dr. Freud for treatment. In the process of her analysis, the feeling now dissociated from all memory returned and was linked to her doctor, creating transference. Freud had not suggested the woman kiss him; her desire to kiss a long-lost love was mistakenly tied to him.

Since transference was not founded on a real interaction between doctor and patient, it freed Freud from the accusations of sexual seduction that long shadowed mesmeric, suggestive, psychic, and hypnotic treatments. "Since I have discovered this," he wrote, "I have been able, whenever I have been similarly involved personally, to presume that a transference and a false connection have taken place." Transference was the fruit of Freud's search for the lawful, mental forces that lay beneath the interpersonal dramas of hypnosis and hysteria. It became a weapon to beat back concerns about the objectivity of a field plagued by simulators, credulous observers, and delusions passing invisibly between doctor and patient. Transference reasserted the boundary between doctor and patient in a way that undercut the growing anxieties that had emerged about the nature of these borders.

Once Charcot's man in Vienna, Freud had ransacked the *psychologie nouvelle*, adopting many of its theories and much of its logic regarding the nature of scientific psychology. As the field came under attack, he devised synthetic positions based on an intrapsychic focus, while aggressively rejecting the proposition that hysteria was due to flawed heredity. Agreeing with Bernheim on the psychological nature of hypnotic states, he took issue with the Nancy doctor's theory of suggestion, opting instead for his model of warfare between desires and inner defenses. By 1895, Sigmund Freud had distinguished himself from other French-oriented psychopathologists with his notions of defense neurosis, mental conflict, psychical analysis, and transference. In the process, he began to refine the ancient dictum: know thyself. If Freud was right, humans could not bear to fully know themselves.

THROUGH A DEEP engagement with French medicine, Sigmund Freud proposed a model that had the potential to redefine the study of psychopathology, but in Paris his ideas won him lifelong enemies, making France hostile to Freudians over the next decades. The man who spearheaded the campaign against Freud was long presumed to be Charcot's heir. Exquisitely trained, with a distinguished pedigree, Pierre Janet completed his studies in philosophy at the École Normale Supérieure in 1882 at the age of twenty-two, and moved to Le Havre to teach. There he stumbled on an old cell of animal magnetists, and with them he began to conduct hypnotic research with hysterics. In 1886, Janet published a series of articles in Ribot's *Revue philosophique* where he laid out his theories on altered states of consciousness. Two years later, he published his *Psychological Automatism: An Essay of Experimental Psychology on Inferior Forms of Human Activity,* a massive, erudite work that unified multiple strands of psychology put forward by philosophers, hypnotists, and alienists.

For Janet, the basic element of psychological analysis was unconscious automatic activity. To conduct experimental studies, Janet adopted the approach pioneered by Ribot. Gestures, language, and bodily signs served as indirect confirmation of psychical states and provided solid ground for an objective psychology. Assembling the vast research that had been done on the shifting states of the "I" in France, Janet postulated simultaneous yet distinct states of consciousness that fluctuated and were at times wholly removed from consciousness. He described multiple centers of automatic activity and parallel selves. These subconscious selves were the result of psychological dissociation. From his researches, Pierre Janet had altered Descartes's famous phrase *I think, therefore I am*, to *We think, therefore I am*, or more curiously, *We think, therefore we are.*

Despite his youth, Janet's psychological work was more nuanced and sophisticated than any of his peers. In 1890, he was summoned to Paris by Charcot and began his medical studies. After completing them in 1893, Janet was promptly made head of the psychology lab at the Salpêtrière. When Breuer and Freud rushed out their "Preliminary Communication" in 1893, Janet had taken notice. In an omnibus review, he remarked that theirs was the most important of a series of new efforts to define hysteria. Important, but not enough to spell the authors' names correctly. Janet referred to "Brener and Frend," and embraced their work as simply confirmation of his own: "We are very happy that the authors in their independent research have been able with so much precision to verify ours, and we thank them for their amiable citation." Janet would get Freud's name right because it would hound him to his grave. In 1895, he would discover that neither author of *Studies on Hysteria* seemed eager to recognize Charcot's heir apparent. In fact, they seemed intent on replacing him.

In 1893, Janet informed his readers that Breuer's notion of a hypnoid state simply reinforced existing French theories. While Breuer and Freud admitted as much in their preliminary communication, by 1895, Breuer distanced himself from Janet's belief that hysterics were inferior degenerates, saying he himself thought it was more likely that they suffered from a form of psychological excess. Still there was no avoiding the fact that Breuer's theory was a version of Janet's. Janet could even claim that Breuer's cathartic treatment was related to his own published therapeutic work, through which he had tried to stitch together broken associations in his patients. When it came to scientific priority and originality, Pierre Janet had little reason to worry about Josef Breuer.

Janet could not so easily dismiss the lesser known Freud, who aggressively rejected degeneration theory, an essential part of Janet's understanding of neurosis. Janet acknowledged that traumatic experiences could instigate the creation of dissociated islands and second selves, but he insisted such dissociations could only happen to those who suffered from degeneration. "As in all other mental maladies," Janet believed bad heredity played a dominant role in hysteria.

By rejecting degeneration theory, Freud distinguished himself from others, but he also lost something terribly valuable. Ever since Ribot, a commitment to heredity had rewarded its believers with a biologically plausible mooring for thoughts and feelings. Those who studied the psyche were no longer in the invisible realm of internal experience; they were not the unloved stepchildren of Auguste Comte's science. Without heredity as the presumed biological cause of psychopathology, Freud would struggle with a long line of critics who saw his endeavors as floating in some metaphysical mind stuff that was cut loose from the material world. For Freud, this was not yet an overriding concern. If the cathartic method and psychical analysis relieved hysterical symptoms, if an idea or a reawakened memory made a paralysis disappear, he knew that was scientific evidence of a dramatic kind. Not only was therapeutic effectiveness of great clinical value for Freud, but it was also the scientific proof he used against skeptics. If ideas cured an illness, who could say ideas had not caused it?

And so despite their common commitments to a psychology of unconscious ideas, despite their commitment to psychotherapy, despite their common lineage from Charcot, or maybe because of all these things, Sigmund Freud and Pierre Janet became bitter rivals. Janet discounted Freud's work as derivative and belittled its critical innovations as flawed. Freud single-mindedly assaulted Janet for his insistence on an inherited feeble-mindedness in hysterics, and positioned his own work as a corrective.

In the end, Pierre Janet's admirers would wonder what happened. This thinker, who by his brilliance and connections seemed destined to carry on French scientific psychology, was increasingly overshadowed by the man he once rather hopefully referred to as "Frend." Janet was not alone. Over the next two decades, French psychopathologists and psychotherapists would increasingly complain that their tradition, their work, their findings were being forgotten. Sigmund Freud had co-opted them, they would insist, and in some ways they were right. Freud had imbibed French notions of scientific psychology and psychopathology only to separate himself from these origins. Those in Nancy confronted in Freud a rival that had a more detailed and scientifically coherent explanation of their own central concept: suggestion. Those in Paris who continued to defend degeneration theory would become discredited on this count, while Sigmund Freud reaped the reward for having pointed the study of psychopathology in a different direction.

With and against Charcot, with and against Bernheim, with and against Breuer: Sigmund Freud moved back and forth in the process creating a distinctive offshoot of French psychopathology. After 1895, if you were attracted to French psychopathology or interested in suggestive psychotherapies, you could pursue either of these by studying the work of Sigmund Freud. When Pierre Janet finally unleashed a full-scale attack on Freud in 1913, it was far too late. The French professor found himself debating a committed Freudian who years earlier followed his interest in the *psychologie nouvelle* to the Viennese doctor.

Furthermore, by 1913, Pierre Janet's charges against Freud no longer held. By then, Sigmund Freud could hardly be dismissed as a derivative French psychopathologist, for he had continued to develop and transform his theory into a body of ideas that was not simply French. After 1895, having embraced the study of psychic causes, Sigmund Freud set out on a dangerous journey that the French had no need to take. For a medical man, the path forward was an odd one, for it seemed headed toward lands usually reserved for novelists and poets:

> Like other neuropathologists, I was trained to employ local diagnoses and electro-prognosis, and it still strikes me myself as strange that the case histories I write should read like short stories and that, as one might say, they lack the serious stamp of science. I must console myself with the reflection that the nature of the subject is evidently responsible for this, rather than any preference of my own. The fact is that local diagnosis and electrical reactions lead nowhere in the study of hysteria, whereas a detailed description of mental processes such as we are accustomed to find in the works of imaginative writers enables me, with the use of a few psychological formulas, to obtain at least some kind of insight into the course of that affection.

Leaving behind French psychopathology, Freud would try and secure his new discoveries, located somewhere between literature and neuropathology, by finding a place for them in a scientifically tenable model of the mind.

———————◼———————

TWO

City of Mirrors, City of Dreams

"A strange picture and strange prisoners."
"No more strange than us," I said.
—Plato, *The Republic*

I.

As FREUD WAS putting the finishing touches on *Studies on Hysteria*, he turned his energies to a new and daunting task. He needed to develop a model of the mind in which his theories of psychic conflict made sense. He needed to understand how a mind could split against itself. How could ideas and feelings create illness? What kind of a mind did that?

These questions forced Freud to take up riddles that had vexed philosophers for millennia. He had no choice but to wade into the perilous waters that ran between mind and body. Of course, one question seemed settled. As a late-nineteenth-century brain scientist, Freud would have rejected any hint of a soul as a dark challenge to scientific reason and knowledge. The brain was the organ of the mind, he would gladly repeat with his scientific confreres. But many troubling questions remained. Was the mind completely controlled by the brain? Could its functions be solely reduced to simpler brain functions? Could poetry, art, and morality be fully explained by nervous physiology?

By emphasizing psychic causes, Freud was pressed to address these conundrums. To do so he turned to long-standing debates in German philosophy and science over the nature of the mind. A hundred years of controversy had left behind a rich series of highly complex, competing models from which Freud could choose, as he began to piece together a theory of the mind in which his ideas about mental conflict made sense.

At the beginning of this tradition stood the philosopher from Königsberg, Immanuel Kant. In 1781, Kant became famous for his "Copernican revolution" in which he integrated two philosophical adversaries: empiricism and rationalism. Empiricists, following Aristotle, believed that the mind mirrored reality. Therefore, the world could be known directly through sensory experience. For many empiricists, the mind was essentially passive, a clay molded by events, not molding in itself. In contrast, rationalists like Descartes saw the "I" as a central constructor and organizer of perception and reality. Each of these positions had long shown its weakness. Strict empiricists were unable to account for phenomena such as visual illusions, and rationalists made the real world little more than a fabrication of the mind so that reality became, in a favorite trope of the time, nothing but a dream.

Was the mind a mirror or a dream machine? In *Critique of Pure Reason*, Kant gave great weight to the solidity of empirical knowledge garnered from the senses, but he also placed a limit on such knowledge. The mind did not simply mirror reality. It relied on a priori forms to organize the ebb and flow of what would otherwise appear chaotic. Space, time, and causality were not simply perceived; they were transcendental categories not derived from experience. These a priori categories structured the "phenomenal" world, allowing for unified conscious experience. By this reasoning, it followed that there was a world outside human perception, a world beyond human knowledge. Kant called it the "noumenal" realm of unknowable things-in-themselves. By Kant's model, the mind was like a funhouse mirror that shaped and twisted an invisible, deeper reality.

Kant's theory had significant implications for science. Empiricists had stressed the role of pure observation in scientific work, which cohered with their associational model of how the mind worked. Kant proposed a different kind of science that made sense with his theory of the mind. He gave a limited but critical role to synthetic a priori principles that organized facts and observations. Scientific knowledge was made up of observations that were actively knitted together by deductive theories that were ultimately not derived from experience, but were metaphysical. To explain, Kant turned to Newtonian physics which, he pointed out, relied a priori on Euclidian geometry and mathematics. The example was telling, because in the end Kant concluded that scientific knowledge could only be found in domains that relied on mathematics. Since psychology seemed unquantifiable, Kant joined the chorus of thinkers who believed it could never be a science.

The Critique of Pure Reason dominated German intellectual life for decades to come; it was avidly read, misread, cited, and appropriated. Perhaps the most influential post-Kantian philosophy to emerge was Friedrich Schelling's Philosophy of Nature. Schelling suggested that Kant's work shouldn't be seen as arguing for two distinct worlds, split into phenomenal and noumenal realms. Rather he believed Kant was drawing a distinction between two aspects of the same world. Subject and object, a child and a rock, consciousness and atoms, the living and the dead, all of these were two facets of the same oneness. Nature itself was a living organism, therefore such antitheses should be thought of together, without attempting to reduce one term to the other. For psychology, Schelling's central message was that the mind and the brain were different ways of approaching the same unified essence. Neither could be explained by the other; both needed to be understood as facets of the same whole, a position that would come to be known as dual aspect monism.

Nature philosophy had a wide-ranging impact on early-nineteenth-century German arts and sciences (not to mention English Romanticism and American transcendentalism). German Romantic poets and writers took up Schelling's work, which encouraged them to forge an intimate link between their own inner lives and the natural world. Nature philosophy also encouraged scientists to study the possible unity of seemingly disparate phenomena. Physicists began to search for unity within forces like electricity and magnetism with striking results. Biologists linked different forms of inorganic and organic life in a grand evolutionary sweep. Carl Carus, for example, concocted schemas by which humans passed between the unconscious world of nature into the consciousness of man. In medicine, Nature philosophy encouraged doctors to experiment with psychic therapies for physical disease. The disciples of Romantic medicine reasoned that since the psychic and physical were intimately connected, the mind might somehow cure the body.

While Nature philosophers pursued the oneness of mind and body, others—to Kant's great dismay—abandoned the transcendental aspects of his thinking and took his work as a catalyst for the development of theories regarding human subjectivity. Kant gave rise to a slew of theoretical positions about the formative character of ideas and mental life. For example, Jakob Friedrich Fries and Friedrich Eduard Beneke founded a theory called psychologism, in which they reduced philosophical questions to subjective, psychological ones. The influential Johann Fichte dismissed Kant's transcendental world as a realm of spooks, ghosts, and gods but avidly embraced Kant to justify his own belief that the study of subjectivity and self-consciousness were crucial tasks for philosophy.

Once shorn of a metaphysical dimension, post-Kantian thinking was but a step away from brain science. The figure who bridged these two worlds was the brilliant misanthrope Arthur Schopenhauer. A former medical student and a student of Fichte's, Schopenhauer had been encouraged by Johann Wolfgang Goethe to take up the study of vision. From his studies, Schopenhauer developed a theory of colors that laid the groundwork for his influential philosophy.

In a generative misreading, Schopenhauer claimed that Kant's a priori categories were not transcendental but material: they were the result of brain activity. The brain projected colors into the world and wrongfully perceived them as existing out there. The brain made yellow and orange: there were no such hues in the setting sun. Four years later, Schopenhauer broadened this conception in his opus, *The World as Will and Representation*. All our mental representations were the result of inner psychic workings that had been projected onto the world. Schopenhauer called the inner force that caused these projections the "Will," which he claimed was Kant's unknowable thing-in-itself. (How Schopenhauer could say he knew what he himself defined as unknowable was a trickier question.) The Will was a blind striving force that was unconscious and biological: it was nature's storm within us. The Will pressed and pushed. It bullied the intellect, and invisibly shaped and skewed our mental representations of the world.

Schopenhauer's philosophy found adherents in the sciences who began to search for distorting powers in mental life. The scientist who did the most to connect post-Kantian philosophy to science was the father of human physiology, the physician Johannes Müller. Influenced by his own hallucinatory experiences, Müller set out to investigate the "fantastic apparitions of vision." In 1833, he published his massive *Handbook of Human Physiology*, where he offered scientific proof of Schopenhauer's contentions.

Müller's experimental findings shocked those who clung to the commonsense notion that their minds provided them with an accurate picture of the world. Müller demonstrated that wildly different stimuli produced the same sensation in a sensory nerve. If you stimulated the optic nerve with electricity, a lit object, or manual pressure, the result was always identical. The subject saw light and color. Therefore, Müller proved the external cause of a sensation was only arbitrarily related to what we saw. He argued that nerves of different sensory organs must have their own specific qualities. The optic nerve registered stimulation with light and colors, whether the stimuli were due to light and colors or not. Müller came to the remarkable conclusion that our so-called empirical knowledge of the world did not have any direct correspondence to reality. Our perception of the world was manufactured in the mind in the same way it made hallucinations, fantasies, and dreams.

II.

IF FREUD HAD been in search of a theory of mind that justified his notions of psychic causes and mental conflict circa 1830, he could have had his pick of the many models put forward by German Romantic physicians, *Naturphilosophen* as well as post-Kantian philosophers and scientists. But in 1895, Freud's path was not so clear. After 1850, Romantic medicine, Nature philosophy, and most forms of post-Kantian philosophy had fallen into disrepute. German metaphysics was subject to a tremendous backlash in the sciences that tarred any philosophical approach to the mind for decades to come. By midcentury, the Frenchman Auguste Comte spoke for many when he called abstract German philosophy false and failed.

The reaction against metaphysics was spearheaded by Johannes Müller's own students. Müller had established a lab for the study of physiology in Berlin. In 1847, his acolytes—Emil Du Bois-Reymond, Hermann von Helmholtz, and Ernst Brücke—met with Carl Ludwig, then a professor at Marburg. Together they formed the Berlin Physical Society and declared their object was to study human life without recourse to metaphysics. They especially targeted Müller's belief in an irreducible life force animating living things and agreed that speculations about vital life forces should be strictly forbidden. They would not assume a life or mind force animated the body, but would proceed as if life was completely determined by mechanistic processes. Life itself could be explained by chemical and physical analysis alone, they believed. With that as their bond, the Biophysics Movement began its research. In growing numbers, a new generation of German scientists attacked problems in human physiology from this point of view and were rewarded with a slew of important discoveries.

Hermann Ludwig Ferdinand von Helmholtz, 1894; German scientist, Kantian, and one of the founders of the Biophysics Movement.

Hermann von Helmholtz was the most famous of the group, though ultimately he was not representative of the whole. A late bloomer who did poorly in school, Helmholtz had been tutored at home by his father, Ferdinand, a Potsdam teacher and avid follower of Fichte. Though he aspired to be a scientist, Hermann took a more practical route and trained as a physician, before he entered Johannes Müller's lab. At twenty-six, Helmholtz burst into prominence. His scrutiny of Müller's belief in an irreducible life source led to questions about the nature of energy. In 1847, through his "Law of the Conservation of Force," Helmholtz showed that force could manifest itself in different forms—chemical, electrical, magnetic—while the overall amount of energy remained unchanged. Nature philosophy's insistence on a unitary life force helped produce this monumental discovery, which ironically then helped kill off that philosophy by furthering the belief that nature's secrets could be revealed if one simply tracked different transformations of energy. It now seemed that inorganic forces could become organic ones. The secret of life was that there was no secret; all of life could be reduced to dynamics, mechanics, and Newtonian laws.

This mechanistic view gained prominence during the 1850s and 1860s and kept squishy speculations about souls, spirits, and life forces at bay. As the leaders of the Biophysics Movement gained prestige, they sought to reorient their culture's assumptions about the nature of knowledge itself. Emil Du Bois-Reymond asked why the German word for "science" (*Wissenschaft*) did not connote the natural sciences (*Naturwissenschaft*) as it did in France, but rather the human sciences (*Geisteswissenschaft*). Human science was a term coined in 1843 by a German historian, who used it to cover disciplines like history, geography, psychology, and sociology, those studies central to the education of the German elite. The biophysicists sought to undermine the prestige of such science. In 1862, Helmholtz noted that unlike natural science, human sciences were subjective and based on the psychology of the knower. Du Bois-Reymond went further, shocking many by suggesting that the only valid cultural science was one that followed the methods of natural science. Over the next two decades, a debate ensued over the character of the different sciences, and the spectacular rise of biophysics increasingly pushed this debate toward the superiority of the natural sciences.

Freud's Vienna boasted three celebrated brain researchers dedicated to the biophysics program: Ernst Brücke, Sigmund Exner, and Theodor Meynert. One of the original founders of biophysics in Berlin, Brücke brought the challenge of this movement to Vienna, where he inaugurated its first physiological laboratory in 1849. He considered all nervous functioning, including brain functioning, to be a reflex action whereby incoming excitation traveled along sensory nerves and was discharged by motor nerves. Psychic events were the side-effects of reflex action and never caused biological events in themselves. Brücke's model had no room for human agency or free will. Man was a reflex-driven, churning machine.

Freud decided to study medicine after he heard Goethe's essay "On Nature" in which the poet romantically imagines nature as an all-embracing mother, but when he entered Brücke's lab in 1876, Freud felt he had found his intellectual home. For the next six years, Freud happily worked on neuroanatomical projects and became closely allied with the community of researchers in and around Brücke's lab. Freud idolized his professor and was deeply impressed by his two brilliant assistants, Sigmund Exner von Ewarten and Ernst Fleischl von Marxow. While Freud was close to the unfortunate, morphine-addicted Fleischl, Exner was the more dominant force in the lab.

Exner had studied with Helmholtz in Berlin, where he proved himself by achieving novel results on vision, and in 1875, he made a splash with contributions to the physiology of hearing. Along with Freud's friend Josef Paneth, Exner became an expert on the localization of brain functions, and in 1894, he attempted to synthesize the burgeoning knowledge on nerve anatomy, physiology, and function in his *Outline of a Physiological Explanation of Psychical Phenomena*. Exner argued that all psychic phenomena were explained by reflex action, the flow of nervous energy, and inborn centers for pain and pleasure. The human being was a doll, and consciousness was a mechanism inside the doll composed of inhibited and excited pathways of electrical stimulation. Exner resolved the mind-brain problem by discarding psychic factors, though he did harbor some regret about the end of free will, since it had such wide implications for matters such as education and criminal law.

After leaving Brücke's lab, Freud wanted to remain faithful to the mechanistic dictums he had learned, and this was not difficult. At the Vienna General Hospital, Freud worked under Theodor Meynert— the professor who later became distressed by Freud's interest in French conceptions of hysteria. When Freud arrived at the hospital, Meynert was famous, for in 1867, he had shown that the laminations of the cerebral cortex were layers of different kinds of neurons, a find that inaugurated the study of cell architecture in the brain. His anatomical research yielded numerous other discoveries, including that myelin formed over nerves during development, which implied neural functions were not set at birth but developed over time. This finding made it possible to imagine that some diseases of the nervous system might not be determined at birth by heredity, but instead might be the result of faulty maturation, a line of thinking Freud would later embrace.

Like Brücke and Exner, Meynert remained dedicated to understanding mental life as the interaction of mechanistic energies and forces. In 1877, he began writing a comprehensive textbook that would sum up his thinking on the brain and mind, and would yield explanations for psychopathology and psychiatric treatment. *Psychiatry: A Clinical Treatise on Diseases of the Fore-Brain Based upon a Study of Its Structure, Functions and Nutrition* was published in 1884. A tour de force, the book offered a complete model of how nerve cells and physiology were organized by automatic reflex actions and how these functions accounted for human psychology and behavior. Meticulously documented and hugely ambitious, Meynert's treatise was intended to cement his reputation as the greatest brain scientist in Europe. It was also an attempt to move psychiatry away from French degeneration theory toward developmental and anatomical thinking. In his introduction, Meynert blasted the "mystical" reliance on degenerative explanations, a position that must have encouraged Freud when he too began to distance himself from this set of assumptions.

Sophisticated, cultured, and astute, Meynert had created a model rich with implication. The brain, he argued, was split into lower and higher cerebral functions, both of which were reflex driven. The automatic, inherited reflexes of the subcortical centers were opposed, controlled, and inhibited by acquired associational reflexes in the higher regions of the cerebral cortex. Meynert asked his readers to imagine a young child who innocently reached out to touch a bright flickering flame. Guided by his lower reflexes, the child stretched out his hand . . . and was burnt. Afterward when these same reflexes guided the child's finger toward the flame, they were opposed and overridden by a learned reflex, that linked the bright, flickering light with sharp pain.

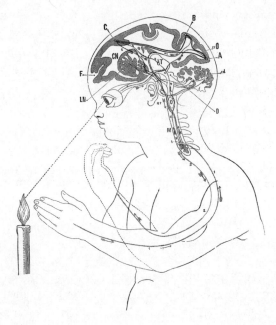

A Diagram of Conscious Movement as the Result of Cerebral Reflexes by Theodor Meynert, 1885.

With this two-tiered model, Meynert believed he had established a kind of psychoanatomy in which psychology could be wholly accounted for by a reflex-driven brain. Fatefully, Meynert pushed these contentions far, by rooting complex psychic functions in particular regions of the brain. Critics wondered if he had simply translated associational psychology into anatomical and physiological language. He had. But Meynert had also done something that set his model apart. Associationalists had long struggled to find a place for inner motivation, desire, and agency in their overly cognitive schemas. Meynert found a way to marry associational psychology with an internal driving force, and in doing so, he banked on the philosophy of Arthur Schopenhauer.

After publishing his great opus *The World as Will and Representation* in 1819, Schopenhauer lived much of his life in obscurity. But just before his death, he experienced a sudden revival. By 1880, his genius was heralded by Friedrich Nietzsche and Richard Wagner. Schopenhauer's fame was further abetted by the wild success of Eduard von Hartmann's *The Philosophy of the Unconscious: Speculative Results According to the Inductive Method of Physical Science*. Hartmann's book spoke to the growing number of thinkers who were committed to science but dismayed by the reduction of human beings into machines. The book was a hit, going through nine editions in fifteen years. In it, Hartmann demonstrated that Schopenhauer's philosophy was coherent with the most recent physiological thinking, offering a model of mental activity that was more refined than those that turned human beings into mechanical dolls.

Meynert agreed. In his lectures, he cited Schopenhauer while putting forward the view that our perceptions and associations were clouded, colored, directed, and transformed by a "Will-impulse." He believed the mind was not a mirror but created inner experience from internal templates. Men and women were driven by this Will in a primal bodily search for pleasure. Foreshadowing Freud's later theory, Meynert described the primal "I" as a collection of pleasurable bodily sensations. Unpleasurable sensations were pushed away from the self by defenses. In addition, a secondary "I" emerged over time and could fall into intense conflict with the primary body "I." Despite these resemblances to Freud's model of mental conflict, Meynert differed in a critical way; he held to the credo that ideas could have no causal role in these proceedings. His was a mind of dueling reflexes.

It was a brilliant synthesis, but to those intent on a solution to the mind-brain problem, there was less in Meynert's model than met the eye. Meynert simply asserted that the creation of ideas took place in specific cortical tracts. Unfortunately, there was no evidence that the assigned nerve tracts had ideas stuffed inside them, a fact not lost upon some readers. The erudite American William James honored Meynert by placing his work at the beginning of his massive, 1890 assessment of scientific psychology. But James rejected Meynert's psychoanatomy and said that any psychology must allow some causal role for ideas and feelings, which Meynert's model did not do. James asked his readers to translate Meynert's edifice back into pure psychology: "We thus get whatever psychological truth the Meynert scheme possesses without entangling ourselves on a dubious anatomy and physiology."

III.

As SIGMUND FREUD approached the daunting work of creating a new theory of the mind in 1895, he could look back on the efforts of his most brilliant teachers. Exner and Meynert were both committed to scientific quantification and mechanistic explanation. Both believed that the mind could be reduced to brain physiology and anatomy. And both, by pushing these beliefs as far as they could, exposed grave problems. Exner's account was publicly condemned by Josef Breuer and, among others, Exner's own brother, for it denied what common sense seemed to demand: some domain for human agency. Meynert found himself dismissed as a "brain mythologist" for his speculative attempt to equate ideas and neurons. Both carried the biophysics agenda into the dark woods of psychological investigation only to return beaten and bloodied.

Freud's training in Brücke's lab and his allegiance to biophysics was formative, but as he was trying to create a model of mind that allowed for the psychic causation of illness, it became clear that biophysics simply offered no help. This was part of a broader failure, for by the 1880s, the biophysics program had run aground. Physicists realized that purely mechanical theories of nature were inadequate, and physiologists turned their energies toward other projects. Nowhere were the limits of biophysics more apparent than in psychology. Biophysics offered no way of understanding psychology except by eliminating the object of study itself, the psyche, and maintaining de facto that all psychic events were identical to underlying brain events. Even the staunchest advocates for biophysics eventually conceded defeat. In a lecture given in Leipzig in 1872, Emil Du Bois-Reymond announced that the problem of consciousness was beyond the reach of science. Eight years later, he listed seven "world enigmas" and included among these imponderables consciousness and free will, issues vital to any psychology. This pessimistic pronouncement from the head of the Berlin scientific community would be reason to despair, had there been no other scientific method to study psychology. But that was not the case. As biophysics revealed its limits, a different scientific approach to the mind emerged. Freud was among a number of scientists who turned from biophysics to this alternative: psychophysics.

The man who pioneered psychophysics was the extraordinary Gustav Fechner. Fechner, a physician, physicist, and occasional writer of mystical philosophy, was determined to establish Schelling's Nature philosophy on scientific grounds. He could not conceive of nature as a dead machine, for how could a dead mother give birth to live children? Fechner adopted a pen name, Dr. Mises, and under that nom de plume published his *Little Book of Life After Death*, in which he outlined a developmental schema that linked all things organic and inorganic from nature to consciousness. Man lived in three stages: in utero, he slept; in life, he alternated between unconscious sleep and consciousness; in death, he entered an eternal wakefulness. Conscious thoughts and motives often derived from unconsciousness, as was made most manifest in madness and clairvoyance.

By the time Fechner published these reflections, the physicist had reason to hide behind a pseudonym. Among scientists, Nature philosophy had become an embarrassment. Fechner didn't care. After recovering from a three-year illness that brought him to the brink of insanity, he felt certain that all of nature was, in its essence, guided by a "Lust" or pleasure principle. The search for pleasure and avoidance of pain guided life, just as Newton's laws governed rocks and air. In 1848, Fechner published *Nanna, or the Soul Life of Plants*, a book in which he argued for such an omnipresent spirit force in nature.

If Fechner had continued in this vein into the materialistic 1850s and 1860s, he would have been marginalized as a throwback and an eccentric. However, in 1850, Gustav Fechner's work took a turn that made him a hero to scientists of the psyche. Fechner's teacher, E. H. Weber, was a physiologist who postulated a rule for the relationship between an outer stimulus and inner sensation. In 1850, Fechner took up the experimental study of external stimuli and their psychic representations. After a decade of ingenious research, he published his landmark work, *Elements of Psychophysics*. This book offered an innovative approach for studying the exact relations between mind and body. Unlike biophysics, psychophysics would not seek to reduce inner psychic experiences to a biological substratum. Following the dual-aspect approach of Schelling, Fechner saw the psychic and the material as two facets of the same entity, much like the convex and concave aspects of the same curved surface. The inner realm could only be experienced from the inside; it was necessarily subjective and psychic. The outer realm was objective and material. Psychophysics would bind the two in the study of their interactions, and track the transformation of energy as it went from physical stimuli to qualitative mental experience.

Fechner spent years measuring the effect of various sorts of stimulation on the inner world. Unlike his alter ego, Dr. Mises, Gustav Fechner employed rigorous scientific method and remained committed to skeptical questioning. Fechner was quick to consider problems with his own hypotheses and ready to cede priority to others. "By no means do I want to say that the contents of this work are absolutely new," he quipped, "and it would be a poor recommendation if they were."

Gustav Theodor Fechner; Nature philosopher and founder of psychophysics.

Fechner was partly right. He and Weber had been influenced and preceded by the philosopher Johann Friedrich Herbart. It was Herbart who suggested that one could measure psychic change by studying "liminal" events in which, for example, a sound finally could be heard. But Fechner operationalized these ideas to experiment with what he called "threshold" effects. For example, Fechner tracked the moment when the stars first became visible as day turned to night. He would measure the precise distance at which a candle disappeared from sight. By this method, he uncovered inner thresholds. Subliminal stimuli, he reasoned, registered in the mind and built until they crossed a line into consciousness.

By 1860, Gustav Fechner postulated the everyday existence of unconscious psychic contents. "As long as the stimulus or stimulus difference remains below threshold its perception is, as one says, unconscious," he wrote. He evaluated the conscious thresholds for vision, hearing, and feeling, and concluded that these barriers to consciousness were crucial for survival because they acted as protection from constant bombardment—the buzzing, whirring, pulsing world. If a mental trace fell below a certain level of energy, it would be unconsciously perceived, but it would not be consciously seen or heard.

Fechner called his groundbreaking studies "Outer Psychophysics," and he expressed the hope that one day a fuller psychology would include the study of how the psyche interacted with unconscious internal stimuli, which he called "Inner Psychophysics." When published in 1860, the *Elements of Psychophysics* was lauded by, among others, the young physicist Ernst Mach. Gustav Fechner had done what Immanuel Kant and Auguste Comte deemed impossible. He had found a way to make psychology—at least the psychology of perception—scientific and quantifiable.

Intellectuals who sought a scientific psychology lauded Fechner. The eminent sensory physiologist and psychologist Ewald Hering encouraged his students, including Josef Breuer, to adopt Fechner's approach. The famed German zoologist Ernst Haeckel adopted Fechner's logic and used it to challenge Du Bois-Reymond's belief that there could never be a science of consciousness. Du Bois's world enigmas, Haeckel retorted, were soluble by a science that did not give in to the reductive principles of biophysics. This other science embraced a dual reality, in which "matter cannot exist and be operative without spirit, nor spirit without matter."

Hering and Haeckel were powerful voices in German science. But Gustav Fechner's most eminent ally was none other than Hermann von Helmholtz. By 1860, the one-time hero of the Biophysics Movement had turned his formidable talents to the study of perception. In the process, he abandoned the reductionism of biophysics and adopted a dual approach to mind and brain. While the outer world could be known by mathematical, scientific study, the world as we perceive it internally was—as Johannes Müller had proved—based on the registering of qualities by our sense organs. Qualitative signs were read in the mind by an unconscious matching process that joined the new sign with the memory of a similar one, a process Helmholtz called "unconscious inference." This process went awry in illusions where the viewer unconsciously confused something from the past with a foreign present.

Helmholtz's theory of perception was deeply indebted to post-Kantian thought, so much so that an elderly Arthur Schopenhauer accused Helmholtz of plagiarism. But to his colleagues in biophysics, Helmholtz's theory of unconscious inference was nothing less than treason. It seemed to reopen the door the biophysicists had hoped to slam shut, a door that they feared led back to Romantic speculations about spirit, soul, and nonphysical mind forces. Still, Helmholtz was not dissuaded from considering the broader role unconscious mental processes might play. Given how richly elaborated and real our dreams feel while we are immersed in them, Helmholtz could not discount the possibility that the most extreme subjective idealists were correct. Was it possible that our minds so powerfully constructed our perceptions that the world as we knew it resembled a dream? One of Germany's most illustrious men of science refused to dismiss this possibility.

Helmholtz hoped to create a new field of study at the edge of natural science and human science, dedicated to comprehending the interactions of physiology and psychology. This borderland would soon grow crowded. In 1874, one of Helmholtz's former assistants, Wilhelm Wundt, published *Fundamentals of Physiological Psychology*, which laid out an ambitious agenda for the new field. Psychophysics, or physiological psychology, would bridge the world of Newtonian mechanics with the world of feelings, ideas, and mental representations. Neither physiology nor psychology would find a solution by being subsumed into the other.

In the last decades of the nineteenth century, psychophysics flourished in Germany and Austria. When Helmholtz died in 1894, Carl Stumpf praised him as the man who had done more than anyone to "bridge the gulf between Physiology and Psychology—a bridge across which thousands of other men now constantly come and go."

NOT LONG AFTER returning from Paris, Freud published neuroscientific papers in which he hinted at the insufficiency of a biophysics model of the mind. In an unsigned 1888 entry on the brain penned for a medical dictionary, the author, assuming his reader had a "fresh brain" before him, dutifully described how this organ worked by processing energy according to the theory of reflex action. Different states of consciousness resulted from these changes, but conscious experience itself, he insisted, was not simply the result of brain reflexes and also needed to be studied through introspection and the inner experiences of sensory perception.

In 1888, Freud further suggested that any theory that did not account for psychic causation flew in the face of common sense. Psychic causation could be demonstrated by such everyday acts as attention and the voluntary act of recalling a name, lifting a thought into consciousness. In its essence, human behavior was psychical. A man reached for a grape, not just because his bodily machinery moved, but because he wanted it.

The unnamed author also noted that ideas and memories could remain unconscious if ethical considerations stood in the way, and he suggested that several psychic intentions might act in competition. To German readers, these notions would have been attributed to some follower of one of the forefathers of psychophysics, Johann Friedrich Herbart. Herbart not only inaugurated thinking of psychological thresholds, he also devised an elaborate theory of mental conflict, in which ideas forcefully opposed one another as they struggled for consciousness. This resulted in the suppression of some ideas while others made it into our conscious awareness. Freud knew Herbart's work, since he had studied Gustav Adolf Lindner's Herbartian textbook of psychology while in the Gymnasium.

In turning away from biophysics, Freud was also following the inclinations of his mentor Josef Breuer, an ardent admirer of Fechner and a former student of Ewald Hering. Freud had been offered a position in Hering's lab in Prague in 1884, and greatly admired the physiologist's essay on unconscious memory, calling it a "masterpiece." Like many of his generation, Freud also revered the work of Hermann von Helmholtz, and his admiration did not diminish after the former biophysicist took up unconscious psychology.

But before embracing psychophysics, Freud had to contend with some serious criticisms. Franz Brentano had pointed out that this method was unable to tackle most psychological phenomena such as desire and actions of the will. Sigmund Exner had also argued against psychophysics. Around the time Freud attended Exner's course on the physiology of the senses, the Viennese brain scientist could be heard declaring that it was unreliable to work with quantitative sensations alongside qualitative perceptions.

Freud considered alternatives to the psychophysics model for studying the mind. In the same medical dictionary, Freud wrote a second anonymous entry on aphasias, illnesses that caused defects in speech or the comprehension of language. After completing this dictionary entry, Freud sank deeper into the study of aphasias and three years later published *On Aphasia: A Critical Study*. This slim work was animated by the author's rejection of Meynert's strictly anatomical approach to this illness. Freud echoed those who believed Meynert's mixing of psychological and anatomical language only fostered confusion. One could not glibly say a nerve housed a word or an idea.

Wilhelm Wundt later considered Freud's work on aphasia an argument for psychophysical analysis, but while studying aphasias Freud actually was probing a different approach to mind and brain. Freud suggested that one might consider mind and brain events to be running in parallel, citing the English neurologist John Hughlings Jackson, who championed "psychophysical parallelism." Hughlings Jackson wanted to simplify neurology by segregating physical and psychic systems into distinct operations that did not affect each other. This lack of intersection in these two realms liberated the neurologist from problems of psychology, and conversely, allowed the psychologist to cease troubling over the mysterious influences of the body. Hughlings Jackson was quick to admit that this was merely a pragmatic distinction meant to facilitate research in an overwhelmingly complex field. He also hoped this segregation would force a moratorium on crude, unsound attempts to simply bypass the problems of how the psychic and physical interacted.

Psychophysical parallelism may have been useful for the study of speech, but in the long run it proved useless to Freud. Sealing off the mind from the body could not help a doctor who was deeply engaged in the study of hysteria. Freud's experiences at the Salpêtrière convinced him that the mind affected the brain and body. He was closely associated with attempts to affect the body through the mind, and had practiced suggestive cures, hypnosis, and psychical treatments to do just that. Freud could not afford to isolate the mind from the brain. He would have to find another way.

IV.

As *Studies on Hysteria* was nearing publication, Sigmund Freud worked feverishly to pull together a grand theory of mental functioning in which there was a role for psychic conflict and defense. This passionate effort became what he called his private "tyrant." After a long day of clinical work, Freud sat down around eleven in the evening and labored deep into the night as he sought to integrate physics, biology, neurology, and psychology. Fueled by cocaine, he hoped to make a psychology for neurologists, a natural science of psychology, a work that came to be known as Freud's "Project for a Scientific Psychology."

Freud's defense of psychic defense grew gargantuan. He found it necessary to take up a panoply of issues: "quality, sleep, memory—in short all of psychology." Undaunted, he forged ahead. As summer turned to fall, he spun out multiple drafts, filling one notebook after another with sketches for a model that reached down to the most basic neural processes and up to the complexities of inner experience. Understandably, Freud relied on many assumptions that had guided his teachers. A September 25, 1895, draft began as if it were an 1850s biophysics manifesto: "The intention is to furnish a psychology that shall be a natural science: that is, to represent psychical processes as quantitatively determinate states of specifiable material particles."

To do this, however, Freud took an unabashedly Kantian approach. By the 1880s, a "Back to Kant" movement had gained momentum in German circles, calling for the rejection of speculative metaphysics as well as materialist speculations that commonly passed under the banner of science. During this period, Freud bought and annotated Kant's *Critique of Pure Reason*, and the influence of Kant is apparent in the "Project." Like the philosopher, Freud allowed that deductive theories would guide the search for empirical evidence, and he organized his scientific psychology around two a priori theorems that would connect facts. These theorems, in turn, would be reshaped and amended in accord with new evidence.

Freud's a priori theorems came from physics and biology. First came Newton's laws of motion and the conservation of energy. In 1892, Freud had recognized the power of applying these thoughts to psychology. Writing to Breuer, he explained that the nervous system strived for constancy by disposing of incoming excitation through association or motor action. Ideas could also disturb the equilibrium of the nervous system, since the affects associated with ideas needed to be discharged. In Freud's "Project," this principle would guide all mental functioning.

Freud's second theorem was founded on the basic anatomical unit of the nervous system, the neuron. These cells were the essential unit for processing neural energy and were charged with energy. Some could dissipate that energy by transferring it, but others could not. Any extrapolations to psychology would have to be consistent with the ways nerves worked.

With these guiding theorems in place, Freud began to build. Just as the nervous system had evolved different motor and sensory neurons, he hypothesized that the brain had developed specific neurons that performed different functions. The permeable "Phi" nerve cells took in new stimuli and were critical for perception. The more impermeable "Psi" cells held their contents tight and explained memory. Lastly, Freud boldly postulated a third system, the "Omega" neurons, which transformed quantities of energy into psychical qualities. These cells were the magical entities that turned energy into inner experience, consciousness, and perception.

But how did this bit of wizardry occur? Freud, like many scientists who came in the wake of Johannes Müller, contended that qualities were not found in the world, but were made in the brain. He proposed that the periodic rhythm and pattern of quantitative stimuli must be perceived as quality. Freud inferred that the Omega system perceived quantitative periodicity and transformed it into signs. Consciousness thereby resulted.

The "Project" also included a dynamic model of the mind. Like Schopenhauer and Meynert, Freud placed a wishful drivenness at the center of mental life. Attention was focused by this internal press for satisfaction. We look for what we want to find. And like Meynert, Freud maintained that higher processes inhibit this driven action. Psychic defenses were encoded, remembered, and reproduced whenever needed. Defenses were not just pathological, but critical for health, making it possible for humans to distinguish between inner desire and outer reality, to rise above animal needs and develop reason. Unfettered wishing persisted in only one place: dreams.

There was much more in this feverish sketch, ranging across psychology and psychopathology. In the end, Freud reported that "everything seemed to fall into place, the cogs meshed, I had the impression that the thing now was really a machine that shortly would function on its own." But Frankenstein's monster would never blink its eyes and rise from the lab table. By late November, Freud confessed: "I no longer understand the state of mind in which I hatched the psychology." It was a "kind of madness." He put the drafts away and never published them.

The "Project" was a trial run that let Freud take the measure of various philosophical and scientific conceptions of the mind. Central to this effort was Freud's experimenting with a Meynert-like linkage of psychology and anatomy, based on his Phi, Psi, and Omega neurons. While such biological grounding seemed to offer tremendous advantages, Freud would ultimately conclude that it glittered like fool's gold. As Breuer warned, if "instead of 'idea' we chose to speak of 'excitation of the cortex,' the latter term would only have any meaning for us in so far as we recognized an old friend under that cloak and tacitly reinstated the 'idea.' " The truth was that there were no such things as Phi and Psi and Omega neurons; they were just made-up, pseudo-scientific names for psychological functions.

In the "Project," Freud also had a chance to evaluate the promise of psychophysics. By maintaining a psychic dimension in his model of the brain, Freud made a place for consciousness, human agency, psychic causation, and psychic defense, and he was able to utilize the data that came from introspection. But this model was quite restricted in that it required a focus on phenomena where psychology could somehow be linked to physical changes. While this framework could make a psychology of perception, as Brentano pointed out, it had little to offer on other psychological matters. Psychophysics had forced a union between inner psychical experiences and physiological changes, whether the parties were ready for such a marriage or not. In 1890 the most prominent spokesman for psychophysics, Wilhelm Wundt, publicly renounced this approach and advocated studying psychology alone. Stung by this betrayal, Hermann von Helmholtz dedicated a good deal of energy during his last years to propping up psychophysics. It was a futile effort. As Wundt's experimental psychology grew in popularity and prestige, it stood as an accusation to the older, more daunting project.

In this shifting theoretical landscape, Sigmund Freud's "Project" was written and abandoned. By 1896, the old lions—Brücke, Meynert, and Helmholtz—had all died. Freud still pursued the dream of making a grand schema of the mind, but he seemed lost. In February of 1896, he wrote: "I am continually occupied with psychology—really metapsychology" and ten months later still reported vainly pursuing his "ideal and woebegone child—metapsychology." While this chase was filled with a youthful desire for philosophical knowledge, it remained focused on the phenomena that dominated post-Kantian psychologies. Freud wrote wistfully, "If I could give a complete account of the psychological character of perception . . . I should have described a new psychology." However there was no end in sight to this project, and Freud knew it. As 1897 arrived, he tried to put on a brave face, exclaiming to his colleague, Wilhelm Fliess: "We shall not be shipwrecked. Instead of the channel we are seeking, we may find oceans. . . . Give me another ten years, and I shall finish the neuroses and the new psychology."

V.

ON OCTOBER 23, 1896, Sigmund Freud's father, Jacob, died. It was a cataclysmic event that shook the forty-year-old physician and inaugurated a period of grief and self-examination. Later Freud would call a father's death the "most important event, the most poignant loss, of a man's life." To tend to himself, Freud, who suspected he himself had a case of neurosis, began a psychical self-analysis. While there was a long and venerable tradition of physicians experimenting on themselves and subjecting themselves to psychological inquiry, this was something stranger. For how could anyone expect to objectively inquire into the unconscious bends of his *own* subjectivity?

If trained self-observation had grave defects for the observation of conscious life, it was, by definition, blind to the disassociated unconscious contents that Freud looked for in his psychical analysis. Romantics who sought out the unconscious recesses of the mind experimented with hashish or other intoxicants, but if Freud did that he would lose any claim to naturalistic study and enter what Baudelaire called *"les paradis artificiels."* Autohypnosis offered possibilities, but without a lab and associates to observe him, what would Freud recall of these lost hours?

Freud turned to his dreams. Extrapolating from the Nancy school's belief that hypnosis was a form of sleep, Freud could consider dream life analogous to unconscious, hypnotic hallucinations. Long before his father's death, Freud had been curious about dreams, and he had begun to take them more seriously as they entered his analyses with neurotics. In July of 1895, while vacationing at the Hotel Bellevue in Kahlenberg overlooking Vienna, Freud had a long, vivid dream he would return to again and again. In it, he was at a party, and conversation with other doctors turned to a treatment that had gone badly. The dream of "Irma's injection," as it came to be called, eventually helped convince Freud that in this way the unconscious could be accessed. Analysis of this inner spectacle confirmed what Freud had postulated years before: dreams were like mirages conjured by men in the desert. They were hallucinations of desire.

After that summer night, Freud began to pay closer attention to his dreams. A year later, on the night after his father was buried, he had another dream. He was in a strange place with a sign that read: YOU ARE REQUESTED TO CLOSE THE EYES. Freud recognized the location: it was his barbershop. On the day of his father's burial, he had been kept waiting there and had arrived late at the funeral home. When he came, his family was angry and offended. Freud interpreted the dream as a command to fulfill his filial duty and a self-reproach for neglecting his father.

You are requested to close the eyes. This command took on more meaning as Freud became less concerned with the psychology of perception and turned to the hallucinatory world of sleep. In the spring of 1897, he started working on a theory of dreams where he felt "so very certain." A few months later, he systematically began his self-analysis. Presuming a willful impulse lay buried in every dream, Freud proceeded to record his dreams after his father's death. To pry open a window into himself, he developed techniques, like writing out these dreams, then rewriting them and analyzing the differences between drafts. The changes, Freud reasoned, would be the result of his own defenses and would point to areas of conflict. It was an ingenious method, a way for a man to think against himself. Freud's investment in his self-analysis grew. "My self-analysis is in fact the most essential thing I have at present," he told Wilhelm Fliess on October 15, 1897.

In the same letter, Freud revealed a primal wish he had uncovered. "I have found, in my case too, being in love with my mother and jealous of my father, and I now consider it a universal event in early childhood . . . if this is so, we can understand the gripping power of Oedipus Rex. . . . Everyone in the audience was once a budding Oedipus in fantasy and each recoils in horror from the dream fulfillment here transplanted into reality." Through his perusal of his own dreams, Freud also understood that childhood jealousy and rivalry had left him with great guilt, especially after his younger brother Julius died.

Freud completed several chapters of a new book on dreams. Shifting his attention to this inner theater led "more deeply into psychology than I had imagined." Indeed. After much writing and rewriting, Sigmund Freud published *The Interpretation of Dreams* in November of 1899. In its finished form, it was really two books. One was a detailed description of a new method for discovering the true meaning of dreams; the second was a psychophysical model of the mind that took dreaming, not perception, as the mind's paradigmatic activity. Unlike anything written by French psychologists, Freud's dream book would be part Kant and Schopenhauer, part Brücke, Exner, and Meynert, part Helmholtz, Hering, and Fechner, and in its final synthesized form, Freud.

THE WORLD OF dreams had long been a source of fascination for philosophers, poets, and writers. Why do we dream? What do dreams represent? Is dreaming a fundamentally different form of consciousness? Why do we remember some dreams but not others? In the *Interpretation of Dreams*, Freud revisited these mysteries, mingling his own reflections with those of philosophers like Aristotle, Artemidorus, Kant, Schelling, and Schopenhauer.

Bedeviling questions had also come up in the scientific study of dreams, for by the time Freud took up his studies, dreams were a well-established arena for empirically studying the undercurrents of mental life. In 1894, when American psychologists compiled a *Psychological Index* listing all scholarly works from English, French, and German contributors, they dedicated one subsection to "Sleep, Dreams, Sub-consciousness." Every year during that period, one could find ten to twenty works on the subject. These studies took different approaches, but the two most important schemas for understanding dreams came from discursive communities Freud knew well: hypnotists and psychophysicists.

Ambroise Auguste Liébeault actively pursued his contention that hypnotism was a form of dreaming, a claim that resonated with the old belief that dreams were a kind of sleeping madness. "Kant writes somewhere (1764) the madman is a waking dreamer . . . ," Freud informed his readership, and "Schopenhauer calls dreams a brief madness, and madness a long dream." If taken seriously, these aphorisms implied that by penetrating dreams, one might also bring to light the mysteries of mental illness.

Dream life had also been taken up by psychophysicists, despite the grave difficulties it presented for them. Psychophysics focused on the way external stimuli registered in perception and consciousness. Dreams seemed to be the very antithesis of that. Here was a vivid experience that occurred while the eyes were closed. Little if any perceptual stimulation seemed involved. Nevertheless, psychophysicists clung to their method and postulated that external stimuli silently invaded sleep consciousness and instigated a flurry of psychic activity. Wundt argued that for a sleeper, a loud wind at night might become a pack of howling wolves. Dreams were provoked illusions, he insisted.

If mental life was scientifically knowable, Freud insisted, it must be lawful, deterministic, and not based on chance events. Dreams could not simply be the result of shutters banging in the night but had to be primarily inwardly determined. Proceeding as he had when confronted by debates over suggestion, Freud asked: What are the internal, lawful causes of dreams? Many had already suggested that inner physical stimuli like pangs of hunger or indigestion might catalyze dreams. But what about inner psychological sources? Wundt denied this possibility, and Freud, in turn, dismissed Wundt and his followers as ideologues who were prisoners of their own beliefs. As psychological experimentalists, they overvalued everything that was open to experimentation. He challenged them and those psychiatrists who shrank from psychological forces, "as though recognition of such things would inevitably bring back the days of the Philosophy of Nature and the metaphysical view of the nature of mind." If scientists could not yet reach down from mental life to find connections with biology, that was no reason to hide from the problem.

Freud was on solid ground with this critique. Wundt and his colleagues merely revealed the limits of psychophysics by insisting on the far-fetched notion that external cues generated all perceptual experiences in dreams. In contrast, Freud proposed a method for the study of inner psychic causes. Much as he had intuited while writing about hysteria in

1895, that method would be a cross between literature and physics, and would require literary analysis as well as a model for the processing of bodily energy.

Freud's approach had developed considerably since his 1895 dream at the Hotel Bellevue of Irma's injection. In fact, he had become so sure of himself, that he imagined a future plaque at that very hotel, proclaiming: "Here on July 24th 1895, the secret of the dream revealed itself to Dr. Sigm. Freud." In *The Interpretation of Dreams*, the dream of Irma's injection functioned as a specimen to be dissected psychologically for his readers. It took Freud less than a page to tell the dream, but another ten to unravel it. As if reading ancient poetry, Freud worked line by line, phrase by phrase, detail by detail, searching for lost contexts and connections. He relied heavily on the feelings, memories, and stray thoughts that the dream provoked in the waking subject, always on the lookout for details that might reveal larger meaning. He proceeded too as if the dream were a hysterical fantasy. His job was to rebuild the broken associations that might make it whole and meaningful.

This exegetical approach was daunting, because unlike a hysterical symptom, dreams were not observable. Worse still, they were often barely remembered and only made it to consciousness in tatters. Freud clung to the idea that the thoughts that disrupted the dream were not random but clues in themselves. He only needed his patients to faithfully report the events as they experienced them. Unfortunately, that too was problematic. To make his method work, Freud required a level of cooperation and openness that was difficult to imagine. In the dream book, Freud suggested a kind of trained introspection, the controversial strategy long advocated by philosophers but called into question by Comte, Brentano, and others. Nonetheless, Freud asked his patients to pay close attention to their inner life, their fleeting thoughts and feelings, and not edit their observations. They were not to criticize or delete, but simply report. To facilitate this inwardness, Freud asked his patients to lie down on his Turkish divan. He requested that they close their eyes to shut off the overwhelming stimuli that came through vision, and facilitate an intense focus on the inner world of fantasy, thought, and feeling. *You are requested to close the eyes.*

Analyzing the story of Irma's injection, Freud showed how the dream concealed a hidden wish to be acquitted of guilt. From that, Freud put forward the starkly simple assertion that a dream was the fulfillment of a primal wish. Manifold kinds of wishes accounted for a multitude of dreams. If you ate salty anchovies before bed, you might dream of drinking. Freud told of a young woman who was sequestered from society while she nursed a sick child. During that period, she dreamt of a festive party attended by dignitaries like Alphonse Daudet. It seemed clear and rather simple. Perhaps too simple.

If dreams were hallucinatory attempts to return to primal experiences of satisfaction, how was going to a party, even a great party, a primal experience? And what about nightmares? Were those wish fulfillments? While the underlying meaning of every dream was the fulfillment of a wish, Freud added this caveat: primal wishes did not make their presence known directly. They came forward dressed in disguise. The mind was like "the political writer who has disagreeable truths to tell those in authority," and can only do so through indirect symbols that evade the censor. Wishes can only be expressed if they elude the psychic censors. What people remember about their dreams was merely the inoffensive, manifest content. Deeper, more disturbing wishes lay beneath innocuous dreams, like the young woman's about enjoying an evening with celebrities. That more profound layer could be reached by a symbolic unpacking that yielded the latent meaning of the dream.

Furthermore, dreams had a syntactic structure of their own. They were not linear and logical but condensed, made up of a knotted, symbolic language. Each of these symbols was the tip of an iceberg, a nodal point in a network of hidden significance. Any one dream element was not just the result of one psychic element but was "over-determined" by numerous other buried sources. Dreams were also structured by what Freud called displacement. High intensity ideas were smuggled into consciousness by attaching themselves to bland but related elements that could sneak past the censor.

Clearly Sigmund Freud had proposed a radically new way of thinking about dreams, but he believed he had done far more. Dreams had led him to "a discovery of general validity" about consciousness and the structure of the mind. In its essence, the mind was a dream weaver. Therefore, the study of dreams could yield what "we have hoped for in vain from philosophy," a valid and wide-ranging model of the mental life.

The second task of the *Interpretation of Dreams* was to make good on that promise, and the result came to be known as Freud's topographic model of mind. His thinking here was not just stimulated by his exploration of dreams but also by reading Gustav Fechner. In 1898, Freud wrote his friend Wilhelm Fliess to complain about the countless dull texts on dreams he had waded through. Yet one author had struck a chord. The only "sensible thought occurred to old Fechner in his sublime simplicity: the dream process is played out in a different psychic territory. I shall report on the first crude map of that territory." In his published account, Freud put it somewhat differently: "the great Fechner puts forward the idea that *the scene of action of dreams is different from that of the waking ideational life*. This is the only hypothesis that makes the special peculiarities of dream life intelligible." Inspired by this insight, Freud created a new theory.

To do so, Freud did not dump the old a priori principle that insisted all mental phenomena be predicated on the physical laws of the universe, but he did drop any attempt to tie psychic functions to any specific neuron or brain region:

> I shall entirely disregard the fact that the mental apparatus with which we are here concerned is also known to us in the form of an anatomical preparation, and I shall carefully avoid the temptation to determine psychical locality in any anatomical fashion. I shall remain upon psychological ground.

Freud conceptualized the unconscious and consciousness, not as located in this or that part of the brain, but rather as organized, psychological structures. His crude map of these psychic localities appeared in the dense seventh chapter of the dream book, where the author proposed that conscious and unconscious psychical domains were separated by a barrier. But unlike Fechner's notion of a simple stimulus threshold, Freud proposed that an active defensive barrier separated conscious and unconscious domains. A "censor" acted as a frontier guard to prevent disruptive elements from invading consciousness.

With that, Freud had done it. He had constructed an elegant model of the mind that made sense of psychical defenses. The mind was not just a passive container of sensations and experiences, but was actively self-regulating and guarded consciousness. Mental contents fell into three categories: they were "Conscious," "Unconscious," or potentially conscious, that is "Preconscious." Mental events could now be understood through the analysis of two opposing psychical forces: wishes and censorship. Following Kant, Freud's unconscious was unknowable in itself and unstructured by considerations of time, space, and causality. It held no opposites, no antitheses, and no logical categories but rather was the stuff of animal passion. The power of this hidden domain was only known indirectly through its effects on consciousness. Nothing from the unconscious reached consciousness without passing the defenses.

Sigmund Freud, like the German Romantics and Nature philosophers, believed the unconscious played a vital role in psychic life, as made obvious by dreams. If this emphasis on dream life smacked of Romanticism, it was countered by Freud's heavy reliance on mechanistic logic. Excitations entered the perceptual system and exited with motor action, returning the system to a lower state of energy. Tension was unpleasurable, and the release of tension was the very definition of pleasure. In sleep, the inhibition of motor release made neural excitation flow backward, leading to a rekindling of unconscious memories. Freud called this "regression." This backward flow of energy led to the excitation of the image of satisfaction preserved from childhood. In sleep, regression led to hallucination, while in waking life, bitter experience had proved that hallucinations were not in themselves satisfying. Since an image of water did not quench one's thirst, these magical aspirations were increasingly held back as painful and frustrating. Censorship prevented such wishes from infiltrating consciousness: the only time they won free passage was during sleep.

For Freud, the mind had become a city of dreams. Dreaming was no longer an odd phenomenon. Rather, it structured and elucidated the problems of consciousness and perception. Subjective vision, illusion, fantasy, dreams, and the distorting force of the Will: for nearly a century German philosophers and scientists had used them to glimpse the psyche's role in structuring our knowledge of the world. In Freud's model, these forces were primary, so primary that they left little role for much else.

And so, Freud faced another problem. Anyone daring enough to make grand claims for either a physiological or psychological unconscious inevitably faced a troubling question: What was the role of consciousness? Freud found help answering this question in the work of a contemporary German philosopher and psychologist. The Munich professor Theodor

Lipps would be remembered for developing the concept of empathy, but he also elaborated ideas on unconscious processes. In the summer of 1898, Freud was amazed to find his own thoughts stated in Lipps. Nearly a month later, Freud noted that he and Lipps had concocted the same notions of "consciousness, quality, and so forth." The seminal idea that Freud found echoed in Lipps was this: consciousness could be seen as a kind of inner eye, a sense organ for thoughts and feelings.

In the dream book, Freud employed this analogy. He wrote that some who accepted the power of the unconscious had grave difficulty finding a proper role for consciousness, but: "We, on the other hand are rescued from the embarrassment by the analogy between our Cs. system and the perceptual systems." Using this framework, Freud defined the interaction of the unconscious and consciousness by basing it on the way perception worked. The unconscious was like the outer world; it presented unknowable quantities to consciousness, which functioned like an eye or ear and turned these quantities into perceivable psychical qualities. The chaotic buzz and whir of unconscious inner life was made coherent and perceivable by consciousness.

Hammering the idea home, Freud wrote: "in its innermost nature it [the unconscious] is as much unknown to us as the reality of the external world, and it is as incompletely presented by the data of consciousness as is the external world by the communications of our sense organs." And so, Freud's theory of mind became doubly complicated. The world out there was unknowable-in-itself; the inner world was dominated by an equally unknowable-in-itself unconscious realm. Consciousness was a feeble lamp that flickered amid darkness inside and out.

Only by studying the distortions of consciousness could one recognize the hidden hand of the unconscious. The unconscious made itself known by subtle infiltrations into consciousness, a process Freud called "transference." Freud had already used the same term, *Übertragung*, in both a nonspecific way that denoted a transfer of neuronal energy and as a psychological concept that described hysterical false connections. Now he yoked both prior meanings together and placed them at the center of his new method. An unconscious wish transferred its energy onto some preconscious idea or perception. In dreams, the preconscious was dense with impressions and recent memories that offered a "point of attachment for a transference." In this description, Freud's theory of misknowledge was not unlike the process of visual illusion described by Helmholtz and Theodor Lipps. But these models were of cognitive miscues more than willful misreadings. For Freud, we get it wrong because we want to.

This was Freud's path to the darkness inside. The interpretation of dreams was his "royal road to the unconscious." Dreams, those airy nothings of sleep, those mysteries conjured in the theater of our unconscious life, had become the very foundation of the human mind. And unlike Romantic poets who may have shared this belief, Freud boiled dreams down to a single, lawful, universal mechanistic cause. Dreams came from primal wishes. The mind was organized around the energetics of wishing and the defenses that rose up against such forces. Wishes were the engine of fantastic dream visions and everyday desire. Only censorship kept us from living in a continual state of hallucination and made way for logic, reason, and more sadly, defense neurosis.

With the *Interpretation of Dreams*, Freud's effort to explain psychological causation and mental conflicts was complete. His studies of hysteria and his engagement with French psychopathology led him to recognize that psychic causes were real, and five years later, after traveling down a long road populated by post-Kantian philosophers, biophysicists, and psychophysicists, Freud arrived at a model of the mind that would ground his ideas for years to come. After experimenting with biophysics, psychophysics, and psychophysical parallelism, Freud emerged with a way to scientifically stabilize the object of his study—the mind-brain—in a manner that made space for psychic causes, without succumbing to either biological reductionism or metaphysics. His model avoided the temptation to make psychology seem more scientific with speculative anatomical reifications, but he still insisted the mind acted in accord with Newtonian laws. By melding ideas with energy, the psychic with the physical, Freud opened up a rich, bounded space for the naturalistic study of inner subjectivity, agency, and intention.

If Sigmund Freud's book on dreams held much in it, few noticed. *The Interpretation of Dreams* sold poorly, and while it did catch the attention of a few in German and Austrian literary circles, it was to the author's great disappointment mostly ignored by physicians and scientists. Still, Freud moved on to questions generated by his theory of a wishful unconscious.

What were these wishes? Did all human wishing come from the same inner source? Was there lurking in the unconscious something universal and lawful that connected all human beings, from aborigines in New Guinea to clerks in London and nuns in Rome? Defining the origins of human desire would be the final step in Sigmund Freud's theory of the mind, but to get there he would have to return to an unsolved mystery that had perplexed him for years.

THREE

The Unhappy Marriage of Psyche and Eros

It is a raging serpent she must wed . . .
　　　　　　　　　　—Apuleius

I.

As THE NEW century began, Sigmund Freud could look back at fifteen years of labor that resulted in dovetailing theories of normal psychology and psychoneurosis. Both theories centered on how the psyche needed to defend itself from its own inner forces in order to maintain a kind of inner stability. But he had still been unable to precisely define those forces that disrupted the mind. What were the unconscious wishes of dream life, or the repressed memories that caused hysteria? In *The Interpretation of Dreams*, Freud suggested these questions could not be fully answered. The unconscious was like the Kantian noumena, unknowable in itself. However, just five years later, the same Sigmund Freud would state definitively (a chorus of critics would say *too* definitively) that he knew the contents of the unconscious. With this claim, he would insert the last piece in the puzzle needed for a whole Freudian theory, one that rooted the mind in the body, human reason in animal passion, and the individual in the species. To find that missing piece, Freud turned to a developing science, one that chased secrets long hidden in bedrooms, brothels, back alleys, secret societies, bathhouses, bars, and morgues. The linchpin that would consolidate Freud's theories of neurosis and general psychology came from the science of sexual life.

Sexuality had long concerned doctors, but in a limited way. Physicians interested themselves in how sexual habits or mishaps played a role in illness. As a young man, Freud became interested in the possible role of sexuality in neurosis, despite the fact that Jean-Martin Charcot and his coworkers had mounted a vigorous campaign to squash the old-fashioned idea that women became hysterical due to sexual frustration. Freud recalled that Charcot and his team considered the very idea of linking hysteria with sexual matters "a sort of insult." Nonetheless, some physicians clung to the idea that hysteria was *une chose génitale*.

One such doctor was the Berlin pediatrician Adolf Baginsky, who believed that masturbation could cause infantile hysteria. On his way home from Paris, Freud visited Baginsky to prepare for a job he had landed as a consultant at Max Kassowitz's pediatric clinic in Vienna. Over the next eleven years, Freud consulted on childhood nervous disorders at Kassowitz's clinic, and during that time he too began to believe that sexual overstimulation in childhood lurked behind some nervous disorders. At the same time, Freud was treating adults with traumatic hysteria, and he began to suspect that these traumas were sexual in nature. As he worked through these ideas, he relied on his confidant: Wilhelm Fliess.

The son of Sephardic Jews, Fliess had been schooled in Berlin, where he studied medicine with Helmholtz and Du Bois-Reymond. In 1883, he opened a general medical practice in Berlin, but his health deteriorated, and in 1886, he decided to take a year off to recover. He traveled to Italy and France before a three-month stay in Vienna in 1887. During that time, Fliess met Freud, and a great friendship commenced.

At first glance, the two men made an unlikely pair. Wilhelm Fliess was an internist who focused on disturbances of the nose, while Sigmund Freud was a man bent on explicating the neuroses and the mind. However, both shared a fascination with the way reflexes in the nervous system choreographed physiological life. Correspondence between the two grew intense around 1893, the same year Fliess published his theory that a vast complex of bodily symptoms could be tied to what he called a "nasal reflex neurosis." Fliess argued that through reflex action, nasal problems could become systemic and affect the sexual organs. Conversely, it followed that sexual disturbances could manifest themselves in the nose. In fact, according to Fliess, some nasal troubles stemmed from neurasthenia, a disorder he and a "foreign colleague" believed was due to "sexual abuse."

Neurasthenia, a psychiatric diagnosis introduced by the American physician George Beard in 1869, was a hodgepodge of symptoms exemplified by nervous exhaustion, malaise, and lethargy. Beard maintained that in these cases excess incoming nervous stimulation led to excess discharge. Behind the resulting debilitation lay some unknown hereditary vulnerability that malfunctioned when confronted by the screaming pace of urban life. Beard also argued for a specific sexual neurasthenia, which manifested itself in sexual weakness and impotence. He warned that early, excessive masturbation as well as unnatural forms of sexual activity, such as prolonged excitation without orgasm, coitus interruptus, and the use of condoms, were potential causes of this disorder.

Beard christened neurasthenia an "American nervousness," but the disease refused to honor national boundaries. In his practice, Sigmund Freud—Fliess's foreign colleague—began to see a number of neurasthenics. Unlike wildly jumpy, dramatic hysterics, these patients appeared enervated. Following Beard's sexual neurasthenia, Freud focused on the role of sexuality in their bodily collapse. Freud and Fliess both concluded that excessive masturbation was the sexual abuse that caused neurasthenia.

Masturbation had long worried physicians, but by the 1870s, it had become commonplace for medical authors to scoff at the list of terrifying ailments physicians once blamed on this practice. At the same time, these authorities often went on to list a host of urological, gastrointestinal, and nervous disorders they believed afflicted onanists, including epilepsy, impotence, neurasthenia, homosexuality, and hysteria. Fliess and Freud now added their voices to this grim chorus.

In late-nineteenth-century Europe, masturbation and other sexual practices were hidden behind a wall of shame, lies, and silences. Freud took it upon himself to knock down this wall and uncover the practices that were harming his patients. After one such discovery, he wrote proudly: "someone who had not searched for it as single-mindedly as I did would have overlooked it." Undeterred by concern that suggestion might distort his results, Freud continued his investigations. Still, he could not help but note his growing isolation. In the spring of 1894, Freud wrote Fliess, complaining about his Viennese colleagues: "They look upon me as pretty much of a monomaniac, while I have the distinct feeling that I have touched upon one of the great secrets of nature."

That great secret was that the cause of traumatic hysteria was not factory accidents, railway rumblings, or random falls, but long forgotten sexual shocks to the nervous system. In May of 1894, Freud sent Fliess a brief outline of his theory called: "On the Etiology and Theory of the Major Neuroses." Inner sexual excitations normally operated via the theory of constancy, but sexual "noxae" disrupted that equilibrium. The disturbance might pass from the body to the mind as somatic tension built and transformed into psychic tension, which led to repression and psychoneurosis.

Freud's explanation owed its language and premises to bio- and psychophysics. It was similar to Josef Breuer's rationale for the cathartic method, not to mention Fliess's nasal cauterizations. The idea was to decrease internal psychic or physical stimulation and thereby provide hysterics with relief. More ominously, German surgeons and gynecologists had taken to performing female castrations to lessen internal stimulation and attempt to treat women deemed neurotic, hysterical, or simply sexually unruly.

This theory of excess stimulation and needed release came together just as Freud was writing his sections of *Studies on Hysteria*. While tempted, he ultimately decided not to add these ideas to that book. Instead, Freud published a separate paper in which he boldly proclaimed that neurasthenia was caused by masturbation. Always. The connection, he suggested, had been long obscured by inaccurate diagnostic practices. Doctors had confused cases of neurasthenia with what Freud now wanted to call "anxiety neuroses." Unlike deadened neurasthenics, these patients were riddled by fear; they suffered anxiety attacks and sudden heart palpitations. While neurasthenia was due to sexual overindulgence, anxiety neurosis struck virgins, prudes, sexual abstainers, and those who practiced coitus interruptus. It was a disorder of the sexually frustrated.

As Freud wrote these words, he was himself experiencing "the horrible misery of abstinence" from his beloved cigars. Alarmed by chest pains, he had put himself in the hands of Breuer and Fliess, who both diagnosed nicotine poisoning and demanded Freud stop smoking. Nicotine withdrawal, however, made Freud feel miserable. And this was not the sole form of deprivation on the doctor's mind. A few months earlier he confided to Fliess that, after having five children in six years, he and his wife were refraining from any sexual activity.

The noxious effects of sexual overstimulation and frustration fit easily into the reflex-arc models of neurophysiology and would have made sense to Freud's colleagues. But his ideas were outrageous in another way. Freud didn't say masturbation sometimes caused neurasthenia, but that it always did. It was unheard of to suggest one kind of trauma invariably caused a particular neurosis. Most nineteenth-century clinicians presented long lists of possible exciting causes, which they loosely connected to a prior heredity taint. Freud's claim was precise, universal, and certain: he even developed a theoretical defense for his approach, called his "thesis of specificity."

This thesis was Freud's adaptation of the approach that made germ theory a medical miracle. In the 1880s, Robert Koch's method, which allowed him to isolate tuberculosis and cholera, set new standards for proof in clinical medicine. Koch's famed postulates were intended to distinguish the specific cause of a specific illness. Among his stipulations was that any isolated causal agent must be found in *every* case of the disease it was believed to cause.

Almost immediately, germ theory saved hundreds of thousands of lives, and it aroused hopes that some diseases that had been written off as hereditary might actually be infectious and preventable. Such optimism found its way to psychiatry through the Frenchman Alfred Fournier, who had become convinced that syphilis was the cause of that dreaded disease filling asylums, General Paralysis of the Insane. (He turned out to be right.) One of Freud's teachers in Paris, the physician Paul Brouardel, was also dedicated to germ theory and extended the rationale of this theory to include the effect of environmental toxins, such as alcoholism, trauma, and sexual assault. Brouardel along with the German Rudolf Virchow and others encouraged doctors to take the responsibility for safeguarding the public from such contagions.

Sigmund Freud saw the secret of sexual trauma in that bright light. His theory offered the hope of displacing worn-out hereditary rationales and potentially curing widespread disorders. In approaching neurasthenia,

Freud followed Koch's rules by isolating one critical environmental factor and claiming it was always present in the disease. If right, Freud would thereby sweep away the unedifying lists of possible influences and replace them with the exact cause that must be prevented. Freud was certain unhealthy sexual practices were a plague that threatened the public. He told Wilhelm Fliess that sexual reform must occur, so that unmarried men and women need not face the Draconian choice of masturbating and getting neurasthenia, having an unwanted pregnancy, or engaging in sex with prostitutes and contracting fatal venereal diseases. In the absence of reform, he bleakly predicted, society appeared "doomed to fall victim to incurable neuroses."

While Freud saw his theory of sexual abuse as a psychiatric equivalent of germ theory, many of his colleagues did not. They had considered sexual factors as possible contributors to mental illness; sexual factors were on their lists. But how could any clinician working with a limited number of patients claim to know that one factor always caused an illness? Such arguments might be possible through lab experimentation, but were they sustainable based on clinical observation? Despite his sympathy for Freud, the distinguished Munich physician Leopold Löwenfeld published a strenuous objection to Freud's thinking. Freud fired back, insisting that neurosis was solely caused by a specific disruption that came from the patient's sexual life.

Privately, Freud began to broaden his theory to other disorders. On October 15, 1895, he wrote Wilhelm Fliess to inform him rather cryptically that "hysteria is the consequence of a presexual *sexual* shock." On January 1, 1896, Freud spelled out his thinking: hysteria was caused by sexual molestation during childhood.

The sexual molestation of children was not unknown among nineteenth-century forensic physicians. In his handbook of forensic medicine, the Berlin expert Johann Ludwig Caspar gathered statistics that showed over 70 percent of all rapes were of children under the age of twelve. Around the same time, French literature on such sexual abuse was initiated by Ambrose Tardieu, who also published horrifying figures documenting the rape of children. In Paris, Freud attended the forensic autopsies of Tardieu's successor, Brouardel, himself no stranger to such cases, for he had been an expert witness in the 1880 trial of Louis Menesclou, a man who raped and murdered a four-year-old.

However, Freud was not just saying that molested children became hysterical. He claimed that childhood molestations quietly festered for years before causing adult hysteria. Like syphilis, this latent disruption broke out in mental illness only later in life. Freud reasoned that the event remained quiescent and only gathered pathogenic force with the emergence of mature sexuality. Only then would the memories of sexual molestation take on meaning, provoke horror, and engender repression. Hence childhood sexual molestation worked by what Freud called "deferred action." This fascinating notion held wide-ranging consequences, for it implied that memories of the past were constantly being rewoven into the present, where they were given new significance and new place.

Elated by his discoveries, Freud fired off three papers to German and French journals, announcing that the riddle of hysteria had been solved. As he waited for the reactions of his colleagues around Europe, Freud began to lose the support of his staunchest ally in Vienna. By 1896, his relationship with Josef Breuer had grown testy. The two fell into open conflict after one of Breuer's greatest acts of loyalty. In October 1895, Freud had delivered a series of three lectures to the Vienna College of Physicians. Before this eminent audience, he laid out his convictions regarding highly specific causes of different neuroses. Afterward, the group met to discuss the talk. Predictably, some physicians refused to believe that hysteria and neurasthenia were due to a specific kind of sexual abuse. For example, Richard von Krafft-Ebing could not accept Freud's diminishment of constitutional factors and his extension of sexual etiology to account for all cases of obsessional neurosis.

The stakes of such a public accounting were high for Freud. With dismissive voices in the air, Josef Breuer rose to defend his young protégé. Breuer confessed that he too had been initially skeptical of his colleague's sexual theories but eventually was won over by the weight of clinical evidence. Had Freud suggested these sexual matters to his patients? Absolutely not. Breuer mocked those men of science who could not bring themselves to imagine sexual abuse, saying they themselves were behaving like hysterics. After delivering this scolding, Breuer allowed that Freud's sexual theories might be overstated, and he let it be known that he did not agree that all hysterical symptoms were sexual in origin. Nonetheless, Breuer hailed Freud's theory as a great advance.

The proceedings were covered in the local newspapers, and it was no small thing that one of the most eminent physicians in Vienna had risen to support such bold notions. Later, Freud thanked his mentor for his timely show of solidarity but was taken aback to hear Breuer mutter: "But all the same, I do not believe it!" Although he accepted sexuality as one critical factor in the neuroses, Breuer could not believe that every hysteria was due to sexual molestation. For Breuer, the case of Anna O. seemed ample proof that some hysterias were due to traumas that had nothing to do with sex.

Over the coming years, Sigmund Freud grew to despise the man who had been his financial, emotional, and professional backer. He would suggest that Breuer prudishly retreated from sexual issues, while Breuer would argue that it was not sex but science that made the two part company: "Freud is a man given to absolute and exclusive formulations," he complained in 1907.

Freud pressed forward, next presenting his theories to the Vienna Society for Psychiatry and Neurology in 1896. This time Krafft-Ebing was more emphatic, calling Freud's schema a "scientific fairy tale." Bitter but undeterred, Freud consoled himself with the belief that his critics were too scared and conventional to address sexual assaults on children. He remained confident that germ theory provided an adequate rationale for his way of theorizing, and continued to expand his thinking into other psychoneuroses. He reasoned that sexual molestation might result in different neuroses if the assaults occurred at different developmental periods. Sexual assaults could lead to hysteria (if they occurred during ages birth to four), obsessional neurosis (ages four to eight), and paranoia (ages eight to fourteen).

Alone in his study, Freud forged forward, but it was hard to ignore the protests coming from the likes of Löwenfeld, Krafft-Ebing, and Breuer. Freud could not fail to notice that unlike his earlier adversaries, these men advocated some role for psychic factors in neurosis and valued psychotherapy; they were his natural allies. And yet they all lined up against his trauma theory. If Freud was to be the Robert Koch of neurosis, he had a good deal of convincing to do. In search of more evidence for his theories, Freud turned to the growing cadre of sexual scientists who had been busy researching the very sexual abuses Freud's theory now heavily relied on.

II.

DURING THE SECOND half of the nineteenth century, the scientific study of sexuality had gained prominence in Europe and North America. Labeled "sexology" in the United States and "sexual science" in Germany, the new field encouraged researchers to bring a naturalist's cold eye to matters long thought indecent, immoral, disgusting, and sinful. Sexologists were encouraged to study varieties of human sexual experience not as vices, sins, or crimes, but as an integral part of the natural world. They would study sexuality in and of itself.

This marked a change. While doctors often considered sexuality in regard to its noxious effects, men of science left general sexual mores to the church and state. But in the last decades of the nineteenth century, the rising tide of Darwinism heightened scientific interest in human sexuality. Darwin's theory highlighted the importance of sexual selection, and made the mating rituals of animals paramount to biology. Evolutionary theory encouraged scientists to wrest sexual matters away from civic moralists and religious leaders, and develop a meaningful science of sexual life.

At the same time, the scourge of syphilis forced doctors and public officials to enter brothels to interrupt the cycle of venereal infection. In order to decrease infection, researchers felt they needed to understand why people would risk their lives by frequenting prostitutes. Early sexologists like Benjamin Tarnowsky and Iwan Bloch entered the demimonde in an effort to arrest the spread of syphilis, and then remained to chronicle the sexual desires and practices long hidden there.

Alongside these early sexual scientists and physicians were a handful of reformers who had become emboldened to oppose openly the criminalization of sexual behavior. Most influential was a gay Hanover jurist named Karl Heinrich Ulrichs, who in 1862 began his crusade to undo sodomy laws in Germany. Using Darwin's argument for constitutional bisexuality and citing documented cases of hermaphroditism in lower animals, Ulrichs argued that same-gender passion was not a vice but a human variant. For Ulrichs, male homosexuality was simply the result of a female soul residing in a male body.

Ulrichs's pamphlets found their way into the hands of psychiatrists like the Berlin physician Carl Westphal, who wrote one of the first psychiatric papers on homosexuality, and a young forensic psychiatrist named Richard von Krafft-Ebing. Born in Mannheim and educated in Heidelberg, Krafft-Ebing became a professor of psychiatry in Graz, where he established himself as an expert in forensic psychiatry. During those years, Krafft-Ebing took note of innovations in France, and he founded a private clinic in which milder cases of nervous illness were treated with a variety of hypnotic and suggestive treatments. For this reason, Freud exulted when Krafft-Ebing accepted one of the two chairs in psychiatry at the University of Vienna in 1889. When Meynert died in 1892, Krafft-Ebing ascended to his more prestigious chair, which included control of the psychiatric clinic. For the next decade, he was the most powerful psychiatrist in Austria.

Postcard of male transvestite from the collection of the Viennese physician and sexologist Richard von Krafft-Ebing. Krafft-Ebing helped legitimize the study of sexual "perversions" in German and Austrian medicine.

Despite his other contributions, Krafft-Ebing's renown came from his work in sexual science. In the 1870s Krafft-Ebing published articles on what he called "perversion." The study of sexuality had long been littered with creative coinages, often in Latin to thwart the ignoble curiosity of nonmedical readers. Krafft-Ebing favored the term "perversion," which literally means "wrong turn." His extensive research culminated in 1886 when he published *Psychopathia Sexualis: A Clinical Forensic Study*, a monumental compilation that would dominate sexual studies for years to come. In it, Krafft-Ebing delineated four major perversions. Sadists, who took sexual pleasure from inflicting pain on others, took their name from the notorious French writer the Marquis de Sade. Masochists, who garnered sexual pleasure from their own suffering, were named after Leopold von Sacher-Masoch, the Graz author of *Venus in Furs*. The anthropological term "fetishist" was adopted by Krafft-Ebing for those whose erotic worship of nonsexual objects (like shoes) resembled primitive peoples' worship of totems. And "inversion" described those who desired members of their own gender.

Over the next seventeen years, Krafft-Ebing stuffed more and more case histories into his compendium, which over the course of twelve editions, ballooned into over three hundred cases detailing diverse sexual acts including incest, bestiality, and necrophilia. Throughout this huge work, Krafft-Ebing proceeded like a botanist eager to organize the flora of the sexual wilds. He also hazarded an explanation for sexual differences, and it was there that the nascent field of sexual science split.

Ulrichs argued that homosexuality was a natural variant and didn't require medical treatment. Krafft-Ebing agreed that homosexuality was inborn, but thought it was a disease caused by the same thing that caused hysteria, neurasthenia, and a host of other ills. Following Charcot and Valentin Magnan in France, Krafft-Ebing resorted to that by now familiar scientific cliché and declared that homosexuality was the result of degenerative heredity. Most sexologists followed the same logic and explained sexual differences through degeneration.

However, over time, Krafft-Ebing changed his position. Banking on embryological studies, which showed that normal development routinely left behind remnants of the sexual organs of the opposite sex, the professor concluded that an inherent, primordial bisexuality had been proved.

Therefore, a variant of development could easily result in female brain regions persisting in a male brain, or vice versa. By 1901, Krafft-Ebing had come around to the view that such sexual difference was most often a natural variation.

Consensus might have coalesced around Krafft-Ebing's view if another theory had not emerged that challenged biological explanations of sexual desire. The Munich physician Albert von Schrenck-Notzing had been intrigued and impressed by a case report, in which Krafft-Ebing used hypnosis to treat a gay Hungarian woman named "Ilma." Ilma lived in a convent, but when she was nineteen she ran away. After fleeing, she suffered from fugue states and peculiar absences. She became a thief, disguised herself as a man, and took female lovers. One of those paramours turned her in to the police. Diagnosed as a hysteric who was afflicted with "congenital perverse feeling," Ilma was treated in Budapest before escaping to Graz, where she was again arrested for stealing and placed in treatment—this time with Krafft-Ebing. He concluded that Ilma had a rare case of acquired homosexuality and began to use hypnotic and suggestive methods to treat her symptoms. One symptom that caused a stir was Ilma's unbridled passion for a sister of charity who worked at the hospital. Krafft-Ebing used hypnotic suggestion to prevent Ilma from making passes at the nun. Reportedly, the treatment worked.

Inspired, Schrenck-Notzing began to experiment with hypnotic and suggestive treatments for a variety of sexual behaviors. In 1892, he presented his stunning findings in *Suggestive Therapeutics in Psychopathia Sexualis*. The Munich doctor announced that he had cured a large number of the seventy cases of sexual perversion he had treated using suggestive and psychic therapies. It was hard to believe. Psychic treatments, everyone assumed, could only cure psychic illnesses. Schrenck-Notzing's findings seemed to prove that perversions were often psychological in nature, thereby contradicting the view that they were either biological variants or degenerative illnesses. If this was so, the biological origins of perversion had been greatly overstated.

To account for his findings, Schrenck-Notzing argued that perversions were due to powerful ideas that took hold in the still undifferentiated child. Because the causes of these perversions were ideas and their associations, they could be cured by psychical means. Marching through Westphal's and Krafft-Ebing's landmark cases, Schrenck-Notzing showed how overlooked psychological factors may have been formative. By 1899, the powerful German psychiatrist Emil Kraepelin embraced Schrenck-Notzing's impressive evidence, which gave doctors new hope for "far reaching improvement and even cure" in such cases.

Throughout the 1890s, numerous authors followed Schrenck-Notzing's formulation by which premature sexual stimulation and chance association in childhood were enough to cause sexual perversion. There was no consensus about the character of the premature sexual stimulation in these cases, but the most common was thought to be sexual molestation. It was said that premature sexual molestation of a boy by a man would make that boy develop associations that would later lead to homosexual desire. Schrenck-Notzing used this reasoning to account for the prevalence of homosexuality in ancient Greece and boys' schools.

As more physicians turned to the study of sexuality in the 1880s, they encountered a simmering controversy at the heart of this nascent discipline. Some sexologists argued sexual differences must be biological variants, others claimed they were the result of degenerative heredity, and others still said these states were due to the psychological aftermath of premature sexual stimulation. Into this debate, enter Sigmund Freud.

III.

As FREUD's THEORIES of sexual abuse crossed into the field of sexology, he followed. There he would find solid support for his claims about the rape and sexual abuse of children. General physicians might scoff, but sexologists recognized the reality of these gruesome events. Sexologists had even profiled sexual molesters. Krafft-Ebing described abusers as alcoholics, epileptics, or morbidly disposed pedophiles, as well as bored libertines, timid onanists, lewd servant girls, governesses, and nursemaids. Freud's first profiles of sexual abusers cohered with these findings. He pointed the finger at strangers, older children, maids, governesses, and relatives. Freud was encouraged when one of his patients returned to his hometown and confronted a former nursemaid who admitted she had sexually seduced him. "The agreement with the perversions described by Krafft is a new, valuable reality confirmation," he wrote.

While aspects of Freud's theory were in harmony with the consensus in sexology, his seduction theory also stood in stark opposition to their seduction theories. Environmentally oriented sexologists believed childhood sexual abuse created perversion, not neurosis. If sexual seduction led to perversion, did it also cause neurosis? If so, Freud's specific etiology was not specific at all. Freud found himself at an impasse.

A year later, he returned to these problems after a breathtaking shift in thought. Freud now concluded that sexual molestations were perpetrated not by bad governesses, older children, and strangers, but always by the child's father. While cases of paternal sexual abuse had been documented, Freud's belief that the father was the culprit in all cases of psychoneurosis simply had no precedent. And on the face of it, it seems needlessly reckless. Why not suggest some or even many cases were due to paternal sexual abuse? Given Freud's limited clinical experience, why accuse the father in all cases? Particular instances of patients who had been abused by their fathers had encouraged, perhaps inflamed Freud to the point where he arrived at a blanket condemnation. He could also see that discomfort among physicians on this subject might lead to an unwitting conspiracy of silence.

But Freud also became excited by the potential explanatory power of this hypothesis. If a father was secretly molesting his children, the whole brood would seem to share a terrible hereditary taint. Their family tree would look like one of Charcot's degenerative lines and follow a pattern of what Freud now called "pseudo-heredity." "It seems to me more and more that the essential point of hysteria is that it results from perversion on the part of the seducer, and more and more that heredity is seduction by the father," Freud wrote to Fliess in December of 1896.

Thus Freud was able to synthesize his seduction theory with preexisting seduction theories. Premature sexual stimulation could result in neurosis or perversion, because men and women processed these experiences differently. Abused boys grew up to become perverse, pedophilic fathers who continued the cycle of traumatization with their own children. Abused girls responded with repression and became neurotic. Genealogies rife with neurosis and perversion were not ridden with bad blood but suffered from the sexual crimes of the father.

Since sexologists believed perversion was more common in men, and psychiatrists thought hysteria—the neurosis par excellence—was more common in women, this added up. But perverse women and neurotic men seemed to have no place in Freud's model. To account for them,

Freud appealed to an omnipresent bisexuality. An inborn bisexuality was a common element in sexological discourse, having been invoked from the very start by Karl Ulrichs and Krafft-Ebing. In 1896, Wilhelm Fliess also began considering how inborn bisexuality might be a determining factor in biological functioning. Freud grabbed onto this theoretical lifeline, arguing that male neurotics and female perverts had to be the result of the female in men and the male in women.

Freud organized these teeming thoughts into a conjecture: *hysteria was the negative of perversion.* This was not a minor matter, for by this reasoning the unconscious memories that caused hysteria became equated with the acts that defined perversion. Since they were supposed to be the same, the contents of the neurotic unconscious could be indirectly determined by empirically studying sexual perversions. Over the next years, Freud repeatedly stressed that perverse actions were exactly the same in content as the repressed fantasies of hysterics. Once, the copiously documented perversions had been stumbling blocks for Freud and his theory of neurosis. With this analogy, they became his Rosetta stone for knowing that seemingly unknowable region, the unconscious.

Of course, all this was highly speculative. In May 1897, however, a Berlin student named Felix Gattel came to study with Freud, and he was set to work testing his teacher's theories. At Krafft-Ebing's clinic, Gattel interviewed a hundred consecutive cases over the course of three months to test Freud's hypotheses regarding neurasthenia and anxiety neurosis. In fact, Gattel's interviews established a correlation between masturbation and neurasthenia, as well as sexual abstinence and anxiety neurosis. But Gattel had disconcerting findings as well. Though he attempted to eliminate all cases of hysteria from his sample, Gattel accidentally included four hysterics and thirteen mixed cases in which hysteria was partially involved. In those seventeen cases, only two confirmed a sexual seduction in childhood. Neither was by the patient's father. By itself that was not so damning, for Freud's theory dictated that these memories of abuse would be unconscious. But the very fact that Gattel's nonhysteric sample ended up with 17 percent who had some component of hysteria, also gave reason to pause. Did that mean the incidence of hysteria was very high? A year earlier, Freud had taken pains to fend off the argument that children were abused far more frequently than the incidence of hysteria, while at the same time rebutting claims that sexual assaults were not common enough to account for all cases of hysteria. Now, he had Gattel's numbers to ponder.

While Gattel's work was in full swing, Freud began to complain of intellectual paralysis. After a vacation to Italy, Freud startled Wilhelm Fliess by announcing that he no longer believed in his sexual trauma theory. He admitted his treatments based on this theory had failed. His patients fled, uncured. He also now realized that the "unexpected frequency of hysteria" had forced him to postulate a great many pedophilic fathers, including his own, which seemed improbable. Making matters worse, there was no clear way to distinguish reality from fantasy in the unconscious. Perhaps some of the clinical histories he had taken at face value were actually fantasies, rather than traumatic events. Finally, he deemed it a bad sign that even when hysterics became totally delirious and uninhibited, the repressed memory of their childhood abuse did not emerge.

Demoralized, Freud felt compelled to acknowledge that despite his efforts, the role of heredity seemed to have made its way back into the picture. But he did not accept such conventional thinking for long. Instead, he abandoned his thesis of specificity, that Kochian pretense that had led Freud to make arguments his clinical knowledge could not support. While recognizing a wider range of traumatizing events, Freud turned away from an exclusive emphasis on sexual abuse and highlighted the role of self-abuse: masturbation.

Many sexologists warned of the dangers of masturbation among youth, but there was a mystery at the heart of such warnings. Young children were supposed to have no sexual drive, so it made no sense that they would want to stimulate themselves. Some sexologists argued that a prior sexual seduction enticed children into such lewd acts. In the fall of 1897, Freud incorporated this logic into his theory. The crime of paternal sexual abuse was replaced by an unspecified cause of sexual excitement that afterward left the child in a state of withdrawal. The child fell into a period of longing, which Freud identified as the most salient characteristic of hysteria. During this intense state of desire, the child concocted sexual fantasies and began to masturbate, stoking an already overstimulated nervous system. When such masturbation was finally repressed, neurosis set in.

This was a desperate attempt to save a theory that was bound for failure. However, almost in passing, these transitional ideas added a critical new element that would take on great import in Freud's thinking. In scientific debates, the Viennese doctor had repeatedly positioned himself against the accidental and external, focusing his inquiry on internal, lawful determinants. He now added a new inner force: fantasy. Sexual daydreams compelled childhood masturbation and thereby played a causal role in illness. Hysterics laid the ground for their illness only if in childhood, spurred by fantasy, they masturbated again and again.

Before that, fantasy played no causal roles in Freud's theorizing. Such reveries simply masked traumatic memories, acting as "protective structures, sublimations of the facts, embellishments of them." But in his new thinking, these psychic concoctions helped create hysteria. Masturbation now stood poised between Psyche and Soma, between mind and brain, between daydreams and neural irritation. Freud grew confident that it was the key to hysteria, writing: "The insight has dawned on me that masturbation is the one major habit, the 'primary addiction.' . . . The role played by this addiction in hysteria is enormous." Enormous perhaps, but obscure. Throughout the ensuing year, Freud struggled with this theory and got nowhere.

In January 1899, Freud received a letter from a sexologist who introduced him to a broader way of thinking about masturbation:

> Something pleasant about which I had meant to write you yesterday was sent to me—from Gibraltar by a Mr. Havelock Ellis, an author who concerns himself with the topic of sex and is obviously a highly intelligent man because his paper, which appeared in *Alienist and Neurologist* (October, 1898) and deals with the connection between hysteria and sexual life, begins with Plato and ends with Freud; he agrees a great deal with the latter and gives *Studies on Hysteria*, as well as later papers, their due in a very sensible manner.

Havelock Ellis would become England's foremost sexologist, but in 1899, he was a relative unknown. Ellis sent Freud an early paper, his 1898 "Hysteria in Relation to the Sexual Emotions," in which the British writer expressed admiration for Breuer and Freud's work and congratulated the Viennese researchers for inaugurating research into hysteria as "a transformation of autoerotism." This itself was quite a transformation, for neither Breuer nor Freud had implied any such relationship. But Mr. Havelock Ellis had. In his 1898 "Autoerotism: A Psychological Study," Ellis sought to isolate the spontaneous sexual feelings that were generated in the absence of another person. According to Ellis, such autoerotism was normal and closely related to daytime reverie. In moderation, auto-erotic masturbation was not harmful, though Ellis warned that too much of it in adolescence might lead to "Narcissus-like" self-absorption. As for hysterics, they were different from the rest, because their autoerotic lives created conflict.

Freud and the English sexologist struck up a warm correspondence. Freud read Ellis's essay on autoerotism and adopted the term, as he began to posit a larger role for sexual impulses and fantasies. However, there was still a central difference dividing the two men. Freud believed sexual trauma roiled a presexual child's nervous system in cases of hysteria, while Ellis suggested hysterics had spontaneous sexual forces that made for conflict. As Freud's masturbation hypotheses began to place increased emphasis on the driving power of fantasy, he found himself inching closer to Ellis.

On December 9, 1899, Freud revealed an underlying intellectual shift to Wilhelm Fliess. He now posited that the earliest phase of normal sexual development was not presexual at all but, following Ellis, autoerotic. Masturbation was no longer a trauma but part of a normal stage of development in which a child took his own body as the object of sexual satisfaction. That phase was followed by a more mature, object-related sexuality, which Freud called "alloerotism (homo- or heteroerotism)."

General formulations regarding the nature of sexual life were not the province of general physicians. By moving from the limited arena of sexual trauma to general questions about human sexual development, Sigmund Freud had crossed a border; he had become a sexologist. As such, he began to consider how sexuality might make sense more generally in his model of the mind.

IV.

WRITING IN OCTOBER 1899, Freud summarized the last decade of his theorizing:

"Psychic apparatus Ψ
Hysteria- clinical
Sexuality. Organic."

"Oddly enough," he told Fliess, "something is at work on the lowest floor. A theory of sexuality may be the immediate successor to the dream book."

Freud increasingly examined the power of intrinsic sexual forces. If his theory of the psyche and his clinical theories were linked to a credible theory of sexual biology, he would have no need for conjectures about brain anatomy or degenerative heredity. Mind would be linked to body, and the psychic would be rooted in the physical. But this step down to the lowest floor was a treacherous one, as Freud's teachers had repeatedly shown. Unfazed, Freud crowed: "I am actually not a man of science, not an observer, not an experimenter, not a thinker. I am by temperament nothing but a conquistador—an adventurer, if you want it translated, with all the curiosity, daring and tenacity characteristic of a man of this sort."

Sexuality would prove hard to conquer. Freud had hoped to cap off his dream book with a chapter entitled "Dreams and Hysteria," which would unify his method of decoding the unconscious and his theory of neurosis. But he could not make everything line up. How could he demonstrate the power of dream interpretation to reveal the underlying sexual determinants of hysteria, when he was quite unsure what they were? By the summer of 1900, Freud's mood sank. "The big problems are still wholly unresolved," he wrote. "Everything is in flux and dawning, an intellectual hell, with layer upon layer; in its darkest corner, glimpses of the contours of Lucifer-Amor."

Then in the winter of 1900, an eighteen-year-old woman named Ida Bauer walked into Freud's consulting room. By the doctor's reckoning, Ida was a hysteric, and with her he would again try to fathom the mystery of this disorder. "It had been a lively time and has brought a new patient, an eighteen year old girl, a case that has smoothly opened to the existing collection of picklocks," Freud excitedly wrote. Those picklocks came from dream interpretation. Dream analysis would be the microscope that illuminated the unconscious causes of Ida's misery.

The result was *Fragment of a Case of Hysteria*, a short, novelistic account that showcased Sigmund Freud's extraordinary rhetorical skills. So powerful and persuasive was the argument that Freud ended up convincing many readers of something he no longer believed himself. The story recounted the travails of an eighteen-year-old who hailed from a prominent Jewish family in Vienna (Ida's brother Otto would go on to be a critical political figure in Austrian politics). "Dora," as Freud called her, had already seen several doctors who tried to treat her depression and shifting bodily ills with, among other things, electrotherapy. After her parents discovered that she had written a suicide note, they brought her to Dr. Freud. Dora suffered from depression, as well as hysterical coughing fits and curious episodes of heavy breathing. It was a case that could easily be categorized, following the French, as a *petite hystérie*. But Freud promised his readers to go beyond labels and lay out the cause of Dora's hysteria, which would centrally involve a disruption of the girl's sexual life.

Sex leapt forward in Freud's account. Dora had been ensnared in a web of unwanted amorous advances and parental betrayals. Her father had formed a close friendship with a certain Frau K., and the woman's husband, Herr K., had taken an interest in Dora. The nature of this interest became clear when Herr K. lured the fourteen-year-old girl to his home, closed the shutters, and forced a kiss upon her. Disgusted, Dora fled. Two years later, Herr K. tried to seduce her again; this time she slapped him, stalked off, and informed her parents. They confronted Herr K., but to the girl's horror, her parents accepted Herr K.'s accusation that Dora had dreamt the whole thing up. Angry and bitter, Dora demanded her father break off his affair with Frau K. He refused. Two years later when a miserable Dora appeared at Freud's door, her father told the doctor that his daughter's troubles had much to do with those unhappy events.

The Bauer family saga was laced with questions that had haunted Freud. What was the role of seduction in hysteria? How could one differentiate sexual fantasies from verifiable memories? Freud now presented his views: Dora's tale was not a fantasy; it was true. Herr K. had tried to seduce the fourteen-year-old. However, there was a twist. Dora manifested hysterical symptoms before turning fourteen. Indicating how wedded he was to an animal, physiological notion of sexuality, Freud claimed that Dora's disgust after Herr K.'s kiss indicated she must have already succumbed to repression. At the time of Herr K.'s advances, Dora was already "entirely and completely hysterical." She was *already* traumatized. If "trauma theory is not to be abandoned," Freud reasoned, "we must go back to her childhood and look about there for any influences or impressions which might have had an effect analogous to that of a trauma."

Freud then prepared his readers for shocking material. Frankness regarding sexual matters was vital for the treatment of hysteria. Readers must put aside their prejudices and recall that the most socially despised of perversions, homosexuality, was highly valued in ancient Greece, a civilization far superior to their own. Sounding every bit the sexual reformer, Freud added that lesser perversions, like oral sex, were widely prevalent, as everybody except medical men seemed to know. Perversions were neither bestial nor degenerate; rather they must be understood through the study of childhood development. Adults who held on to childish sexual modes were deemed perverts by society.

At this point, the author seemed to have digressed. Why was he taking such pains to discuss perversion in a case study of hysteria? The answer was forthcoming. Hysterics, like all psychoneurotics, Freud asserted, suffer from strong, repressed perverse tendencies:

> Consequently their unconscious *phantasies* show precisely the same content as the documentarily recorded actions of perverts—even though they have not read Krafft-Ebing's *Psychopathia Sexualis*, to which simple-minded people attribute such a large share of the responsibility for the production of perverse tendencies. Psychoneuroses are, so to speak, the *negative* of perversion.

In their unconscious fantasies, hysterics imagined doing what perverts *did*. Unfulfilled perverse desires lurked in Dora's unconscious. To defend that assertion, Freud turned to the young woman's dreams.

Dora told Freud about a recurring dream. In it, her home was on fire and her father was standing by her bed, waking her. As she dressed, Dora's mother could be heard saying she wanted to stop and save her jewel case. Her father refused, saying he would not let his family die for that. It was a dream, the young woman confided, that she first had in the house where Herr K. tried to seduce her.

Freud proceeded with a close hermeneutic dissection, asking the girl to associate to each element in the dream. In the end, he announced that Dora had summoned the love for her father as a defensive transformation

of Herr K. Her father was in the house, where she had almost been seduced, saving his daughter from the burning flames of sexual desire. At a
lower frequency, however, the dream represented something older and
odder—it was a veiled accusation of masturbation.

Dora had a history of thumb sucking as a young child. When she was
broken of the self-stimulating habit, Freud believed the girl was left in a
frustrated state of aroused longing. Years later, when she began experiencing the stirrings of sexual desire for others, she fantasized about
having oral sex with her father. These fantasies were not pathological in
and of themselves, but in Dora's case, they were charged with excess
yearning. Therefore, the girl masturbated. At age eight, she repressed her
conscious, incestuous masturbatory fantasies and succumbed to hysteria.
Dora didn't become ill because of her Oedipal love. Rather, masturbation
pathologically reinforced her all-too-human love for her father. When
Freud informed Dora of his conclusions, she denied masturbating. But in
a gesture Freud took as symbolic confirmation, she began to play nervously with a small purse, opening it, sticking her finger in it, and then
shutting it.

Soon thereafter, an uncured Ida Bauer stalked out of Freud's office
never to return. Nonetheless, the doctor was thrilled. He began to write
up the case and on January 25, 1901, told Fliess: "I finished 'Dreams and
Hysteria' yesterday, and today I already miss a narcotic." The case contained "glimpses of the sexual-organic foundation of the whole" and was
"the subtlest thing I have written so far." Five days later, Freud added a
few more words, saying the case held glimpses of bodily, erotogenic zones
and a frank recognition of bisexuality. This description would prove be-
fuddling for, in fact, the published account of this case devoted only one
footnote to bisexuality and Dora's purported love for Frau K. The confusion probably stems from the rapidly fraying bond between Sigmund
Freud and Wilhelm Fliess.

While expanding his work on the nasal reflex, Fliess began down a
theoretical path that Freud could not share. Fliess had become intrigued
by the menstrual cycle and devised an ambitious theory of male and
female periodic cycles. In 1897, Fliess published *The Relations Between the
Nose and the Female Genital Organs, Presented in Their Biologic Significance*, arguing that a masculine sexual period of twenty-three days existed

alongside a feminine period of twenty-eight days. These inborn cycles determined numerous fluctuating biologic processes, including date of birth, gender, and periodic illnesses like migraines, anxiety attacks, epilepsy, gout, and asthma outbreaks, not to mention the precise date of one's death. Astonishingly, Fliess declared: "The date of death is menstrual."

Fliess insisted that both masculine and feminine cycles existed in all human beings due to a universal bisexual disposition. The Berlin doctor proposed that the dominant period would determine an embryo's gender. He mapped out a mixture of twenty-three- and twenty-eight-day events, and by employing these two numbers along with their combined sum and difference, Fliess believed he could account for the precise date of many occurrences in a person's life.

When Freud announced to Fliess that the Dora case would include glimpses of bisexuality, he was making a friendly gesture in the direction of Fliess's theory, something he had done before. In 1896, while developing the seduction theory, Freud had made use of Fliess's theory. During a meeting between the two men in 1897, Fliess had suggested that repression might be engendered by a battle between masculine and feminine periods. Men repressed their feminine periods, and women their masculine ones, he ventured. When the two friends later met in Breslau, Fliess revealed that he had expanded this theory to cover whether a person was right- or left-handed. Fliess's bisexual-bilateral or "bi-bi" theory was a bit much for Freud, who had voiced his doubts. Nevertheless, the Viennese nerve specialist remained excited about Fliess's thoughts about sexuality, declaring in the summer of 1899: "But bisexuality! You are certainly right about it. I am accustoming myself to regarding every sexual act as a process in which four individuals are involved."

While fascinated and at times persuaded, Freud ultimately could not share his Berlin colleague's enthusiasm for male and female periodicity. By Fliess's reckoning, these inborn periods predetermined a good deal of biological and psychic development as well as pathology. If that were true, other causes of illness—like the ones Freud proposed—were simply irrelevant. In effect, Fliess's theory completely undermined Freud's. While excitedly experimenting with Fliess's periods in private letters written to his friend, Freud was circumspect about showing any enthusiasm for them in print. In February 1897, he assured Fliess that he would not be using his friend's ideas on periodicity, writing: "The truth is I have long since given up my attempt, never intended seriously, to play on your flute."

Despite Freud's attempts to be diplomatic, around the time he began to see Dora, conflicts between the two men began to boil over. In the fall of 1900, the doctors had agreed to meet in Achensee in the Tyrolian Alps, but during this meeting they fought bitterly. Fliess was enraged that Freud would not accept his grand theory. During their time in the Alps, Fliess derisively dismissed one of Freud's interpretations, saying: "The reader of thoughts merely reads his own thoughts into other people." This was a variant of the damning critique Comte and a legion of philosophers had thrown up against those who hoped to objectively know another's inner life. For Freud to hear this from his most avid supporter was devastating. The animosity between the two men became so bad that Fliess later accused Freud of physically attacking him. Decades later, members of Fliess's family would recall that Wilhelm believed Freud wanted to kill him.

The meeting in Achensee had long-lasting effects on Freud as well. Long afterward, Fliess's condemnation rang in Freud's ears. Almost a year later, in August 1901, Freud lambasted his friend for dismissing him as a mind reader. A month later, Freud again returned to Fliess's criticism, saying: "If as soon as an interpretation of mine makes you uncomfortable, you are ready to agree that the 'reader of thoughts' perceives nothing in the other, but merely projects his own thoughts, you really no longer are my audience either and must regard my entire method of working as being just as worthless as the others do."

Sigmund Freud had fashioned his derivations of French psychopathology and German psychophysics into a scientifically plausible model of the mind. And yet he still had to contend with the indictment that his science was all a figment of his imagination, a projection of his own subjectivity. Worse still, Freud's theory encouraged such speculation, for it was predicated on the belief that human beings must defend themselves from unpleasant self-knowledge. Like Nietzsche and Schopenhauer, he believed we must mistake ourselves. Therefore, Freud could not expect easy confirmation of his conjectures from the objects of his study; rather, his model required that patients resist and reject such disturbing knowledge. And yet, without external confirmation, wasn't Freud's method prone to subjective distortions of all sorts? In the Dora case, there seemed ample evidence of this. Freud cajoled, suggested, and arm-twisted the young girl, getting her to, if not agree, at least not actively dissent, as he peppered her with interpretations. Fliess knew that the baroque etiologic formulation Freud seemed to discover in Dora's dreams was exactly the same working theory he had drawn up before he ever met the woman. Was that fortuitous, or had Freud simply read his own thoughts into her?

As Freud's relationship with Fliess began to sour, the Viennese doctor sent the Berliner a new work, *The Psychopathology of Everyday Life*. Freud had turned his interpretive methods to small, everyday occurrences, quite ingeniously focusing on minor linguistic errors, trivial bungled actions, and misperceptions that he saw as symbolically laden eruptions from the unconscious. He worried Fliess would dismiss the book as nothing more than his own projections. If that were the case, Freud advised his one-time friend to toss the manuscript into the garbage. For more than a decade Freud had tried to secure a place for an empirical observer of the psyche, but still he seemed vulnerable to the taunt—suggester! mind reader! Having lost the confidence of Wilhelm Fliess, Freud became unsure of himself. He would hold back his completed analysis of Ida Bauer for four years.

As the relationship with Fliess died, Freud made a last-ditch attempt to save their collaboration and strengthen his theory. He wrote:

> And now, the main thing! As far as I can see, my next work will be called "Human Bisexuality." It will go to the root of the problem and say the last word it may be granted me to say—the last and the most profound. For the time being I have only one thing for it: the chief insight which for a long time now has built itself upon the idea of repression, my core problem, is possible only through reaction between two sexual currents. I shall need about six months to put the materials together and hope to find that it is now possible to carry out the work. The idea itself is yours. ... So perhaps I must borrow even more from you; perhaps my sense of honesty will force me to ask you to co-author the work with me; thereby the anatomical-biological part would gain in scope, the part which if I did it alone would be meager. I would concentrate on the psychic aspect of bisexuality and the explanation of the neurotic.

If Freud could give lawful definition to the organic basis of his psychological theories, then perhaps his interpretive work could find the stability it needed. If that lawful definition was Fliessian, the two men might reconcile. Unfortunately, Fliess was not impressed by this offer to collaborate on *his* discovery. Fliess reminded Freud that his own innovations would link biorhythmic periodicity with bisexuality, and chastised his colleague for not supporting him in it. It was true, Freud admitted, he could not follow Fliess on these matters, which he rather weakly attributed to his lack of mathematical skill. There was little left to say. In confusion and anger, this passionate collaboration died.

V.

FREUD DID NOT borrow Fliess's theory of bisexuality to account for repression, but he continued to move his sexual theory away from trauma toward inner fantasy and biological forces. He questioned his prior emphasis on unhappy events, wondering now if that was the result of a kind of paranoia on his part. After his distressing meeting with Fliess in the Tyrol mountains, Freud began to interrogate his own thinking, including his pronounced tendency to hold superstitious beliefs. He wrote Fliess: "If we attribute significance to an external accidental happening, we project to the outside our knowledge that our inner accident is invariably intentional (unconsciously)." As far back as 1895, Freud had postulated that such projections could be used for self-defense. Now he wondered if his trauma theories were just that, defensive. Was his theory of bad things happening to good people obscuring a deeper, more disturbing vision of human nature?

Sometime after 1901, Sigmund Freud decisively broke away from theories of trauma that he had relied on since his encounter with Charcot, and shifted toward a view that psychoneurosis arose from conflicts stemming from a universal sexual drive. He did not dismiss the pathogenic effects of sexual molestation, but no longer made it the central cause of repression. After years of silence, Freud's new theory would appear in 1905, and it would astound reviewers.

Three Essays on the Theory of Sexuality was a thin book with a misleadingly modest title. With lightning speed, the author took apart, reframed, and re-presented the major questions facing sexology. What made for sexual difference? What was inborn and what was the result of experience? What constituted normal sexual development? What kinds of sexual desires were abnormal? What was love? In three terse, tightly argued essays, Freud presented crushing critiques of prevailing theories of sexuality, and then lifted out of that rubble a synthesis centered on a theory of the sexual drive, which he called "libido."

The notion of an inborn sexual drive was not the sole property of Wilhelm Fliess and his periodic whirligig. In both *Origin of Species* and *Descent of Man*, Charles Darwin advanced the theory that natural and sexual selection were essential laws of the natural world. According to natural selection, the fittest of a species successfully reproduced and survived. Sexual selection gave an evolutionary advantage to those who were successful attracting a mate. Nature was geared for one thing: reproduction and species survival. After Darwin, the vicissitudes of animal sexuality became a central fulcrum for natural history.

As a boy, Freud had been entranced by Darwin's writings. In the fall of 1873, when he arrived at the University of Vienna, he immediately sought out Carl Claus, a renowned naturalist and Darwinian, who handed his student an embryological riddle. A Dr. Syrski had announced that he had located the testes in a supposedly hermaphroditic eel. In 1876, the young Freud was sent to a field station in Trieste, where he dissected eels to check Syrski's claims. In the end, Freud did not find this research compelling, but he left Claus's lab with a deeper appreciation of sexuality as understood, not through the lens of social convention, but through scientific analysis and evolutionary doctrine.

Sexual scientists would march under much the same banner. With them, Freud seized the right to investigate sexuality with the indifference of a man dissecting eels. Adopting the cool, detached voice of a natural scientist, Freud announced in 1905 that popular conceptions regarding sexuality offered an extremely unreliable picture, shot through with error and inaccuracy. In opposition, the author would rely on the scientific work of sexologists like Krafft-Ebing, Albert Moll, Paul J. Möbius, Havelock Ellis, Albert von Schrenck-Notzing, and others.

However, any reader who expected a reverent summary of sexual science was headed for disappointment. In fewer than a hundred pages, Freud addressed the claims of sexologists, then proceeded to rework, rename, and reposition so much of their findings that in the end, he had spun their data and doctrines into something alien. As an opener, Freud insisted that what most sexologists called sexual drive needed to be broken up into component parts and analyzed. Only through this crucial division could one move past the tired dichotomies of nature and nurture, the normal and the perverse.

To prove the value of such an analysis, Freud jumped into the great debate at the center of sexology—homosexuality. Freud cautioned that same-gender love included a complex set of behaviors, not a single entity. Some people were homosexual from earliest childhood, while others became so only after later experience. Some people were exclusively attracted to the same gender, while others vacillated. Some were comfortable with their orientation; others repulsed. This human variety, Freud warned, made it easy for sexologists to wear blinders and focus exclusively on one small subset to support a favored but flawed theory. The man who once believed he had solved the riddle of hysteria based on his experiences with twenty or so patients now acknowledged the danger of such narrowly founded claims. He positioned himself as the fair-minded arbiter, ready to guide the reader through a number of partial theories while he searched for a more convincing explanation.

Degeneration did not cause homosexuality, he began. In fact, degeneration had become a meaningless term, indiscriminately employed in almost any case that was not obviously traumatic or infectious. It was a farce. How could it be used to describe homosexuals when they were otherwise normal? This group could hardly be accused of diminished capacities, given that some of the most cultured and intellectually advanced members of the human species could be counted among them. Perhaps homosexuality was an innate variant, but if so, what about those who became homosexual later in life? Furthermore, sexologists had shown that early impressions like seduction and same-sex social settings could encourage homosexuality. And homosexuality had been cured by suggestion, proving it could be an acquired psychological trait. Freud admitted that he too had first come to the conclusion that sexual difference was acquired, only to then discover that the traumatic cause he envisioned, sexual molestation, affected some but not all.

With this intellectual jujitsu, Freud exposed all three prevailing solutions as partial and wanting. Later, he observed that debates between innate disposition and accidental factors were "a case in which scientifically thinking people distort cooperation into an antithesis." Libido was not either inborn or acquired. Those false dichotomies needed to be exchanged for a theory that showed how inborn factors met environmental stimuli "half-way." To make such a refined theory, Freud suggested breaking down libido into three parts: the impulse, the object, and the aim.

Constitutional theories proposed that sexual impulses were either heterosexual or homosexual. According to this logic, homosexuality was a psychological form of hermaphroditism—a female brain in a male body. Given the scant evidence for such claims, Freud rejected this leap from embryology to human psychology. In the end, he said, such theories were confused, because they assumed that the object of desire was inherent in the sexual drive itself. Homosexuals did not differ from others in their biological sexual impulse, but rather in their choice of sexual objects. The sexual impulse, Freud continued, was never dependent on its object. It was quite possible to have a normal inborn sexual drive and a very atypical sexual object. Consider pedophiles. The terrible frequency of the sexual molestation of children by teachers and servants was not due to an abnormal sexual drive but rather easy opportunity. And think of zoophilia, that passion that crossed over to other species. Freud deemed it quite common among farmers. In these variations, the sexual impulse was constant, while environmental factors explained the selection of a deviant object of desire. Finally, Freud distinguished sexual aims from sexual impulses. With this decoupling, one could study anatomical transgressions, such as the sexual utilization of the lips and mouth, or the use of a fetish.

If in the Dora case, he had maintained a clear distinction between normal and perverse sexuality, those distinctions had now collapsed. Freud pushed hard on the boundaries sexologists had erected and asserted that perversions were quite common. Under the right circumstances, such abnormal acts might appear in many, for perversions were normally only inhibited by shame and moral concerns. That is, they were *repressed*. In extreme cases of perversion, no such self-restraint existed. "The conclusion now presents itself to us that there is indeed something innate lying behind the perversions but that it is something innate in everyone, though as a disposition it may vary in its intensity and may be increased by the influences of actual life."

Twenty-five pages into the *Three Essays*, it became radiantly clear that this book was a bombshell, which sought nothing less than a complete re-definition of the categories critical to sexological discourse. Freud's appraisal rapidly exposed the sterility and deficiencies of competing models, and declared that both sides of the debate on sexual difference were partly wrong. This prepared the ground for new conceptual distinctions that offered greater clarity: in short, Freud's new theory. In addition, the Viennese doctor aimed for nothing less than a co-optation and integration of earlier sexual theories into his preexisting theories.

The opening for this integration emerged when Freud noted that perverse inclinations were normally repressed, and then he launched into a discussion of how the sexual impulse was critical not just in perversions, but also those illnesses of repression, the psychoneuroses. Neurotics repressed their own perverse sexual desires, so that many, if not all, the common perversions appeared in their repressed mental life. Going further, Freud declared that all of us were somewhat hysterical, which now meant we were all a bit perverse too.

Everyone a bit perverse? How could such an outrageous statement be defended? The answer was forthcoming. Perversion was neither degenerative, nor a biological variant. It was the result of a universal sexual impulse. To understand this inborn impulse, Freud suggested his readers turn their attention not to fetishists or sadomasochists, but . . . to children.

This must have struck some readers as adding provocation to provocation. Most nineteenth-century sexologists did not even believe children were sexual. Any premature manifestation of sexuality in a child was thought to be the result of something terrible, familial degeneration or sexual abuse. As the Italian sexologist Paolo Mantegazza wrote, puberty was "the passage over the fatal bridge" from innocence to adolescence and sexual life. However, a few of Freud's colleagues disagreed. Wilhelm Fliess's theory demanded that sexual periods commence at the beginning of life. And the Berlin sexologist Albert Moll, whose work Freud knew well, published a two-volume work on sexual libido in 1897, proposing that the sexual impulse began early on. The "contrectation" drive awakened in childhood and created the desire for physical contact: it compelled youngsters to cuddle, touch, and fantasize about love objects. With puberty, Moll suggested that this drive was melded with a genital drive, and together these made for mature sexuality.

Freud went further; he was not suggesting that children were vaguely stimulated to touch each other and imagine romance. Shockingly, he was arguing that the little innocents were perverse. Perverse sexuality had been very carefully constructed to refer to socially maligned groups, not little Rosa and Otto. Fearlessly, however, Freud marched ahead, steadied by a passionless rhetoric that referred to skin, mucous membranes, and irritations, as if he were still a zoologist writing on the gonads of some sea creature.

That too was the message. Childhood could not be grasped by sentimental reminiscences; it was an alien place, a lost land hidden from memory. That did not mean nothing of lasting import remained from those years, but rather that the disturbing remnants of our prehistory were repressed. Childhood amnesia, Freud ventured, was the bit of hysteria in all humans. The once shameless child repressed his earliest memories, and as a grown-up created comforting myths of childhood as pure, kind, and good. For Freud, we become adults in part by forgetting.

To reveal what adults must forget, Freud turned to data from different, converging fields. From sexology, Freud drew on Moll, Krafft-Ebing, and others who believed normal children had a number of bodily zones that were highly arousable: erotogenic zones. Freud had already incorporated some of these ideas into his sexual trauma theory by arguing these normally quiescent zones could be awakened by premature stimulation and thereby become perversely alive. Now he allowed that these zones awoke as a child matured, and over time folded into genital sexuality. In perversion, however, erotogenic zones remained as they were in childhood.

For Freud, children were perverse, and perverts were still children. He explained that the most basic form of infantile sexuality derived from stimulation of the mouth and lips—the oral zone. Here, the Viennese doctor echoed an array of sexologists, as well as a Hungarian pediatrician, S. Lindner. Sucking was a primal childhood pleasure that began at the breast and led to pleasures such as thumb sucking. Freud called this earliest form of the sexual impulse "autoerotic." The sexual object was one's own body, and pleasure was derived from stimulating its arousable areas. Another erotogenic zone was the anus. That zone, Freud continued, in an almost unearthly tone, was a source of pleasure for children, as demonstrated by their desire to retain feces and manually stimulate themselves.

Finally, Freud counted the urethral zone and the stimulation of the mucous membranes during urination as a bodily pleasure.

Freud reminded his recoiling readers that the child was above all shameless. It had not yet taken in society's proscriptions. It allowed itself all the pleasures of feeling, looking, touching, and cruelty. It wanted no object other than itself. But in the course of development, these infantile impulses and aims were usually repressed, sublimated, and transformed into object-directed sexuality. Only when genital sexuality was repressed, did a central damming result in the flooding of these older, dried-up tributaries. Then erotogenic zones would come back to life and fill the unconscious with perverse fantasies from infantile sexuality. Conversely, if these earliest forms of sexuality were never repressed, then they would live on unencumbered, creating exclusively oral or anal sexual desires— the so-called perversions.

And so Freud wove neurosis, perversion, and normal development into a model that centered on the complex interaction between an awakening sexual impulse and the vicissitudes of repression. He laid out a separate developmental line for the choice of sexual objects, arguing that during puberty the child's desire critically shifted from himself to others. The search for an external love object had been prepared, starting at the mother's breast. For Freud, the finding of a satisfying object always entailed a refinding of that first, primal object. By this account, finding a lover included an element of nostalgia, a vain attempt to recover a time when the mother's stroking, kissing, and snuggling with her child first awakened the sexual impulse within.

Although the giving of love from a parent entailed a meticulous avoidance of any overt sexual stimulation, Freud posited that the child's first sexual feelings were for his parents. To illustrate this, he would refer to a story familiar to his audience, the Sophoclean drama of Oedipus Rex, the boy-king driven to murder his father and marry his mother. For Freud, it was critical that this childhood passion be frustrated, for only then could a child break away from home and find sexual love in the world. Later, couples might delight in cooing childlike to each other, thereby blindly giving succor to unconscious incestuous fantasies. That was neither perversion nor neurosis. It was love.

Before leaving his readers, Freud returned to the most startling of his claims. All humans had an innate, polymorphously perverse disposition.

Freud asked his readers to consider this claim from the perspectives of anthropology and evolution. In the polymorphously perverse child, he declared, it was impossible not to recognize the origins of man. The unrepressed child was a vestige of our primitive forebears; it was the Cro-Magnon man and woman in all of us. Polymorphous perversity was in the human unconscious, for it was the repressed prehistory of civilization.

This assertion was plausible only because Freud and many of his readers—following the famed biologist Ernst Haeckel—believed that all human history lay somewhere in our minds. Haeckel believed he had discovered a fundamental law: ontogeny was the recapitulation of phylogeny. This meant the historical development of the individual from conception to adulthood exactly mirrored the evolution of the species from its most primitive origins to its present state. Every person in the course of their development reenacted the entire evolution of the human species. Thanks to Haeckel's mind-brain position, this applied to the mind as much as it did the lungs. And so, it followed that children had the minds of cavemen.

It was difficult, however, to imagine how the history of all psychic life could possibly be inherited. In 1870, Breuer's mentor, Ewald Hering, came up with an explanation: unconscious memory. Psychic experiences became nervous excitations that were transferred to germ cells and passed down to the next generation. If a parent developed deeply ingrained skills like playing the piano, an unconscious memory of that musical ability might be inherited. Whether it involved spiders spinning webs, birds making nests, or Mozart's compositional gifts, learned behaviors could be passed down as acquired instincts through unconscious memory. The simplest examples of unconscious memory were hunger and the sexual instinct, those forces that had exerted their power over all organic creatures through the ages. For Hering, these life-or-death instincts were guarded and safely passed down from individual to individual thanks to their storage in the unconscious.

An enthusiastic admirer of Hering's theory, Freud chose a similar definition of unconscious memory. Unconscious memories of sexuality had been highly conserved. The pleasure principle was an evolutionary ploy to make sure these animal needs were met without requiring the fickle aid of consciousness. The oral zone was a sucking drive that was requisite for an infant's nourishment. Oral and anal zones could be justified by moving up and down on the evolutionary ladder, and linking them to the needs of more primitive forms of life, like amoebas. With the advent of human civilization, these ancient drives became repressed. They only remained manifest in the shameless child, the pervert, the primitive, the savage, and, Freud contended, sharing a misogyny that would cripple his thinking in the future, in those not fully civilized human beings—women.

By linking his sexual theory to evolutionary biology, Freud reinforced his account and made it more plausible to those in the natural sciences. He also turned for support to anthropological evidence discovered by Iwan Bloch. In stark contrast to Freud's ideas, it was commonly believed that European civilization had degenerated and fallen prey to a high incidence of sexual perversity. To test this hypothesis, Bloch did fieldwork and discovered perverse sexuality was no more common in so-called advanced civilizations than in those deemed primitive. In fact, he concluded that primitive societies were frequently based on sexual perversity and promiscuity.

In 1914, when Freud wrote a preface to the third edition of the *Three Essays*, he baldly stated that human "disposition is ultimately the precipitate of earlier experience of the species." Carrying inside them the cumulative weight of evolution, a child was driven by archaic, perverse impulses. In civilized societies, these forces were hidden and rendered unconscious. But under the veneer of progress, there lay a primal sexual force that was vital to life itself. For years afterward, Freud would look to children, savages, and primitives, arguing they all lived free from repression. In these beings, one could circumvent the problems of knowing what was normally forgotten and impossible to apprehend by interior observation. In them, it was possible to grasp the mystery of the unconscious.

VI.

BETWEEN 1885 AND 1905, Freud took up French psychopathology, German biophysics and psychophysics, and sexology in an attempt to make sense of the bewildering psychic forces that seemed to cause illness. He entered each of these fields, mastered the details of each discourse, and then set forth solutions, modifications, or amendments to pressing problems that faced each discipline. Freud's solutions were organized by two long-standing commitments that had been articulated by Théodule Ribot in 1870: he refused to eliminate subjective, interior experience from psychology, while insisting that psychological knowledge somehow be made scientific. These sometimes competing concerns organized and informed the new contours that Freud gave these fields and, in the end, became the backbone for a new Freudian field.

As early as 1899, Freud had seen the outlines of this extraordinary synthesis looming before him. After completing the book on dreams, he announced to Wilhelm Fliess that he had nearly finished "the first third of the large task." He then listed the elements of that task: "(1) The organic-sexual; (2) the factual-clinical; (3) the metapsychological." By 1905, Freud had completed the job. Libido caused the psychic disruptions in psychoneurosis and was the source of unconscious wishing in dreams. Both arose from the press of sexual drive. The unknowable-in-itself unconscious had been newly defined in a way that clarified the foundations of Freud's model. The unconscious did not need to be discovered on a case-by-case basis, for it was said to be universally inhabited by a highly conserved drive, which Freud defined in a manner coherent with evolutionary and anthropological data. The problems of knowing another's inner world diminished: all humans were driven by libido.

Libido was the source of the mind's energy and the cause of perpetual psychic conflict. The strange dissociated mental states studied by French psychopathologists could be redefined: they were due not to degenerative heredity but rather to ruptures in consciousness caused by repression and unconscious sexual drives. The subjective distortions and illusions that psychophysicists researched could be studied through an internal psychophysics in which unconscious libido distorted consciousness. Freud's libido also clarified debates over the role of nature and nurture in sexual difference, and massively expanded the domain of sexology. Rather than merely being the investigation of overtly sexual actions, behaviors, and desires, Freud had made sexuality central to the workings of all psychology.

Sexual theory was Freud's missing link. *Three Essays* was not just an addition to Freud's prior work: it was a series of answers to questions left hanging by his earlier theories. Libido was the foundation upon which Freud now placed the rest of his theoretical edifice. "No one," he wrote in 1905, "probably, will be inclined to deny the sexual function the character of an organic factor, and it is the sexual function that I look upon as the foundation of hysteria and of the psychoneuroses in general."

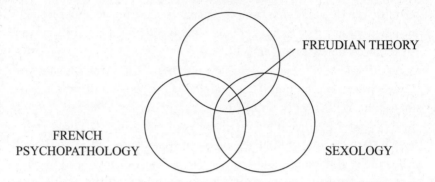

Psychosexuality, Freud's synthesis, took the notion of psychical causality pioneered by the French, placed this in a mind/brain model that cohered with Newtonian theory, and combined it with an underlying sexual drive that sexologists and Darwinians saw as central to life.

By straddling mind and body, libido theory also linked these two realms. There was no need for degeneration theory or psychoanatomical speculation. These strategies had proved themselves lacking and now could be avoided. Sexuality was clearly universal; it was also both biological and psychical, an interaction of drives and mental representations. As Breuer had quipped, it was ridiculous to say that all erections came from ideas, but it was also ridiculous to say that no erections came from them. Libido allowed Freud to posit a credible, underlying biological force that had impact on the psychic world. While proceeding in a purely psychological manner, he suggested that someday chemical substances might better account for the source of libido, but since this was a psychophysical phenomenon, he could afford to focus his full attention on its psychic manifestations alone.

Libido theory also held great advantages for the scientific status of Freud's model. Model building in science often proceeds by a search for coherence. If ideas in one scientific domain are harmonious with evidence from another, even if hard proof is absent, they gain credibility. Like Ribot, who had shored up his study of inner subjective states by rooting them in heredity, Freud deployed sexuality in a way that cohered with evolutionary theory and contemporary models of biological functioning.

Finally, the advent of libido theory immediately altered Freud's thinking about his method of psychical analysis. The same year he published the *Three Essays*, he also finally published his analysis of Ida Bauer. It was a mysterious decision because the case write-up, now four years old, was based on his now repudiated trauma theory. Freud did not rewrite the entire manuscript, or eliminate the role of trauma, but he went through the manuscript, penned "Preliminary Remarks," and added some new citations. Most importantly, he appended a "Post-script" in which he mulled over the failure of this case, and in so doing he articulated a new way to analyze another's mind.

Trauma theory turned the psychical analyst into a sleuth who had to discover painful buried memories. This method raised questions about the reliability of historical memory and the stability of these discoveries, given the cunning way humans fictionalize their own lives, not to mention the distorting influence of the doctor's demand that there was necessarily a trauma in the past that must be recalled. Libido theory shifted Freud's attention from memory retrieval to the constant power of desire in, not just the past, but also the present as revealed in the transference.

A few months before meeting Ida Bauer, Freud had started to recognize the potency of such a shift: "I am beginning to understand that the apparent endlessness of the treatment is something that occurs regularly and is connected with the transference," he told Fliess. In his early draft of the Dora case, however, Freud did not directly address this issue. In an anachronistic footnote Freud neglected to delete four years later, he confided: "At this point my interpretation touches for a moment upon the subject of 'transference'—a theme which is of the highest practical and theoretical importance, but into which I shall not have much further opportunity of entering in the present paper." With the new clarity afforded by libido theory, however, Freud returned to transference in 1905:

What are transferences? They are new editions or facsimiles of the tendencies and phantasies which are aroused and made conscious during the progress of the analysis; but they have this peculiarity, which is characteristic for their species; that they replace some earlier person by the person of the physician. To put it another way: a whole series of psychological experiences are revived, not belonging to the past, but as applying to the person of the physician at the present moment.

Transferences resurrected the past in the present. Hidden in quiet fantasies, stray associations, small behaviors, and explosive interpersonal dramas, transferences manifested the patient's forgotten libidinal ties to early objects. Freud suggested that this awakening of the past should become the center of the psychoanalyst's attention. The moment a neurotic entered psychoanalytic treatment, the formation of neurotic symptoms began to be replaced by these new mental structures. Transferences were revenants of a lost time, but their force was still to be reckoned with, and so the physician became a stand-in for those archaic prototypes of loved objects—mother, father, brother, and sister. Freud confessed that he had bungled Dora's case, because in a single-minded search for painful memories and trauma, he neglected the very rich drama taking place between him and his young patient. Transference cut straight from the here-and-now to the long forgotten.

ONCE CLASSICAL POETS sang of the marriage between Psyche and Eros. In 1905, after years of work, a Viennese physician presided over a similar union. By merging dynamic French notions of the psyche with sexology, Freud created a biologically rooted model of the mind that allowed for psychic causes and intentions. A universal sexual drive gave lawful definition to the contents of the unconscious and drove consciousness. In Freud's view, children prod, poke, suck, and stare as if they are prehistoric savages, but as they grow older, their mental life develops the capacity to do battle with these animal parts of their inner selves. That battle would never end: the Freudian marriage of Psyche and Eros would be one filled with conflict and discomfort.

Since Aristotle, science has proceeded by attempting to reduce great complexity to simpler, essential entities. By that criteria, Freud—if his synthesis was right—had pulled off something spectacular. He had taken up complex debates in psychology and psychopathology, quarrels over the mind and brain, nature and nurture, inner cause and outer experience, perversion and neurosis, normal and abnormal human behavior, individual development and evolution, fear and desire, and in the end, he could sum up his answers in one highly theorized word: *psychosexuality*. Ribot, Charcot, Bernheim, Janet, Brücke, Helmholtz, Exner, Meynert,

Breuer, Fliess, Fechner, Hering, Lipps, Wundt, Krafft-Ebing, Schrenck-Notzing, Moll, Möbius, Haeckel, Lamarck, and Darwin. All of these men were in the scramble to answer some of these same questions. But after repeated effort, dramatic failures, and finally, a series of deft theoretical engagements, it was Sigmund Freud who had taken the massive complexity of these problems and synthesized a theory of the psyche that was rich in implication and explanatory power, yet breathtakingly simple: *psychosexuality.*

Between 1895 and 1905, Freud published three major works on hysteria, dreams, and sexual theory. After 1905, all three could be seen as arguments for psychic causation and determination, unconscious wishing, and theories of neurosis and perversion based on psychosexuality. Framed in this way, these works pulled together a new approach for the study of mental life. Freud's synthesis was much more than a way to make a hysterical twitch disappear; it was a new discursive space that brought *Geisteswissenschaft* into *Naturwissenschaft*; it broadened natural science so that it could take up the great questions of human interiority, that space explored by the great psychological novels and poetry of the French, the Russians, and the English; the studies of character in theater from Aeschylus and Shakespeare to Ibsen and Schnitzler; the eye-opening lessons of human history, and the chronicle of human fantasy and belief in religions, fairy tales, and fables. Through this integration, it seemed science could be rescued from an embarrassing poverty and the humanities could be understood according to universal laws.

Psychosexuality situated a full range of studies of the human heart and mind, so that they made sense in a Newtonian universe, in Darwinian biology, and in a world where truth was decided by the epistemological demands of science. And as Freud's synthesis was busy being born, a number of competing views were waning. Despite having a sophisticated model of mental life, French psychopathologists remained doggedly committed to degeneration theory, which limited their capacity to expand into general psychology. Suggestive therapists struggled to establish a credible scientific ground for their treatments. Freud could explain the same dynamic unconscious psychic states, according to a scientific model that cohered with evolutionary biology. The biophysics reduction of the mind to brain had been a dead end in regard to consciousness, and those who launched psychophysics found themselves limited by the extremely difficult promise embedded in the very name of their field. Freud's model of mind included psychical qualities alongside energetic quantities but proceeded as a purely psychological method. Sexologists had become divided between theories of mindless bodies performing sexual acts, and bodiless minds in search of pleasure. In their search for scientific terra firma, these sexual scientists had foundered when it came to modeling the interactions of body and mind. Libido theory offered such a synthetic account. Sigmund Freud had proposed solutions to internal problems in all of these disciplines then pulled together these more local solutions into an overarching new theory with new terminology and new methods. From these older disciplines, a new field of study was being born: a Freudian one.

From these same origins, Freud would make many enemies. His science of the mind would offend religious leaders who saw it as a materialistic degradation of the everlasting soul. He inherited the enemies of French psychopathology, psychophysics, and sexology, for those psychiatrists who resisted dynamic psychology, those who remained committed to a strict brain-based neuropathology, and those who spurned sexology, would also reject Freud. Furthermore, he found himself pitted against fervent opponents who came from the very fields he had mined. French psychopathologists bitterly complained that Freudian theory had been plagiarized from them, despite the fact that Freud's theoretical development after 1895 had little to do with the French. German academic psychologists and psychiatrists who followed Wilhelm Wundt from psychophysics to empirical psychology, vehemently argued that Freud's work lacked the scientific credibility they had found in lab experimentation. When confronted by these attacks, Freud did not underestimate the power of his own ideas. These psychologists, he would comment, were like the giants in *Orlando Furioso* who continued to strike out at their enemy, long after their own heads had been chopped off.

As enemies came forward, so did allies. Many roads now led to Freud, so followers began to find their way to him. If you studied sexuality, hypnosis, hysteria, dreams, or memory, any of these might bring you to the work of Sigmund Freud. If you wanted to be a psychopathologist and treat psychogenic illnesses with psychotherapy, you could turn to Freud. If you wanted to be a sexologist who studied and treated perversions, you could find guidance in Freud. If you wanted to study psychophysics and parse the relationship of the unconscious to consciousness, and the mind to bodily energies, you could look to Freud. By following any or all of these paths, recruits entered a new Freudian nexus that linked psychopathology, normal psychology, psychotherapy, sexology, evolutionary biology, and the hermeneutic study of memory, symbolism, and dreams. Before long, a small community would consolidate around these ideas and approaches. They would come to be known as the Freudians.

PART TWO

Making the Freudians

Vienna

Where are the new physicians of the soul?
—Friedrich Nietzsche, 1881

I.

THE FIRST INTELLECTUAL community to come together around Freud coalesced in his hometown of Vienna. Psychiatrists, neurologists, medical therapists, writers, cultural critics, and social reformers all found their way to this
thinker, who in his wide-ranging synthesis spoke to some aspect of their concerns. However, as they would discover, many of these men differed greatly
in their assumptions, their aims, and their methods. During their first years,
these Viennese followers engaged in vociferous, at times vicious debate about
who they were and what exactly they hoped to achieve.

As a docent at the university, Sigmund Freud had been delivering Saturday evening lectures to small groups since the mid-1880s. In 1902,
Freud's academic prestige lifted when he was awarded the honorific title
professor, yet his lectures still drew but a scattered few. During that same
year, however, Freud found serious interest in his work coming from a
few local physicians. In the winter of 1902, he sent out invitations to four
colleagues, and asked them to join him at his home to discuss matters of
psychology and neuropathology. In honor of their chosen meeting day,
the group would call itself the Wednesday Psychological Society (*Psychologischen Mittwoch-Gesellschaft*), and join the many other small groups,
associations, clubs, and parties that flourished in turn-of-the-century
Vienna. Alongside teetotalers, feminists, ethnic groups, socialist physi-

cians, Fabians, anti-Semites, pan-Germanists, educational reformers, and Secessionist artists, the Wednesday Psychological Society brought together a few Viennese who shared common interests.

Forming the Society was not Freud's idea. The impetus came from a local doctor, Wilhelm Stekel. Stekel had studied at the University of Vienna Medical School, where he had worked under Krafft-Ebing. After leaving academia, he opened a medical practice but felt unequipped to deal with his patients' problems. Bored, he threw himself into writing feuilletons—short, often personal essays that were printed at the bottom of Viennese newspapers. Stekel met another doctor, Max Kahane, who told him about Freud's Saturday evening lectures. Depending on whose account one believes, Stekel's subsequent transformation occurred either after he read *The Interpretation of Dreams*, or was successfully treated for a sexual problem by Freud. Stekel then composed a glowing review of Freud's dream book for the *Neues Wiener Tagblatt*, in which he announced the dawning of "a new era in psychology." He began writing more feuilletons on Freud, so many that his beleaguered editors demanded he write about something, anything, else.

Max Kahane was also invited to Freud's home at Berggasse 19. Kahane was an old friend whose career paralleled Freud's. The men attended the same secondary school and university; as young doctors, both were fascinated by French psychopathology and took up the task of translating some of these French works into German. (Kahane produced a volume on Charcot and Janet, and reworked Freud's Bernheim translation for a second edition.) Both men were interested in childhood diseases and served as consultants at Max Kassowitz's Institute of Pediatrics, before finally going separate ways. But Kahane stayed abreast of Freud's work by attending his friend's university lectures. Critical of the smug elitism he found among academic physicians, Kahane dedicated himself to the practical imperatives of medicine and was intrigued by Freud's therapeutic innovations. When he published a textbook of medical practice in 1901, Kahane included mention of Freud's new psychological treatment. That year, he founded a sanatorium, and a year later became a founding member of the Wednesday group.

While the first members of the society were all Jewish by birth, Rudolf Reitler was the son of a prominent, wealthy Jewish family that had converted to Catholicism, a not uncommon method of accommodating Vienna's anti-Semitism. Like Kahane, Reitler was one of Freud's childhood friends. He too had studied medicine at the University of Vienna, and he kept an eye on Freud's innovations by attending the Saturday evening lectures. The same year Kahane opened a sanatorium, Reitler opened his own institute devoted to thermal cures.

The youngest member of the group was Alfred Adler. While at the Vienna Medical School, Adler had been impressed by Krafft-Ebing and French psychopathology and, like the others, attended some of Freud's evening talks. A socialist, Adler shared Max Kahane's disdain for the abstractions of academic medicine and was committed to a medicine that worked for the betterment of society. In 1898, Adler wrote a book on the occupational ills of tailors and followed that with a series of feuilletons that demanded the university establish a chair for social medicine, as well as take up problems caused by poor hygiene, poverty, and ignorance. Adler opened a general medical practice in 1899, and not long after that he consulted Freud on one of his cases.

Stekel, Kahane, Reitler, and Adler. All were medical clinicians who struggled to make their way in practice. All were economically and, in some cases, ideologically committed to therapeutics. And when they joined Freud, they all faced a common trouble. They had been trained at Vienna's medical school, the elite mecca of European medicine, where doctors were taught to think like high-minded scientists and then, rather unceremoniously, were tossed out of the ivory tower with precious little to guide them when it came to the life-or-death matters they confronted in their daily work.

In 1900, VIENNA was arguably the greatest center for medical study in Europe. Its medical school boasted an extraordinary roster of famed scientists and discoverers. The university had risen to prominence alongside other German-speaking medical schools which, during the second half of the nineteenth century, took the lead in basing medical studies on the natural sciences. Unlike the French, who trained physicians in hospital-based clinics, German schools were located in universities where medical students were broadly integrated into chemistry, physics, physiology, and other scientific curricula. While bedside teaching remained the favored method in France, in Germany medical training took place in the lab and lecture hall. By 1870, thanks to numerous scientific achievements, German scientific medicine became the envy of the Western world.

In Vienna, this turn from medical pragmatics to science was effected by the anatomist and dean of the medical school, Carl Rokitansky. He insisted that pathological anatomy be placed at the center of all specialties. Dissections and the microscopic study of diseased organs and tissues were to be the bedrock of medicine. With Rokitansky at the helm, researchers in Vienna made great scientific advances, but these advances did not come without a cost. As professors eagerly focused on their lab work, Vienna's hospitals fell into disrepair. In 1887, the Viennese deputy Engelbert Pernerstorfer denounced the city's hospitals for criminal neglect of their patients. A decade later not much had changed.

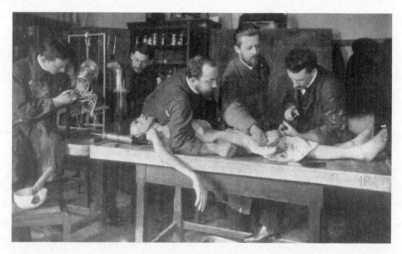

University of Vienna medical students in the dissecting room of the Institute for Anatomy, 1884.

This was no accident. German scientific medicine held therapeutics in low regard. Most therapies could not be scientifically proved and were based on little more than hearsay and the scattered experiences of seasoned doctors. Skepticism regarding the possibilities for rational therapies became extreme in Vienna, where an attitude of therapeutic nihilism took hold. From 1850 to 1880, prominent leaders at the medical school suggested that therapeutics were at best worthless, at worst dangerous. These beliefs were championed by professors like Joseph Dietl, who argued that the present state of knowledge made it impossible to offer any scientifically based treatments. He suggested physicians solely concentrate on understanding diseases, not treating them. Dietl's oft-repeated phrase was that nature alone could heal.

If Dietl hoped research might someday narrow the gap between science and medical practice, that was little consolation for the sick. Nor was it much help for the vast majority of medical graduates spit out into practice each year. As Freud discovered, few students ever went on to secure positions as scientists in the university. After extensive lab training and years of academic lectures, most German-trained doctors went into practice unequipped to deal with the clinical problems they faced. Physicians were forced to quickly shed their identities as students of nature and adopt a variety of treatments that, proven or not, would sustain their practice and provide some relief for their patients. As a result, two quite disparate medical cultures existed side by side in fin de siècle Germany and Austria. A small but prestigious university faculty insisted on furthering science, while most medical practitioners did what they could to master often unproven therapies. Commenting on this situation, Rudolf Virchow dryly observed: "It is said of the academic physician that he can do nothing, and of the practitioner that he knows nothing."

At the University of Vienna, however, there were a few exceptions, and these more therapeutically minded teachers would be crucial to the first Freudians. In 1850, Johann von Oppolzer was appointed to the faculty against the professoriate's wishes. Oppolzer was a therapeutic activist who had earned the scorn of men like Rokitansky. But patients flocked to him, making him one of the most widely sought-after consultants in Europe. Oppolzer experimented with different kinds of hydrotherapies, but perhaps most importantly, he refused to give up on his patients. His energetic approach proved popular with some students, including the young Josef Breuer, who became Oppolzer's assistant in 1867. In 1871, Oppolzer died and Breuer went into private practice, where he would dream up an extraordinarily ambitious therapy for the ailing Bertha Pappenheim.

In the 1880s, a few other champions of medical therapeutics found their way to Vienna. Hermann Nothnagel had been trained in Berlin, where he was first tempted to become a research pathologist. Instead he turned to the clinical practice of medicine, and in 1870 he wrote a practical guide to pharmacotherapy. In 1882, Nothnagel arrived at the University of Vienna and began to compile a huge handbook for pathology and practical therapies. Nothnagel believed the relief of suffering—not the scientific study of disease—was the highest ethical obligation for doctors. Following Rudolf Virchow, Nothnagel saw disease as life under changed conditions. By carefully studying the individual, he believed a physician could come to understand the specific nature of his pathology, and then determine the correct circumstances under which that individual might effectively cure himself.

After Freud introduced himself to Nothnagel in 1882, the older doctor proved to be a faithful ally. Some of Freud's most dedicated followers came to him after imbibing Nothnagel's therapeutic ideals. Alfred Adler called Nothnagel his greatest influence in medical school. Stekel was Nothnagel's student, patient, and friend. But Stekel would also discover that despite Nothnagel's fine words about the obligation of physicians to heal, his teacher had a frightfully limited repertoire of treatments. When in 1901 he sought out Nothnagel's care for relief of a headache, the great teacher applied leeches to bleed him.

The arrival in Vienna of Richard von Krafft-Ebing in 1889 heralded a more activist stance for the university's department of psychiatry. Rokitansky's protégé, Theodor Meynert, had vigorously aligned his psychiatry department with the study of neuroanatomy, despite the protests of therapeutically minded asylum doctors. When Krafft-Ebing was appointed over Meynert's objections, he brought a more clinically inclined perspective with him. At Graz, Krafft-Ebing had employed hypnosis and a suggestive cure he called psychotherapy. When Krafft-Ebing ascended to Meynert's chair in 1892, suddenly Vienna, unlike most other German-speaking psychiatry departments, could boast of a chairman not wedded to anatomical research. His lectures turned the interests of numerous medical students in a more clinically minded direction.

While Viennese medicine was generally known for its therapeutic pessimism, the doctors who joined Freud in 1902 had all studied with the likes of Nothnagel and Krafft-Ebing and were curious about treatments, such as heat, air, water, and electrical cures. A number of them had shown serious interest in the dynamic treatments that the French had pioneered; it made sense for them to seek out the Viennese representative of such psychical treatments. Freud offered them hope for another way to cure.

He also offered these men something else. While there were many odd medical therapies floating around in 1900, few were as highly theorized as Sigmund Freud's. By 1900, Freud had linked his ideas not just to the work of Charcot and Janet, but also to that of Brücke, Helmholtz, and Fechner. He had a detailed, deterministic explanation for how his therapy worked, and a complex model of the mind rooted in descriptive psychology, biophysics, and psychophysics. Thanks to this ambitious, nuanced model, Freud offered his followers not just increased clinical effectiveness but also a new lease on their identities as natural scientists. By following Freud, these castoffs from the ivory tower could claim to be both clinical activists *and* natural scientists. They could treat hysteria and be scientists of the mind.

Thus, a handful of doctors made their way to Berggasse 19 in the last months of 1902 to join Freud in a discussion of scientific psychology and neuropathology. Others soon followed. In 1903, Paul Federn joined the group after being sent to Freud by Nothnagel. In 1905, Federn's friend, Eduard Hitschmann, joined the band, and in 1906 three others arrived: an internist named Alfred Bass, a physiotherapist, Adolf Deutsch, and Isidor Sadger, a hydrotherapist. They came looking for new treatments for their patients and sometimes themselves. Schooled in the German tradition of scientific medicine, they took up Freud's method as a way of curing as well as studying the psyche. And so the little group began to grow, as more doctors came to Sigmund Freud's home in search of something new.

II.

THE PHYSICIANS IN the Wednesday Society would be joined by nonphysicians from the larger swirl of Viennese culture for, by 1902, Freud's medical theories and his positions on degeneration, repression, and sexuality had begun to filter into broader debates on Austrian political and social life. Many concluded the Austro-Hungarian Empire was sickly, and that radical measures needed to be imposed before this illness proved fatal. Freud, they concluded, might have a treatment for their culture's malaise.

Unlike these fin-de-siècle reformers, the middle-aged Freud himself was an 1860s liberal who had come of age in a time when this political philosophy had held great hope for rational progress and enlightenment. That hope began to flourish after Emperor Franz Josef established the first pan-Austrian parliamentary structure in 1861. Optimism ran even higher after the rise of the Burgerministerium of 1867. The same year, the Jews of Austro-Hungary were granted civil equality, which effectively allowed them to move about freely and brought many of them to Vienna. In 1868, Catholic laws on marriage were repealed, further marking the power of liberal reform and the retreat of the church.

Many men and women of Freud's generation believed society could be changed by the power of reason and science. When Hermann von Helmholtz gave public lectures dismissing knowledge based on traditional authority as not only empty but immoral, his message rang out in church and court. He argued that science alone provided a legitimate means to understand the natural, social, and political world. As evolutionary theory made its way into Germany and Austria, this challenge was heightened. By the late 1860s, Ernst Haeckel was busily attacking the church, and ruing the way superstition and religious belief retarded social progress.

In the last decades of the nineteenth century, however, evidence of rational progress in the Austro-Hungarian Empire began to grow scant. The collapse of the Vienna stock market in 1873 resulted in a wave of bank failures and bankruptcies, followed by years of economic recession. During this bleak time, antiliberal, mass parties formed and appealed to their constituents' ethnic and national identities, rather than to individual liberty and rationality. Often anticapitalist and anti-Semitic, some populists blamed the Vienna stock market crash on the Jews, not to mention the liberal policies that brought Jews to Vienna as citizens. Increasingly, hate mongers like Georg Ritter von Schönerer propounded a virulent form of anti-Semitic, reactionary politics.

The bright hopes of old school liberals began to fade as social reform was met with populist rage. Even the university was no haven from the rising tide of ethnic hatred. In 1876, the famous physician Theodor Billroth questioned the practice of admitting Jewish medical students from Galicia and Hungary, characterizing these students as vain, untalented, and not German. In 1889, university fraternities excluded non-Aryans from membership. A few years later the general assembly of Austrian fraternities forbade Jews from dueling, contending they had no honor to defend. In 1896, the Anti-Semitic League was formed in Austria to protect artisans from being displaced by Jewish craftsmen. Blood libel cases continued to be brought against Jews for sacrificing Aryan children, with a guilty verdict coming down in 1899. To top all this off, the anti-Semitic, Christian Socialist politician Karl Lueger was elected mayor of Vienna in 1897.

In this increasingly hostile atmosphere, educated liberals of Freud's generation found themselves embattled. Many had harnessed their political hopes to work as functionaries and experts. As government workers, lawyers, judges, doctors, and scientists, they were committed to the belief that reason would lead to reform. By 1900, their confidence had been undermined. This atmosphere formed the backdrop to Freud's furious attempts to win a professorship from the medical school. It was not until one of Freud's rich, noble patients—whom he cynically nicknamed the "Goldfish"—intervened with government officials, that he was granted the title. This bitter episode demonstrated that despite its emphasis on merit and intellectual achievement, liberalism did not effectively counter either anti-Semitism or the old Habsburg methods of doing business via aristocratic wealth, access, and power.

What had gone wrong? By the fin de siècle, liberals tried to make sense of their failed project. One explanation for the collapse of reason came from psychiatry: degeneration theory. Its most famous exponent was the Hungarian Jew Max Simon Südfeld, who took the nom de plume Max Nordau. Trained as a doctor, Nordau studied with Charcot and settled in Paris, where he churned out political reviews for German-language newspapers, as well as a stream of novels, plays, and works of cultural criticism. Nordau was committed to the belief that science would lead to social progress, and his despair that this progress had been thwarted animated his 1892 book, *Degeneration*. A work of social criticism, *Degeneration* read more like a coroner's report.

According to Nordau, the European fin de siècle was characterized by degeneracy as manifested by the writers and artists who represented the culture. They embraced the irrational and mystical, a most dangerous trait, because knowledge of reality was critical for adaptation and survival. The degenerates of aesthetic modernism suffered from a kind of moral insanity; their sexual hedonism was infantile and ego-maniacal, their prophets sick men like Ibsen, Tolstoy, Wagner, Oscar Wilde, Zola, and most centrally, Friedrich Nietzsche. Thanks to these apostles of madness, degeneracy had become a severe mental epidemic among the urban educated classes. If not arrested, Nordau predicted that this plague would spread murder, suicide, sexual perversity, drug abuse, and illiteracy throughout Europe.

In his sweeping indictment, Nordau did not hesitate to impugn his own race. Along with Charcot, he believed Jews were disproportionately degenerate. To ameliorate this hereditary curse, Nordau lamely advised the practice of gymnastics. Others would devise more extreme solutions; Nordau's critique was taken up by right-wing anti-Semites who also saw modern degeneracy as a Jewish disease. They called for a *Lebensreform* in which the German people would renew their inner beings as Aryans, and reject modernity, industrialization, and above all, the sickness carried by Jews.

Hence, when the young Sigmund Freud positioned himself against degeneration theory in the service of casting more light on the problem of neurosis, he was also taking a position in a larger political debate over the collapse of European liberalism. Freud's opposition to degeneration brought the old liberal some unlikely allies, including the very modernists that Nordau accused of sowing disease. These writers and artists rejected the notion that they were the problem and turned to the work of Freud to in part explain the disarray around them.

The Viennese moderns saw a decaying culture hidebound by unnatural, sickening traditions. Long the guarantors of the status quo, the Habsburgs had been swiftly losing authority. While Franz Josef won many hearts by insuring equal rights for minorities, the suicide of his sole heir, Crown Prince Rudolf, in 1889 dashed any hope for the Habsburgs to maintain long-term control of what was becoming an increasingly chaotic society. By 1900, Austrians may have shared a fondness for their dotty, elderly monarch, but they shared very little else. When universal male suffrage was instituted in 1907, the parliament that emerged dramatized the fact that Austria was home to an ethnic stew of Germans, Italians, Poles, Czechs, Slovaks, Romanians, Magyars, Ruthenes, Croats, and Slovenes. Parliament consisted of over thirty different parties, including pan-Germans, Zionists, Socialists, Christian Socialists, and Social Democrats. Members addressed the body in ten different languages. Each representative was allowed to speak in his native tongue, though incredibly, no translators were provided. Liberal reforms had led to a parliamentary government that resembled the Tower of Babel.

To some this was symbolic of a broader dilemma. It was said that to be Viennese was to be a question mark. Issues of nationality, religion, language, and politics were all contested. It did not help that prominent voices in Austrian science suggested the human self had no *there* there. Ernst Mach had posited that the self was nothing more than a stable bundle of disparate sensations, a logical extension of the reductive methods of biophysics. By reducing psychic qualities to their energetic quantities, Mach had made human identity an illusion. This idea was taken up by Mach's student the Austrian novelist Robert Musil, who famously described a man without qualities:

> The inhabitant of a country has at least nine characters, a professional one, a national one, a civic one, a class one, a geographic one, a sex one, a conscious, an unconscious and perhaps even a private one; he combines them all in himself, but they dissolve him, and he is really nothing but a little channel washed out by all these trickling streams.

Musil added a tenth character, an empty invisible space which made all of the others unreal. Insofar as the unreality of the self could become distinct and apparent, Musil mused, it had become so in the Austro-Hungarian monarchy.

Café intellectuals and apostles of modernism blamed this state of affairs on the weight of religious and social convention, as well as scientific positivism, which appeared to yield only a dead, sterile hyperrationality. This critique was most powerfully expressed in the explosive work of Friedrich Nietzsche. After losing his religious faith, Nietzsche fell under the spell of Arthur Schopenhauer and began to conceive of himself as physician to a diseased culture. Animated by the belief that moral regeneration would only come about from an infusion of the emotional and irrational into an otherwise conventional, half-dead way of being, Nietzsche began decimating social and intellectual beliefs. For Nietzsche, enlightened men were self-deceiving creatures who constantly sought what they called truth, which was in fact little more than simplified and falsified notions meant to provide them with reassurance in a chaotic world. At the foundation of rational knowledge was hungry, unspoken psychological need. In *The Gay Science*, Nietzsche attempted to construct a science that attended to that need and restored the health of European culture:

> I am still waiting for a philosophical *physician* in the exceptional sense of that word—one who has to pursue the problem of the total health of the people, time, race, humanity—to muster the courage to push my suspicion to its limits and to risk the proposition: what was at stake in all philosophizing hitherto was not at all "truth" but something else—let us say, health, future, growth, power, life.

Nietzsche reminded his readers that humans were animals driven to maintain their species. The imperatives of the species required little of consciousness. Unconscious thought, feeling, will, memory, and action drove human behavior. "Man, like every living being, thinks continually without knowing it; the thinking that rises to consciousness is truly the smallest part of all this." In 1890, Nietzsche collapsed in Turin due to tertiary syphilis; he would die in an asylum ten years later, but during those years his work began to grow in the public's estimation. German and Austrian socialists, anarchists, and progressives all took up Nietzsche's call for liberation and embraced the ethic by which the health of individuals came before conventional rules, science, or the morals of the church.

In Vienna, a leading advocate for such liberation was Hermann Bahr. Born to a liberal Austrian Catholic family, Bahr traveled to Paris and came back a changed man. Rather than critique what others called decadence, Bahr embraced this as indicative of a turn inward, away from pseudorational attempts to portray the world mechanistically. Decadence was a shift toward the sexual, the psychological, even the psychopathological self. Scientific positivism and religious belief both needed to be replaced by what Bahr called a "nervous romanticism." By focusing inward, Bahr hoped to map out the modern nervous self and its universe.

Bahr returned to Vienna and held court at the Café Griensteidl, nicknamed by some the Café Grössenwahn, or Café Megalomania. His critical and literary work was avidly followed by the cultural elite. Young Alma Schindler, soon to be the wife of the composer Gustav Mahler, found one of Bahr's novels hollow and obscene, but quite enjoyable. "I never experienced anything like the physical thrills of his smuttier episodes," she wrote in her diary. Around Bahr, there congregated a group of writers including Arthur Schnitzler, Hugo von Hofmannsthal, Felix Salten, and Richard Beer-Hoffmann, who collectively became known as the *Jung Wien* writers, and together they made Austrian literary modernism famous.

While Bahr called for a turn inward, another *Jung Wien* writer attacked the world that made such a retreat necessary. Peter Altenberg (né Richard Engländer) had taken a stab at law and medicine, before he was declared neurasthenic and unfit for work. Freed from responsibility, Altenberg became the ultimate bohemian, spending his days and nights in cafés, drinking and cavorting with prostitutes. He wrote short prose sketches, as well as attacks on the unnatural qualities of Viennese culture.

In 1897, Altenberg published a collection entitled *Ashantee*, inspired by the establishment of a West African village in the Vienna Zoo. These village dwellers became his archetype of man and woman living in harmony with nature, free from the pathological effects of a gilded Habsburg past. Altenberg sought to cast off conventional ethics and return to a natural primitivity; toward this goal, he advocated a panoply of health measures aimed at a liberation from clothing, especially women's undergarments. His motto was: "One cannot wear too little!" One winter he caught pneumonia and died.

Bahr and Altenberg were inspirations to a generation that had grown cynical about the aristocracy and the church, and had also lost faith in their parents' liberal ideals. "Nobody knew exactly what was on the way," Robert Musil wrote of this time, "nobody was able to say whether it was to be a new art, a New Man, a new morality or perhaps a re-shuffling of society. So everyone made of it what he liked. But people were standing up on all sides to fight against the old way of life." Some of these restless troops embraced an erotic, nervous, modern self, and in so doing hoped for a more natural, healthier way of life. In this effort, they would find support, justification, and guidance in the work of Sigmund Freud. As Freud's psychoanalysis began to take root in Vienna, Vienna began to take root in psychoanalysis.

III.

FREUD BELIEVED OPPRESSIVE conventions regarding sexuality could create illness, and he was alarmed at the sexual dilemmas young people faced because of society's demands. Therefore, he found common cause with a growing band of sexual reformers in Vienna. The Austrian writer Stefan Zweig noted: "It did not take us long to discover that all those authorities in whom we had previously confided—school, family, and public morals—manifested an astonishing insincerity in this matter of sex." When it came to pushing sexual reform, ethicists, feminists, bohemian aesthetes, and critics were all drawn to Dr. Freud, and they used his ideas to give scientific weight to their case.

They needed bolstering because the reformers were up against a formidable foe in Austria: the Catholic Church. Catholic doctrine defined sex outside of marriage as a sin, but the open secret was that extramarital sex was a thriving business in Vienna. The city hosted what Zweig called a "gigantic army" of prostitutes. Some argued that secretive sexual practices were the only thing that kept the pretense of a moral, monogamous middle class afloat. In 1912, a poll asked Viennese medical students to reveal their first sexual partner. Four percent named a woman who was a potential spouse, 17 percent a maid or waitress, and 75 percent confessed that they had been initiated into sexual life by a prostitute.

Debates over sexual ethics also expanded to take in the so-called Woman Question. Austrian liberalism had offered men the promise of autonomy and self-fulfillment, but in Catholic Austria, women had not been given the same opportunities. Women could not enter professions, receive university educations, or vote. They were to marry, raise children, and be monogamous, no matter what their husbands did. And they were not to show much interest in sex. Middle-class custom forced women to conceal their bodies under layers of clothing, and enforced an ignorance of sexuality in girls, all in the hope of maintaining the notion of a sexless, virtuous woman. Women who acknowledged their erotic longings were seen as aberrant. An anguished young Alma Schindler wrote in her diaries: "A curse on me and my sensual temperament . . ." then later: "Why am I so boundlessly licentious? I *long* for rape!—Whoever it might be." Freud's young female hysterics came from the same airless world.

By the end of the nineteenth century, the restrictions on women began to be challenged. In 1897, the University of Vienna opened its doors to women, though Catholic women were discouraged from the study of medicine, where they would have to view men naked. At that same time, women's emancipation movements began to spring up, led by Viennese such as Rosa Mayreder, Grete Meisel-Hess, the Eckstein sisters, and Bertha Pappenheim, as well as the Berliner Helene Stöcker. Some early feminists were proletarian radicals who sought to emancipate women by calling for a new socialist order, while others were middle-class reformers. Grete Meisel-Hess was a socialist who would argue that capitalism and contemporary sexual mores were making men and women ill. More traditional liberals like Stöcker and her League for the Protection of Mothers, as well as Bertha Pappenheim and her Jewish Women's League, were out to stop the sexual exploitation of women via prostitution and white slavery. All saw the traditional roles of women as unjust. Some like Meisel-Hess would look to Freud's idea of repression to aid their arguments. Two other prominent feminists knew Freud's theories well, for they had played a role in their creation. Bertha Pappenheim was Breuer's "Anna O.," and Emma Eckstein was one of Freud's early patients.

The *Jung Wien* writers also took up the cause of women, but from their own vantage point. Peter Altenberg envisaged a new relationship between the sexes in which women were not sexually shamed, repressed, or exploited. In his 1896 story called "The Primitive," he described picking up a woman who spoke of her desire without shame, as if she were "earth in the Cretaceous Age." Unabashedly, she stripped off her clothes, saying: "I love my body and I treat it as something holy." For the first time, the narrator exclaimed, this woman experienced a man in freedom and understanding rather than exploitation. Why? Because the narrator understood that the sexual, the bodily, the primitive, resided in the unconscious of all people. He had given the " 'beautiful unconscious' a philosophical basis, a psychological interpretation. He had 'discovered' the primitive!"

This embrace of the sexual, the natural, and the primitive easily led to Freud's medical claims regarding the dangers of sexual repression, and soon the doctor found himself in the midst of efforts for reform. Divorce was strictly limited in Austria, as were contraception and abortion. Critics argued for new laws to end this unhealthy state of affairs. A tireless proponent for such reform was Freud's friend Christian von Ehrenfels. For Ehrenfels, civilization's repressive demands created a double consciousness. Civilization made a pathological split between day and night, above and below, the socially sanctioned and the sexual. For women, the problem was even worse, since civilization required her to renounce all sexual urges, lest she be called a whore. Men, on the other hand, were persecuted by the cultural belief that monogamy was natural and therefore made to feel that their polygamous sexual urges were evil or sick. Pretending to be virtuous by day, they would slink off to prostitutes at night, where they contracted venereal diseases. Ehrenfels wrote that this psychical splitting had been described by Breuer and Freud, and it could be cured by their method, but he also wondered how much more could be done through sexual reform.

In 1907, Ehrenfels went further in his *Sexual Ethics*, a book-length critique of contemporary sexual mores. Ehrenfels called for a new ethic based on polygamy. Citing Freud, Ehrenfels made the case that men and women had been placed in an impossible bind. If men complied with civilization's demands for sexual abstinence, they became ill. Doctors were compelled to tell young men to have sex or risk becoming neurasthenic. But moral young men were often loath to seduce an upstanding virgin, so after great torment, they ended up in the "cloaca of prostitution." Instead of coming down with a nervous disorder, they contracted syphilis. A new moral code was required based on natural selection, which would be aided if the most virile men reproduced with many women. Therefore, nature (and Ehrenfels) favored "polygyny," in which the most vigorous men fertilized several women, either in serial monogamous relations or simultaneously.

Ehrenfels sent an inscribed copy of *Sexual Ethics* to Freud, who annotated the work carefully. In response, Freud would publish his own 1908 reformist essay, "Civilized Sexual Ethics and Modern Nervous Illness," making clear that he shared concerns about the widespread dangers of repression.

Essential to Ehrenfels's critique was a vision of the natural state of sexual relations. Banking on Darwin's primal horde, Ehrenfels argued that polygyny was nature's way, but others argued it was women who were meant to be polygamous, not men. It was said that woman in her natural state engaged her sexual desire freely, a line of thought that made the prostitute a natural woman and therefore a figure of some veneration. Felix Salten, the beloved author of the children's book *Bambi*, devoted a good deal of energy to writing anonymously *The Memories of Josephine Mutzenbacher,* a pornographic work that detailed, often in heroic terms, the sexual escapades and traumas of a courtesan.

The figure in Vienna most associated with the idealization of the prostitute was Karl Kraus. Kraus began his literary career in the *Jung Wien* circle before breaking with the group, and founding his vitriolic weekly, *Die Fackel* (The Torch) in 1899. From this editorial perch, Kraus surveyed Viennese culture, mercilessly attacking opponents like the old-line liberal *Neue Freie Presse*. In the fall of 1902, Kraus took up his pen to expose the corrupt sexual morality of his culture. Homosexuality, adultery, prostitution, and extramarital sex should be legalized, he argued. Abortion and contraception should be available. Middle-class women should not have to suffer the hypocrisy of a tradition that allowed them no sexual freedom, while men were permitted to feign moral purity by day and whore by night.

The Austrian writer and publisher of *Die Fackel*, Karl Kraus, 1908.

While some of Kraus's hatreds were idiosyncratic, his views on women and sexuality were not. Like Schopenhauer, Nietzsche, and others, Kraus saw men as the embodiment of rationality and women as the embodiment of emotion and animal desire. Kraus opposed feminists because he viewed woman as man's savior: the natural, sexual woman would redeem arid men. This glorified figure of deliverance was not the mother or sister or daughter of the bourgeois family, but the courtesan.

Similar views were elaborated in the startling work of Otto Weininger. Like many reformers in Vienna, Weininger was the son of a middle-class Jewish family. He studied at the University of Vienna and became interested in philosophy and psychology. Before taking a trip to Paris with his friend Hermann Swoboda for the Fourth International Congress of Psychology, he sat in on Krafft-Ebing's lectures. When the two friends returned from Paris, Swoboda entered psychoanalytic treatment with Freud. At some point, Swoboda told Weininger that his doctor believed all humans were bisexual. With that Weininger was said to have clapped his hand to his head and gone off to write his masterpiece.

After the publication of *Sex and Character* in 1903, Weininger became a literary sensation. To many, it seemed that he had solved the riddles of human sexuality and gender identity. In his book, Weininger argued that all men and women were part male and female, and that the ratios of these qualities determined both a person's character and his or her object of sexual desire. All people sought an object that worked in proportion to their own masculine and feminine qualities, so as to make one whole male and female out of a couple. Using this theory, Weininger also mounted an argument for the naturalness of homosexuality and prostitution. The homosexual and prostitute were simply different constitutional amalgams of the male and female, who searched for objects based on the needs of their characters.

It was difficult to discern what male and female meant, now that Weininger had freed these concepts from gender. For Weininger, however, the true female was pure sexuality. Sex for women was natural. Chastity was unnatural and impossible. Hence: "the disposition of and inclination to prostitution is as organic in a woman as is the capacity for motherhood." Two types of woman—mother and whore—may be mixed, but women as a rule were polygamous, while men were monogamous. Women had no capacity for conscious morality or reason, but in their fully achieved state were sexual, irrational—a whore.

Weininger argued that his ideas were shocking only because society had had a perverting influence. Not free to live naturally, women had been traumatized. Citing Breuer and Freud's notion that hysteria was the result of sexual trauma, he added that the trauma was caused by the way women were forced to imbibe male notions of their own sexuality. Civilized women learned to hate themselves for secretly being lusty. In the end, Weininger (much like Schopenhauer) recommended a complete renunciation of sexuality even for propagation, published his magnum opus, and promptly committed suicide later that year. After his tragic end, Weininger became wildly popular among European intellectuals.

Kraus read Weininger's book and was profoundly impressed. He began to do battle with critics who failed to understand women and their polygamous nature, and in this spirit, in 1904, Kraus took up the Hervay case. In the provincial town of Mürzzuschlag, Frau Hervay had been accused of polygamy; during the ensuing public scandal, her humiliated husband committed suicide. After publishing a scathing commentary on the case, Kraus received a visiting card, congratulating him on his "insight, courage, and the ability to perceive the larger implications of a small affair." The card was from an older Viennese physician who was quite interested in sexuality himself.

SIGMUND FREUD'S ENGAGEMENT with Kraus and other Viennese sexual reformers came at a critical juncture. In 1904, Freud had an academic reputation and a small following of physicians, but he was not famous. After his break with Breuer and then Fliess, Freud felt isolated. The doctors that met at his home and attended his lectures were his sole audience. His books had fared poorly: *Studies on Hysteria* took thirteen years to sell out an edition of 626 copics, while *The Interpretation of Dreams* took eight years to sell its 600 copies.

Then rather suddenly, this middle-aged liberal doctor found himself embraced by laymen who dreamed of a new culture. Their world was not the university, the clinic, or the lab, but the Vienna coffeehouse, the meeting place for artists, writers, reformers, and utopians where the daily meal consisted of the many newspapers these cafés provided. Thanks to his book on dreams, Freud's name began to appear in these newspapers during the first years of the twentieth century, but the doctor seemed unimpressed. He had great hopes for scientific recognition and was disappointed by the lack of reaction his work had generated among his colleagues. In 1902, he complained: "I have received only two reviews in professional journals." Newspaper reviews were little consolation. In 1904, Freud began to write a number of summarizing essays intended to win over the medical community, but he still professed a proud unwillingness to pursue the masses directly. When Fliess suggested his Viennese friend write a short article on the dream book for the *Neue Rundschau* in Berlin, Freud said he would not stoop to such measures.

And yet, Freud did begin to approach the Viennese public. In 1903, he agreed to publish short pieces in the *Neue Freie Presse,* contributing a total of five reviews and one obituary over two years. In one review, Freud tellingly positioned himself as the voice for a new generation, dismissing one author by saying: "It really serves as a useful warning, if one is shown how much that was described by people of an earlier generation as 'self-evident' or as 'nonsensical' ranks with us today conversely as nonsensical or self-evident." Freud's card to Kraus may have also signaled an increased willingness to engage nonmedical audiences. For within the cultural elite of Vienna, reading Kraus was de rigueur. In her journals, Alma Schindler returned again and again to the tumult surrounding Kraus and his latest diatribe; his insults against the painter Gustav Klimt, the lawsuit filed against him by Hermann Bahr, and his repeated savaging of prominent figures. Farcical, brutal, and often hilarious, Kraus's likes and dislikes were devoured by the intellectuals of Vienna.

One of his favorite targets was Viennese psychiatry. After his essay on the Hervay case, Kraus went on a tear against forensic psychiatry. He was horrified by the indiscriminate rationales these doctors used to commit individuals to mental institutions. On the case of Louise von Coburg, a princess deemed mad because she did not love her powerful, aristocratic husband, Kraus wrote: "The police and the military have a new function: to channel sexual drives in desired directions. As a result there exists in Austria an office that may be called the Ministry of Jealousy. The Ministry does not rule by naked force; it does not poison or strangle. It uses psychiatry instead." Since Freud also scorned traditional psychiatric thinking on these matters, Kraus's attacks created an allegiance between the two men.

More vitally, they shared an interest in sexual reform, a fact that was underscored when Freud gave public testimony advocating a change in the marriage laws. In 1904, the Austrian government had appointed a commission of inquiry to consider reform. The Political-Culture Society, a group of Fabians dedicated to a synthesis of socialism and capitalism, decided that their group would conduct a parallel investigation. Lawyers and prominent feminists including Grete Meisel-Hess were asked to appear before this other commission, and experts on the sexual and medical matters were invited, including Wilhelm Stekel and Sigmund Freud.

Dr. Freud's answers made an effort to remain above the political fray. He declined to answer any questions that were not directly related to his expertise as a physician. In a series of written responses, however, this medical expert did not hide his convictions. Was successive polygamy consistent with the demands of civilizations? Freud replied that it could be, and that such a change in law would only sanction the conditions that already existed. Would morality be better served by legalizing extramarital sex? Freud believed it would. A greater measure of sexual freedom was the only way to actually encourage morality, he ventured. As for women, Freud suggested that full equality of the sexes was impossible because of their different biological roles in reproduction. But he argued that the indissolubility of marriage did a great disservice to women, and argued for female representatives in all marriage courts.

Freud was most expansive on the health consequences of sexual continence. Many physicians avoided this issue for fear of disrupting the existing sexual order, but the majority of men found it impossible to maintain abstinence over long periods. The physical harm of such continence was substantial, and included a disposition to nervousness. Psychic damage was even more damaging; it undermined "self-confidence, energy and daring." Women might appear capable of sexual abstinence, but that was an illusion. Women paid an even greater psychological price (which prompted Freud a year later to claim that there was scientific justification for "women's emancipation"). Finally, Freud answered explosive question number thirty-eight. If abstinence was not physically and psychically possible, could adultery in sexually dead marriages be legitimate? It was, Freud answered. Since abstinence was pathological, "the existence of a marriage is in itself no grounds for sexual obligations when the marriage no longer fulfills the task of satisfying normal sexual instincts." Freud's reply may have been buttressed by personal experience: after 1895 his marriage to Martha had grown cold, and it seems likely that sometime after 1898 he conducted an affair with his sister-in-law, Minna Bernays.

For Karl Kraus, Sigmund Freud had become one of the few voices of reason among a horde of corrupt and ignorant psychiatrists. Readers of *Die Fackel* soon found Kraus opposing the general immorality of psychiatrists with the good sense of this doctor: "One must agree with Professor Freud that the homosexual belongs neither in prison nor in the madhouse," he wrote. In 1906, Freud addressed Kraus warmly, writing: "That I find my name repeatedly mentioned in the *Fackel* is presumably caused by the fact that your aims and opinions partially coincide with mine."

Six weeks later, Freud had even more reason to be pleased. On December 21, 1905, *Die Fackel's* readers were treated to the first review of Freud's new book, *Three Essays on the Theory of Sexuality*. The reviewer was not a doctor, but rather a young writer interested in sexual ethics. In his review, he juxtaposed Freud's *Three Essays* with a work by the Zurich psychiatrist Auguste Forel. The aging Swiss psychiatric authority had recommended that young men insure their health by doing their best to rid themselves of sexual thoughts. "Forel's treatise," the reviewer commented, "has not quite reached the level on which the modern science of sexuality stands; Freud has left it far behind." Freud, the critic continued, questioned the conventional morality and metaphysics that dominated European political and ethical culture, and hence he provided the first comprehensive account of the physiology and psychology of love.

The twenty-three-year-old who composed this glowing review was Otto Soyka. Soyka had already written a critique of psychiatry for Kraus, and soon he would contribute another on sexual morality for *Die Fackel*. In 1906, he published *Beyond the Boundaries of Morality*, the first book that used Freud's psychosexual theory as the foundation for a new sexual ethic. According to Soyka, man's only true morality was the expression of his desires and drives. Citing Freud, he insisted that sexual desires were not just procreative, and that sexual perversions were not diseases. He encouraged homosexuality as a useful educational instrument between teacher and student, and asked for the acceptance of masochism and sodomy. That year, Soyka joined Freud's Wednesday Psychological Society after being nominated by Philipp Frey, a schoolteacher who in 1904 had just published his own treatise, *The Battle of the Sexes*. While Soyka would soon drop out of the group, he was followed by others who found their way to the Wednesday Society through the debates on sexual ethics.

Soyka's praise for *Three Essays* was shared by a leading feminist, Rosa Mayreder. Writing in the *Wiener klinische Rundschau*, she announced that Freud had not just made a contribution to the problems of sexuality but also had created "solid fundamentals for a novel theory." Mayreder used her review to reiterate the point she had made in her coolly devastating *Towards a Critique of Femininity*. Most notions of woman had one thing in common, she observed: they were written by men. These abstract, contradictory, and often absurd sketches were all male projections, fantastic concoctions of woman as sexual fetish object.

In her view, Freud was different. "Very few observers have been able to free themselves from conventional prejudices and give their studies true scientific value," Mayreder commented. She praised Freud for resisting "conventional pseudo-psychology and its twin sister, conventional pseudo-morality." Freud offered deep insight into female sexuality by suggesting that repression characterized female psychology.

Along with the enthusiasm of reformers, Sigmund Freud's *Three*

Essays on the Theory of Sexuality was warmly endorsed by his medical colleagues. "All parents and educators should make it their duty to become familiar with the author's theories," wrote one. The respected sexologist Albert Eulenberg offered the comment that a brilliant Freud had brought forth a "rich harvest." The Swiss-American psychiatrist Adolf Meyer called the thin book essential. With the *Three Essays*, Freud impressed neurologists, psychiatrists, and sexologists, as well as cultural critics and sexual reformers. This little book showcased his ability to enter debates, seek out the flaws in different positions, and then present startling new syntheses. He had done this before in other domains, but this was his most dramatic achievement, and its success cut across professional lines to reach a broader readership.

The publication of the *Three Essays* made Freud a hero of the Viennese coffeehouse scene. It placed him at the center of a network of artists, writers, journalists, feminists, and reformers who believed the decay of Habsburg Vienna was not due to degenerate heredity, but rather centuries of unhealthy rules and regulations. Embraced by those who wanted to cure their culture, Freud began to respond to their overtures. He considered recruiting more intellectuals to the Wednesday Psychological Society. Max Graf recalled that Freud even hoped to interest the leader of the *Jung Wien* writers, Hermann Bahr. This never came to pass, but Freud was soon in correspondence with an equally famous member of that group, Arthur Schnitzler.

A former physician, Schnitzler had turned his energies to writing and the pleasures of life as a café bohemian and ladies' man. His plays fearlessly probed the hypocrisies and pains of conventional sexual ethics, and often provoked outrage. In 1906, Schnitzler publicly acknowledged the influence of Freud on his work, and the Professor was thrilled. He wrote the playwright: "For many years I have been conscious of the far-reaching conformity existing between your opinions and mine on many psychological and erotic problems; and recently I even found the courage expressly to emphasize this conformity. . . . Now you may imagine how pleased and elated I felt on reading that you too have derived inspiration from my writings."

With physicians coming to Freud for new therapeutic tools and sexual reformers welcoming his psychosexual studies as key to social reform, it seemed that Freud's wait for recognition was over. Then came scandal. In the summer of 1904, Wilhelm Fliess suddenly wrote his former confidant after years of silence. He had recently come across Otto Weininger's *Sex and Character*, where to his horror he discovered his theory of bisexuality put forward by the author. Since Weininger's friend Hermann Swoboda had been Freud's patient, was Freud somehow involved in this plagiarism? Had he ruined the Berlin doctor's chance for fame? Freud replied that he had used the idea of bisexuality in his treatment of Swoboda. If Swoboda passed the idea on to his friend, what could Freud have done about that?

Fliess refused to drop the matter. Wasn't it true, as their mutual friend Oskar Rie had told him, that Freud had read Otto Weininger's book in manuscript? the Berlin doctor demanded. Trapped, Freud admitted it was true, saying he had forgotten the reading, and acknowledged this forgetting represented an unconscious wish to steal Fliess's ideas. With that admission, the correspondence between Sigmund Freud and Wilhelm Fliess ended. Still Fliess would not let the matter die. In 1906, as Freud was winning plaudits from the likes of Karl Kraus, Arthur Schnitzler, Rosa Mayreder, and Albert Eulenberg, Fliess published a book-length blow-by-blow account, attacking his former friend as a liar and a plagiarist and backing up the accusation by publishing Freud's self-incriminating, final letter.

It was a dangerous moment for Freud, for he was a man already marked by more than his share of scientific scandal. But now, he had powerful friends in the press. Freud wrote to Karl Kraus, hoping to find a sympathetic audience. When the affair broke, Kraus covered Fliess's accusations and supported Weininger and Freud. Later, Freud admitted that Karl Kraus had been vital to building the psychoanalytic movement, though the Professor would have ample reason to rue his affiliation with the acerbic Kraus for years to come.

IV.

As FREUD'S REPUTATION spread, the Wednesday Psychological Society grew. In 1906, the group hired a secretary to take formal minutes. By then the Society was no longer the cozy, informal group that Wilhelm Stekel portrayed in 1902—five physicians meeting at the home of a prominent neurologist, leisurely bantering about cigar smoking and sexual matters.

In 1906, the group's membership had mushroomed to seventeen, and over the next few months it would reach twenty-two. The men now assembled for a scientific presentation followed by a formal discussion, in which every member was expected to participate. The last word was reserved for Freud, but otherwise these meetings were not exclusively centered around the Professor. Unlike his lectures at the university, where Freud would command the stage, lecturing and shooting questions at the audience, the Professor did not hold forth at the Society, nor did he present his own work very often. Instead, the Wednesday Psychological Society allowed a number of men to actively participate alongside Freud, thinking out loud, debating and engaging in what had become a wide-ranging consideration of all things psychological.

The new recruits included doctors like Adolf Deutsch, Paul Federn, and Eduard Hitschmann, as well as writers and intellectuals like Max Graf, David Bach, and Hugo Heller. The most influential society member in the greater world of Viennese culture was Heller. A publisher and bookseller, he was a strong supporter of the *Jung Wien* writers and the Secessionist artists. Later, when Freud needed a publisher to launch psychoanalytic publications, he would recruit Heller.

When the new secretary, Otto Rank, took up his pen in 1906, he recorded both the free flow of ideas and the equally uninhibited flow of venom that charged the meetings. Ad hominem attacks electrified the society's atmosphere. Years later, Freud mourned the fact that the group had not been able to put aside personal hostilities and work smoothly together. But such hostilities were continually enflamed by confusion about the group's methods and purpose. Was this a group dedicated to *Naturwissenschaft*, or *Geisteswissenschaft*, or some undiscovered third path? What kinds of knowledge were they seeking, and what kinds of practices would they support? Furthermore there was no consensus as to Freud's role. Was this a group that had come together to learn from the Professor, were they his students, or was this a group of independent researchers who had come together because of interests they shared with Freud?

The lack of agreement was dramatically revealed when the Society was asked to define itself. In January of 1907, Max Eitingon came to Vienna as an emissary from the renowned Burghölzli clinic in Zurich, where an interest in Freud's writings had taken hold. Eitingon hoped to flesh out the views of the Vienna School. Before the gathered members of the Wednesday Society, he asked for their thinking on the cause of neurosis, the essence of psychoanalytic treatment, and the kinds of outcomes that could be expected from this clinical method. As he listened to the answers offered, Eitingon grew increasingly dismayed.

The vast array of responses that showered forth from Society members was nothing short of stunning. In Charcot's clinic, a visitor could expect to hear Charcot's views elaborated by junior physicians. A doctor of Freud's generation, Auguste Forel, described how in medical school, students were very careful not to contradict their professors. Eitingon came from Zurich to hear Professor Freud's ideas explicated by members of his school. But at the Professor's home, everyone had his own particular view. Psychoanalysis was prophylactic, or perhaps curative, or maybe educational. As for the etiology of neurosis, one pointed to constitutional disposition, another to unnaturally increased perverse libido and familial syphilis, a third to defective organs, a fourth to sexual trauma, a fifth to general psychic conflict; and a sixth to the psychical residuals of birthing. The visitor from Zurich grew so exasperated he began to argue with his hosts, suggesting they were diluting specific and potentially powerful aspects of Freud's theory.

The Professor was given the final word. He reiterated his view that psychosexual conflict caused neurosis and did not comment on the fact that most of the Wednesday Society either did not accept or understand his position. Freud then turned his ire toward his visitor: "Mr. Eitingon's question betrays the theoretical disavowal of the sexual etiology of the neurosis, a disavowal which has not always been maintained by the Zurich School." If it was true that Eitingon's simple question as recorded did not assume sexual etiology, neither did it deny a role for sexuality, as many of Freud's Viennese followers had explicitly done in that same meeting. Acting as if there was consensus within his own group, Freud went on the offensive with the stranger, reinforcing a contrived sense of unity among the Viennese group.

But there was no hiding from this: the visitor had revealed that the Wednesday Society was a loose confederation of heretics. The members didn't share the same theories, and they did not even advocate the same methods of inquiry. Freud had set up—and was still modifying—a novel and inventive framework for studying mental phenomena. It was perhaps the most unusual and original part of his contribution to date, but his Society members had their own ideas about methodology. Some said the psychoanalytic approach consisted of strengthening the psychic field via a kind of psychic training; others insisted on making the patient his own therapist; some followed Freud by laying emphasis on transference, while others believed analysis washed away unhealthy deposits like a psychic water cure.

These disparities among the Society's members must have been particularly disappointing for Eitingon, because in 1907 divining Freud's clinical technique was no easy task. In Vienna, Society members had the advantage of being able to attend Freud's Monday consulting hours, where they could present cases to him for supervision. But in the Society meetings, psychoanalytic technique was not a regular focus of discussion. Wilhelm Stekel believed matters of technique could not be turned into laws and should be tailored to the patient. Alfred Adler expressed the view that psychoanalytic method could never be taught. Freud disagreed, expressing the hope that psychoanalytic method could be learned and would become set "once the arbitrariness of individual psychoanalysts is curbed by tested rules."

The Wednesday Society was a hodgepodge of arbitrary opinions alongside few tested rules. Moreover, no clear guideline existed to check various opinions, so they might be impartially insisted upon or rejected. They had no stable method and could not develop explanatory principles that did not constantly require a great deal of explanation themselves. The problems of developing a common method were buried in the excitement of discovery. Freud was avidly working to solidify his theory of psychosexuality, and in the case of a boy he named Little Hans, for example, the doctor did not worry too much about the potential conflict involved in the fact that Max Graf was not only the child's analyst but also his father. Similarly in the case of Ernst Lanzer, Freud did not feel compelled to discuss the extraordinary technical problems that arose. As Freud attempted to interpret Lanzer's psychosexual dynamics, the patient heaped "the grossest and filthiest abuse" on him, spewing fantasies about defiling the doctor and his daughter. Lanzer's diatribes were like dirt that kept getting on the lens of Freud's microscope; the scientist just kept wiping it off.

In the rush of exploration, how one traveled seemed less important than getting to the destination. Freud and the others in the Society ran off in myriad directions, into the wilds of psychopathology and back into the pasts of their patients. And for them, as with many revolutionaries, the means were less important than the ends.

WHILE MATTERS OF method would remain loose and undefined for some time, the wide-open character of the Wednesday Psychological Society in Vienna began to change between 1906 and 1908. A set of rules, some implicit, emerged to govern what kind of claims were acceptable in this community. These standards would come forward after a series of open conflicts, the first of which involved Wilhelm Stekel. Stekel was one of the few members who actually practiced clinical psychoanalysis. His clinical experience created a crisis in the Society for it posed a direct challenge to Freud's authority, and forced the group to decide exactly how far that authority extended.

Stekel was a prolific writer, and in 1907, he wrote a popular pamphlet on the cause of nervousness. When the Wednesday group read the book, they discovered Stekel claimed all neuroses were based on psychic conflict. What might have seemed *plus Freud que Freud* was, as one member tartly noted, a contradiction of the Professor's beliefs. Freud of course contended that unlike the psychoneuroses, the actual neuroses (neurasthenia and anxiety neurosis) were due to purely sexual disturbances, and had no psychological involvement. In the Society, Stekel's contradictory view was met with disdain. Inaugurating a critique that like the proverbial genie would be difficult to get back in the bottle, Rudolf Reitler declared that Stekel's theory of neurosis was a symptom of *his* neurosis. By denying real sexual causes, he had succumbed to repression.

Freud did not pick up on this personal attack, but then, as Stekel's onetime physician, he could not have done this without breeching his patient's confidentiality. Instead, the Professor accused Stekel of not keeping up with his advances. Neuroses—whether actual or psychoneurosis—were all based on sexual matters, but only some were based on psychic conflict, Freud patiently explained. Defeated, Stekel confessed to having written the brochure in a state of grave depression.

During the next meeting, Stekel went to great lengths to prove his bona fides. When Alfred Meisl presented a paper on instinctual life in which he discriminated between instincts of hunger and love, Stekel slammed the author for not acknowledging Freud's priority and added that what was good in the paper wasn't original. Repeating the line of attack Reitler unleashed on him, Stekel argued "personal repressions of the speaker lie behind Meisl's postulation of an 'asexual' component of preserving the species." Freud was finally moved to defend poor Meisl, who never presented to the group again and not long after, resigned his membership.

The harshness and the invasive quality of such attacks had an effect. Some members never dared present to the group. Others did their best to avoid commenting during the discussion. This was no simple matter. The Society had a custom of passing around an urn filled with numbered pieces of paper. Each member was expected to withdraw a number from the vase and then discuss the presentation in order. As the urn made its appearance, so too did the backs of a number of reticent members who hurried off into the night before being forced to speak.

The minutes of the Society spare Freud much blame for the vituperation. They show a Freud who—given the authoritarian standards for medical professors in his culture—was reasonably charitable in the face of dissenting views and often curious about novel, even oddball suggestions. In the debate with Stekel, however, Sigmund Freud showed another side, going to great lengths to squash his rival's theory. And despite being chastised, Wilhelm Stekel refused to back down from his contention that all neuroses, including the so-called anxiety neuroses, were due to psychic conflict. Furthermore, he believed he had evidence to back up his claim.

An anxious bank cashier had come to see Stekel sometime around 1907. The man was afraid of crossing a public square. While standing before it, he would find himself flooded with anxiety and would turn back. According to Freud's theories, this phobia and its attendant panic had to be due to a sexual dysfunction in the cashier's present life, perhaps sexual abstinence or coitus interruptus. After vainly searching for such a cause, Stekel began to root about for a psychological origin. He discovered that the man had never fallen in love, which led him to presume a parental fixation. He then discovered that the object of that fixation, the man's mother, was gravely ill. Tormented by the fact that he could not afford to pay for the treatment her doctor prescribed, the cashier developed a psychic conflict due to his impulse to steal the money from his bank and save his beloved mother, and then flee, not just across the square, but across the ocean to America. By making these unconscious struggles conscious, Stekel reported that he had cured the man of his phobia. Stekel also claimed to have effected similar cures in two other cases.

In a lab, even the most dictatorial professor might be forced to alter his views: Forel recalled performing an experiment that refuted one of his professor's theories, and the professor, while enraged, after repeating his student's experiments could not help but be convinced. In clinical medicine, however, there was no such possibility. A doctor's word that he had cured a patient otherwise deemed incurable was the closest thing to scientific validation for a therapy available. Stekel's cures shifted the Society's debate from debating theory to weighting evidence. Since Freud considered the actual neuroses incurable, he could not counter Stekel's cures with those of his own. It seemed that Stekel had generated clinical proof that would *compel* acceptance from his audience.

Excitedly, Stekel told Freud that he wanted to write a scientific work on anxiety. Freud promised to provide a preface, but as Stekel recalled, there was one condition: "we should go through the whole book together, lest I should write something that was not in accordance with his theory." Stekel agreed. "I was proud to explain to Freud the psychic mechanism underlying these cases," he remembered. If Stekel was contradicting one aspect of Freudian theory, he was extending another and expanding the therapeutic range of psychoanalysis. In sum, it seemed welcome news for those interested in pressing forward psychoanalytic therapeutics.

Nevertheless, Sigmund Freud was not pleased. He insisted Stekel's cases must have been mixed cases of anxiety and hysteria, and that his extraordinary cures were therefore the result of a quite ordinary misdiagnosis. Stekel went away discouraged. When he next returned to Freud, the Professor offered a compromise. "I will give you a royal present," he told him. "We shall call all cases where anxiety has a psychological root, 'anxiety hysteria,' while cases where anxiety can be traced back to injuries of the sex life will be called 'anxiety neurosis.'" With that, Freud began to review Stekel's work in progress. For twenty Sundays, the acolyte came to Berggasse 19, manuscripts in hand.

Stekel also brought his ideas to the Wednesday group, where he did not hide his differences with the Professor. Psychic conflict, he insisted, was essential to all anxiety neuroses; these disorders were the result of a conflict between a life and death force, which he called "Eros" and

"Thanatos." Stekel shrank from a direct confrontation with Freud by noting that the case he had presented to the group included a history of coitus interruptus, but still his paper was poorly received. Federn found it inconceivable that all phobias could be based on repressed sexual ideas. Others complained that Stekel had thrown them into confusion. Freud offered this simple clarification: Stekel's cases were all misdiagnosed.

A few weeks later, Stekel was back, lecturing on anxiety. Again, his presentation went poorly. A little more than a month later, he rose to speak to the Society yet again, but by now, his tune had changed. He presented the case of a cantor who could not sing, and gave the man a diagnosis of anxiety hysteria. The response from the Society was dramatic. One of Stekel's worst critics, Eduard Hitschmann, applauded him. Freud also trumpeted Stekel's virtues, complimenting him on his ability to penetrate into the essence of a case.

When Stekel published *Conditions of Nervous Anxiety and Their Treatment* in 1908, it would be pored over by German readers who sought guidance on the nuances of psychoanalytic theory and method. He described the goal of treatment as the lifting of repression and presented Freud's free-associative process, the problem of resistance, and the rules of dream interpretation. The center of the book was now—*as per* Freud—on the differences between anxiety neuroses and anxiety hysterias. Stekel obediently proposed that psychic conflict only caused anxiety hysteria. However, he could not bring himself to write an introduction to the section on anxiety hysterias, and later claimed that Freud—rather appropriately under the circumstances—wrote those pages for him. Stekel's clinical discoveries had been recategorized, redefined, and made meaningless. Freud had established that hysteria was due to psychic conflict back in 1895; to say that some mixed hysterical entity was also due to psychic conflict was nothing more than obvious. Stekel's humiliation, however, was not over. When it was time for Freud to supply a preface, he produced something so lukewarm that Stekel rejected it not once but twice. Then just before publication, Freud asked Stekel to delete the entire section on anxiety hysteria, in other words, to delete the only part of the book that in any way diverged from Freud's published theory. Unable to directly confront the Professor, Stekel managed to get his publisher to say it was too late for such a change.

Freud had shown himself willing to tolerate dissenting theoretical views in the Society. But theoretical differences were just that, theoretical. They had no empirical weight and could not demand acceptance by others. Stekel's challenge was different. His cures were precisely the kind of evidence most clinical medicine relied on. This was no longer a dispute between theories. If left unchallenged, the Society would have to accept Stekel's views.

Furthermore, if Stekel was right, a whole host of patients Freud said were untreatable had been written off incorrectly. If Freud was wrong about these clinical matters, what would it mean about his other observations? Professor Freud vigorously and skillfully defended his authority as master clinician and objective observer by shifting the debate away from Stekel's evidence to matters of clinical judgment. Identifying the difference between anxiety neurosis and anxiety hysteria was such a judgment. The Professor was the man who had separated the anxiety neurosis from the vague category of neurasthenia in the first place, and he was the most esteemed clinician in the group. Stekel was a generalist, a popularizer, and a man whose vitriol had already made him enemies in the Society. He could not muster evidence to prove his cases were not mixed cases of anxiety and hysteria, and with the debate reframed in this way, the outcome was assured.

Freud didn't stop there. He co-opted Stekel's innovations and incorporated them into his own theory. Again and again, over the coming years, Sigmund Freud would employ the same strategy: when opposed, he would fight bitterly to hold his ground, and then after rebuffing a foe, he would quietly incorporate those aspects of the challenge he most admired into his ever-expanding models. The Freudian field grew fat on a host of vanquished opponents, Wilhelm Stekel among them. In 1909, Freud announced that all phobias—previously attributed to anxiety neurosis and real sexual dysfunction—were actually cases of anxiety-hysterias. "I suggested the term to Dr. W. Stekel when he was undertaking a description of neurotic anxiety-states," Freud wrote. He hoped this diagnosis would come into widespread use, now believing that anxiety-hysteria was the most common of all the psychoneurotic disorders. What was once impossible had become omnipresent.

Stekel's claims for curing anxiety neurosis had been woven into Freudian theory via a creative relabeling. Freud had seamlessly pulled what was new in Stekel's work into his own theory, and in the process, given up nothing. His theory of actual neurosis remained unchallenged. His clinical expertise was unmarred. For Freud, this was a win-win outcome. The same could not be said for Wilhelm Stekel.

STEKEL'S ROUGH AND tumble treatment helped define what was acceptable and what was not among the Viennese Freudians. One could not—it seemed—question Freud's clinical authority, his veracity as an empirical observer, or his therapeutic claims. But more abstract, theoretical challenges were another story, as became clear from the Wednesday Society's reception of Alfred Alder.

After medical school, Adler had opened a practice in the working-class Second District of Vienna, where his clientele differed from the wealthy patients that frequented Breuer and Freud. Attracted to socialism, he wrote essays on environmental factors in illness, and in that context, first referred to Freud's work. When he joined the Wednesday group at Freud's behest in 1902, Adler still harbored hopes of returning to academic medicine.

In the fall of 1906, Adler was slated to present on a topic listed as "On the Organic Basis of Neuroses." In the end, the secretary's minutes recorded the title as "On the Basis of Neurosis," but in fact, the key word was "organic." For this paper was a prelude to Adler's soon-to-be-published *Studies on the Inferiority of Organs*, in which he would insist on the organic cause of neurosis.

Adler considered his thesis to be more in the tradition of Sigmund Exner and Ernst Haeckel, rather than Sigmund Freud. Beginning with the partial drives Freud attributed to erotogenic zones, Adler assigned such drives to various bodily organs and posited a conflict among *all* organ drives and the cultural demands placed upon them. In this model, many organ drives might fall into conflict and cause neurosis. Adler made a weak attempt to give priority to sexuality by saying that every diseased organ was also secondarily linked to a diseased sexual organ, but it was hard to escape the conclusion that Adler had fundamentally rewritten Freud's theory.

Further, Adler's theory was predicated on degenerative biology—the doctrine Freud had fought against for over a decade. Adler postulated that degenerate organs could be remedied by what he called cerebral "over-compensation." In this way, the stuttering Demosthenes became the greatest orator in ancient Greece. Musicians, Adler argued, often had ear anomalies, and 70 percent of painters had optical peculiarities. By this reasoning, childhood organ defects and cerebral overcompensation were the rule in neurosis.

It also followed that libido theory was an error. Inferior sexual organs created hypertrophied psychic structures that made the sexual drive persist strongly into adulthood. Such an intense sexual drive was not normal; it was an overcompensation. Adler did not hide from the implications of this explanation:

> I must state that the interesting psychic phenomena of repression, substitution, conversion, which Freud demonstrated in his psycho-analyses and which I also found to be the most important constituents of the psychoneuroses, develop upon the above-described formation of the psyche in the case of the inferior organs. In a like manner, the usual statement of "sexual basis" of psychoneuroses is cleared up.

Simply put, Freud's psychosexual theory was wrong.

One might imagine that after Adler presented such views to the society, the lions of the Wednesday group would begin to roar. But in fact, Adler's presentation was warmly received. One physician remarked that he too always advocated searching for organic factors in neuroses. Freud himself praised Adler's effort. "To judge from immediate impression," he said, "much of what Adler said may be correct." Hugo Heller called it an "impressive intellectual achievement," a "continuation and supplement of Freud's ideas." Only a few naysayers refused to join in the praise.

Strikingly, Adler did not face an obvious, ad hominem line of attack. He had already brought up the fact that many medical men had suffered illnesses in their childhood and overcompensated by becoming physicians. Adler himself had been a sickly youth, who suffered from debilitating rickets as well as breathing and speech problems. While it is not clear that the members of the group knew this, it is likely that they knew no less about Adler's illnesses than they knew about Stekel's sexual life. But no one asked if the creator of this scientific theory was secretly also its subject. At least not yet.

Adler's theoretical presentation intrigued his audience, insofar as it offered a bridge to clinical medicine. His case presentations fared less well. For example, Adler described a wealthy young Russian student who had come to treatment with an "inferior alimentary tract," who had compulsive desire to hold his head under water while bathing. Several members considered the interpretation of the case nonsensical. Freud noted that it didn't even cohere with Adler's own theory. In the Society, however, Adler continued to press his view on inferior organs, insisting that the writer Jean Paul suffered from renal inferiority and searching out connections between symbols and inferior organs.

Over time, Adler's interpretations began to be met with a touch of exasperation, but in general, he was afforded great leeway. Adler's theory could be seen as a massive expansion that would bring psychological thinking into a wide range of medical matters. For a fledgling discipline with few supporters in academia, such potential advantages could not be quickly discounted. Perhaps Adler's emphasis on pathological localization, his reliance on defective heredity, and his biological assumptions were the price to be paid for a full psychobiologic theory of medical illness. In every disease, Adler claimed, biological weakness and cultural demand resulted in symptomatic psychological structures, by which the mind attempted to compensate for anatomical weakness. Despite his opposition to libido theory and the primacy of psychic conflict, Adler's work received respectful consideration.

In part, the difference between the reception of Stekel and Adler had to do with the character of each challenge. Alfred Adler had no real data to back up his claims. His was an abstract theory that could be entertained, but nobody felt forced by his strained evidence to adopt his views. Adler's work could be seen as a fascinating theoretical physiology that did not directly compete with psychoanalytic theories of mind. In fact, Adler's ally Carl Furtmüller recalled that Freud praised Adler's work as physiology. Adler could be seen as adapting a partially Freudian model for a broad range of medical illnesses. Was it tenable to give up intrapsychic conflict and psychosexuality for a dynamic model that combined degenerative changes with psychic forces? Adler asked members of the Wednesday group to consider this question, and for many the answer was maybe. There was no pressing need to decide.

V.

ALONGSIDE THE SCIENTISTS and physicians in the growing Wednesday Psychological Society were humanists, critics, and social reformers who, in the parlance of the time, were students of cultural or human science. By 1907, these intellectuals numbered roughly a third of the Society, and they joined a healthy number of the physicians who were also seriously interested in such matters. Among these thinkers were provocative revolutionaries, who packaged Freudian thought and lobbed it like a Molotov cocktail at sacrosanct social and cultural ideals, and their presence alongside those primarily interested in medical treatments made for friction. While lectures on diseases and cures relied on the evidentiary rules of clinical medicine and natural science for their authority, these other Freudians explored culture, history, and society in presentations that aspired to persuasive logic and broad explanatory power, not scientific proof. Soon the Society found itself at odds over these competing notions of what was real or imagined, sensible or nonsensical.

Otto Rank, the young secretary of the Society, was among the cultural researchers, and his path to Freud was distinct from the physicians interested in treating neuroses. Born Otto Rosenfeld in 1884, Rank had been forced by economic necessity to attend technical school rather than university. Unhappy, at times desperate, the nineteen-year-old Jewish boy changed his surname to Rank and his religion to unaffiliated. Filled with self-loathing and contempt for others, he suffered from a sense of emptiness and often contemplated suicide. Hope came when he discovered great writers. "I vowed never to forget the man Schopenhauer," he wrote in his diaries. The young Otto next proclaimed Nietzsche the "nourisher" of his generation. He read Darwin avidly and found that Otto Weininger took his breath away. Privately, he railed against his culture's sexual hypocrisy, equated woman's essence with sexuality, and saw the economic exchange in marriage as no different from prostitution. Over time, Rank's enthusiasm for Weininger and Nietzsche cooled as he questioned the psychological needs that created these philosophies. These thinkers "admire what they don't have and despise what they do," he wrote. He turned from philosophy to French psychology and in 1904, tried to secure funds to go to Paris, to learn French and translate some psychological works into German.

In his neo-Romantic intellectual passions, Rank mirrored the interests of many of his peers in turn-of-the-century Vienna. They too had moved from politics and natural science to matters of aesthetics, the self, and psychology. Rank's journey took a turn when in October 1904 he noted in his diary: "Meistersinger (Freud—Interpretation of Dreams) dreamsong." From then on, Freud the Meistersinger appeared again and again in Otto's diaries. The young man tried to synthesize his prior intellectual idols with this one, welding the Freudian unconscious to the Schopenhauerian will, and concluding that sexuality was the essence of all psychological life. In May 1905, he wrote: "sexuality is the power that drives everything."

Otto's doctor was Alfred Alder, who introduced Rank to his new hero. The young man came armed with a manuscript that impressed Freud, for the book ambitiously took up the psychology of artistic creativity, territory where most scientific psychologies feared to tread. Sigmund Exner and Theodor Meynert could create scientific models to account for automatic mental functioning, but they floundered when confronted by human intention and choice, most dramatically exemplified in art. How could one of Exner's doll-like automata ever write a great sonnet or compose a symphony? The young Rank attempted to show that in this regard Freudian psychosexuality could succeed where these other models had failed. He reworked this text under Freud's supervision, and in 1907 published *The Artist: Towards a Sexual Psychology*. The artist was like the neurotic, Rank argued, but artists had the capacity to know their own unconscious by recognizing its force on the world. Rank's theory of knowing one's inner being by recognizing its projections came right out of Schopenhauer, and could easily be transposed into Freud's theory of transference. In his diaries, Rank wrote: "If the world is my projection, so is becoming conscious of this projection, my birth." Rank ended his book with a vision of the artist as superman, who through knowledge of his own projections guided his own sexual drives with clarity and control.

On October 10, 1906, Rank joined the Wednesday Psychological Society and became its salaried secretary. To mark his admission, Otto presented his own work. At eight-thirty in the evening, he stood before a room of older doctors and critics and presented his thinking on sexual psychology and the arts. From the recorded discussion, it appears that Rank took a strong position on the ubiquity of sexual instincts in creativity, and the presence of Oedipal and incestuous themes in art. Facing the crowd of more educated men, the inexperienced Rank could at least take comfort in the fact that the Professor had worked closely with him on his theories. Before the Professor's students, he should have little to worry about.

Yet many in the Wednesday Society did not consider themselves the Professor's obliging students. And so, no doubt to Rank's shock, his reading was attacked for being *too* Freudian. "The tendency of the paper is to interpret everything according to Freud's method, and that therefore too much has been read into material and interpreted into it," one member complained. Another cautioned against interpreting such broad themes from a "far-fetched angle." A third felt Rank was going too far in his interpretations, stretching Freud's ideas, "over-extending an elastic band." When Rank presented a second time, Wilhelm Stekel complained that "everything in this book is seen through spectacles colored by Freudian teachings, without going beyond Freud."

The critics—including clinicians like Wilhelm Stekel, Max Kahane, and Eduard Hitschmann—rejected Rank's use of psychosexuality as an expansive concept. Although these men also gave psychosexuality import, they recoiled from the assumption that it could be universally applied to every unconscious factor and every cultural product. They also made it clear that they were free to think for themselves and were not participating in the society simply to learn and apply Freud's notions.

Rank had no such hesitance. He was Freud's pupil and was not interested in debating the worth of Freud's theories; he was interested in expanding this hermeneutic into art and culture, a domain free of issues like clinical improvement or failure, cures or disasters, and empirical proof. Armed with Freud's interpretative techniques, Rank and a number of other humanists tracked the hidden manifestations of libido like hunters. It was from this troop of followers that Freud's psychosexual synthesis would find some of its most vociferous adherents.

HUMANIST APPLICATIONS OF Freudian theories followed a wave of psychiatric writing on social, cultural, and political matters in late-nineteenth-century Europe. Ambitious, well-known psychiatrists often offered their pronouncements on matters of social import, even when the questions did not easily yield to categories of bodily disease. Some tried to reduce cultural questions to scientific ones by pressing complex phenomena into a diagnostic category. In this fashion, neurologists and psychiatrists declared a host of complex phenomena in art and politics, degenerative or neurasthenic, a result of this or that purported brain disease. The rest— and that terrain was mammoth—was left to philosophers, writers, and artists.

Psychiatrists also weighed in on cultural matters through a new style of biography. "Psychobiography" was inaugurated around 1870 by one of Freud's competitors, Paul Möbius. Möbius believed that no one could possibly understand a historical figure without a medical and psychiatric evaluation. Influenced by the idea of degeneration *supérieur*, Möbius analyzed the lives of diseased geniuses and at the time of his death in 1907, had produced biographies of Rousseau, Goethe, Schopenhauer, and Nietzsche. Other doctors followed with accounts of great men and women seen through the lens of degenerative heredity. Journals were founded to foster this kind of work. Freud's friend and rival Leopold Löwenfeld joined Hans Kurella to publish pathographies in a book series, featuring writing on Otto Weininger, Ibsen, Berlioz, Guy de Maupassant, and Tolstoy.

Freud offered a new approach to psychobiography. His theory did not reduce psychology to brain and biology but created a bridge between *Naturwissenschaft* and *Kulturwissenschaft*, by articulating the laws of psychic life in a biological world. By securing a scientific approach to dreams, fantasies, and thought, Freud offered a way to study not just illness, but also minds, selves, and personalities and their impact on a culture. As early as 1906, Freud tried such a study when he wrote a short, brilliant application of his theory to drama, in part inspired by a play written by Hermann Bahr. Although Freud declined to publish "Psychopathic Characters on the Stage," others would not hesitate to let their work loose on the public.

Consider Isidor Sadger. By the time he joined the Wednesday Society in November 1906, Sadger was an old hand at the traditional approaches to psychobiography, having published studies on Henrik Ibsen, Nikolaus Lenau, and that hero of the German people, Goethe. Sadger had been in contact with Freud for over a decade, having attended the Professor's university lectures between 1895 and 1904. Upon admittance to the Wednesday group, Sadger revisited the case of Lenau, reiterating his belief that the man's mania was due to heredity. Sadger's work was treated respectfully, though Freud balked at the emphasis on hereditary taint. To this, Dr. Sadger patiently reminded the group that "certain phenomena cannot be explained by psychosexual arguments."

Over the next months, Sadger doggedly reaffirmed his belief in traditional degenerative factors while adding psychic, at times erotic, ones into his psychobiographical formulations. He warned the group not to overestimate the significance of Freud's teachings, reminding them that the importance of sex for psychology and the unconscious could be overemphasized. On December 4, 1907, Sadger presented a paper on the Swiss writer Konrad Ferdinand Meyer and again mixed degenerative and erotic influences.

This time a vitriolic discussion ensued. Max Graf said Sadger had glibly relied on heredity and not plumbed the inner world of the man. Federn was furious that Sadger had said nothing about the writer's sexual development. Wilhelm Stekel, who had already warned of the incendiary potential of psychobiography, was aghast and feared the author would bring great harm to the Society. Turning beloved cultural heroes into degenerates risked provoking widespread hatred. Stekel already criticized Möbius for doing that, and now Sadger was doing the same thing, potentially compromising the Society. He asked the author to withhold his work from publication. To this avalanche of criticism, Sadger said bitterly that it added up to nothing more than personal insults. That night, Sadger's nephew, Fritz Wittels, was in the audience, and he vainly tried to defend his uncle. He left, a witness to the hard treatment afforded those who appeared to pose a risk to the Society with their cultural applications of Freudian thought. Soon, Wittels would learn that lesson firsthand.

Like a number of Freud's adherents, Fritz Wittels was both a physician and a writer. After attending the University of Vienna, Wittels trained for four years at the General Hospital of Vienna, during which time he met Freud by appearing at the Professor's lectures. During the same years, Wittels turned his talents to literary and cultural criticism, and found himself catapulted into the limelight due to his relationship with Karl Kraus.

In December 1906, Wittels wrote Kraus, audaciously suggesting that his short stories were better then Strindberg's. Wittels enclosed a sample; Kraus liked and published it. Wittels was welcomed into Kraus's coterie and became a regular *Die Fackel* writer. That summer a "personal experience" prompted Wittels to take up his pen. In a polemic entitled "The Greatest Crime in the Penal Code," Wittels argued against the criminalization of abortion. Like other sexual reformers, he relied on Freud to claim that methods of contraception such as coitus interruptus caused anxiety neurosis. He went further, arguing that the criminalization of abortion led to the murder of newborn infants, and that the forces of capitalism, the military, and the church supported this status quo, so as to amass cheap labor for their factories, cannon fodder for war, and more believers to fill the pews. Fearing for his medical reputation, Wittels published this explosive piece under a pseudonym, Avicenna.

Freud knew Wittels had written the piece. After a Saturday lecture, the Professor approached the young firebrand and said: "Did you write this? It is like a brief and I subscribe to every word of it." Wittels was invited to join the Wednesday group. It was a decision that the Professor would come to bemoan. The twenty-six-year-old Wittels had tasted notoriety, and it was to his liking. He immersed himself in the coffeehouse world, as part of Kraus's circle, where he attentively listened to his editor expound on his theories of women. Wittels then took up his pen.

Wittels combined Kraus's misogyny with Freudian thinking, and cooked up a polemic on the role of women in European society. On March 27, 1907, he presented this concoction to the Wednesday group. Ostensibly, the topic was female assassins who, the author contended, were hysterics who killed after they had been spurned sexually. But the talk was also a general condemnation of woman and hysterics, for whom the brash Wittels confessed "his personal distaste."

It was not an auspicious beginning. Stekel opened the discussion by remarking that "the speaker has projected the unpleasant self-knowledge of his own insignificant hysteria onto a quite harmless class of people." This was the first in a series of criticisms that challenged not just Wittels's argument, but also his motives. Because his polemic was so transparently a product of his own seething psychosexual dilemmas, Wittels forced his audience to address the larger issue: whose psychology was central, Fritz Wittels or the female hysteric? The problem was profound and familiar, running back through the debates over Stekel's theories of anxiety, Fliess's accusation of Freud as a "mind-reader," the French debates over the nature of suggestion, and Auguste Comte's denunciation of psychology as unscientific. If it couldn't be contained, then all psychological statements could flip back and forth between the proposed objects of study and the investigators themselves.

Wittels's presentation was condemned as outlandish and superficial. Freud himself agreed that the talk was marked by the author's neurosis. On May 3, 1907, Wittels published a polemic against women doctors in *Die Fackel* and less than two weeks later, stood before the Wednesday group again. Women had only begun to be admitted to German medical schools; Wittels disapproved. Since women were (as per Kraus) exclusively sexual beings, their interest in education must be due to hysterical repression. Woman doctors were dangerous; patients would have no confidence in them. As for female psychiatrists, they could never understand a man's psychology. Wittels did not pause to consider the logical consequence of this belief, which would undercut his own authority to speak so confidently about the inner worlds of women.

Others were not so myopic. Wittels's diatribe forced the society members to focus on how such personally-driven cultural uses of Freudian theory should be handled. One possibility was to concentrate less on what Wittels said and more on his motives. Max Graf noted the intense feeling that accompanied Wittels's presentation and proposed that this stemmed from his anger at women who rather than wanting to have intercourse (with him), preferred to pursue their studies. Professor Freud reproached Wittels: he dealt in half-truths, lacked courtesy for women, and had no sense of justice with regard to their entering the medical profession. Wittels expressed a childish viewpoint, that of a young man who, out of his own sexual needs, distorted and debased female sexuality. Still, Freud added his own prejudice, that a woman could not reach a man's cultural achievements.

After that the recorded minutes grow confused. Much of the ensuing discussion is directed at Wittels's paper on female doctors, but other comments concern a text Wittels delivered at the next meeting. Entitled "The Great Courtesan," its subject matter was, one scholar has speculated, too scandalous for the Society to record. If so, it was not too scandalous for *Die Fackel*, where it was published six weeks later. Modeled on Kraus's young lover, Irma, Wittels wrote of a woman so sexual that she began her erotic life as a child and remained highly sexual for the rest of her days. Thoroughly unrepressed, she was serene and unneurotic, though her sensual nature condemned her to be persecuted by society as a whore. This woman was not a prostitute but embodied the ideal of the ancient Greek Hetera. In support of this glorified vision, Wittels cited Sigmund Freud.

After commenting on Wittels's attack on female doctors, Freud shifted his attention to the topic of the Hetera:

> The ideal of the courtesan (Hetäre) has no place in our culture. We endeavor to uncover sexuality; but once sexuality is demonstrated, we demand that the entire repression of sexuality becomes conscious and that the individual learn to subordinate it to cultural requirements. We replace repression with healthy suppression. The sexual problem cannot be settled without regard for the social problem; and if one prefers abstinence to the wretched sexual conditions, one is abstinent under protest.

Freud went on to say that a woman who acted like a courtesan was not trustworthy and was worthless, a *Haderlump* or tramp. Given a chance to reply, Wittels said he was so shocked by Freud's calling his ideal courtesan a *Haderlump* that he was rendered speechless.

Freud's annoyance with Wittels was even more intense in private. He asked Wittels to read him the paper on courtesans before publishing it. When Wittels did, Freud could not contain his dismay. He made it clear that it was not his intention to lead the world to an uninhibited frenzy, but rather to allow men and women to consciously decide what to do and what not to do. But the fifty-one-year-old liberal spoke across a cultural divide to the twenty-seven-year-old sexual radical. This clash of worldviews would mark Freud's relationships not just with Wittels, but also with others who had jumped on his bandwagon. Freud understood the value of personal liberty as primary, but also something necessarily restrained by reason and the institutions that embodied it, legal and sociopolitical. Extremists like Otto Soyka and Fritz Wittels did not share that view. Wittels called for a sexual revolution, one that would eradicate the old structures and reinstate, not reason, but nature into its rightful place. After hearing Freud's critiques, Wittels revised himself a bit, then published his essay. Later, he recalled that the piece created outrage among Freud's followers but was a hit in the coffeehouses. He added: "we knew from Freud that repressed sex instincts made men neurotic to such an extent that an entire era was poisoned. What we did not know then was that former puritans running wild would not help either."

In the winter of 1907, Kraus offered Wittels a chance to put together a special issue of *Die Fackel*. Wittels dedicated the issue to the fight against venereal disease. Prior to publishing his essay on the topic, Wittels gave it a reading at the Wednesday meeting. Again, Rank apparently declined to take notes. In the essay as it appeared in print, Wittels took no prisoners. He assailed the ruling class and the church for using venereal diseases as a way of maintaining puritanical morality and their own power. Quoting Kraus, Wittels wrote: "As Sancho Panza rides behind Don Quixote, so syphilis behind Christianity." At the Society, Hitschmann immediately honed in on Wittels's own neurotic motives for writing this paper. Three things thwarted Wittels's sexuality, he dryly concluded: impregnation, the chastity of female medical students, and venereal disease.

Wittels wasn't through. On March 11, 1908, he presented yet again, this time spinning a dystopian fantasy that began with a free, polymorphously perverse, sexual Eden and ended with the sorry present, in which women wanted to be men. Wittels's uncle Isidor Sadger had supported the younger man through other trying times, but even he could bear it no longer. Sadger asked if the paper was meant to be taken seriously. Stekel denounced the essay as a fantasy, and Hitschmann dismissed Wittels's work as that of a neurotic reactionary.

JUST AS DEBATES over Stekel's and Adler's work helped to define the boundaries of the Society's clinical discussions, the reception of writers like Rank, Sadger, and Wittels began to lay out rules for a specifically Freudian cultural science. A major concern was that this work, which opened up into a wider public discourse, might injure the group's standing. Incendiary journalism struck many as dangerous at a time when Freud was just starting to win followers. The Society warned those who pursued a Freudian cultural critique to censor themselves accordingly.

Wittels's presentations forced another critical question front and center. If Freud's method offered an understanding and delineation of another's subjectivity via psychosexuality, it could only do so if the author's psychosexuality did not mar the field of inquiry. The risk of projecting one's own unconscious into another's was great in clinical work, but it was perhaps even greater in cultural, political, and sociological writings. Wittels was ridiculed because his broadsides appeared to be dressed-up expressions of his own sexual urges. This young man's invective and fantasies served notice to the group. These theories were not to be cavalierly used as an excuse for personal likes and dislikes.

With his inflammatory speeches, Wittels also forced the group to consider if its goals were strictly clinical and psychological, or also political and reformist, committed to social activism. If Freud believed that conventional sexual ethics caused neurosis, it was logical to think that the group might attack these conventions. Would the Society engage in such a fight, and if so, how far would they be willing to go?

Many in the Society, like Freud, were interested in both clinical treatments and sexual reform, but Wittels highlighted the potential conflict between the two. He had made his position clear when he denounced "the psychologist's arrogance shown by some of the gentlemen, who consistently overlook the facts and are interested only in theoretical-psychological aspects." For Wittels, ideological political commitments took precedence over attempts to build a scientific psychology. He did not worry that polemics and social activism might run away with the hopes for a new science.

But Wittels's call for a Freudian politics would not prevail. The messenger was clumsy, but more importantly his message violated a position dear to many in the group. Sigmund Freud had gone to great lengths to construct an approach to the psyche that could claim to follow scientific principles. A highly politicized Wednesday Society had the power to nullify any such claims. Despite sympathy for the sexual reform movement that had taken up his clinical theories, Sigmund Freud doggedly insisted that scientific studies should dictate social reform, not the other way around. The reformers who came to the Wednesday Society with a ready-made agenda turned to Freud because he gave them ammunition for beliefs about society that they were more deeply committed to than a science of the mind. For many of the Society's members, Fritz Wittels represented this inversion of commitments. Despite his many attempts to sway the group, he was ostracized and denounced.

VI.

THE WEDNESDAY SOCIETY housed men of varying interests. Some were attracted to the French Freud of 1895 who studied hysteria and used psychical treatments; among them were those who were skeptical of Freud's later ideas, his psychosexual synthesis in particular. Others, fascinated by the dream book and the interpretative method it offered for myth and literature, had little interest in the requirements of scientific epistemology. Others still yearned for social and sexual reform but were less invested in clinical psychology. In short, Freud lured people interested in the very fields he had plundered. These Society members mixed and matched their Freud with a conglomeration of their own ideas on dynamic psychology, degeneration theory, brain science, and sexology. From this jumble, conflicts inevitably arose. Fritz Wittels, Eduard Hitschmann, and Alfred Adler all had goals that differed from one another and from Freud. The essential question became: How should such differences be resolved? Who had the final word as to what was and what wasn't acceptable in this community?

By 1908, running debates had taken place over the work of some of the group's most creative members, and these conflicts began to define tentative borders. Direct challenges to Freud's clinical authority were not allowed, but members could theorize freely about the nature of the mind. Slavish overextensions of Freudian psychosexuality were chided, but some recognition of sexual factors was required. Social critiques founded on scientific claims were encouraged, but polemical calls for revolution were not countenanced. Rhetoric and argumentation indebted to both natural science and cultural science would be accepted, but acute vigilance was required to ward off claims driven by irrational, subjective desires. Sigmund Freud, Paul Federn, Wilhelm Stekel, Eduard Hitschmann, and other members of the Society tried to police these boundaries, with stern warnings, personal pleas, and ferocious attacks when the lines were crossed.

As these debates went on, the Wednesday Psychological Society began to rethink its mission. In 1908, a motion was made to change the ground rules: Alfred Adler suggested they abolish the dreaded urn and make participation voluntary. In addition, he also asked that new members be elected by secret ballot. Until then, the Professor and a few others had in-

vited members to join and put them up for an open vote. All the votes recorded in the minutes were unanimous, a fact that suggested members
might not have felt comfortable making their reservations public. Despite
a few minor qualms, Adler's proposals were warmly received. Max Graf
said: "These proposals to reorganize stem from a feeling of uneasiness.
We no longer are the type of gathering we once were. Although we are
still guests of the Professor, we are about to become an organization." In
recognition of this changing identity, Graf suggested the group no longer
meet at the Professor's home, but at a university location.

The proceedings seemed headed for a changed Society where Freud's
power would become shared. But Paul Federn stepped in, declaring that
the group should also put an end to what he called "intellectual communism." He heatedly demanded that an idea not be employed without the
clear consent of its author. Federn's target was not named, but that did
not obscure the fact that his charges were levied against Stekel, whom he
believed was using Freud's ideas in his journalism without crediting the
Professor. Stekel tried to defend himself, and Wittels jumped in to support him, saying he too was incapable of writing anything without using
the Professor's ideas and could not always cite his source.

While Adler's proposals pushed the group to consider a more democratic, academic structure, Federn's motion accused members of underplaying their dependence on Sigmund Freud. What was Freud's
intellectual property, and what were the common working assumptions
of the group? In striving to build a community that shared certain terms
and concepts, the Wednesday Psychological Society had become an intellectual collective of sorts. Federn's term "intellectual communism," however, raised the concern that social cohesion might rob individuals of their
rights. Communism—which in 1908 referred to the French Commune—
implied collective ownership. No one quite knew what was shared property in the Wednesday Psychological Society, and what belonged to Freud
alone. The quandary forced members to go back in time and consider
Freud's intellectual origins and debts. After all, much of his terminology
and theory was derived from others. Was this to be a group of dedicated
Freudians, or a society, while seriously indebted to Freud, that organized
itself around a common theory arrived at from multiple sources?

Ultimately these questions were postponed for a gentlemanly solution.
The Society would not regulate intellectual property, but leave it as a

matter of individual honor. The person most familiar with intellectual property issues, the publisher Hugo Heller, defended Stekel and argued he should be free to use the ideas expressed at the meetings, warning that a resolution to the contrary would seriously restrict creativity. Stekel maintained that many ideas of the group were "in the air," and not specific to one man. With the generosity of the wealthy, Freud seconded Heller's views and added that "each person might himself state how he wants his ideas dealt with." As for the Professor, he "personally waives all rights to any of his own remarks."

Given his desire for public recognition, his bitterness about the failure of others to acknowledge his work, and his hawklike defense of his own priority in matters of intellectual property, Freud's magnanimity would seem to be insincere. And yet his statement offered his comrades the flexibility needed to develop a common set of assumptions and a common identity, as opposed to the far more restrictive role of serfs under the tutelage of a king. Stekel performed a mea culpa, and Federn withdrew his motion. Subsequently, an amended motion was unanimously passed: ideas put forward in the Society could be used by others, unless they had been explicitly forbidden to do so by the author.

With this matter put to rest, Adler's proposals were raised again but not voted upon. Instead, in the first signs of bureaucratization, these suggestions were deferred to an ad hoc committee. A week later, the committee delivered its decisions. The urn was abolished. But the proposal to elect members by secret ballot was mysteriously shot down, despite its earlier strong support. New members were to be accepted using the old rules. "(T)his assembly," the committee explained, "is something in between a group invited by Professor Freud and a society; therefore, whoever is acceptable to the Professor must be acceptable to the others." The meetings would continue at the Professor's home.

It would be hard not to infer Freud's intervention in the committee decisions. Two of the three members—Adler and Hitschmann—had already openly favored secret balloting. Hitschmann remained silent about his change of heart, but Adler felt compelled to dissent from the committee's report. He requested that meetings be held at a university location, and that a two-thirds majority be required for the election of members. In the end, he withdrew his proposal.

Adler had emerged as the most vocal spokesman for change in the Wednesday Psychological Society. An ardent Social Democrat, his concerns about democratic rule were founded on sincere convictions. In addition, his proposal to adopt a university location would have created more space within the Society for his divergent ideas. And it would not hurt too, if membership was not in the hands of the Professor, so that men and women interested in Adler's ideas might be more easily welcomed into the group. After his challenge failed, in May 1908, Adler tendered his resignation. Freud convinced Adler to stay on, saying he had the sharpest mind in the group. But a critical divide had been crossed: the Society was to remain in Freud's home under his direction.

The group that bewildered Max Eitingon with its range of opinion a year earlier became more focused. An overarching theory of psychosexuality began to organize debates in the Society. Many warily circled this idea, focusing increasingly on the way it linked mind and body, psyche and soma, Kant and Darwin, Charcot and Krafft-Ebing, Helmholtz and Schnitzler. They began to more deeply explore the way Freudian psychosexuality transformed French psychopathology, psychophysics, and sexology into a possibly uniform, new field.

During a Society meeting in 1907, a total commitment to Freudian psychosexual theory was first broached, albeit flippantly. Discussing the work of his rival Paul Möbius (who had himself emphasized the ideogenic in hysteria, employed sexuality, and believed most psychic processes were unconscious), Freud complained that Möbius never recognized or reviewed his work. In the discussion, Max Kahane made the following blunt summary: "There are only two possibilities in relation to Freud's doctrines: adhering to them or ignoring them." At the time, this either-or position made little sense in a Society populated by partial adherents of many stripes.

In the spring of 1908, Society members took another giant step toward consolidating their identities. Members discussed going public with a questionnaire about sexuality, in conjunction with the Berlin sexologist Magnus Hirschfeld. In supporting this questionnaire, the Society would take a public stand as a single voice for the first time. In preparation for this, the group decided to change its name. By majority vote on April 15, 1908, this Viennese community that had for six years met as a Psychological Society transformed itself into a "Psychoanalytic Society." The change was noted without comment in Rank's minutes, but it was critical. Psychoanalysis was Freud's word. As early as 1894, he had publicly launched the term "psychische Analyse" or "psychical analysis," and two years later he first used *la psychoanalyse* in French and *Methode der Psychoanalyse* in German. Between 1904 and 1905, Freud wrote a number of short essays consolidating and distinguishing his postcathartic methods as psychoanalytic. The new name acknowledged the Society's commitment to a particular psychological method and theory founded by Sigmund Freud.

As the Society in Vienna became Psychoanalytic, others around Europe made it known that they too wanted to join a community of Freudians. In December 1907, Freud announced to the Society that Dr. Carl Jung of Zurich had proposed a congress in Salzburg, Austria, for all of Freud's followers. Twelve days after the Wednesday Psychological Society renamed itself, the members of the group traveled to Salzburg to attend the congress. The First Congress for Freudian Psychology was directed not by the Society members from Vienna, but by Freud's Zurich followers. For some Zurich doctors had embraced Freud, though in a manner dictated by their own prior commitments and their own histories.

FIVE

Zurich

I.

AROUND 1900, IT would have been hard to mistake Zurich for Vienna. Vienna was the center of a failing empire, a swarming multiethnic metropolis barely held together by Catholic and monarchist authority. Zurich was the stable, Protestant capital of one of twenty-two Swiss cantons, each of which enjoyed a good degree of autonomy and democracy. In Zurich (as Freud would discover), communal self-rule was fiercely prized. Like Vienna, Zurich was German speaking, but Switzerland was unique in that it unified linguistically French, Italian, and German citizens by allowing each to have their own region. After 1870 when bitter nationalism forced the French and Germans apart, Switzerland remained an open marketplace of ideas where those cultures mingled. Aided by these advantages, the Swiss would play an inordinately large role in attempts to synthesize French and German ideas of the mind, and they would be vital to the acceptance of the French-inflected psychology of the German-speaking doctor Sigmund Freud.

In the second half of the nineteenth century, Switzerland had become a haven for free thinkers, especially those who fled in the 1840s when the liberal revolutions in France and Germany failed. Partisans left the barricades for Swiss borders, and a number of intellectuals, including some of Europe's leading psychiatrists and psychologists, found their way to Switzerland. One such refugee, August Zinn, submitted a plan to the Zurich city council in 1863 for the reform of the overcrowded, poorly run psychiatric hospitals, throwing his support behind the University of Zurich's professor of medicine, Wilhelm Griesinger, a famous German émigré psychiatrist who had also been advocating change. Their plan for a small university-based clinic was defeated by the council, which decided to build a larger hospital called the Burghölzli. In 1870, the Burghölzli opened its doors, but by then a disgruntled Griesinger had left Zurich. The hospital was placed under the leadership of another distinguished German brain researcher, Bernhard Aloys von Gudden. He was not particularly interested in running a hospital. After two years, he left Zurich and was replaced by Eduard Hitzig.

The Berlin-born Hitzig was also primarily committed to cerebral research—in his case, experimenting with electrical stimulation on dog brains. In Zurich, he found himself in the rather absurd position of speaking only High German, which meant he could not understand the Swiss-German dialect spoken by his patients. Apparently this caused him little distress, for Hitzig didn't spend much time talking to patients. By 1878, he too was preparing to leave, and by then the hospital was such a mess that attempts to recruit a new director failed. Eventually, this job fell to a young, upright Swiss physician, the thirty-one-year-old Auguste Forel, who assumed the directorship in 1879, and soon transformed the Burghölzli into a leading center for psychiatric teaching, research, and care.

Born of a Swiss father and French mother, Forel became an avid naturalist at a young age. Like many of his contemporaries, Forel was permanently altered by his reading of Darwin. He attended medical school in Zurich, became a psychiatrist, and went to Vienna to study brain anatomy with Freud's teacher Theodor Meynert. Arriving in Vienna, the young man was appalled by the licentiousness and moral filth around him. His experience with Meynert was hardly better, for the schemas of the great brain anatomist struck the young man as fantastic. As Forel dryly put it: "I could not always see what Meynert saw."

In 1873, Forel moved to Paris, where he was again shocked by the widespread drunkenness and whoring, and soon he turned to the sanctuary of Bernhard von Gudden's lab in Munich, where he quietly pursued the study of brain anatomy with his mentor. When the Burghölzli job was offered to him, Forel took it, despite the fact that he had very limited clinical skills. Perhaps he knew that this deficiency did not much distinguish him from his predecessors.

At the Burghölzli, Forel continued neuroanatomical research, furthering his scientific reputation, but unlike some of those who had held the directorship before him, Forel also showed a zeal for clinical care. His desire to improve his patients' lives was aided by his capacity to easily communicate with his patients. One of his most central therapeutic tenants came out of a conversation with a villager who somehow managed to cure a local alcoholic. The director asked this humble townsman how he had done it, and the man replied that he could get others to stop drinking because he himself abstained. From that time on, Auguste Forel became a fierce enemy of alcohol, enforcing abstinence with his staff, his patients, and himself.

"The Morning Report" at the Burghölzli, circa 1890. *Left to right*, Dr. Delbrück, Dr. Auguste Forel, Dr. Bach, and Dr. Gottschal.

In his search for novel treatments, sometime around 1885 Forel turned to the French research on hypnotism. In 1887, he visited Hippolyte Bernheim and before long began to lecture to huge audiences on hypnotism at the University of Zurich. Forel and his doctors employed hypnotic suggestion on one another and opened an outpatient clinic to provide suggestive and hypnotic treatments for physical ailments. Forel became deeply committed to these psychic treatments, editing a journal and publishing widely on the subject. During this period, he and his junior doctors also experimented with a number of psychotherapeutic practices, including trying to interpret the symbolic meaning of delusions in paranoid patients at the Burghölzli. Forel's public endorsement of French psychic, suggestive, and hypnotic treatments meant much in Austria, Germany, and German-speaking Switzerland. If a great "German" neuroscientist believed there was something to suggestion and hypnotism, others could not glibly dismiss this work as French silliness.

Through Forel's zeal, the Burghölzli found itself at the cutting edge of both brain research and clinical therapy. Forel's assistants were recruited to be directors for other asylums, prestigious professors visited Zurich, and the Burghölzli became a center for training in psychiatry. But in 1898, the man who effected this transformation abruptly resigned. Though only fifty years old, Forel lost his appetite for hospital administration; the directorship was turned over to one of his former students, Paul Eugen Bleuler.

Bleuler was born in Zollikon, a small farming village outside Zurich. In 1874, when Eugen was seventeen, his sister developed catatonia and was hospitalized at the Burghölzli. During her hospitalization, the family was enraged by the haughty remove of then director Hitzig, who could not understand a word the sickly Bleuler girl said. Family legend had it that this experience so impressed Bleuler's mother that she instilled in her son a desire to be a psychiatrist who could understand and help those whose lives had been torn by mental illness.

Bleuler graduated from the University of Zurich Medical School in 1881 and then like so many others, he took a scientific grand tour, visiting Europe's most prestigious labs and clinics. In Paris, he studied with Charcot and Janet, and was impressed by the power of hypnotism and unconscious mental processes. For six months, he worked on brain anatomy in Gudden's lab in Munich, and in 1885, he followed Forel to Zurich to become his assistant.

The job was only offered to Bleuler after he promised to swear off alcohol, a demand that had become de rigueur for Forel. But at the Burghölzli, the two men who vowed to practice self-restraint experimented with a technique that demonstrated how limited human self-control could be. They hypnotized each other. In 1887, Bleuler published his first paper on hypnotism and, two years later, followed it with a first-person account of being put under.

After only a year as an assistant at the Burghölzli, Bleuler was tapped to become director of the asylum in Rheinau. Housed in a former monastery on an island in the Rhine, Rheinau was the last destination for the Burghölzli's incurables; it held over eight hundred of them. No one was better suited than Bleuler to take up this overwhelming task. He had been trained by Forel to be curious, searching, and optimistic about therapeutic treatments. He was a bachelor and therefore not distracted by family obligations, and he had a personal wound that encouraged him to never give up on his patients. Bleuler threw himself into his job. He worked hard, too hard. Not long after he arrived at Rheinau, he informed Forel that he had come to think of sleep as no more than a bad habit. Soon, Bleuler collapsed from exhaustion. Forced to accept sleep as a necessary evil, a determined but rested Bleuler dove back into his work.

For the next twelve years, Eugen Bleuler lived alongside psychosis. He observed his patients' behavior closely, their likes and dislikes, their ability to reason and their forms of madness. Eventually, he concluded that the reigning pessimism about insanity was simply wrong. In 1896, Emil Kraepelin had presented his influential theory of dementia praecox. As the name suggested, Kraepelin believed this kind of psychosis entailed an early and inevitable decline into dementia, with an increasing loss of cognitive capacity. Forel also believed psychotic patients were doomed to imbecility, but at Rheinau, Bleuler saw that many of his psychotic patients were not demented. He redoubled his efforts to find out what was wrong with them.

By closely attending to his insane patients' utterances and behaviors, Bleuler grew certain that their mental lives had not been extinguished. Each patient had a particular intellectual and emotional range, and each reacted psychologically to his or her environment. A favorite aunt or detested brother elicited different responses. A member of the opposite sex sometimes encouraged a wild patient to suddenly become socially appropriate. That had to mean that certain symptoms were psychological reactions, and that the illness could not be totally reduced to an invariant, organic process. Never doubting that his patients had a biological illness, Bleuler decided psychology also played a role in their dysfunction. Therefore, he looked to psychological therapies to provide some relief and developed a form of psychological shock therapy. For example, Bleuler might decree that a regressed patient suddenly be discharged, or invite a violent patient to a formal dinner at the director's home.

Eugen Bleuler in 1902, as recently appointed director of the Burghölzli Asylum.

During these years, Eugen Bleuler began to formulate his theory of psychosis. He distinguished primary signs of the illness directly related to biological pathology, from secondary signs that were psychological reactions to both the disease and the environment. As Bleuler immersed himself in the study of new models of the psyche that emerged in French psychopathology, hypnotism, and hysteria, his theory began to coalesce. As a book reviewer for the *Münchener Medizinische Wochenschrift*, a weekly medical newspaper, Bleuler kept abreast of French writings on altered states of mind, and in that role he reviewed many central works, including Josef Breuer's and Sigmund Freud's work on hysteria. Bleuler was impressed and deemed the book "one of the most important additions of recent years in the field of normal and pathological psychology."

When summoned back to the Burghölzli in 1898, Bleuler brought these formative experiences with him. Upon his arrival, he found an institution on its way up, and in his first decade of leadership Bleuler accelerated that climb. The hospital had a budget of 8000F in 1900 and by 1913 commanded ten times that amount. In 1900, the medical staff numbered four; Bleuler more than doubled that number in thirteen years. Burghölzli admissions grew from 203 to 578 in the same period.

Upon his arrival, Bleuler laid out his clinical philosophy. Much of it was conventional wisdom, except for a few sentences in which Bleuler suggested hypnotism might be effective for symptomatic relief and could even effect some cures. Those few words became a bridge between the new Zurich director and Sigmund Freud, but the two men also shared many other perspectives. While at Rheinau, Bleuler had worked hard on the problem of how psychology could be a natural science, outlined his opposition to metaphysical philosophical concepts, and argued that unconscious and conscious mental contents were not qualitatively different. Like Freud, Bleuler concluded complex mental processes could be unconscious, and that this psychological unconscious was implicated in disassociated actions, fugue states, hysterical symptoms, and cases of split, double, or multiple consciousness.

Along with Freud (though to a much lesser extent), Bleuler also engaged in debates about human sexuality, in part spurred by his former boss, Auguste Forel. For many years, Forel had been one of Zurich's leading advocates for a moral approach to sexuality; in 1888, when morality leagues were founded by Zurich's elite, they included not only city leaders and Protestant clergy, but also him. Although these groups assailed the double sexual standard for men and women, they prescribed a solution that was exactly the opposite of most Viennese reformers. Rather than loosen the restraints on female erotic life, the Zurichers wanted to clamp down on male sexuality and enforce abstinence. The Zurich Morality League descried theater and dance but focused on the brothel as the true root of evil and advocated a strict, criminal approach to prostitution. When one doctor argued that legalization and state hygienic intervention would be more effective as a way of limiting the lethal spread of venereal diseases, Forel supported the counterreport that said he was wrong.

After his retirement in 1898, this upstanding Swiss doctor had a gradual change of heart. He continued his lifelong study of insects, which led him to evolution, sexual selection, and the debates over human sexuality.

Forel began to ally himself with those who called for a new sexual ethic based on natural science. After years of study, Forel published *The Sexual Question* in 1905. While considered tame by Viennese radicals, in Switzerland the book was dynamite. A founding member of the Zurich Morality League now argued that sexual abstinence was unnatural and doomed. With Freud and other sexologists, Forel concluded that while the sexual drive had to be controlled, complete abstinence could damage one's health. Sex, Forel asserted, was a basic human right as well as a source of creativity and happiness.

The Sexual Question was applauded in the medical press and by progressives in the *Neue Züricher Zeitung* and the *Züricher Post*. Bleuler also leapt in to support Forel, but the Morality League members and clergymen were furious and denounced the book as a moral aberration. In some parts of Switzerland, Forel was banned from public speaking.

Through his interest in hypnotism, French psychopathology, evolutionary biology, and sexology, Auguste Forel traversed some of the same roads as Freud, and his familiarity with that terrain prepared the ground in Zurich for an open appraisal of Freud's work. Forel's student Eugen Bleuler had imbibed his teacher's enthusiasms (including not imbibing) and had taken charge of the Burghölzli after being well versed in hypnotism, suggestive cures, French psychopathology, and sexology. He was committed to the natural scientific study of the psyche, including unconscious psychical processes, and he became an eager reader of Sigmund Freud. But Bleuler also had research interests of his own, and it was the integration of those interests with Freud's theories that would bring the Burghölzli international acclaim.

II.

ALONG WITH A number of academic psychiatrists of his generation, Eugen Bleuler had given up on the hope that anatomical research would yield much for the clinical understanding and treatment of mental illness. Instead, he turned to scientific psychology in the hope that it might offer more immediate gains. The emergence of experimental psychology provided the possibility for a new scientific foundation for psychiatry based on quantifiable measures.

Early in his career, Bleuler tried his hand at psychological research by studying synesthesia, that odd mix-up in which a subject saw sounds or heard colors. A favorite of Romantic poets, this phenomenon was also intriguing for scientists who endeavored to understand how the mind synthesized (or failed to synthesize) experience. At Rheinau, Bleuler came to believe that psychotics had a similar difficulty. They suffered from abnormal processes of association due to an inherent inability to knit connections together. When he got to the Burghölzli, Bleuler set out to study this and other mental aberrations with word-association experiments.

Charles Darwin's cousin, the eccentric and brilliant British polymath Francis Galton, had inaugurated the word-association experiment. In 1879, Galton wrote: "My object is to show how the whole of these associated ideas, though they are for the most part exceedingly fleeting and obscure, and barely cross the threshold of our consciousness, may be seized, dragged into daylight and be recorded." Once recorded, these shadowy thoughts provided an understanding of deeper individual motivations. "No one can have a just idea," he wrote, "before he has carefully experimented upon himself, of the crowd of unheeded half-thoughts and faint imagery that flits through his brain, and of the influence they exert upon his conscious life." His technique for making his own barely conscious associations more manifest was clever; he would suddenly flash a printed word, allow a few ideas to emerge in response, and then record those ideas before they faded from mind.

While Galton invented the word-association tests, Wilhelm Wundt standardized and popularized the method. In Leipzig in 1879, he set up the first laboratory for experimental psychology, where researchers adopted a simple method to study associative processes. The Experimenter called out a word and the Subject replied with the first word that came to mind. Wundt categorized these associations and, from that data, pieced together mental operations. In *Outlines of Psychology*, Wundt tracked associations as they fused into more complex compounds, arguing that imagination, understanding, illusions and hallucinations, dreams and hypnotic phenomena could all be thereby explained. The mind could be understood by breaking down mental phenomena into its elements, analyzing how the elements were connected, and determining the laws of their relations.

Wundt's methods spread. Psychology labs were founded in Göttingen, Berlin, Bonn, Munich, Geneva, and Copenhagen, generating interest among psychopathologists and psychiatrists, including Emil Kraepelin. Kraepelin began as a brain researcher, but after working with Wundt he decided psychopathology could be better understood by precise, methodical, psychological study. After settling in Heidelberg, Kraepelin assembled a research group, and in 1895, they began to publish their findings in their *Psychologischen Arbeiten*. In the preface of the first volume, Kraepelin paid tribute to Wundt and predicted psychological experimentation would be the future for scientific psychopathology.

Kraepelin's ideas quickly attracted academic psychiatrists, because they offered a solution to a crisis that plagued the field. Despite decades of work, the search for precise brain lesions specific to mental diseases had reaped little. In his *Textbook on Psychiatry*, Kraepelin's student Theodor Ziehen was emphatic: psychology, not neurology, must be the foundation for the study of psychopathology. To investigate mental illness, Kraepelin and his colleague Gustav Aschaffenburg turned to the word-association test. Aschaffenburg distinguished internal associations based on meaning (i.e., "bat-cave-dark") and external associations based on sound (i.e., "bat-rat-cat"). In so doing, he found that fatigued or ill subjects were more prone to external associations. He postulated that psychopathological states might also be characterized by different associative patterns. Others pursued this line of thought, most importantly Ziehen, who in 1898 began to publish his studies on associations in children. He showed that associations which awakened unpleasant memories took longer to emerge. Ziehen also discovered that one could thematically link these delayed responses and define the obstructing mental contents, which he considered an emotionally charged idea-complex.

In 1901, Eugen Bleuler was ready to adopt these experimental methods to study psychosis. He sent one of his new assistant doctors, Franz Riklin, to Munich to study with Kraepelin's team. When Riklin returned, Bleuler asked him to team up with another doctor to administer the word-association tests to the Burghölzli patients. The other doctor was Carl Gustav Jung.

Jung would play a significant role in Freudian psychoanalysis before leaving to found his own dynamic psychology. Born in Kesswil, a town in northeastern Switzerland, in 1875, Carl was the grandson of the Basel physician Carl Gustav Jung the Elder, who had distinguished himself as an ardent democrat and a Romantic fascinated by psychological models of mental illness. Carl's father, Paul Achilles Jung, was a university-trained philologist and linguist who became a Protestant clergyman. His mother, Emilie, came from a prominent Basel family, the Preiswerks.

The Jung family was an unhappy one. Carl's father was prone to depression and fits of rage. His mother's intimate confidences gave Carl the feeling that he had taken his father's place. Carl himself was moody and prone to episodes of grave demoralization. In the 1890s, the teenage boy discovered Arthur Schopenhauer and was captivated by the philosopher's tragic vision of man driven by blind desires. Turning to Kant, Carl believed he understood Schopenhauer's wrong turn: the philosopher had endowed the unknowable thing-in-itself with particular qualities, an epistemological error. The thing-in-itself, Jung understood, thanks to Kant, was as mysterious as divinity itself. Long afterward, Carl Jung would preserve a realm for the unknowable in his psychological thinking.

In 1895, Jung began medical school in the German-speaking Swiss city of Basel. Basel's university was a liberal bastion in a city that was also home to apocalyptic brands of Protestantism. On the weekends, Carl studied Kant and that philosopher of the unconscious—Eduard von Hartmann. He also absorbed the work of Friedrich Nietzsche, who made a great impression. In 1896, while Carl was in medical school, his father died. Paul Jung had been sick, and in his final years the clergyman had had a crisis of faith. Hospitalized and dying, the man had turned to books on hypnotic suggestion to understand what plagued him.

Even after his father's death, Jung, or "The Barrel" as he was known to his drinking buddies, remained a buoyant and active member of the university and its Zofingia Students Association. A friend later recalled that Jung was always ready to do battle with another university fraternity, nicknamed the League of Virtue, an allusion to the morality leagues that had attacked sex outside marriage and other vices. As a member of the Zofingia Association, Jung gave lectures that revealed a neo-Romantic desire to limit science and cordon off a domain for mystery and human subjectivity.

These youthful speeches dramatize the divide between Carl Jung and earlier generations of medical students. While his older colleagues had been fueled by a desire to roll back religious dogma and champion a mechanistic, rational, and scientific philosophy of life, Jung and many of his generation positioned themselves against the excesses of hyperrational science. Jung mocked those who would parrot "Papa Du Bois-Reymond." Influenced by neo-Romanticism and schooled by "our great master" Kant, as well as Schopenhauer and Nietzsche, not to mention his own religious convictions, Jung was eager to open a space for the irrational and subjective in an otherwise mechanized world.

To uncover this territory, Carl began to conduct some research on his own, picking up a line of inquiry from his family. The Preiswerks were interested in spiritism, a movement that emerged from the great American awakening of the 1840s and became popular in Europe soon after. The popular craze, which included receiving messages from the dead, mysterious turning tables, and mediums, was not restricted to lay people. It had also attracted scholars and scientists like F. W. H. Myers and William James.

In his first year in medical school, Jung began to attend family séances. The meetings began with a parlor game of table-turning. His aunts and cousins sat around an old oak table and waited for rumbles, knocks, and other signs from the beyond. Soon signs came in abundance through Jung's cousin, the fifteen-year-old Hélène Preiswerk. Helly, it seemed, could speak for the dead. After his father's death, Carl was often in attendance at the séances, but after four years his interest dimmed, especially after Helly was caught cheating. By then, Carl had decided to study psychiatry. When he set out to write his medical thesis, he chose his cousin and her strange voyages into the beyond as his subject. But believing no one at the university would be receptive to such work, eventually Jung set it aside.

On December 11, 1900, the young medical graduate was hired as an assistant doctor at the Burghölzli. In Bleuler, Jung found a teacher who was fascinated by bizarre unconscious and psychotic mental phenomena. Bleuler encouraged his staff to study the most recent literature on these subjects. Six weeks after his arrival in Zurich, Carl Jung fulfilled this duty to his director by presenting his colleagues with a synopsis of the latest work on dreams by Sigmund Freud.

Bleuler also encouraged Jung to immerse himself in a wide range of psychopathological studies. Since "The Barrel" had taken a vow of abstinence and was holed up in the hospital with few distractions, he took it upon himself to read through fifty volumes of the *Allgemeine Zeitschrift für Psychiatrie*, one of the oldest psychiatric journals in Germany. Unlike those journals that remained narrowly anatomical and physiologic, this one was open to work in sexology, hypnotism, hysteria, neurosis, and a range of therapeutics. In it, Jung found articles by Max Dessoir on the psychology of the sexual life, Eugen Bleuler's argument for unconscious mental phenomena, and a piece by Krafft-Ebing on zoophilia.

In this more accepting setting, Jung went back to work on his dissertation. In 1902, he completed it in a way that showed a marked change from his medical school days. "On the Psychology and Pathology of the So-Called Occult Phenomena" made no claims about the soul, nor did it allow for supernatural happenings. Instead, it was committed to demystifying the strange events of spiritism via natural scientific explanation. Jung presented the case of a fifteen-and-a-half-year-old medium whose family history was rife with madness. Using the pseudonym "Miss S. W.," Jung described his cousin's somnambulistic episodes, her visions, and her voices. He detailed séances in which Helly spoke from the beyond, read others' minds, and revealed the cosmic order of life. Jung relied on the French psychopathological literature for theoretical support, sprinkling his text with references to Charcot's "*grande hystérie*," Janet's psychic "*désagrégation*," and Charles Richet's work on unconscious processes. He noted that most German psychiatrists had ignored these unconscious states, though he singled out Freud and Löwenfeld as exceptions.

But the authority Jung turned to most was a fellow Swiss, Théodore Flournoy. Flournoy was a scientist who had tried to make research into spiritism respectable. After studying with Wundt, Flournoy went to Geneva and took the chair of psychology in 1891. In 1892, he opened the first experimental psychology lab in Switzerland, and eight years later published his classic, *From India to the Planet Mars*. The book, a case study of a medium named Hélène Smith, attributed the woman's altered states to hysteria, which was a result of emotional shocks, psychic trauma, and hypnoid states. In these ways, Flournoy's account was quite harmonious with Josef Breuer's etiological thinking.

Jung's dissertation followed Flournoy by positing that his cousin had an unconscious, second personality that had split off from the conscious "I" and could therefore perform complex tasks on its own. The girl's mind reading was the result of her extraordinary unconscious receptivity, a gift she owed to her hysteria. "The somnambulist not only incorporates every suggestive idea into himself," Jung wrote, "he actually lives himself into the suggestion." Helly had unconsciously lived her way into the suggestions that surrounded her. Jung spoke in passing of repression, hysterical forgetting, hysterical identifications, and Freud's theories of dreams, but in the end, his view of her hysteria was more conventional: Helly had a pathological constitution.

After finishing his dissertation, in October 1902, Jung escorted his cousin Valerie Preiswerk to Paris and there met up again with his mind-reading cousin, Hélène. During that winter in Paris, he attended lectures by Pierre Janet at the Collège de France, before returning to Zurich in 1903 to resume his association experiments alongside Riklin.

In his winter lectures of 1902, Pierre Janet discussed how fatigue could reveal aspects of psychopathology and allow lower mental states to emerge. Under Jung's guidance, the Zurich association experiments shifted to focus on similar forces that might reveal lower states of the mind, especially Theodor Ziehen's notion of an emotionally charged complex. Jung asked himself whether such complexes, if they were unconscious in a Freudian sense, might act like fatigue and disrupt associations. He pointed the Zurich association experiments in that direction. When a hostile reviewer of Jung's thesis derided its indebtedness to French psychopathology in 1904, Carl Jung snapped back that his mode of analysis was not dependent on French thinking, but rather on Freud's studies of hysteria. The reviewer was undeniably right about Jung's 1902 thesis, but by 1904 Carl Gustav Jung had become a follower of Sigmund Freud.

"ASSOCIATION IS A fundamental phenomenon of psychical activity," Eugen Bleuler declared in his preface to the remarkable findings gathered by the Burghölzli researchers in 1906. Therefore by studying associations, one could grasp the complete psychology of a man. Furthermore, by analyzing associations one could understand the effect of illness on the mind, and even characterize various mental illnesses. Bleuler announced that Carl Jung, Franz Riklin, and the Burghölzli group had done all that. If the empirical classification of mental illness was psychiatry's Holy Grail, a quantifiable method to penetrate inner subjective states was psychology's. Bleuler declared that both these long-sought-after solutions had been found. And while acknowledging a debt to Wundt and others, Bleuler made it clear that these breakthroughs had also been greatly influenced by a vision that came from outside academic psychology. Sigmund Freud had revealed part of this new world, Bleuler announced.

The Zuricher's excitement about Freud had risen as they conducted their experiments between 1903 and 1906. In the beginning, in order to study psychosis, Bleuler attempted to establish a baseline for normal associative patterns. Jung and Riklin took on the job and began testing themselves and other hospital staff. In 1904, they published their surprising results. After gathering the responses of thirty-eight normal subjects to four hundred stimuli words, Jung and Riklin found they could classify different associative patterns according to different character types. Instead of sorting associations, they began to sort people into what they called objective and subjective personalities. The subjective ones were subdivided into those who evaluated objects in terms of themselves, and those for whom the stimulus word provoked strong emotional "complexes." In discussing the latter group, Jung and Riklin transformed Ziehen's concept of an emotionally charged complex by claiming complexes could be unconscious due to repression, "in the sense of Breuer and Freud, to whose work we are indebted for valuable stimulus in our investigations."

The association studies brilliantly integrated Freud's theory of repression with empirical psychological findings. But the bombshell came when Jung and Riklin claimed they had found unconscious complexes throughout a normal population. Were we all suffering from the repression of affect-laden ideas? Did we all have what one of their patients called a "small soul in the large one, poisoning it when it awakes"? The implication was clear: Sigmund Freud's hypothesis that repression was ubiquitous appeared to have found scientific confirmation.

II. *Test* :	Right	none, it doesn't exist
I. „	False	women : 5 seconds
II. „	False	man, S. (in the sense of man is falsely organized)
II. „	False	corruption of man, S. (in sexual sense)
III. „	Correct	not everything, S.
I. „	Unjust	Russia : 2·8 seconds

Franz Riklin's 1906 Zurich study included word associations from the case of Catterina H., whose spurned Russian boyfriend committed suicide, sending her into a hysterical state. "S" indicated the presence of a sexual complex.

By 1905, Jung embraced Freud's ideas even more explicitly. He studied the reaction times of his subjects and discovered some were longer than others. These delays were due to unconscious repressed complexes, but going further, Jung argued that these same complexes were so central that they characterized the psychology of these individuals. Jung described a married woman whose erotic complex occurred in 18 percent of her associations, and a young man whose family complex occurred in 54 percent. These unconscious complexes marked each person. Thanks to the Zurich experiments, it seemed that one could now experimentally locate, define, and quantify the makeup of a man's or woman's unconscious world.

Between 1904 and 1905, the Zurichers published a string of association studies that sought to define the unconscious complexes in hysteria, epilepsy, and idiocy. While some of these complexes were understood as psychological manifestations of a biological disorder, Jung himself believed that in psychogenic diseases the complex itself caused the illness.

The implications of this flurry of findings were large. When Bleuler stepped back to consider what they meant for theories of the mind, he began with a pointed critique. Skepticism about a psychological unconscious had caused Germans to leave the scientific study of these processes to the French. The origin of this problem was Wilhelm Wundt. Training virtually all German practitioners of empirical psychology, Wundt had imparted the presumption that associations were by definition conscious, making it impossible to conceive of unconscious processes and study disassociated mental states. Bleuler reminded his readers that for over two decades he had argued that unconscious and conscious mental contents were qualitatively the same, a belief that he noted was shared by both Hermann von Helmholtz and Sigmund Freud.

The Zurich association studies were published as one explosive volume in 1906. The reader who opened this work expecting to enter the experimental world of Wundt and Kraepelin would close it after having reached foreign land. Jung's summation, "Psychoanalysis and the Association Experiments," marked the Zurichers' voyage from the clinics of Paris and the labs of Leipzig and Munich to an office located on Berggasse 19 in Vienna. The Zurichers had become followers of Sigmund Freud. Describing Freud's theories, Jung warned that the Viennese doctor's clinical method did not provide a secure framework for gathering data. Creating that framework was the great triumph of the Zurich Freudians.

THE ZURICH EXPERIMENTS took place during a critical period for scientific psychology. After 1900, psychiatrists and psychologists began founding labs, academic journals, and organizations at a feverish pace, giving rise to a new community. When the first international Congress of Experimental Psychology was announced in 1906, word spread over the same network of journals that would pick up news of the Zurich findings.

For Jung and his colleagues, the most important forum for their work was the *Journal für Psychologie und Neurologie*, a Swiss publication edited by Forel and a Burghölzli alumnus, Oskar Vogt. The journal had changed its name in a way that mirrored the shifting perspectives of those eager to study the mind. Founded in 1893 as the *Zeitschrift für Hypnotismus, Suggestionstherapie, Suggestionslehre und verwandte psychologische Forschungen*, the journal dropped its affiliation with suggestive therapies in 1896, and became the *Zeitschrift für Hypnotismus, Psychotherapie sowie andere psychopathologische und psychopathologische Forschungen*. In 1902, the editors dropped all references to hypnotism and psychotherapy and went forward as the *Journal für Psychologie und Neurologie*. Vogt explained that the journal had expanded to cover studies in normal, pathological, and comparative psychology, and neurobiology, while maintaining a special emphasis on psychic etiologies and therapies.

Jung and Riklin published their first studies in the journal in 1904 and, in the process, ushered Sigmund Freud into the world of empirical psychology. Freud had no lab and no quantifiable results. In 1904, most experimental psychologists and psychiatrists would have connected Freud to his 1895 work on hysteria, but the Zurich studies changed that by giving Freud's broader theories of a repressed unconscious quantifiable scientific verification.

In French-speaking Geneva, the *Archives de Psychologie* carried news of the Zurich association tests. Founded in 1901, the journal was edited by Flournoy and Édouard Claparède. The editors reviewed Freud's dream book, Jung's thesis, and Freud's *Psychopathology of Everyday Life*, before coming upon Jung and Riklin's "very important" association studies. For French-speaking psychological scientists, Sigmund Freud suddenly mattered.

In Germany, Jung and Riklin's experiments were generally warmly received. Take for example the *Zeitschrift für Psychologie*. It too had struggled with shifting affiliations during the last decade of the nineteenth century, as both psychology and psychiatry sought to define themselves. In 1890, the journal was called *Zeitschrift für Psychologie und Physiologie der Sinnesorgane* and was the flagship journal for psychophysics. In 1906, the journal split. One journal retained the old title and the old mission, while the other, *Zeitschrift für Psychologie*, dropped all references to physiology following Wundt. Edited by the experimental psychologist Hermann Ebbinghaus, the later journal paid close attention to the Zurich association studies, printing six extensive, though not uncritical, reviews.

The diagnostic association studies made Freud's ideas relevant to those interested in Wundt and experimental psychology, those scientists of the psyche who were committed to a scientific study of inner life. They also made the thirty-one-year-old Carl Jung renowned overnight. By 1906, Jung's writings and research had become widely recognized. His notion of the repressed, unconscious complex entered academic psychology and psychiatry.

News of the Zurich work crossed the Atlantic. In the newly established *Psychological Bulletin*, the Swiss-American psychiatrist named Adolf Meyer had adopted the view that psychiatry must find its scientific footing in psychology, and he excitedly reported that the Jung-Riklin paper on unconscious complexes in normal subjects was "the best single contribution to psychopathology during the past year." At the end of an eight-page review, Meyer wondered if Breuer and Freud's work would achieve the stature that it seemed to deserve now that Jung and Riklin's studies had provided such powerful validation. The American would help answer his own question by enthusiastically reviewing five of the Zurichers' papers in the same issue of the *Bulletin*.

Afterward, Meyer began corresponding with Carl Jung, who presented himself as a research psychologist trying to adapt Freud to this more scientific discipline. "I have been working for some time trying to find empirical rules of psychoanalysis, which is not an easy task," he wrote. "With psychoanalysis, one can take the wrong path, but the association experiment is an absolutely sure guide." Jung went on to warn that while delayed reaction-times revealed complexes, the *meaning* of the stimulus-word was still subjective. Therefore, the association tests yielded only the outlines for conducting psychoanalysis. Despite these warnings, Meyer sensed the promise of Jung's lab-based method to objectively ratify Freud's claims. To E. B. Titchner, a prominent American psychologist, Meyer defended Freud from the accusation that his psychology was "two generations" old, by pointing out that Jung's theory of the complex brought Freud into contemporary, experimentally based psychology.

Around Europe and America, the Zurichers helped Sigmund Freud get a hearing among scientific psychologists. "Now that you, Bleuler and to a certain extent Löwenfeld have won me a hearing among the readers of the scientific literature," Freud would later write to Jung, "the movement in favour of our new ideas will continue irresistibly despite all the efforts of the moribund authorities." The weight of the Zurich studies was increased by their connection to a prestigious, well-positioned institution, a center that linked the French and the Germans, published in both languages, and welcomed students from both communities. Auguste Forel remained one of the most visible psychiatrists in Europe, writing, lecturing widely while commanding the loyalty of a battalion of former colleagues and students. Bleuler had grown in stature, and his students, Carl Jung and Franz Riklin, had helped make the Burghölzli an international hub for the study of the mind.

In 1904, Bleuler added a new doctor to his staff, Karl Abraham of Bremen. In 1906 Ludwig Binswanger of Kreuzlingen, the nephew of the well-known psychiatrist Otto Binswanger, joined the hospital staff, as did a young Russian medical student, Max Eitingon. Alongside with Jung and Riklin, these new doctors would begin to agitate for Freudian psychoanalysis.

III.

EUGEN BLEULER AND Sigmund Freud had a lot in common. They were political liberals dedicated to the power of science; they both began their careers as brain anatomists and then became fascinated by French psychopathology and hypnotism. Most importantly, both men became convinced that complex, unconscious psychic processes could explain the mysteries of the mind. But while Freud was a Viennese practitioner who earned his living treating private patients, Bleuler was a Zurich professor, the director of a laboratory and the leader of a large hospital that took care of severely ill patients. As the doctors drew closer, these differences would become more pronounced.

In the first extant letter between the two men, dated June 9, 1905, Bleuler wrote to Freud as an admirer who was not afraid to air his reservations. Having just read Freud's book on wit and the *Three Essays*, Bleuler praised the theory of jokes but voiced doubts about his thinking on sexuality. The *Three Essays* lacked the detail and proof Bleuler had found in Freud's other work. The Zuricher was reluctant to accept Freud's use of material garnered from the analyses of adult neurotics as verification for theories regarding normal children. He suggested that Freud's sexual theories would be strengthened by a phylogenetic and teleological argument, and recommended a more critical attitude toward Theodor Lipps. Bleuler closed by saying he hoped Freud did not consider his blunt objections as arrogant. With that, Eugen Bleuler introduced himself as an admirer and confident expert, a colleague quite willing to question Freud's judgment and logic.

Yet as Bleuler's next letter revealed, the Zurichers were not just interested in debating ideas with Freud. Since 1903, they had been energetically trying to practice psychoanalysis. One of the first case accounts published from Zurich was Franz Riklin's case of "Lina H.," a 1903 analysis of a twenty-nine-year-old woman with a history of hysterical conversions and mood swings. Using hypnosis, Riklin traced the girl's symptoms to sexual abuse at the hands of her father. Along with hypnosis, he reported using association tests and symbolic interpretations to unearth Lina's charged, repressed emotional complexes. In sum, Riklin's case study revealed an amalgam of methods.

Feeling he could benefit from practical instruction, Bleuler asked Freud for assistance. The director had been struggling with the analysis of a rather difficult patient: himself. Like the other Zurichers, he was trying Freud's psychoanalytic method on his own dreams and those of his colleagues. Bleuler's self-analysis was not going well. "Although right after the first reading I recognized your Dreambook as correct, I only rarely succeed in interpreting my own dreams," he confessed.

This confidence marked the beginning of a long-distance, epistolary analysis. Bleuler enclosed four dreams in the letter, appealing to the "Master" to show him the way. Five days later, Bleuler thanked Freud for his analysis. He had to respond immediately, he confessed, "as a catharsis," to rid himself of the weight of Freud's criticisms. It wasn't true that he had been unable to interpret *any* of his dreams; he had succeeded nu-

merous times. Once, however, his answer referred to an event that occurred after the dream, so it had to be wrong. A second time, Bleuler had
presented a dream to a group of doctors and their wives, but the group
was unable to untangle it. However, when he left the room and returned,
the others had come to a conclusion that was obviously wrong. The director's wife presented the group's interpretation, and Bleuler reassured
Freud that it reflected the play of his wife's unconscious complexes,
not his.

The familiar shadow that darkened the efforts of one mind to know
another had returned, and yet Eugen Bleuler was unconcerned. Erroneous interpretations were easy to brush aside, he believed, while correct
ones hit like lightning. Unfortunately, Bleuler felt that Freud's interpretations did not carry electric force. In his letter, he abruptly shifted the discussion from his own analysis to reservations he had about Freud's sexual
theory, seeming to imply that sexual meanings were part of Freud's incor-
rect interpretation. For Freud, the implication was that Bleuler was
unable to think about Freud's sexual theory, in the very same way he was
unable to rationally consider Freud's sexual interpretations of his dreams.
Bleuler prided himself on his intellectual openness and went to some
lengths to assure Freud that sexuality caused him no shame, and that he
was not in the least repressed.

The Swiss doctor followed this assertion with a classic Freudian slip. "I
myself have not been seduced in childhood," he wrote. "My sexual drive
however was clear to me from very early on, and I believe with my
~~Mother~~ wife to clearly perceive that my 2¾ year old boy makes a distinction between the sexes." Bleuler had first typed in "Mother" then crossed
it out, for he meant to write "wife." Proceeding in the same letter, Bleuler
now doubled back and acknowledged he had the complex Freud supposed, adding that the typing mistakes that marred his letter must be the
result of his complexes. Still, the director persevered, enclosing a series of
associations for Freud to interpret, adding that Freud need not send the
materials back unless "something very special comes out of you." Another
error. He had intended to write "them."

Freud could not resist the temptation to interpret these slips. After
reading the Professor's thoughts, Bleuler returned to the books: "I just
read the *Three Essays* again. I still believe my resistance versus certain specific deductions is not an emotional resistance." Bleuler defended his ear-

lier critique and his own psychological makeup, saying neither was marked by repression. As if to stress the point, his next letter revealed sexual dreams about hospital attendants and his wife's sister.

A few weeks later, Bleuler thanked Freud for sending him the "brilliant" Dora case, which he and his colleagues were eagerly devouring. The case was the most extensive demonstration of how Freud conducted an analysis to date, and the Zurichers pored over it for clues. But Bleuler began to weary of his own analysis by post. He wrote shorter letters and less often. He seems to have no longer asked Freud to analyze his dreams. After January 1906, the director of the Burghölzli stopped sharing dreams of pistols and maidens, and asked for no more news about his own unconscious.

MEANWHILE, CARL JUNG was undergoing a transformation. The young Romantic who once descried the overextension of scientific rationalism now found himself famous for the scientific quantification of nothing less than the deepest strata of human interiority. Boldly, he set out to bite the hand that fed him and reposition himself not just as an empirical researcher, but also as a depth psychologist. As he would later say, he wanted to move from seeing from the outside to seeing inside.

In April 1906, Jung sent the completed first volume of the *Diagnostic Association Studies* to Freud. The Viennese doctor wrote back, saying he had been so eager to read the work that he had already rushed out and bought it. "I am confident," Freud wrote, "that you will often be in a position to back me up, but I shall also gladly accept correction."

Freud needed backing up almost immediately. Although publication of the Zurich studies gave Freud increased scientific legitimacy among academic psychologists and psychiatrists, it also brought forward powerful new enemies. For men such as Gustav Aschaffenburg, Theodor Ziehen, and Emil Kraepelin, the Zurich studies posed a threat. Before the association studies were published, these academics found it easy to ignore Freud. The Zurichers, however, used the methods of these scientists to verify something the students of Wundt deeply opposed: Freud's idea of unconscious motivation. They were poised to be killed with their own swords, and were not about to idly sit by and let it happen.

In 1899, Emil Kraepelin dismissed Breuer and Freud's theories of hysteria without a second thought. But in 1906, Kraepelin's associate Gustav Aschaffenburg unleashed a fierce attack on Freudian thinking. It was Aschaffenburg who had taught Franz Riklin how to perform the association tests; he was one of the leading experts on this method. Rather tellingly, when this critique was published, Sigmund Freud had never heard of Aschaffenburg. The Viennese doctor and the Munich psychiatrist once existed in separate spheres, but the Zurich studies made those spheres collide. Awakening to what was happening, Freud wrote Jung: "Here we have two warring worlds."

The Zurichers were not eager for war. They were on cordial terms with the Munich group and Aschaffenburg, who Jung maintained an active correspondence with, lately devoted to explaining Freud's theories. These elucidations were for naught. On May 27, 1906, speaking before an audience of German neurologists and psychiatrists, Aschaffenburg struck again, this time with a more detailed attack on Freud. The lecture was a sustained attempt to demolish Freud's scientific standing and was published in the *Münchener Medizinische Wochenschrift*.

According to Aschaffenburg, Freud's sexual determinism was patently absurd. Hysteria was a psychogenic disease, but it was not psychosexual. Aschaffenburg was outraged by "Fragment of a Case of Hysteria," in which a young girl's sexual life became the focus of a doctor's attention. Furthermore, the scientist lambasted Freud's method. The physician was not discovering sexual complexes but putting sexual ideas into his patients' heads by suggestion. It was Freud who linked sexuality to the thoughts of his patients. To avoid doing harm, doctors should not discuss sexuality. The Munich expert also noted that the experimental approach advocated by Jung should be dismissed. To get from the recorded associations to their unconscious contents, Jung indulged in leaps of inference. His method was unsound, a tautological trap that offered nothing.

After that, the Zurichers seemed ready to accept the fact that this was their battle. While Freud declined to respond to Aschaffenburg, both Carl Jung and Eugen Bleuler did. They knew Aschaffenburg's premises, they knew his audience, and they were the ones who started this fight. In November 1906, Jung's reply to Aschaffenburg appeared in the *Münchener Medizinische Wochenschrift*, where he argued for making a clear distinction between Freud's sexual theory and his psychology, to "not throw out the baby with the bathwater." In a letter to Freud, Jung previewed his defense: he would admonish Aschaffenburg for not seeing the value of Freud's general psychology and accuse him of "harping" on Freud's theories of sex, a matter that Jung would concede, he too had questions about. Freud replied that he hoped his younger colleague would grow closer to his own position on sexuality over time.

In his published rebuttal, Jung added another defense: Freud's achievements could not be challenged by those who had "never taken the trouble to check Freud's thought-processes experimentally." This was his trump card. Men of science were supposed to accept or reject findings not according to their likes or dislikes, but only after a careful attempt to reproduce experimental results. After all, what man of science would want to act like those who refused to look through Galileo's telescope? "We hear a great deal about 'experiments' and 'experiences,' but there is nothing to show that our critic had used the method himself." All Aschaffenburg needed to do, Jung continued brazenly, were . . . association tests! "In earlier writings, I have already pointed out that the association experiment devised by me gives the same result in principle, and that psychoanalysis is really no different than the association experiment." Since it was obvious that suggestion was not part of the association experiments, why did Aschaffenburg think it was operative in psychoanalysis? Jung delicately reminded his audience that Aschaffenburg had made excellent contributions to the association tests and that those same tests had verified the Freudian unconscious.

Jung concluded his rejoinder to this powerful psychiatrist by harshly rebuking someone else entirely. In passing, Aschaffenburg had cited Walter Spielmeyer's critique of the Dora case. Filled with mockery, this review focused on Freud's interpretation of sexual symbols and relied on sarcasm more than reasoned rebuttal. Spielmeyer was not a prominent ally of Wundt and Kraepelin, but merely a young doctor from Freiburg. He made an attractive target for Jung's central point: "When a person reviles as unscientific not only a theory whose experimental foundations he has not even examined, but also those who have taken the trouble to test it for themselves, the freedom of scientific research is imperiled." Spielmeyer's critique was beneath contempt for a scientist.

The parry by Jung was tactically brilliant. In the beginning, he gave up a bit of ground on sexual theory, but after questioning the empirical basis for Aschaffenburg's critique, Jung returned to sex and took back terrain he had originally conceded. His blistering summation was not directed against the much-admired and well-connected Aschaffenburg, but a relative unknown, and his indignation was not on behalf of himself or Freud, but rather the corruption of scientific ideals. An attack on the scientific value of Freud's theories had been transformed into an attack on the scientific values of Freud's critics. Bleuler also joined in the fray by condemning Spielmeyer.

By taking on their academic colleagues and seeking to forge a relationship with Freud at the same time, Bleuler and Jung were trying to manage their way between enemy camps. As scientists from the University of Zurich, it was critical for them to defend their independence and scientific objectivity. At the same time, they had to defend Sigmund Freud, at least in part, because aspects of their conclusions were predicated on Freudian ideas. The Zurichers pulled this off by distinguishing between the *psycho* and the *sexual*. They declared their allegiance to the first half of the Freudian synthesis but remained wary of libido theory. In this way, they appeared to reaffirm their balance and independence. The Zurichers could point to experimental data that offered support for the Freudian notion of unconscious ideation and its symptomatic action and take a strong stand, girded by their science. But they had nothing to definitively link sexual complexes to all neuroses, and they freely said so.

At the same time, the Zurichers were anxious not to alienate Freud, whom they were running toward and away from at the same time. On December 29, 1906, Jung sent Freud a paper, asking the Viennese professor to remember that Bleuler was writing for a German academic audience, and that Aschaffenburg's critique had "whipped up a storm of protest against you." Jung asked Freud to consider that his differences might be due to their different patient populations, his lack of experience, as well as "my upbringing, my milieu, and my scientific premises," which were quite different from the Professor's.

Different upbringing. Different milieu. Different scientific premises. In that mouthful, Jung pierced to the heart of the differences between the Zurich and Viennese Freudians. One group was Jewish, the other Protestant. One group resided in sexually open Vienna, the other in buttoned-down Zurich. One had a foot in biophysics and psychophysics, and the other was committed to the new kind of lab-based study of psychology that had emerged from a rejection of those approaches. Jung made it clear that he had no intention of abandoning his good standing in academic psychology and psychiatry, at least not yet. So he happily reported to Freud that he had introduced a new concept he thought Freud would not like, but would help facilitate a link to Wundt's psychology.

Freud seemed to take it all rather well. He didn't interpret Jung's reticence as emotional resistance. In January 1907, he politely asked Jung to pay less attention to the critics. "The 'leading lights' of psychiatry really don't amount to much," Freud wrote. "The future belongs to us and our views." He asked Jung not to give up anything essential to assuage these authorities, and not to stray too far from his own views. If not, Freud warned, "we may one day be played off against one another." Freud gave Jung one more piece of advice. "My inclination is to treat colleagues who offer resistance exactly as we treat patients in the same situation." It was, perhaps, a subtle warning.

The Professor was not so sanguine with Bleuler. On January 10, 1907, the Zurich professor sent Freud a postcard asking if sexual trauma was the cause of hysteria. Many had not yet recognized that Freud had abandoned his sexual trauma theory, but Bleuler had read and reread the *Three Essays* and had corresponded with Freud on sexual theory. Freud accused Bleuler of willing incomprehension and brought up Bleuler's inadequate defense when he went after Spielmeyer. Bleuler replied that "you will however excuse me, when you realize what difficulties defending you presents." Behind the scenes, Freud's irritation with Bleuler was stoked by Jung, who reported on the "frosty bachelor's" reserve toward libido theory, which was an efficient way to take the focus off Jung's own reservations on that very subject.

Jung's desire to get closer to Freud prompted a trip to Vienna, accompanied by his wife and his young assistant, Ludwig Binswanger. The group arrived in Vienna on a Sunday in March 1907. Binswanger recalled that Freud invited them to his house, and not long after their arrival asked both men to share their dreams. He interpreted Jung's as a disguised wish to displace Freud, and Binswanger's as a desire to marry Freud's daughter. Jung later recalled that he and Freud went on talking for thirteen hours, and he left impressed by Freud's commitment to sexual theory. Jung was particularly disturbed by Freud's interpretation of psychosexuality, which made it seem that all of culture might be reduced to an animal farce. According to Jung, Freud asked him to promise never to abandon sexual theory. "You see we must make a dogma of it, an unshakeable bulwark," Jung recollected Freud saying. When Jung asked what this was to be a bulwark against, Freud replied: "Against the black tide of mud . . . of occultism." Jung, whose family was deeply immersed in that black tide, was being asked to protect psychoanalysis from it.

On March 6, the two Zurichers attended the Wednesday Psychological Society meeting in Vienna. Freud pulled Binswanger aside, scanned his Viennese followers, and said derisively: "Well, now you have seen the gang." None of the Viennese had any academic stature: no one could compare to Bleuler, Jung, or Riklin. Next to the Zurichers, the Viennese seemed like rabble. That evening, Alfred Adler launched into a discussion of organ inferiority. Somehow the debate moved from organ inferiority to the size of Jewish and Christian penises. With the experimental psychiatrist from Zurich sitting by, Otto Rank said the numbers seven and forty-nine represented the small and big penis. Freud chimed in: the number three represented the Christian, seven the small, and forty-nine "the large Jewish penis." Most experimental psychologists in their labs and white coats would have shuddered at this strange hermeneutic.

A few weeks later, Bleuler wrote Freud. Commenting on the visit of his Zurichers, he said it must have been pleasant to be in the company of students who were still thinking a bit. The cutting implication was clear: Jung must have reported that the Freudians of Vienna were simply following the Professor. Ten days later, Jung wrote Freud to criticize Otto Rank's full acceptance of libido theory. With Rank, he said, there was an uncomfortable feeling of a disciple who *jurat in verba magistri*, or swears by the master.

By 1907, THE Freudians had two distinct centers, in Vienna and Zurich. While all hoped for mainstream medical and psychiatric acceptance, the Viennese Freudians counted on the support of private clinicians and a network of social and cultural reformers committed to sexual reform, while the Zurichers were part of a growing community of academic psychiatrists who were committed to the scientific study of psychopathology. These differences became pronounced when the Zurichers came under attack and chose to defend Freud's unconscious psychology, while not embracing his sexual theory. In their reluctance to embrace libido theory, the Zurichers were not alone, for the Wednesday Society was filled with theorists who quietly did not fully endorse libido theory. But the Zurichers made their hesitance part of a very public battle over Freud's scientific worth. As they publicly pivoted against this aspect of his thinking, Freud became increasingly concerned that his grand synthesis would come undone. He began to increase pressure on his most prominent backers to close off this line of retreat.

During this jockeying, all concerned parties struggled to somehow contend with the concern that they were either resistant to Freud's ideas or true believers primarily so as to serve their own emotional needs. Was there a way to enter into an evaluation of a theory of unconsciously driven motives and not have one's own motives in question? How could one take up these ideas without getting tangled up in a net of emotional refusal or childish credulity?

For Freud, resistance when confronted by psychosexuality was just as predictable from doctors and scientists as it was from patients. For that reason, he urged Jung to treat scientific critiques as, in essence, emotional. From that moment on, Freud increasingly relinquished the responsibility to distinguish valid criticism from rationalizations compelled by personal disgust. At the same time, Bleuler began to seriously consider his own refusals and resistances and develop a theory that specified the roots of such behavior. In the process he would bring the Viennese Freudians more into his clinical world, the world of psychosis.

IV.

IN 1900, THE Burghölzli admitted 203 patients; only one was deemed to be a hysteric. None were diagnosed with obsessional neurosis, and only two were labeled paranoid. While Freud was treating psychoneurosis in his office, the Burghölzli hospital filled with men and women who had constitutional, organic, epileptic, and intoxicant forms of insanity. The largest group were diagnosed with "acquired idiopathic psychosis," which meant no one could even take a guess at the cause. For Freud to truly enter their clinical world, the Zurichers would have to adapt his methods for these patients.

What was wrong with the madmen and women on Bleuler's wards?

The Zuricher's engagement with Freud gave him a new perspective on this quandary, which had possessed him for years. Simply finding a stable category for psychoses was difficult enough and had greatly occupied late-nineteenth-century European psychiatry. In 1883, Wundt's student Emil Kraepelin sought to rationally categorize mental disease the way his teacher had categorized normal psychology. Reviewing the literature, he found a sprawling mess of terms and concepts, and he embarked on a re-organization, based on the belief that illness could be characterized by cause, course, and prognosis. Since the cause was rarely known, Kraepelin mostly relied on disease course and prognosis. His term, dementia prae-cox, described a psychosis that led to an early and permanent cognitive deterioration. Eventually he folded catatonia and dementia paranoides into that same category, packaging hallucinations, delusions, and other odd behaviors into this one illness.

According to Kraepelin, the disorder was to be distinguished from two others: manic-depressive insanity, in which mood dysregulation was the defining problem and no long-term cognitive deterioration could be found, and paranoia. The latter was a big loser in Kraepelin's reshuf-fling, for this term had once been loosely bandied about to cover a great number of ailments. Kraepelin restricted paranoia to a small subset of psychotics characterized by a fixed thought disorder, one that did not decline over time or manifest affective disruptions. Paranoia was more related to psychological causes and was more amenable to cures and spontaneous remissions.

Skeptical of Kraepelin's belief in the inevitable cognitive decline of most psychotic patients, Bleuler began to wonder about a curious phe-nomenon exhibited by some on his wards. They were deeply resistant to the wishes of their doctors. This psychotic "negativism" could dominate patients, making them refuse the simplest of requests. Were those patients demented, or were they in a state unreceptive to suggestion, "negative suggestibility" as the hypnotists might say? Bleuler wondered if these two states of negativity might be related.

At the same time, Bleuler struggled with his own negativism. Accord-ing to Freud, Bleuler's refusal to accept his sexual interpretations were the result of emotion, not intellect. With this accusation ringing in his head, Bleuler reread Freud's *Three Essays* and came upon the paragraph in which the author declared that certain impulses for sexual perversion were regularly found as pairs of opposites. Upon reading this, Bleuler had an epiphany. Excitedly, he wrote to tell Freud: ". . . our entire life is regulated by an interplay of contrasting forces. We find this in the chemical, as well as the nervous and psychic areas." Consider fine motor moyements where resistance was critical to any coordinated action, Bleuler went on. Consider the powerful sexual drive and the powerful resistance it encountered. Consider exaggerated states of intense negativism and utter suggestibility. These were all due to a common, underlying process that Bleuler would later name "ambivalence." He meant it literally: all mental phenomena were, like chemical elements, subject to a positive and a negative charge.

Bleuler tried to show how experiences of negativity cast light on the way mental processes were, as Jung put it, a "play of contraries." Neurotics experienced opposite feelings side by side, but psychotics had lost the glue that kept such feelings together. They succumbed to unalloyed positive or negative states. Negativism was a state in which contrary positive feelings had split off and disconnected. Seeing this as fundamental, Bleuler would name the consequent ailment "schizophrenia," literally split minds.

Bleuler expanded his thoughts on ambivalent mental processes in a 1906 monograph. Since affects drive the mind, they also control seemingly cognitive matters like suggestibility. Suggestion was a wildfire of affects that spread from person to person. Its evolutionary value was clear to anyone who had ever watched an awareness of danger spread through a herd of gazelles. Suggestibility was simply a preverbal form by which affects were communicated, a form that commenced with the bonding of mother and child. It was a normal transfer of feelings from one to another. An idea from a beloved person was easy to accept, for it traveled on the wings of love. Notions from a disagreeable person were easy to reject for the same affective reason. Suggestibility could be broken down into the poles of utter credulity and blind refusal. Negativism, negative suggestibility, and positive suggestibility were all extreme states due to breaks in the normal equilibrium between negative and positive affects.

Bleuler's line of thought could be read as a personal rejoinder to Freud: I don't like you and that's why I refuse to submit to your authority. But it was more than that. After contemplating different lines of thought on negativism, Bleuler had wrestled with Freud, examined his conflicts with the Viennese doctor, and in the end, created a new theory of negative emotional states. In 1908, Freud believed suggestibility was due to the transference of unconscious sexual feeling, and resistance was reactive, secondary to these repressed sexual feelings. Bleuler suggested such negative states might be primary and could better be explained by a broken dance between two primal affects, love and hate.

Bleuler's theory offered a new approach for a hopeless illness. If psychotic negativism was not due to brain degeneration but rather the splitting of affects, then perhaps psychological treatments could do something to help. Bleuler had taken a similar stance at Rheinau, and now he had a fuller theory to back up his beliefs. He knew the challenges to implementing his theory would be formidable. Many psychotics were impervious to their doctors, and performing psychotherapy with such patients required somehow reaching them and penetrating their remove.

To do that, the Zurichers honed in on the concept of transference. With the exception of Isidor Sadger, most Viennese Freudians weren't so impressed by this innovative idea. But Bleuler and Jung quickly recognized that it might provide a key to approaching withdrawn, disturbed patients. In his first letter to Freud, Jung asked about the means of cure in psychoanalysis. Transference was the mechanism, Freud replied, and "the chief proof that the drive underlying the whole process is sexual in nature." Six weeks later, Freud told Jung that psychoanalytic cures were effected by a fixation of libido on the analyst. Without it, patients would not make the effort or listen to their analyst's interpretations. Hence, transference was critical for curing neurosis, but unfortunately psychotics did not generate transferences, only resistance. That seemed to leave little hope for the patients at the Burghölzli.

In January 1906, Freud sent Bleuler a psychoanalysis of a case of paranoia. Bleuler thanked Freud, disagreed with him on the diagnosis, and casually added that it would be easy to employ Freud's dream theory and work on hysteria to account for the symptoms of dementia praecox. If dreams were a form of sleeping madness, then Freud's methods of dream analysis could be used to unravel insanity. In 1906, Bleuler wrote "Freudian Mechanisms in the Symptomatology of Psychoses," in which he announced that paranoid delusions and hallucinations were nothing more than disguised wishful dreams, in which reality was symbolized as a persecutory Other. Freud dissented from this analysis but added: "my experience in this field is meager. In this respect therefore I shall try to believe you."

The Zurichers soon produced a number of Freudian studies on psychosis, and in 1907, Jung published a major work on the subject that utilized association studies and Freudian theory. In *The Psychology of Dementia Praecox*, Jung argued for an overlap between the symptoms of hysteria and dementia praecox, which legitimized the extension of Freud's views on hysteria into psychosis. Like hysteria, dementia praecox was based on an unconscious complex. But unlike hysteria, dementia praecox was caused by a toxin—an "X" factor. Toxic psychoses were well-established entities no psychiatrist could fault. In fact, the best example of such a psychosis was caused by the bête noire of the Burghölzli: alcohol. Jung suggested that dementia praecox was a form of hysteria dramatically altered by an unknown, internal poison.

Jung cited Sigmund Freud as well as Emil Kraepelin; Bleuler wrote the Professor promising the study of dementia praecox would soon become a "panegyric" to the Viennese doctor's ideas. But neither Kraepelin nor Sigmund Freud were buying this proposed merger. After Jung's book came out, one of Kraepelin's assistants, Max Isserlin, published a biting critique. "It will soon come to your attention that an assistant of Kraepelin's has slaughtered me," Jung wrote to Freud. "At least they have now started using the heavy artillery." At the same time, relations between Bleuler and Freud were also growing cool. Jung informed Freud that Bleuler had rejected the concept of autoerotism and would develop his own competing concept of "autism" in Gustav Aschaffenburg's textbook. This was stunning news. It meant Bleuler had turned against an aspect of libido theory and had agreed to write a monograph announcing his rejection for one of Freud's most vehement critics. The man who would make dementia praecox a panegyric to Freud's theory was dancing with the enemy.

And yet, Eugen Bleuler was undeniably the single most important member of the Freudian movement. As director of the Burghölzli, he had the power to keep the doors of the institution open to Freudian thinking, or shut them as he pleased. It was Bleuler who had supported the diagnostic association tests, and Bleuler who had sent a Freudian, Franz Riklin, to run the Rheinau asylum. Bleuler had supported Jung, and thanks to

Bleuler and his colleagues, it seemed possible that Freud's thinking might become a legitimate part of European hospital psychiatry.

In the fall of 1907, Bleuler's import grew even more when along with Jung and Riklin, he founded a Freudian Society of Physicians in Zurich. When the society held its second meeting in November of that same year, Professor Bleuler opened the proceedings with doggerel mocking Freud's critics. Twenty-five people were in attendance, more than could be expected most Wednesday nights in Vienna.

In the course of a few years, the Zurichers had taken up Freudian theory and done a great deal to foster it. Their association experiments forced many academics to pay serious attention to Freud's ideas. Jung's notion of the complex was an elegant synthesis that brought Freudian repression into the lexicon of experimentalists. Bleuler's understanding of negativism and his commitment to analyze madness as if it were a dream had the potential to push Freudian thought into asylums everywhere.

After Forel's groundbreaking acceptance of psychic theories and cures, Eugen Bleuler and his Zurich colleagues embraced Sigmund Freud and established a thriving Freudian hub in Switzerland. From that base, they also disseminated Freudian thought throughout Europe. The Burghölzli was a stop on the itinerary of medical students and recent graduates who traveled to cutting-edge labs and clinics, before going home and settling into their careers. When they came to Zurich, they were exposed to Freud's thinking, not just as a body of ideas, but also as something that had been experimentally verified and clinically put to use. In August 1907, Jung informed Freud that the Burghölzli had hosted six Americans, a Russian, a Hungarian, and an Italian in just three weeks. That same summer, Carl Jung received three applications from students who wanted to travel to Zurich to do research with him; they hailed from Budapest, Boston, and Switzerland. A number of these foreign students would come to Zurich and return home with Freudian theory in their heads. And these itinerants, more than anyone else, would transform the Freudians into an international movement.

SIX

Freudians International

I.

THE SMALL FREUDIAN movement was growing. Sigmund Freud had built a theory that drew its ideas across disciplines, attracting interest from dynamic French psychopathologists and psychotherapists, sexologists as well as sexual reformers, medical clinicians looking for treatments, and neo-Romantics schooled in Nietzsche, Weininger, and Kraus. The Zurichers had brought crucial, new audiences to Freud, thanks to their application of his psychology in the lab and the hospital. In June 1907, Jung wrote to Freud, exulting, "all the lines are converging on you."

But in their attempt to weld Freud and academic psychiatry, the Zurichers incited some in German psychiatry to attack. A charged contest of ideas was waged in academic journals and at conferences, and for the Freudians the stakes were high. In 1907, the monumental First International Congress of Psychiatry, Neurology, Psychology and Alienists was held in Amsterdam. In the spring of that year, Jung informed Freud that he had been invited to discuss theories of hysteria, along with Aschaffenburg. Jung braced for a dogfight with his former ally. Freud too had been contacted by the organizers of the conference. "Apparently a duel was planned between Janet and myself," he wrote, "but I detest gladiatorial fights in front of the noble rabble. Now, you will have to measure yourself with Aschaffenburg. I recommend ruthlessness; our opponents are pachyderms, you must reckon with their thick hides."

In the weeks before the Amsterdam conference, Jung grew jittery. He would have to publicly defend Sigmund Freud before an august, influential, and skeptical audience. In preparation, he asked Freud for a clarification: Did he really believe that only sexual complexes made for neurotic symptoms? Freud replied that the determining role of sexual complexes was a theoretical necessity that was not yet proven. With that wiggle room, Jung went on his "apostolic journey," hoping he would not be the only Freudian in all of Amsterdam.

The First International Congress of Psychiatry, Neurology, Psychology and Alienists brought together many of the finest psychiatrists, neuropathologists, and psychologists from around the world. The conference's unwieldy title was indicative of still unsettled disciplinary boundaries, and the hope that all these workers might come together in the same mind-brain field. Breakthroughs in neuron theory and pathological anatomy were presented alongside discussions of clinical psychology and lab-based psychological studies. Over six hundred participants from twenty-nine countries attended, including official delegates from forty-six medical, neurological, psychiatric, hypnologic, alcohol, asylum-related, and psychological societies.

On the morning of September 4, 1907, the Psychiatry and Neurology Section convened a panel on "Modern Theories of the Genesis of Hysteria." The speakers were Pierre Janet, Aschaffenburg, Jung, and G. Jelgersma, the president of the congress. The morning began with the Frenchman, Janet, who employed his notions of dissociation, fixed ideas, and the subconscious to characterize hysteria. Next was Aschaffenburg, who presented Breuer and Freud's 1895 work and launched into a critique that condemned both Freudian and Jungian perspectives as modes of thought that were empirically unwarranted and completely misguided in their emphasis on sexuality.

Carl Jung then rose to the podium. He replied at length, great length, for he had not timed his presentation well. After his allotted thirty minutes, Jung was cut off, his defense incomplete. In the lecture as it later appeared in print, Jung argued that the questions about Freud could be reduced to one empirical question: "Do the associative connections asserted by Freud exist or not?" It was a question on which Jung had a good deal of evidence, which he presented. The Zuricher then gave an example of a contemporary Freudian analysis, something that went far beyond Breuer and Freud's earlier theories and was predicated on repressed infantile sexuality creating an unconscious complex that manifested itself through transferences.

During the discussion, protests by the pachyderms began. The first objector was a competitor in the psychotherapy trade, Dumeng Bezzola. A Swiss leader of the abstinence movement, Bezzola had borrowed Freud's method then turned around to advocate his own "psychosynthesis" treatment. Next, a German psychiatrist, Professor Alt, condemned Freud, proclaiming that he would never refer a patient to a sanatorium that had any Freudian doctors. This was greeted by applause and compliments from the chair. Professor Heilbronner of Utrecht then asserted that the association experiments were the only thing of value in Freudian theory, and Jung's proofs of the Freudian unconscious were fraudulent. No doubt to Jung's relief, these condemnations were interrupted by a Zurich doctor, Ludwig Frank, and one from Munich named Otto Gross who defended Jung and Freud. At a later session, Gross even employed Freudian and Jungian theory in his own paper.

While the German and Swiss doctors were embroiled in what Jung called the "great Freud debate," the French were supremely unimpressed. The French doctors focused on French accomplishments in hysteria, without any comment on the disturbances across the Rhine. However, since Pierre Janet had the final word, he was obliged to take stock of the entire proceedings. Janet ventured to say that Breuer and Freud's work was "an interesting contribution to the oeuvre of the French physicians who for fifteen years had been analyzing the mental states of hysterics via hypnotism or automatic writing." The French had shown the value of subconscious ideas, but the Viennese had taken the fact that some hysterics had an "idée fixe" in regard to sexuality and generalized wildly on the subject. In the end, Freud's theory was a derivative work of French psychopathology that had been marred by the all-too-German penchant for grand philosophical speculation. With that, the congress was finished with Freud and Jung.

More than any before, the Amsterdam conference laid out the new lines of opposition that the fledgling Freudians faced. They were not assailed by psycho-anatomists and neuropsychiatrists who dismissed the very notion of a psychological science, psychic causes, or psychic treatments. The French had won over a contingent of European doctors on those points. Freud, Jung, and their colleagues now faced German and Swiss competitors, men like Bezzola, who pushed his own theory of psychic cause and cure, as well as French physicians who dismissed Freud's theories, maintaining that what was good was not his, and what was his was not good. The French were confident that their theories and therapies were the standard for the field and that Freud was but a footnote. A panel at the congress devoted to the different varieties of psychotherapy agreed. Discussion lingered on the advantages and disadvantages of numerous psychotherapies that had emerged from French suggestive treatments, but nowhere did the panelists mention Freudian psychoanalysis.

Along with this serene French disdain were loud protests from German academic psychiatrists, especially the psychologically minded followers of Wundt and Kraepelin. Due to the Zurich association experiments, these psychiatrists saw their own empirical methods used to prop up Freud's work. If Freud's work was nonsense, then their methods would be discredited. They attacked Freud's sexual theory and attributed his empirical findings to suggestion. Furthermore, they chastised their colleagues, Bleuler and Jung, for an egregious loss of judgment. In the months ahead, these opponents repeatedly assaulted Jung. "One thing is certain," Jung wrote Freud, "the cause will never fall asleep again. The worst thing is being killed by silence, but that stage is over and done with." As the siege continued, Freud assured Jung: "Many enemies, much honor."

II.

JUNG WAS RIGHT: there would be no death-by-silence for Freud and his adherents. As medical students and doctors arrived in Zurich for training; as readers sank their teeth into the writings of Freud, Bleuler, Jung, Riklin, Adler, Sadger, Wittels, and Stekel; as psychiatric conference attendees were forced to take note of their controversial theories and methods, the Freudians were on the map of European psychiatry.

But while Freud's profile had been raised, he still had little to offer recruits. When a would-be student from Breslau contacted him, Freud was forced to admit that he could offer no ward of patients, no lab, and no lecture series, very little really. He demurred, telling the man that he couldn't teach him much in a short visit. By contrast, the Burghölzli was an academic teaching hospital that hosted a steady stream of young doctors who, after their stint in Zurich, went home to London, New York, Budapest, Jena, Geneva, and Munich, bringing what Freud jocularly referred to as "a focus of infection" to their hometowns.

These far-flung Freudians began to clamor for an international congress. On November 30, 1907, Jung wrote to Freud saying that an Englishman and two Hungarians had brought up the idea of a congress of Freudian followers. Freud was delighted, but the idea lay dormant as Jung and the Englishman, Ernest Jones, turned their attention to the founding of an international journal dedicated to Freudian studies. A journal could be named for psychopathology and psychoanalysis, Freud suggested, or "more brazenly" just for psychoanalysis. It would be "a matter of life or death for our ideas," he wrote excitedly. Jung responded enthusiastically but countered with Archive for Psychopathology, a title that conspicuously dropped any reference to Freud's creation.

Since an international journal could hardly get by on the small number of Freudians, Jung imagined consolidating a group of journals run by psychotherapists and psychopathologists with common interests. He traveled to Geneva to ask Claparède and Flournoy whether they would be interested in folding their *Archives de Psychologie*, which was losing money, into a common international journal. They agreed, despite worries that the audience in France for a Freudian journal would be small. Jung then approached Morton Prince in America, suggesting he fold his *Journal of Abnormal Psychology* into this new international journal as well. Prince countered with an offer to make it a dual publication, a prospect Jung found troublesome.

With negotiations for a journal at a standstill, Jung turned his energies back to organizing a conference. Soon thereafter, a flyer went out in Jung's name:

From many quarters the followers of Freud's teachings have expressed a desire for an annual meeting which would afford them an opportunity to discuss their practical experiences and to exchange ideas. Since Freud's followers, though few in numbers at present, are scattered all over Europe, it had been suggested that our first meeting should take place immediately after this year's 3rd Congress for Experimental Psychology in Frankfurt (22–25 April), so as to facilitate the attendance of colleagues from Western Europe. The proposed place of meeting is *Salzburg*.

The meeting would be called the First Congress for Freudian Psychology. Jung's chosen name caused some consternation with Ernest Jones, for the title implied an allegiance to a man rather than a subject: "I remember vainly protesting against a wish to call it a Congress for Freudian Psychology, a term which offended my ideas of objectivity in scientific work," Jones later recalled. He proposed naming the congress the "1st International Psycho-analytical Congress," but Jung sent off the flyers unaltered, proudly telling the Professor that Jones had been dismayed by the "pretentious title of the circular."

Freud might have wondered why Jung thought the title pretentious, though in some ways that was obvious. In 1908, most of the Zurichers and many of the Viennese had theoretical commitments that were not fully in line with Freud's. In addition, most of the international members had been trained in Zurich, where they had absorbed Jung's and Bleuler's modifications of Freudian theories. For many who might attend, a conference billed as Freudian misrepresented or even compromised their interests. To alleviate these concerns, Jung stressed the "completely private nature of this project." Later, when one participant asked how he should cite his own presentation, Freud told him that the congress was not to be mentioned in public at all.

About forty people met on April 27, 1908, at the Hotel Bristol in Salzburg to attend a private, international meeting of Freudians. The international character of the meeting turned out to be modest; the meeting resembled a spring outing for the Viennese and the Swiss. There were exactly twelve Freudians who came from outside Austria and Switzerland, but each one of these men would become an indispensable emissary for Freudian thought in places like Budapest, Berlin, New York, London, and Munich.

One of the initiators of the Salzburg meeting was the Welshman Ernest Jones. As a young doctor, he had begun studying French psychopathology, when his friend, the surgeon Wilfred Trotter, mentioned a review in *Brain* that detailed Freud's studies on hysteria. By 1906, Jones had tried his own brand of psychoanalysis in London. After attending the Amsterdam International Congress and presenting a paper on hysteria heavily indebted to Janet, Jones made Jung's acquaintance. He informed the Zuricher that he was quietly trying to practice psychoanalysis and soon was invited to visit the Burghölzli. While in Zurich, Jones attended the second meeting of the Zurich Freudians, and it was there that he and Jung's "Budapest friends" hatched the idea for an international conference. Upon returning to London, Jones was encouraged by Jung to establish a Freudian society there. On April 27, 1908, Jones and Wilfred Trotter arrived in Salzburg. The surgeon left unimpressed, but Ernest Jones returned to London and dedicated himself to the cause for the rest of his life.

Carl Jung's two friends from Budapest also clamored for a meeting where they might build relationships with other Freudians. Phillip Stein was a Hungarian psychiatrist and a prominent crusader against alcohol. At the Tenth International Anti-Alcohol Congress in Budapest in 1905, Stein met his fellow teetotaler Eugen Bleuler. Not long after, he was at the Burghölzli, working alongside Jung in the lab. Stein returned to Budapest as an enthusiast for the Zurich method.

Sándor Ferenczi had marched down the roads of French psychopathology, sexology, and academic psychology. While at the University of Vienna medical school, he attended lectures by Krafft-Ebing and took up the study of French literature on hypnosis and hysteria. He encountered Freud's name, but it made little impression on him. Upon graduating, he returned to Budapest in 1897 and took a job at the Prostitutes' Ward of St. Rókus Hospital. He continued to research hypnotism, automatic writing, and suggestive therapies, and like others of his generation, he chafed at the reductive materialism he found in scientific colleagues. Like Jung and Flournoy, he became interested in the experimental study of spiritism as a manifestation of unconscious mental life. And because of his work with prostitutes, Ferenczi was naturally drawn to sexology and sexual reform. He became the Budapest representative for Magnus Hirschfeld's International Committee for the Defense of Homosexuals, a group that sought to make homosexuality legal.

While Ferenczi's intellectual journey in many ways mirrored Freud's, he remained unimpressed by the Viennese doctor. He even turned down an offer to review Freud's dream book. But in 1906, after reading Jung's association tests, Ferenczi ran out to buy similar equipment and became a convert. After a visit to Freud in Vienna, Jung traveled to Budapest, where he stayed with Stein and met Ferenczi. In 1907, Ferenczi came to the Burghölzli and was in some way analyzed by Jung. Subsequently, Jung wrote a letter of introduction to Freud on behalf of Stein and Ferenczi. The Hungarians met the Professor in February 1908, and Freud was so impressed he immediately suggested Ferenczi present a paper at the fast-approaching Salzburg gathering. Within months, Sándor Ferenczi was lecturing, writing, and generally advocating for Freud and Jung with Budapest's physicians and literati.

The American at the Salzburg congress was Abraham Arden Brill. Born to a Jewish family in Austria in 1884, Brill arrived in America at the age of fifteen. While working odd jobs, he managed to attend City College and then graduate from Columbia medical school. He began to work as a pathologist but tired of this pursuit, and decided to travel to Europe to prepare himself for clinical practice. Brill's first stop was Paris, where he worked under the neurologist Pierre Marie. Demoralized, Brill was encouraged by the American Frederick Peterson to travel to the Burghölzli, which he did in the summer of 1907. Brill arrived at the moment the Zurichers were riding the crest of their diagnostic association studies. For nine months, he held the position of Third Assistant Physician, attended meetings of the Zurich Freud group, learned the association method, and treated patients with these new methods. Brill published an account of his treatment of a patient with dementia praecox, in which he used association tests and Freudian interpretation. He concluded that the patient had repressed sexual complexes, which alone caused his psychosis, a purely Freudian position that neither Bleuler nor Jung were yet willing to consider. Brill also began translating Jung's book on the psychology of dementia praecox into English, a task that would lead to his translating Freud.

In 1908, Brill returned to New York, bringing psychoanalysis with him. Unlike Adolph Meyer, Morton Prince, and others who had only read Freud, Brill could boast of personal instruction in Europe. He tutored two proponents of psychoanalysis in America: James J. Putnam and Smith Ely Jelliffe. Jelliffe found that Brill's experience had deepened his understanding so that his explanations had a way of "assembling and crystallizing a large background of general medical experience. I felt that bottom rock had been reached and we could then build with confidence." With Brill, Putnam, and Jelliffe, psychoanalysis gained powerful advocates in the United States.

The wealthy Russian Max Eitingon, of Buczacz Galizien, came to Zurich in 1906 as a student. That same year, he began a correspondence with Freud, after an attempt to refer a patient to the Professor. In Zurich, Eitingon was nicknamed "Oblomov" after the slothful figure in the Russian novel who promised much and delivered little. But once he settled in Berlin in 1909, Eitingon delivered a great deal indeed and dedicated his fortune to establishing psychoanalysis in that city.

The future leader of psychoanalysis in Berlin would be Karl Abraham. Born in Bremen to Jewish parents, Abraham admired Bleuler's thought and grew determined to work at the Burghölzli. Three years after medical school, Abraham secured a post in Zurich, where he was introduced to Freud's work. Unlike many students who were enamored of Jung, Abraham kept his distance while passionately steeping himself in Freud's ideas. In 1907, Abraham wrote a paper on sexual trauma and dementia praecox and sent it to Freud. Freud congratulated the young man despite his rather embarrassing use of a completely repudiated theory. "I particularly like that you have tackled the sexual side of the problem," he wrote, "the side that hardly anybody is willing to approach."

In October, the German-born Abraham told Freud he was leaving the Burghölzli: "As a Jew in Germany and as a foreigner in Switzerland, I have not been promoted beyond a junior position in seven years," he complained. In the belief that he had no future in hospital-based university psychiatry, Abraham wanted to move to Berlin and set up a private practice. He asked for Freud's support through patient referrals. Freud coyly replied that he was often in need of a doctor to refer patients to in Germany, and if he could call Abraham "my pupil and follower," then he would be all too happy to send him patients. Freud also remarked: "if my reputation in Germany increases, it will certainly be useful to you." Abraham was delighted, and when he arrived in Salzburg, he was already planning a "propaganda" campaign to win Berlin for the Freudian corps.

Another doctor Jung invited to Salzburg was a man who had been coura-geous enough to stand up for Freud in Amsterdam. In 1908, Otto Gross was perhaps the most promising of the young recruits. A child prodigy, he was the son of a famous criminologist, Hans Gross. After medical school in Graz, Austria, Gross took a position as a doctor on a liner sailing to South America in the hopes of studying plant life in Patagonia, and there he became first acquainted with the coca plant. Afterward, Gross returned to Germany and worked in Professor Gudden's lab before he became an as-sistant under Meynert's former student, Gabriel Anton, in 1901. By the time he joined Emil Kraepelin's team in 1906, the twenty-nine-year-old was on the academic fast track and had produced a number of ambitious papers.

Gross took the propositions of the new psychologically oriented psy-chiatrists and used them to break down and classify pathological mental and perceptual states in an original way. Like Bleuler, he turned to the problem of negativism, and in a series of papers, he sensitively explored the various states of patient noncompliance, postulating that these refusals were specifically predicated on an affect of helplessness. In 1904, Gross classified and distinguished the forms of negativism in neurosis and psy-chosis, and associated neurotic negativism with Freud's concept of uncon-scious repression. In 1907, Gross boldly extended Freudian thought into Kraepelin's turf. While the Zurichers sought to remake dementia praecox for Freud, Gross tried to transform Kraepelin's other main diagnosis: manic-depressive insanity. Gross, like the Zurichers, positioned himself against the sexological Freud to advocate for a more general notion of psychic conflict. Gross's academic success and intellectual reach suggested that he would become a leading member of Freud's circle. But as both Freud and Jung knew, all was not as it seemed with Otto Gross.

Another international follower with an impressive pedigree was not at Salzburg. Because of his admiration for Bleuler, Ludwig Binswanger had eagerly joined the Burghölzli staff in 1906. Born in Kreuzlingen, on the Swiss side of Lake Constance, Ludwig was the son of Robert Binswanger, the director of the town clinic where Josef Breuer had hospitalized Bertha Pappenheim. In addition, Binswanger's uncle Otto Binswanger was a highly respected professor of psychiatry in Jena who had written on, among other subjects, hysteria. After studying in Lausanne, Zurich, and Heidelberg, Ludwig chose to return to the Burghölzli, where he employed the diagnos-tic association method under Jung's supervision. In 1907, Binswanger left the Burghölzli to work for his uncle in Jena and attempted psychoanalysis on asylum patients. As the Freudians prepared to meet in Salzburg, Binswanger was getting married, which perhaps precluded his attendance. A few months later, he joined his father at the Bellevue hospital in Kreuzlingen, where he would quickly develop into one of the pivotal figures striving for an integration of academic psychiatry and Freudian thought.

Despite Binswanger's absence, there were still a band of apostles of Freudian psychoanalysis who left Salzburg and established Freudian communities in their own cities. Some established institutions, while others wrote, lectured, cajoled, or propagandized. The common thread among the international Freudians was their debt to the Zurichers. Thanks to the empirical validation of Freud's work achieved in the Burghölzli, these men left Zurich confident that their belief in the repressed unconscious was scientifically defensible. The dynamic unconscious had been proved in Zurich, they believed, and they carried this conviction out into an often uncomprehending world.

IF THE 1907 Amsterdam congress flushed out the kinds of external opposition the Freudians faced, the Salzburg congress of Freudians illuminated potential conflicts within the nascent community. In Salzburg, there were Zurichers and Viennese, gentiles and Jews, academic psychiatrists and private practitioners, physicians and sexual reformers, and this great diversity would make for trouble.

Even before the Salzburg meeting convened, Freud's desire for an international community had changed his relationship to his Viennese champions. Freud worried that the locals would embarrass him in front of the academics from Switzerland. Jung tried to reassure the Professor but secretly confided to Jones that he thought it a pity that Freud was surrounded by a "degenerate and Bohemian crowd." Jones visited the Wednesday group himself and concluded that Jung's view was unfair and possibly anti-Semitic, though he seconded the belief that the Viennese were not impressive. Another Zurich trainee, Karl Abraham, commented on the sad state of affairs among the Viennese, deeming most of them insignificant.

Another factor played a role in Freud's wish to distance himself from Vienna and move closer to Zurich. In 1906, Freud's Viennese followers were almost all Jewish. Whether that was simply the result of the demographics of Viennese liberalism or the result of racism, it was clear to Freud that if psychoanalysis was seen as a Jewish science, it would face stiff resistance. The addition of the Zurichers to the Freudian community undercut the racist label. Bleuler, Jung, Riklin, Jones, Binswanger, Flournoy, Claparède, and Prince were all Christians, whose capacity to move Freud's theories out into the world would not be thwarted by anti-Semitism.

Nevertheless, if Freud simply ceded the movement to the Zurichers, his psychosexual synthesis might be in jeopardy. Bleuler and Jung had both taken public positions against what they saw as the excesses of Freud's sexual theory. Therefore, the Salzburg congress became an opportunity for the Professor to bring the Zurichers around. Jung had asked Freud to deliver a clinical presentation to open the proceedings, an obvious request, because the Freudians who did not live in Vienna had little idea how to conduct an analysis. Freud chose to present the case of a twenty-nine-year-old lawyer with an obsessional neurosis (the case of the "Ratman," to which we will return). For Freud, the case proved that infantile sexual conflicts caused neurosis, and it was Freud's chance to sell libido theory to the Zurichers.

His presentation was like cold water to thirsty men. When Freud suggested he was ready to stop, he was persuaded to continue, and by the end, the Professor had spoken for nearly five hours. There are no records of the presentation, but when Freud discussed the same case at the Wednesday meetings, he talked about free association as a clinical technique and positioned it against the directed technique used in the Zurich laboratory. In analyzing the case itself in Salzburg, Freud described the "alteration of love and hate in respect to the same person, the early separation of the two attitudes usually resulting in the repression of hate." This thinking must have struck Eugen Bleuler as familiar, for a year earlier he had presented this theory to Freud in their correspondence. It's unclear whether Freud acknowledged Bleuler's contribution in Salzburg, but in his early published accounts of the case, Freud never mentions the Swiss doctor.

Carl Jung's presentation was no surprise to the Salzburg audience for it was a reiteration of his published view that the unconscious complex in dementia praecox worked in combination with some "X" toxin, and was therefore different from psychoneurosis. This measured opinion, shared by Bleuler, should have provoked little debate. Who at Salzburg could claim to know more about dementia praecox than Jung and Bleuler? Surprisingly, Dr. Karl Abraham did. Or was it Sigmund Freud speaking through Abraham? Freud had floated the idea that dementia praecox (or rather "paranoia" as Freud preferred to call it) could be caused by a psychosexual conflict in the earliest stage of development. Some of the younger doctors like Brill had pursued this purely psychological view of psychosis. But Jung could not accept it without losing credibility in academic psychiatry. He knew dementia praecox was a terrible, raging disease with symptoms that reeked of massive brain damage. And he had already been exposed to withering attacks for merely hinting that the phenomenology of this illness could be explained psychoanalytically.

Unlike Jung, Karl Abraham had nothing to lose. A year earlier, he had begun to investigate the relationship of sexual factors to dementia praecox. At first, he too argued symptoms alone could be attributable to sexual trauma. When Abraham sent a reprint of the paper to Freud, the Professor replied that while the Zurichers had been busy unearthing similarities between hysteria and dementia praecox, he had been thinking about their psychosexual differences.

Abraham took this lead and in the winter of 1907, visited Freud and had a long discussion on hysteria and dementia praecox. "Freud," Abraham wrote to Max Eitingon, "very much wants me to deal with this subject very soon." Abraham had a case of psychosis that he was treating, a man Freud had also once seen in consultation whom the two referred to as "G," or "the Görlitz patient." "The case of G. seems very important to me for our concept of dementia praecox—that is, its differentiation from neurosis," Abraham wrote to Freud. "G." displayed negativism in his sexual rejection of the world. In January, when Abraham received an invitation from Jung to speak at Salzburg, he prepared a paper entitled "Psycho-sexual Differences between Hysteria and Dementia Praecox."

Jung got wind of Abraham's correspondence with Freud and let the Professor know he disliked his former colleague, whom he considered an untrustworthy, competitive loner. Nonetheless, Karl Abraham presented his paper in Salzburg and made no mention of Freud's suggestions, much less the groundbreaking work of Eugen Bleuler or Carl Jung on dementia praecox. Freud had explicitly invited Abraham to take ownership of his suggestions, perhaps understanding that espousing this line of thought openly might imperil his relationship with Bleuler and Jung. But Jung was furious at the slight. "In Salzburg," he fumed, "I was able to prevent a scandal only by imploring a certain gentleman, who wanted to shed light on the sources of A's lecture, to abandon his plan." When it came time for publication in the new international journal, Jung put his foot down and told Freud the journal would not publish such plagiarized work. In the end, Abraham added footnotes that credited Jung and Bleuler's influence, as well as oral communications by Freud, but then he thumbed his nose at Jung by publishing the piece elsewhere.

In the published account, Abraham laid out a Schopenhauerian vision of a world bathed in libido. Libido created the human affective responses to people and things, and projected sexual transferences everywhere. The absence of such transferences characterized dementia praecox and negativism. This emotional disconnection from the world was the result of a developmental failure to progress through the stage of autoerotism, caused by an abnormal psychosexual constitution. Abraham took time to cast doubt on the "recently discussed toxin theory." He also implied that a modified psychoanalytic method might reinstate developmental progression and cure psychosis.

Although the seeds of a rivalry had been planted, it was not enough to disrupt the Salzburg festivities. Nor was the tension between the medical men and the sexual utopians. When Otto Gross toasted Freud as a moral revolutionary, the Professor dryly replied: "We are doctors, and doctors it is our intention to remain." Still that was only an awkward moment in what otherwise was a grand success. An international Freudian community had come together. Sigmund Freud had taken the opportunity to press his views of psychosexuality on a group that included many who had been trained by the skeptics in Zurich. "I have never seen a more remarkable triumph than that secured by Freud at this congress," Fritz Wittels recalled. "The Swiss, who are cautious folk, had a good many objections to raise; but Freud carried the critics off their feet by the impetus and the clearness of his utterances." Jung himself reported that Bleuler had left the congress nearly convinced of the psychogenic cause of dementia praecox. Freud left the congress certain that, in Wittel's words: "through the growth of the Zurich School, his teachings would be given a footing in the domain of general science."

After Salzburg, Freud conducted correspondences with dispersed followers like Max Eitingon, Ludwig Binswanger, Karl Abraham, Sándor Ferenczi, A. A. Brill, and Ernest Jones. What might have been a pretentious title for the gathering in Salzburg became closer to the truth once it was over. During the Salzburg proceedings, Jung, Bleuler, and Freud met with Jones, Ferenczi, Brill, and Abraham to lay the foundation for an international Freudian journal to be published in German with Bleuler and Freud as codirectors, and Jung as the hands-on editor. As the men struggled to find the right name for the journal, unarticulated questions of identity were forced into the open. While Jung's "Archive for Psychopathology" made no mention of anything Freudian, Freud countered with "Yearbook for Psychosexual and Psychoanalytic Researches." Jung then proposed "Yearbook for Psychoanalysis and Psychopathology," which dropped the reference to psychosexuality but explicitly recognized psychoanalysis. Finally, it was agreed that the journal would be called the *Jahrbuch für psychoanalytische und psychopathologische Forschungen*, or "Yearbook for Psychoanalytic and Psychopathologic Research." This new periodical, linking psychoanalysis and the general study of psychopathology, gave the Freudians a forum where, for the first time, they could develop their ideas among themselves.

III.

THE BEHIND-THE-SCENES CONFLICT between Abraham and Jung signaled that the Salzburg euphoria might not last. Some Freudians accepted only limited aspects of Freudian libido theory, while others fully embraced psychosexuality. These pure adherents included three recruits with ties to Zurich: Jones, Ferenczi, and Abraham. They launched an offensive on the partial Freudians like Bleuler and Jung, baldly seeking to advance themselves with the Professor. Freud was taken aback by some of these ardent warriors and told Jung that Sadger was a "congenital fanatic of orthodoxy," and Jones a zealot who "denies all heredity; to his mind even I am reactionary." A few months later, Freud assured a new acolyte: "we do not demand thoughtless parroting of our views, and that all my followers had reserved judgment until convinced by their own work."

Yet Freud played both sides. While he comforted some, he encouraged the zealots to keep pressure on the Zurichers regarding sexual theory. And Freud did not hesitate to keep score. After a visit to Vienna, Karl Abraham informed Eitingon that Freud had developed different grades for his followers:

> Freud has come to divide his followers into three grades: those in the lowest have understood no more than the Psychopathology of Everyday Life; those in the second the theories of dreams and neuroses; and those in the third follow him into the theory of sexuality and accept his extension of the libido concept.

Like an ingratiating schoolboy, Abraham reported: "he includes me in the third grade, which is very gratifying for me."

The most powerful adherents of Freudian theory, Eugen Bleuler and Carl Jung, now found themselves squeezed by German academics on one side and Freudian flame throwers on the other. Before Salzburg, Freud had scolded Jung for his timidity and restraint regarding libido, while assuring his colleague that a partial acceptance of the theory was fine. Through Abraham, Freud was able indirectly to keep the heat on the Zurichers, who had rapidly accrued the power to make Freud over in their image. Their Freud had won scientific recognition and attracted new followers, so their reticence on the issue of sexuality had the power to rapidly multiply.

This was a delicate game. Freud did not want to press Jung too hard. He admonished Abraham for his attempt to force Jung into open opposition and asked him to remember that Jung was not acting independently as medical director of the Burghölzli. Jung was also the son of a pastor, Freud wrote, implying that a once devout Christian would have to overcome great internal obstacles in order to accept Freud's psychosexual vision. At the same time, Jung was critical for the movement, for he prevented psychoanalysis from the slur that it was a "Jewish national affair."

Freud's Young Turks did not relent. After Salzburg, Jung again defended his views of psychosis to Ernest Jones:

> When Hegel had judged that the reproach that his philosophy didn't match the facts, he said "too bad for the facts" (Tant pis pour les faits!) I say the opposite: if I am not able to set up a clear psycho-

logical theory of Dementia Praecox, still I endeavor to adapt myself
to the impressions which this illness gives us.

The debate about psychosis had become not so much a discussion about
severe mental illness, so much as a test of loyalty to Freud. After all, Jung
did *not* reject libido theory in cases of psychoneurosis. He only backed
away from the claim that it always caused psychosis. Jung wrote Freud
and warned him that his devout followers would not advance the movement. Meanwhile, Abraham continued to complain about Jung's hesitance. Freud worked to tame both parties, writing to Jung: "We must not
quarrel when we are besieging Troy." At the same time, he reprimanded
Abraham: "there are still so few of us that disagreements, based on personal complexes, ought to be excluded amongst us."

The Salzburg congress stimulated great rivalry among the Viennese as
well, who felt humiliated when they were left out of both the conference
planning as well as the meetings that created the new journal. These men,
some of whom were Freud's oldest advocates, were not pleased. After
Salzburg, Freud returned home to find a letter of resignation from Alfred
Adler.

Adler's disaffection was probably caused by the recent consolidation of
the Psychoanalytic Society in Vienna around Freud, as well as Adler's diminished role at the gathering in Salzburg. But the two colleagues also had
theoretical divisions. Only days after Freud urged him to stay in the Society,
Adler presented a paper that firmly broke with libido theory. Adler argued
that every drive came from an organ activity, and the aggressive drive existed in a distinct form from the sexual drive. Freud suggested that Adler
had left behind psychology in order to "make a connection with medicine,"
and argued that what Adler called aggressive drive was none other than
libido. Otto Rank's notes of the meeting only say that a long debate ensued.

An international Freudian community had started to take shape, but
its boundaries were being negotiated. The "second graders," who did not
accept Freudian psychosexuality included Freud's three most prominent
exponents: Bleuler, Jung, and Adler. All of them found themselves on the
defensive after Salzburg. At the same time, serious problems arose for the
third graders, those who fully embraced Freudian psychosexuality without any doubts.

Standing to the right of unidentified man, Otto Gross,
physician, libertine, Freudian, and anarchist.

OTTO GROSS WAS not just a psychiatrist; he was a sexual revolutionary. His brilliant psychiatric writings only hinted at his growing dedication to a utopian vision in which drug use fostered the loss of inhibitions, free sex, and a wholesale rejection of monogamy and the patriarchal order. Gross came to these beliefs in part through Freud, but also from his exposure to the Zurichers, for in 1902 Otto Gross had been institutionalized at the Burghölzli due to his spiraling drug addictions.

When Gross quit Emil Kraepelin's lab in 1906 and became part of Munich's thriving counterculture, he advocated revolution in the name of Freudian psychoanalysis. He spent his days and nights seated in a café, his clothes smeared with cocaine, seeing patients whose fees supplied his drug habit. Like some radicals in Vienna, Gross championed the view that a society free of repression would cure neurosis.

Charismatic and spellbinding, Gross's call for change soon found attentive listeners among artists and intellectuals. He lived his own credo and conducted open sexual relationships with patients and disciples. Gross entangled himself in a love quadrangle that included his wife along with Else and Frieda von Richthofen. "You were born for freedom and only for freedom," he urged Frieda, who would later marry and exert a strong influence on the English writer D. H. Lawrence.

At the 1907 Amsterdam psychiatric congress, Gross met Carl Jung, and told him about his adaptation of Freud's psychoanalytic method. Jung later reported to Freud:

> Dr. Gross tells me that he puts a quick stop to the transference by turning people into sexual immoralists. He says the transference to the analyst and its persistent fixation are mere monogamy symbols and as such symptomatic of repression. The true healthy state for the neurotic is sexual immorality. Hence he associates you with Nietzsche.

At the 1908 Salzburg meeting for Freudians, Otto wrote Frieda that he would announce his life's work, the application of psychoanalysis to culture. "My road is clear, the giant shadow of Freud no longer lies across my path." But Gross's road was far from clear. Cocaine and morphine had worked their ways on him. After Salzburg, the brilliant young Freudian revolutionary was again hospitalized at the Burghölzli.

During his time at the hospital, Gross was not only detoxified. He also underwent analysis with Carl Jung for up to twelve hours a day. During this grueling treatment, the tables would sometimes turn, and Jung found himself analyzed by his patient. One day, Gross jumped the Burghölzli walls and fled back to Munich, where he continued to spread his version of a Freudian utopia among residents of that city. Freud wrote Abraham: "there seems to be a similar centre of infection in Munich, and it seems to have affected the craziest artists and people of that kind." The Professor awaited the consequences of Gross's action with foreboding. To Jung, he wrote: "he is addicted [again] and can only do great harm to our cause."

Little did Freud know that Gross had infected not just crazy artists but also Jung. After Gross absconded from the Burghölzli, Jung confessed that his patient's departure wounded him deeply: "for me this experience is one of the harshest in my life, for in Gross I discovered many aspects of my own nature, so that he often seemed like my twin brother—but for the Dementia praecox." It would not be long before Freud discovered which aspects of Jung's own nature he discovered in Gross. In March 1909, Jung wrote that he had difficulties with a former patient, who demanded that their relationship become sexual. Jung added the cryptic remark that he once had "a totally inadequate idea of my polygamous components." In May, Freud received a letter from a female intern at the Burghölzli asking if she could discuss a matter of great import. Freud asked for an explanation; he got an earful.

Sabina Spielrein had been hospitalized at the Burghölzli from August 1904 to June 1905. Placed under Jung's care, her treatment had run smack into the problem of transference. Sabina fell in love with her doctor. In September 1905, Jung wrote his first letter to Freud seeking counsel on this matter. In it, he described the erotic complex and the unfortunate transference, but he never sent the letter. Sabina told Professor Freud that after he had served as her doctor, Jung had become her friend and then her "poet." According to Spielrein, the married Jung had approached her, preaching the gospel of polygamy, telling her "with strong emotion about Gross, about the great insight he has just received (i.e., about polygamy); he no longer wants to suppress his feeling for me." Jung's letters to Spielrein in the spring and summer of 1908 urge her to embrace a radical freedom and not suffocate due to banal conventions. Spielrein also wrote that Jung saw many other women besides herself. The letter included the disturbing admission that she had contemplated suicide, and recounted a gory episode in which she had attacked Jung with a knife, cut herself, and ended up covered in her own blood.

Jung defended himself to Freud, saying, "during the whole business Gross' notions flitted about a bit too much in my head." While Gross had advocated curing transference through sex, Freud cautioned Jung to do the opposite and overcome what he called "counter-transference." Countertransference was reawakening of the analyst's libidinal past. It was a "permanent problem for us," Freud confided, and needed to be controlled. This critical recognition was paramount, but for years it would not be discussed in public for fear it would nourish those eager to write off Freudian theory as all suggestion, just the sorry delusions of mind readers who projected their own thoughts onto their patients.

"You have been oscillating, as I see, between the extremes of Bleuler and Gross," Freud wrote to Jung. *Between Bleuler and Gross*—these were the poles of the Freudian community. Bleuler accepted psychic determinism but remained uncommitted to psychosexuality. Gross eagerly embraced Freud's sexual theory so emphatically that it had become a call to arms, not a medical treatment. In 1909, Eugen Bleuler and Otto Gross defined the margins of the Freudian field.

As for Freud, he made it clear that while he disapproved and dreaded Gross's actions, he ultimately preferred them to Bleuler's skepticism. The Professor would have opportunity to rethink this preference, for as news of Jung's problems made their way to Freud, he had also heard disturbing news about Ernest Jones, who had abruptly left London for the New World. After Otto Gross jumped the walls of the Burghölzli, Jones met with him in Munich and became enthralled. The prophet of free love became Jones's ideal of a Romantic genius, and his first instructor in the technique of psychoanalysis. Now came reports that Jones had taken up with a harem of women. Jones was no newcomer to sexual scandal, for he had twice been accused of child molestation while in London.

Freud was well aware that sexual scandals could make the Freudians anathema in scientific and medical circles. Unfortunately, there was more trouble brewing among Freud's sexual shock troops. After the Salzburg gathering, Freud returned home to find a public appeal from Karl Kraus, who asked the Professor to rein in his followers. While maintaining respect for Freud, Kraus had been publishing increasingly negative assessments of the Freudians. The targets of Kraus's ire included Isidor Sadger, the pathographer of revered German writers, but his greatest wrath was reserved for his one-time sidekick, Fritz Wittels.

Despite the scorn heaped upon him at the Wednesday meetings, Wittels chose to publish his sexual polemics after the Salzburg meeting. *Sexual Misery* appeared in print in 1909, and it was dedicated to Sigmund Freud. Kraus attacked Wittels' book as a vulgarization of his own ideas. In retaliation Wittels presented a paper to the Wednesday Society on Karl Kraus, portraying him as a deformed neurotic. Not content, Wittels began to plan a full-scale counteroffensive, a roman à clef that ridiculed Kraus— whom he renamed "The Giant Snout"—and mocked his sexual philosophy as nothing more than compensation for his own hideousness.

Learning of Wittels's plan, Kraus asked his lawyer to approach Freud and inform the Professor that if he didn't intervene with Wittels, Freud would be spared no mercy in *Die Fackel*. Freud castigated Wittels, told him he had no right to endanger the movement with such silliness, and added the threat: "you are impossible in my circle if you publish this book." Wittels promptly resigned from the Society.

The Freudians had become a swelling band of renegades, whose ranks swung between the views of Eugen Bleuler and Otto Gross. At their center was one man, getting on in years, who labored to right this strange ship. He tried to keep some from mutiny and, in 1909, first tossed a follower overboard. Bleuler and Gross represented the two wings of the Freudians, each of whom did not share the others' views. One side denied Freud's psychosexual synthesis, the other distorted it. Bleuler would keep the psychic and minimize the sexual; Gross and Wittels celebrated sexual freedom and had little use for psychic mastery. Meanwhile, Freud's followers in Vienna looked upon the Zurichers with resentment, as the Jews and the gentiles warily took stock of each other. And all these internal difficulties occurred as German academic psychiatrists continued their warfare against Freud, Bleuler, Jung, and their supporters. And yet despite these fissures in their community, the number of Freudians began to grow at an explosive rate.

IV.

BETWEEN 1908 AND 1911, the Freudians consolidated into an international movement. For the leaders of the movement, Sigmund Freud and Carl Jung, it was a time to tend to possible allies and beat back competitors and enemies. But as early as 1909, it was already clear that allies were arriving in greater numbers than adversaries.

In Vienna, the increase in Freud's prestige and the growth of the number of his followers was obvious. In December 1908, Freud was invited to receive an honorary degree at Clark University in Massachusetts. The invitation came from the psychologist G. Stanley Hall, who had once been a student of Wilhelm Wundt's. He informed Freud that Janet had given a similar course at Clark and had "a profound influence in turning attention of our leading and especially our younger students of abnormal psychology from an exclusively somatic and neurological basis to a more psychological basis." Hall suggested that the ground had been prepared for Freud, and that a visit by him might usher in a new day for psychology in the United States.

Freud turned Hall down. Writing to Jung, he complained of economic and scheduling difficulties, but he also confessed to a hesitation when it came to Americans: "I also think that once they discover the sexual core of our psychological theories they will drop us." But after Hall made some changes in his proposal, Freud reversed himself. After another honoree at the gala event canceled, Carl Jung was also invited to receive an honorary degree and be one of the twenty-nine distinguished lecturers to speak at the festivities.

In September 1909, Freud and Jung arrived in Worcester, Massachusetts, accompanied by Ferenczi, whom Freud had invited along, and A. A. Brill, their New York ally. After meeting with Hall and William James, Freud delivered a series of five lectures that lucidly tracked his theoretical development. He countered Janet's degeneration theory with his own theory of hysteria, described his topographic theory of consciousness and unconsciousness, and in the last two lectures discussed sexuality from an individual and social point of view. Jung delivered two lectures on the word-association tests and a third on the psychosexuality of a normal four-year-old girl, who was in fact his daughter, Emma. Jung's first two lectures were intended to demonstrate that the unconscious complex was scientifically verified, his third that infantile sexuality was a reality.

Freud discovered after their trip that he had won over Hall, as well as the Harvard University neurologist James Jackson Putnam. Prior to Freud's visit, Hall commented, "many, if not most psychopathologists have leaned upon the stock psychologists like Wundt, your interpretations reverse the situation and make us normal psychologists look to this work in the abnormal and borderline field for our chief light." Freud was thrilled.

When he arrived home, Freud found five letters from Switzerland either referring patients or asking for information, and a large number of foreign requests, eleven of which Freud answered his first day back. Freud received inquiries from Russians, Italians, and Americans. His books were suddenly selling. Writing to Brill in New York, Freud noted that "according to the publishers, the demand for all of the books is extraordinary." Between 1909 and 1910, Freud published second editions of *Studies on Hysteria*, *The Interpretation of Dreams*, and the *Three Essays on Sexuality*.

In Zurich, the sweet smell of success was also in the air. Jung's letters to Freud were filled with news of bountiful recruits and new colonies of supporters. He had received visitors like the Swiss-American Adolf Meyer, and new students such as Otto Juliusburger of Berlin and Wilhelm Strohmayer of Tübingen. New Zurich followers stepped forward like Alphonse Maeder, Johann Jakob Honegger, and the pastor Oskar Pfister. Dr. Leonhard Seif of Munich had approached Jung to study with him, and he was joined by myriad Moscovites and Budapesters. And what could top the news that a Freudian advocate, Jaroslaw Marcinowski, had received a referral for analysis from Sir Richard Semon, physician to King Edward VII of England?!

Jung brimmed with confidence. In January 1909, he wrote: "little by little your truth is percolating through to the public." Ten months later, he was jubilant: "One must let the forest fire rage, there's no stopping it now." This sense of inevitable victory came as problems between Jung and Bleuler also began to boil. In the fall of 1908, Jung resigned his post as head physician at the Burghölzli but retained his leadership of the lab. Freud understood the gesture to signify a deeper commitment by Jung to the Freudian cause. Jung was severing himself from the imperatives of academic psychiatry and "going unreservedly with us." Jung then left the Burghölzli altogether and opened a private practice. He did not seem to suffer for this change and reported that four assistants had joined him, along with wealthy patients like the American heiress Edith Rockefeller McCormick. While Jung remained prominent in Swiss psychiatric circles, he was now more clearly Freudian. After managing to win over the Swiss psychiatric association, Jung crowed: "Your (that is our) cause is winning all along the line. . . . In fact," he continued, "we're on top of the world."

During this period of quick expansion, Jung kept watch over his enemies in academic psychiatry, for they had hardly disappeared. Jung took some comfort in the fact that Kraepelin (the "Great Pope of Psychiatry" as Freud called him) had moderated his views toward Freud in his last textbook. At a conference of Swiss psychiatrists in Zurich, Jung sensed his enemies were weak and dared not show their faces.

Growing acceptance was also apparent in that former bastion of resistance, Germany, where Freud wrote "the affirming voices definitely have the upper hand." Alongside Vienna and Zurich, Berlin became a third hub of the Freudian movement. Karl Abraham had built his practice aided by the fact that his cousin was married to the famed neurologist Hermann Oppenheim, who sent Abraham patients on the condition that the young man not use Freudian methods. Still, Abraham persevered and told Freud about a propaganda campaign conducted at the home of the professor of psychiatry at the Berlin Charité, where Abraham had taken on the men in one room, while his wife argued for Freud with the women in another.

Abraham found willing allies among Berlin sexologists like Iwan Bloch, Magnus Hirschfeld, and Albert Moll. In August 1908, he started a Berlin Psychoanalytic Society with these men and two others—Heinrich Koerber, the president of the local Monist League, and Otto Juliusburger, a doctor who had studied with Jung and now worked in a private clinic. Abraham presented his psychoanalytic work to the Berlin Psychological Society and the Berlin Society for Psychiatry and Nervous Illnesses, where he was surprised at how well his talks were received. Members expressed interest in his work and relief at not having to sit through another anatomical presentation. Freud's ideas, Abraham confirmed, were spreading through lectures, avid students, and grateful patients, including a headmaster who had presented his school doctor with an exam, to make sure he actually understood Freudian sexual theory.

With Moll, Hirschfeld, and Bloch on board, the Berlin group was heavily weighted with men who studied the biology of libido without a particular interest in psychology or psychotherapy. Their emphasis was the opposite of those in Zurich who were fascinated with unconscious states, but suspicious of sexual theory. Jung had a low opinion of Magnus Hirschfeld, who had become one of the world's leading sexologists, and he chastised Freud for thinking about publishing an essay in Hirschfeld's journal. The Professor, however, reiterated his hope that sexologists would be part of the Freudian movement, and wrote to Jung that Hirschfeld had visited him and left a favorable impression: "He is moving close to us and from now on will take our ideas into account as much as possible."

The Freudians were not the only ones who gained adherents. Psychotherapists of all varieties were established in the first years of the twentieth century. In 1909, an International Society for Medical Psychology and Psychotherapy convened for the first time, in of all places, Salzburg. The society's goal was to be a kind of clearinghouse for psychotherapies, including French methods, the Swiss doctor Paul Dubois's persuasion therapy, Oskar Vogt's rational therapy, and the Swiss physician Ludwig Frank's psychosynthesis. The group's inaugural meeting took place while Jung, Ferenczi, and Freud were all in America, and so it was not until his return to Zurich that Carl Jung received Forel's invitation to join this society. Jung declined, arguing that he and his fellow Freudians would not be welcome. After some debate, Freud, Bleuler, and the reluctant Jung agreed to join, encouraged by the idea that they might sway members to join their ranks.

As the international Freudian movement flourished and competing psychotherapeutic schools also took root, the question that had haunted the Wednesday Psychological Society took on renewed urgency. What was required to become part of the Freudian school? How tightly defined would that school be? Was it to be a confederation that made room for multiple psychotherapeutic, sexological, and psychopathological theories loosely associated with Freud? Or would it outline more stringent requirements?

Jung and Freud believed the time had come for the Freudians to form their own international organization dedicated to the definition and defense of their cause. As preparations for a second conference of Freudians began, Freud asked his Budapest colleague, Sándor Ferenczi, to present a proposal at that gathering calling for an international psychoanalytic organization. The specific character of this organization would provoke an outcry within Freudian ranks and, in the end, prove fateful.

SEVEN

Integration/Disintegration

The 1911 Weimar Congress of the International Psychoanalytical Association, the first after the critical 1910 Nuremberg Congress.

I.

IN VIENNA, ZURICH, Berlin, and beyond, the degree of adherence to Freudian theory varied enormously. Freudians freely disputed, disagreed, created their own amalgams, and still held a place at Freud's table. How could it have been otherwise? Freud's sense of isolation, whether real or in part imagined, encouraged him to embrace almost any potential follower. In its early years, the Wednesday Society did not turn away any applicants for membership. Students, journalists, book publishers, hydrotherapists; Freud welcomed more voices, even when some broke off to sing their own tunes.

Soon this would change. Around 1910, Freud and some close advisors began to reconsider the loose boundaries of their community. Freud had redefined and incorporated aspects of psychotherapeutics, hypnosis, physiologic psychology, and sexology, not only to redirect and reform those fields, but also to merge them into a single, new discipline. Psychosexuality was to be the linchpin of Freud's new psychology and psychopathology; it could neatly define and unify the Freudians. But such self-definition was complicated by the fact that many of Freud's strongest supporters such as Eugen Bleuler, Carl Jung, and Alfred Adler were Freudians who refused to fully embrace psychosexuality.

By 1910, however, Freud believed he was no longer desperate for any and all support. When they gathered that year, the Freudians were growing steadily in number. Their academic adversaries were in retreat. The reach of Freudian theory had spread from the neuroses to the psychoses, from marginal private practitioners to prestigious academics, from medical circles to literary and political ones, from Vienna to other major cities in Europe and America. Freud had begun to worry less about attracting new followers and more about controlling what would pass under his name. Along with Jung and Ferenczi, he hatched a plan to consolidate the movement.

On March 30, 1910, more then fifty people convened in Nuremberg to attend the second private meeting of Freudians. Freud had planned to feature three interwoven talks at the congress. The Professor would discuss the future of the field; Carl Jung would speak on the development of psychoanalysis in America; and Sándor Ferenczi would propose the founding of an international psychoanalytic group that would safeguard the future through "organization and propaganda."

At eight-thirty in the morning, Sigmund Freud opened the congress. Before his ragtag array of followers, he called for standardization in the field. He argued that psychoanalysis, as a medical therapy, should aspire to a kind of uniformity. Like other medical practitioners, psychoanalysts ought to set guidelines for therapy to aid the training of newcomers and protect patients against gross malpractice.

There was a problem with such invariance, but Freud did not dwell on it. The desire to maintain sameness from practitioner to practitioner made sense for clinical work, but it ran counter to the requirements of scientific research, which called for freedom of inquiry, experimentation, and innovation. Since the clinical situation was in fact the only research forum for Freudians (save for a few Burghölzli researchers), a demand for clinical standardization could imperil future advances. Adding to the problem, Freudian technique remained very much a work in progress. In his Nuremberg address, Freud reviewed the development of his technique as it moved from the cathartic method, to the patient freely associating while the analyst cajoled and urged, to his latest vision, in which the analyst no longer pushed the patient forward but presented "anticipatory ideas" that helped the patient find repressed thoughts inside himself.

This new method hardly seemed like the final word on psychoanalytic technique; in fact, it was deeply troublesome. Freud had foreseen the objection his new technique would face: many would worry those anticipatory ideas would act like suggestions. If the analyst offered up such ideas, wasn't psychoanalysis a suggestive therapy in the French tradition? Freud said the answer to this and other questions would be forthcoming in a textbook on the technique of psychoanalysis.

Freud had been promising his followers this primer for two years. After the Salzburg Congress, he excitedly informed Ferenczi, Brill, Jung, and Abraham of a book project entitled *A General Exposition of the Psychoanalytic Method*. Freud knew his followers were in dire need of instruction, but uncharacteristically, he had encountered insurmountable obstacles writing this work. As Freud wrestled with his textbook, he told Jung that their Hungarian ally, Ferenczi, had written something that closely approximated a section of *A General Exposition*. Ferenczi's work mirrored Freud's thinking for good reason: the Budapester had written it with Freud looking over his shoulder as they vacationed together.

For the audience in Nuremberg, Ferenczi's 1909 paper, "Introjection and Transference," offered the most comprehensive account of analytic technique to date. The paper began with a broadside against suggestive therapeutics that included a devastating critique of the theory of suggestion itself. Using clinical examples, Ferenczi illustrated that suggestion was, at root, based on the Freudian process of unconscious transference and had its origin in repressed parental complexes. Patients accepted suggestions because, unconsciously, their doctor had assumed the role of all-knowing parent. This explication made it possible for the Freudians to distinguish themselves from their French competitors, and claim they could explain and manage suggestion better than suggestive therapists.

Building on this critique, Ferenczi emphasized the centrality of transference theory, insisting this phenomenon was typical of neurosis and that
psychoanalytic technique should focus on it. In his Nuremberg speech,
Freud also stressed the unraveling of silent projections and displacements
in the transference. Transference was by far the most powerful force at
the analyst's disposal, he advised, a point he promised to clarify in his
never-to-appear textbook. Going further, Freud urged his listeners to
heed "counter-transference" as well. Uttered in public for the first time at
the Nuremberg Congress, the clumsy technical term contained dynamite
for the Freudian cause. Libidinal transferences coming from the analyst
had to be surmounted, Freud warned. He had witnessed how an analyst's
passions could be set on fire; this doctrine was introduced to douse the
flames. Unanalyzed countertransferences could lead to therapeutic failure and encourage others to chase after that pied piper of carnality, Otto
Gross.

Countertransference theory was not just a safeguard against sexual
malfeasance. It also reopened a philosophical can of worms that never
would fully shut. If transference was a core neurotic process that distorted
perception of others, countertransference implicitly acknowledged that
the doctor might be subject to his own illusions. If both doctor and patient
were half-blinded by their own transferences, how could one ever claim
to know the other? Rather than wade into these currents, Freud tersely
limited his Nuremberg discussion by insisting that countertransferences
were the result of the patient's influence. For now, countertransferences
were only provoked responses to the patient's transference.

But in his ongoing work, Freud did not duck the larger implications
raised by the presence of the analyst's unconscious. Countertransference,
he believed, could affect not just clinical treatments, but also scientific
work as well. It could cause misperception and mar theoretical construc-
tions. If Bleuler's refusal to accept libido theory was due to his own resistance, and if Adler's need to see inferiority everywhere was a projection,
how could psychoanalytic knowledge be stabilized?

One answer was that the analyst must be analyzed himself. In this way,
the problem of countertransference led Freud to consider standardized
training for analysts, which he next took up in Nuremberg. The analyst
must conduct a self-analysis; if he could not, Freud declared, such a
person should not be an analyst. A psychoanalyst will only go as far with

his patients as he had gone with himself. The doctor's complexes and resistances must be conquered in order to contain the irrational distortions of his inner life.

Another way to clean up the epistemological morass made by the analyst's subjectivity was to establish invariant knowledge regarding the cause of neurosis. If a universal theory existed, even the most inexperienced and poorly analyzed practitioner could apply it. He would not need to uncover subtle unconscious forces through close individual observation, inference, and interpretation, but would know of their existence in advance. In Nuremberg, Freud presented his hope that a general theory of neurosis (he says nothing of psychosis) might be forged.

In Vienna, Eduard Hitschmann had been working to standardize a Freudian theory of neurosis. Nearly a year before the Nuremberg Congress, at a Vienna Society's meeting, he suggested it was high time for the Freudian School to produce such a textbook. Hitschmann correctly noted that international acceptance had produced papers from around Europe and America that were supposedly Freudian, but on inspection turned out to be based on an outdated 1895 version of Freud. A new work of propaganda, as Hitschmann called it, should be written for physicians. Freud thought this was a marvelous idea and confessed that his publisher had been begging for such a book. But the author would have to resist the temptation to present Freudian theory as closed and finished, Freud warned. He must draw attention to the limits of present knowledge, including the still dark origins of repression.

Six months after the Nuremberg proceedings, Hitschmann published *Freud's Theories of the Neuroses*. In his introduction, he explained that resistance to Freud's theories was spawned by disgust for sexuality. He stressed that the rules of psychoanalytic technique were not yet finalized, but he explained the most up-to-date theories of resistance, interpretation, and transference. Hitschmann avoided the controversy over the psychoses but made it clear that the Freudian method had not yet been modified for these illnesses. Freud's response was enthusiastic, and soon Hitschmann's book became a standard for young German and Austrian doctors.

In Nuremberg, Freud knew this primer was in the works and he knew it would necessarily be incomplete, for he himself had only recently started to imagine a unified theory of neurosis. Citing advances that might lead to such a theory, Freud mentioned Jung's complex, which he acknowledged as an invariant mental process, and Stekel's idea that the meaning of symbols appearing in dreams, myths, and folklore were constant and universal. Initially, Freud had interpreted dream symbols individually, but eventually he came to accept the argument that these signs must have been highly conserved over the millennia. If the meaning of these universal symbols could be established, psychoanalysis would achieve another piece of bedrock.

Those advances led Freud to hope for "a succinct formula of the factors regularly concerned in constructing the various forms of neurosis." While the old "choice of neurosis" problem had long remained a mystery, Freud now offered the inklings of a solution. During his self-analysis in 1897, he had been seized by the idea that the Greek myth of Oedipus Rex—the foundling fated to kill his father and marry his mother—might hold a universal truth about childhood development; three years later he had written about the Oedipal drama as a part of normal psychology. In the Dora case, Freud had elaborated on Oedipal themes while still relying on physiological traumas like masturbation to account for the girl's neurosis. As Freud dropped trauma theory, he increasingly turned to the symbolic meaning of the Oedipus legend to guide his general theory of neurosis. At the Vienna meetings and in letters to followers, Freud spoke more about a child's Oedipal struggle, and the way love and hate for one's parents might determine later neurosis.

In 1909, Freud published two case histories that relied on Oedipal explanations. "Little Hans" developed fears of bodily mutilation as a result of his jealous love for his mother: "Hans really was a little Oedipus who wanted to have his father 'out of the way,' to get rid of him, so that he might be alone with his beautiful mother and sleep with her." The "Ratman" (the Russian patient that Freud discussed in Salzburg) suffered from oscillations between love and hate for the Father. In his 1909 Clark University lectures, Freud went further and announced that the Oedipal narrative was a core complex in psychosexual development.

In Nuremberg, when Freud hazarded a guess that the defining aspect of neurosis would be the "Father Complex," he was referring his followers back to these other intimations. After this congress, Freud sat down to write the first of a series of papers on the psychology of love life and coined the term: the Oedipus complex. Melding Freudian theories of psychosexual conflict with Jung's complex theory, the Viennese doctor argued that among all the array of complexes cataloged by Zurich experimenters, this *one* caused psychoneurosis. If true, a well-informed initiate, no matter how blinded by his own neurosis, would know the source of his patient's neurosis.

After calling for standard methods, training, and theory, Freud wound up his Nuremberg address by turning to impediments that might thwart psychoanalysts in the future. Since it destroyed sicknesses that society in part created, psychoanalysis would be seen as the public's enemy. "Because we destroy illusions, we are accused of endangering ideals," Freud warned. Psychoanalysts, it seemed, were fated to be gadflies and outcasts, carrying a message no one wanted to hear. But Freud asked his listeners to imagine what would happen if, instead of opprobrium, social acceptance came the analysts' way.

For Freud that future was dizzying. If societies came to accept the revelations of psychoanalysis, the value of the neurotic symptom would disappear. Imagine a picnic in which the ladies who needed to urinate excused themselves by saying they were going to "pick flowers." What would happen, Freud asked his audience, if everyone had been informed that "I am going to pick flowers" meant the ladies were going to pee? Of course, no one would say such a ridiculous thing again. It would be without value. Instead, "the ladies will have to admit to their natural needs without shame." While acknowledging that this might sound utopian, Freud encouraged his followers to consider the radical power they held in their hands. By treating their patients, analysts were working for science, giving patients the best medical remedy available, and reforming a hypocritical society. If radicals like Wittels and Gross envisioned a world without repression, Freud envisioned a world where repression would be of little value, and where conscious choice would hold sway. It was a liberal's dream: increasing rational control over unreason and furthering individual emancipation. And it called for nothing more than the practice of psychoanalysis: it would be a revolution from the couch.

Standing before the Freudians, Carl Jung addressed the future in America, and he picked up on Freud's final theme. A culture of intense repression would engender massive resistance to their field in places such as America, Jung warned. In America, whites live so close to the wild

Negro that they have had to redouble their efforts at repression. Jung's racist belief that African-Americans were primitive and unrepressed was not uncommon among his listeners. But his recourse to racial psychology would not help quell whispers among the Jews, who suspected him of being anti-Semitic too.

Freud and Jung touched on how psychoanalysis would adapt in communities where it was accepted or reviled. These themes dovetailed with the most important presentation, delivered at the end of the first day by Sándor Ferenczi. A favorite of Freud, Ferenczi had won broad admiration with his theoretical work. He also held a more dubious distinction. Unlike leading Freudians in Berlin, Zurich, Munich, and New York, Ferenczi had been unable to bring together a group of Freudians in Budapest. He had produced a good deal of propaganda, but few serious followers. Ferenczi even managed to fall out with one of the few other Freudians in Budapest, Phillip Stein. Ferenczi had criticized Stein for maintaining too close an alliance with academic psychopathology. Under Freud's questioning, Ferenczi admitted that these attacks were due to his own competitive nature. He also complained that colleagues slandered him and left him scientifically isolated. Freud's quick analysis was that Ferenczi had a "Brother-Complex," which made him so rivalrous that he destroyed potential alliances. The Professor observed that Ferenczi made little progress with male recruits but managed to sustain close relationships only with intelligent women.

Freud's choice of the impolitic Ferenczi to deliver this crucial lecture was telling. The firebrand was slated to offer a tough-minded, new vision of a Freudian community, determined to protect itself. He had lobbied for a defense organization, telling Freud that someone at the congress should speak about "the most expedient method of *propaganda* for our psychological movement." After mulling it over, Freud wrote back: "What do you think of a tighter organization with a small fee?" Ferenczi agreed and added that the group should maintain strict guidelines, like the Vienna Society, so as to exclude "undesirable elements." Of course, the only strict guideline in the freewheeling Vienna Society was that the Professor approve each candidate.

Freud and Ferenczi drew up plans for an organization and jointly prepared the Hungarian's Nuremberg speech. After discussing their plans, Ferenczi wrote the Professor: "I do not think that the psa. worldview leads to democratic egalitarianism; the intellectual elite of humanity should maintain its hegemony." As liberals, these men looked for reason and virtue in an educated elite; neither placed much faith in the democratic whims of the masses. Their proposals were meant to inspire the group to grant ruling power to an enlightened few.

"Psychoanalysis, looked at objectively, is a pure science, the object of which is to fill in the gaps in our knowledge of the laws that determine mental events," Ferenczi began. By pursuing its aims, this science disrupted the family and the church, thereby eliciting revulsion and attack. Dispassionate study in such an atmosphere was impossible. "(W)e were thus, very much against our will, involved in a war, and it is well known that in war the muses are silent." To defend the cause, Ferenczi called for an "Internationale Psychoanalytische Vereinigung," or "International Psychoanalytical Association" (hereafter the "I.P.A.").

Who exactly were the enemies of the I.P.A.? The academic psychiatrists who wanted to discredit psychoanalysis were on the horizon, but Ferenczi, like Freud and Jung, believed that the war had already been won. Ferenczi was more concerned about losing the peace to internal enemies. The opponents of psychoanalysis drew strength from what Ferenczi ominously called "our irregularity." This new international organization would constrain members whose selfish tendencies would be kept in check by "mutual control." The group would maximize personal liberty under a leader whose authority was based solely on merit, and whose pronouncements were subject to criticism. But to be part of this group, members would have to alienate some intellectual freedom for the greater good of the cause.

Ferenczi gave thinly veiled examples of who the troublemakers might be. The organization would put the brakes on the clinician who was unsuited to theoretical questions, but nonetheless concocted scientific theories. The organization would constrain the member who mistakenly generalized from his own subjective experience and tried to rewrite this young science. Names were not named, but Freud had openly complained that Stekel had no head for theory, and that Adler had grossly inflated his own ideas in an attempt to reshape all of psychoanalysis. Ferenczi went on to say that such an organization would not eliminate these types, merely hold them in check. The organization would also protect psychoanalysis from unqualified members who spouted nonsense. As an example, Ferenczi denounced psychosynthesis, while the founder of that theory, Ludwig Frank, sat in the crowd. By repudiating such idiocy, psychoanalysis would maintain a more unified front and gain respect in the outside world.

The I.P.A. would be run by a president who operated from a central office, which would tightly control membership and ensure "that Freud's own psychoanalytic methods were being used, not methods cooked up for the practitioner's own purposes." The organization would limit membership and not let in those who were not clearly committed to the same principles. "Profitable work is possible only when agreement prevails on fundamental matters," Ferenczi reasoned. The president would also be in charge of a great deal, including congresses, editing the *Jahrbuch*, and a monthly newsletter. Finally, Ferenczi proposed that the president be elected for a life term. Carl Jung was nominated for the job.

Carl Jung had planned the first Freudian congress in Salzburg, and when he arrived at that gathering, he could have reasonably expected to be greeted as the scientific champion of the cause. Instead, Jung found himself outflanked by Karl Abraham and forced to defend his loyalty to Freud. Since then, Jung had left his post at the Burghölzli and thrown himself in with the Freudians. This was his reward. As head of the I.P.A., entrusted with extraordinary power to decide who was and who wasn't a Freudian, it seemed Jung would never be outflanked again. For life.

Appointing an enlightened leader to a life term was not strange for late-nineteenth-century liberals like Freud. The Viennese clung to a fond memory of Franz Josef, a leader not subject to the illiberal whims of the rabble. For older liberal Zurichers, their council had run on lifetime appointments. But many younger members of Freud's community were democrats committed to a politics of equality. A lifetime appointment appalled them.

As Jung's coronation began, there was an outcry. The Viennese were not going to passively watch the center of the movement move to Zurich for the duration of Carl Jung's life. Wilhelm Stekel rose to denounce the proposal and insist the new science would be destroyed if it was not completely free. Adler seconded Stekel's objections. The motion was put up for a vote. By Stekel's count, all the Viennese and German analysts voted against the I.P.A. statutes. Freud asked for a postponement of the final vote until the next day. Immediately afterward, the Viennese convened a secret meeting at the Nuremberg Grand Hotel. About twenty analysts gathered and listened to Stekel argue that the Viennese were about to be permanently marginalized. Psychoanalysis must be an independent science, and Vienna must be independent of the whims of Zurich!

Freud burst into the room. He pleaded with the Viennese to agree to the I.P.A., predicting hard times and bitter opposition from those in the scientific establishment. He grabbed his coat and cried: "They begrudge me the coat I am wearing." Freud insisted the Viennese cede his brainchild to Zurich, for as Jews they had no hope of winning acceptance in the world of science. The Viennese refused to cave in. The next day the proposal for the I.P.A. passed, but only after key modifications had been made. Carl Jung was still elected president, but his term would be two years.

THE NUREMBERG PROPOSALS were a watershed event for the Freudians. By moving the center of power to Zurich, Freud hoped to undermine anti-Semitic responses to his science. He created the I.P.A. in part because he had come to believe that psychoanalysis was too closely identified with himself. A recent attack by the German academic Alfred Hoche suggested Freud had founded a religion, where he was worshipped and viewed as infallible, a cutting reference to the Catholic doctrine of papal infallibility. Because he had sat through scores of meetings where his Viennese compatriots had wandered off in all theoretical directions, Freud felt this was absurd and unfair. He concluded that "the time has come to withdraw personally and show our opponents how foolish it is to believe that psychoanalysis rests upon my personal experience and will pass with me."

But the idea of building a movement bigger than Freud conflicted with another imperative: the father of the field felt it was critical to control his followers. "I see two imminent dangers," Freud wrote to Bleuler, "firstly, that some of the followers would react unwisely to the personal attacks and that others would present to the general public some of their constructs as psychoanalysis, which could bring discredit to this expression." A central office was needed to conduct public relations and adjudicate what was and what was not psychoanalysis. But how was the group to develop a psychoanalysis not centered on Freud, while its central office was policing ideas that differed from Freud's?

For Ferenczi, the desire for a broader psychoanalytic movement simply had to be sacrificed. In times of war, as he warned, the muses must be silent. An open, cacophonous psychoanalytic community was impossible. and irresponsible, given the attacks the movement faced. They needed to hold fast, commit themselves firmly to central tenets, enforce commitment among wayward followers, and be prepared to fight.

Freud concurred. He wanted an organization with a "foreign policy" directed at hostile academic psychiatrists and competing psychotherapists. He wanted propaganda to support the cause, not unlike the campaigns launched by sexual reform and abstinence advocacy groups. Members of those communities were expected to be devotees not skeptics. Nuremberg initiated an attempt to narrow the range of ideas that could be called Freudian. Freud and his closest allies sought to mandate a number of central beliefs that would force partial adherents to fall into line. In a letter to Karl Abraham, the Professor wrote: "defence is required not only against enemies, but also against rash fellow-workers."

In the end, despite great turmoil, Freud seemed pleased with the results of the Nuremberg meeting. Jung was pleased too. But almost immediately, it became apparent that the Nuremberg proposals had drastic, unforeseen consequences.

II.

EUGEN BLEULER was in bed with appendicitis during the Nuremberg Congress, and he jovially wrote Freud to assure him that his illness was not an unconscious resistance. The joke was no longer amusing a month later, when Freud heard the startling news that Bleuler had refused to join the I.P.A. as the head of the Swiss group. The Zurich professor believed the association's rules were too biased and too exclusive, and he was loath to join up with everyone in the group. Bleuler suggested that the Zurich Freudians meet in joint sessions with his Burghölzli staff, which enraged Jung, who privately vowed to one day kick out the whole lot.

Freud attributed the conflict in Zurich to jealous attacks on Jung, but he also worried that with the Nuremberg proposals, "we forged ahead too fast, perhaps we should have waited for things to ripen a bit." At least in Vienna, however, Freud assured Jung, the Nuremberg rules had been quite helpful. It was a view he would revise.

In June, after much delay, Jung announced the founding of the Zurich Society. Bleuler had not joined. Since the statutes mandated that no president or secretary of the I.P.A. could head a local branch, Ludwig Binswanger was elected president. To Jung's despair, Binswanger promptly announced he would only accept the title if nonmembers were allowed to participate alongside members. Jung demanded a vote on Binswanger's proposal: Binswanger won. Freud immediately wrote Jung and insisted this arrangement was impossible, and he chastised Jung for allowing such a "stupid" situation to develop.

The proposals for the I.P.A. had sent the Zurich Freudians into disarray and opened up schisms among Jung, Bleuler, and Binswanger. Freud and Jung both conceded their plans for consolidation had proceeded too swiftly. In September 1910, Freud wrote Bleuler, asking the prodigal to return. Freud admitted that he had expected trouble from the Viennese, but not "from the other side and least from you who would have become the leader of the Zurich group." "If it is your intention," Freud admonished Bleuler, "to preserve the bridge of communication between psychoanalysis and academic psychiatry, then your absence from the society" would do just the opposite.

Bleuler reminded Freud that he had been in favor of forming an organization in Salzburg, but that the statutes of the new organization offended him. They reeked of an exclusivity, which he himself did not share. In discussions with Jung, Bleuler had confirmed that this exclusivity was intentional: the organization was built to have no competitors like Ludwig Frank and no academic critics like Kraepelin's associate Max Isserlin.

Ludwig Frank had openly complained to Bleuler that he had been snubbed by Freud at the congress. The case of Isserlin, however, was filled with intrigue. Before Nuremberg, Jung wrote Freud to report that this professor, one of Jung's most severe academic critics, wanted to come to the 1910 meeting. Bleuler later discovered that Jung (or one of his assistants) had sent Isserlin an invitation as a hoax. Isserlin accepted. Without hinting of the prank, Jung wrote Freud that Isserlin was a "bastard" and a slanderer, and that he did not want such filth around. Blindly,

Freud assented and refused the man admission. After Nuremberg, Emil Kraepelin visited Zurich and lambasted Bleuler for the thoughtless treatment of his associate. Bleuler later told Freud that he could easily understand not inviting Isserlin to the congress, but once the man had been invited, it was exceedingly rude to bar him from attending.

Eugen Bleuler did not want to join an I.P.A. that admitted no critics and no competitors. If "one wants a scientific discussion, and if one wants to appear as a scientific association to the outside world, then one can not from the beginning render an opposition impossible," he reasoned. Freud reiterated that his motives were practical and had "nothing to do with stifling opinions within our group." Furthermore, he saw nothing wrong with a psychoanalytic organization that only accepted those members who themselves accepted psychoanalysis.

But what was a psychoanalyst in 1910? Was it possible to be a psychoanalyst in the broad, generic sense that Bleuler would have accepted and still be a member of the I.P.A.? This organization was meant to force more eccentric Freudians into a newly forged center, whether they liked it or not. On October 18 and 19, 1910, Bleuler wrote Freud a long, eight-page letter, where he withheld nothing. The warlike tone of the I.P.A. statutes cried out against any partial opposition. To put it simply, it seemed that members who maintained their intellectual independence were no longer welcome. Furthermore, Bleuler was concerned about associating himself scientifically with certain types. Preempting Freud's retort, Bleuler denied that this was due to anti-Semitism or academic snobbery. Rather, he recoiled from the idea of being chained to a group of people who had been chosen "solely depending on whether they have adopted your theories or not." Bleuler ended his letter with a farewell. You, he told Freud, are an intellectual giant, a Copernicus or Darwin for psychology. But your truth is still one among many. In closing, Bleuler chided Freud for having turned from the pursuit of science to the politics of getting his theory accepted.

Freud was a proud man, and the direct criticism must have offended him, but he could not let the matter drop. Losing Eugen Bleuler's support was too monumental for the movement. "(W)e have a cause to defend before the public," he wrote Jung, and therefore must take up the "witches 'Politics' and 'Diplomacy.'" "My guess is that the insides of other great movements would have been no more appetizing if one could have looked into them," he observed. Freud asked Bleuler to list the changes in the association that would be required for him to join. To further assure Bleuler, Freud informed him that Alfred Adler had been appointed president of the Vienna group, the same Adler who was mainly interested in biology and whose psychological ideas were so counter to Freud's that they made him fume on a weekly basis. Freud would stick to his own views, but no one should mistake "steadfastness for intolerance."

Meanwhile, Carl Jung bitterly noted the irony of Bleuler's criticism. This leader of the abstinence movement was demanding open inquiry in scientific communities. And yet the same man insisted all doctors at the Burghölzli completely abstain from drinking, though there was little scientific data to support the belief that moderate amounts of consumption were unhealthy. Would Eugen Bleuler accept drinkers into his abstinence groups? Jung wondered. If not, why should psychoanalysts take in anti-psychoanalysts?

In December, Freud tried another tactic to win back the most famous supporter of Freudian theory. Knowing that Bleuler was in the middle of composing a position paper on Freudian doctrine, the Professor urged him to take on Freud's detractors. Worried about what Bleuler might say, Jung and Freud also offered him space in the *Jahrbuch*, in the hope that publishing his work in that forum might dampen his dissent. Bleuler's rebuttal of Freud's critics appeared in the second issue of the *Jahrbuch*, alongside an account of the Nuremberg proceedings. He accused critics of basing their opposition to sexual theory on personal emotions and conventional ethics, rather than on scientific observation. He defended sexual theory in relation to neurosis, and stated that the Oedipus complex was a fact. Bleuler's defense further pointed out that repression, the sublimation of sexuality, transference, and the unconscious were not new ideas. Instead, Freud had brought these ideas together in a new way. For the thinking person, Bleuler insisted, much of Freud was self-evident. The rest needed to be tested by science.

After reading this, Jung was stunned and Freud jubilant. The Professor caught the train to Munich to meet Bleuler and discuss a rapprochement, for Abraham had passed on word that Bleuler was eager for a reconciliation. Abraham had spied Bleuler at a psychiatric conference, where the Zuricher openly defended Freudian perspectives against Kraepelin and Aschaffenburg. In December 1910, Freud and Bleuler sat down together in Munich, and soon the Swiss doctor joined the I.P.A. In Vienna, Freud happily announced that Professor Bleuler had published a "magnificent *apologia* for psychoanalysis" and joined the Zurich society. Bleuler, however, would not stay for long.

FOR YEARS, BLEULER had promised his colleagues a big work on psychosis. In the fall of 1911, he published *Dementia Praecox or the Group of Schizophrenias*, which was quickly recognized to be a towering achievement. In this landmark book, Bleuler renamed the disease that Kraepelin tied to dementia, "schizophrenia," and he characterized it anew by core psychopathological features. A good deal of Bleuler's theory contained remnants of his debates with Freud. For instance, he called the psychotic's inward isolation "autism," Bleuler's desexualized version of Freud's autoerotism. Schizophrenic "ambivalence" was based on the theory of contrary but simultaneous feelings that Bleuler developed while arguing with his Viennese colleague. He also borrowed aspects from French psychopathology to portray schizophrenics as people with split mental associations. Finally, Bleuler distinguished himself from Kraepelin by arguing that besides these cardinal characteristics, much remained intact in schizophrenia, from processes of perception to memory and consciousness.

In his preface, Bleuler acknowledged the work of Sigmund Freud and optimistically asserted that his readers surely understood the great debt owed to the Viennese innovator. Bleuler, however, only referred to Freud's work during his description and theory of symptoms. Freud, Bleuler implicitly suggested, was an excellent phenomenologist, but his libido theories could not explain the cause of such calamitous disturbances. Bleuler concluded that in schizophrenia some brain disease was most likely at work, a rationale that made it reasonable not to mention psychoanalysis in his section on therapeutics. Bleuler sent the work to Freud and anticipated a response of "respect-filled horror." To preempt Bleuler, Freud had already stormed the bastion of psychosis and written an analysis of the memoirs of the mad jurist, Daniel Paul Schreber. Bleuler puzzled over Freud's interpretations that regarded paranoia as the result of conflicts over homosexuality and narcissistic fixations, a theory the Zuricher found quite dubious.

Then on November 27, 1911, Bleuler resigned his place in the I.P.A. for good. He had attended the Zurich Society's meetings and had brought a group of young assistant doctors from the Burghölzli with him. This invasion annoyed Jung, who later, along with Alphonse Maeder, asked Jung's Burghölzli replacement, Hans Maier, to join the Freudian group. When Maier expressed reservations, he was told he should enlist or stop attending meetings. For Bleuler, Maier's treatment amounted to a serious breach of conduct. Looking back at the reconciliation with Freud in Munich, Bleuler concluded that he had been wrong to believe that Freud did not "want to close the doors.... The 'who is not for us is against us' and 'everything or nothing' is in my opinion necessary for religious groups and useful for political parties. Thus I can understand the principle as such, but I deem it noxious to science," Bleuler wrote.

Bleuler predicted that such a polarizing view of psychoanalysis would be its downfall in Zurich. The Zurich Society could have been the largest and most robust of all, but exclusivity had already driven away friends and made enemies. Nuremberg had changed the culture of the Zurich Freudian movement and made it impossible for him to continue his allegiance with it. If it had been a scientific association like any other, he would have been eager to stay. But the I.P.A. was different:

> Instead of striving for having as many shared views with the rest of science and scientists, it has closed itself off to the outside world with a prickly skin that is hurting friend and foe. And that seemed wrong to us from the very beginning. Unfortunately the events have proven our fears right, in fact surpassed them. Everything here in Zurich was favorable to Psa. The doctors were generally interested ... One had everything one could wish for for the next 10 years. That has been destroyed by the manner of the Association and can not be rebuilt ... Hoche's malicious term of calling it a sect, which at the time was inappropriate, has been turned into truth by the psychoanalysts themselves.

In a devastating reversal, Bleuler maintained that Freud's attempt to defend the Freudians from their enemies had resulted in sectarianism that made the worst accusations real. Bleuler would never be a Freudian

again. In a December 1911 letter to A. A. Brill, Freud reported the bad
news: Eugen Bleuler, whose interest in Freudian theory had single-
handedly done more for the Freudian movement than that of any other,
had now resigned from the I.P.A. Ten days later, Freud wrote Brill again
and repeated the bad news, as if he could not fully bring himself to
believe it.

With Bleuler at the helm, the Burghölzli had lured interested doctors
in and spit out Freudians. Eugen Bleuler's prestige, training center, and
laboratories had forced German academic psychiatry to take Sigmund
Freud seriously. While Freud's heightened reputation attracted hostility,
it also made his ideas impossible to avoid. Increasingly, a thoughtful European psychiatrist had to take a position on Sigmund Freud. Karl Abraham merrily reported that an obviously irked Gustav Aschaffenburg
reported that even in America, the first question on every physician's lips
was: "What do you think about Freud?" Bleuler had been instrumental
in creating the urgent relevance of Freud to psychiatry.

But Bleuler had different commitments that would ultimately make
him choose not to be a Freudian. A powerful member of European academic medicine, his own reputation would be threatened by a full-scale
adoption of Freud's thinking. Had his call for a calm, dispassionate scientific Freudian organization been heeded, he would have been able to fend
off his disapproving colleagues by retorting that it was unscientific of them
to belittle a method they did not know or practice. But the I.P.A. was not
primarily a scientific group; it had been born to defend the cause. It had
little room for scientific skeptics, doubters, and critics. Just as Bleuler's
abstinence group could bar those who were quite interested in drink because they were drunkards, the I.P.A. would exclude those interested in
Freud who were dissidents.

Bleuler's interest in French psychopathology and the study of unconscious psychological processes had led him to Freud and nurtured a version
of psychoanalysis that included these interests. But the Nuremberg Congress, with its calls for a standard theory and method, codified psychoanalysis in purely Freudian terms. Professor Bleuler had nothing to gain by
joining such a group, particularly at the moment when his magnum opus
on schizophrenia was about to be published. In the end, he retained his
primary commitment to German academic psychiatry and left the Freudians behind.

As early as 1905, Sigmund Freud resolved to discount disciplinary conflicts and treat Bleuler's theoretical arguments as if they were no different from neurotic symptoms. Freud's cry of emotional resistance clouded the relationship between the two men long after Bleuler stopped asking Freud to interpret his dreams. Freud's stance made it impossible for Bleuler to have legitimate scientific questions that did not revert to his unconscious motives. The Viennese's final plea to Bleuler again brought up the idea of resistance, but by then his colleague had long stopped listening.

Although Freud's ad hominem tactics were unfair, they were not inconsistent. The mind of the patient and the mind of the scientist were one; the unconscious and repression were always at work. Yet the Professor was inconsistent in failing to consider his own resistances to intellectual differences, paternal authority, and fraternal competition. Bleuler told Freud that turning scientific differences into personal ones—no matter how theoretically sound—would destabilize any scientific community. Even if a theoretical opposition was in part emotional, it had to be argued or rebutted on its merits. Otherwise all criticism would be dismissed without a second thought, while submissive parroting would be blindly accepted. What kind of a community would the Freudians become if they did not admit skeptics or dissenters, if they treated scientific difference as always a rationalization for distaste, disgust, and resistance?

III.

NUREMBERG HAD UNINTENDED repercussions in Vienna too. Though they were meant to restrain and marginalize the Viennese, Ferenczi's proposals backfired. To assuage the hurt feelings of his Viennese colleagues, Freud felt he had to cede control of the local Society. He handed the presidency to Adler and the deputy chair to Stekel. The transfer of power was intended to demonstrate Freud's commitment to changing the group into one less closely associated with his person.

The two men the Nuremberg proposals sought to marginalize now controlled its oldest Society. And they had not forgotten the indignities suffered in Nuremberg. Adler criticized Freud for overstating the danger the movement was in and encouraging an exaggerated reaction. Others complained that the high-handed Zurichers were of another "breed."

Adler, the committed socialist, declared that the Vienna group—unlike the I.P.A.—would have free elections, and he envisioned strengthening the movement by inviting more outsiders to attend, by giving public courses, and providing training in scientific matters. Toward that end (as well as to undermine Jung's power), Adler announced that he and Stekel had formed a new psychoanalytic journal to be published in Vienna called the *Zentralblatt für Psychoanalyse*.

The creation of the I.P.A. had prompted the Viennese to rally around an anti-I.P.A. platform. Freud worried that in the hands of his angry colleagues, this new journal might be turned against him. Parrying Stekel and Adler's move, Freud informed them that he had contacted two publishers, both of whom were willing to take on the *Zentralblatt*, as long as Freud was editor. "What kind of a guarantee can you give me that this journal will not be directed against me?" he asked them pointedly. Adler and Stekel professed their loyalty, so Freud suggested a modus operandi. Every paper published by the new journal would have to be approved by all three of them: each man had the right of veto. Primarily concerned about Carl Jung's power, not Freud's, the Viennese agreed. Not long after the decision was made, Freud exercised his newly won veto power . . . over one of Stekel's own submissions! And unbeknownst to Adler and Stekel, Jung had arranged to have Freud veto any article he disliked as well.

Despite all the infighting, the Vienna Psychoanalytical Society grew rapidly. Around 1906, Society membership had stagnated around twenty members, but in 1910 interest surged. The Society admitted sixteen new recruits in 1910 and twelve more in 1911, making those years the greatest period of growth ever. Suddenly, there were plenty of novices. Following the edicts of Freud's Nuremberg speech, they would be trained in a standardized, if still incomplete, Freudian theory of Oedipal psychosexuality.

But the new president of the Society had other ideas. Alfred Adler had matured considerably as a psychological theorist. The biologically minded doctor who once saw organ inferiority everywhere had transformed himself into a psychologist. In 1909, Adler announced his own unified theory of neurosis, which emphasized an aggressive instinct, which when repressed led to *feelings* of inferiority and neurosis.

Just before the Nuremberg meeting, Adler presented another paper to the society that showcased his move from speculative biology to a purely psychological approach. The key to neurosis was the child's inner experience of inferiority engendered by his confrontation with sexuality and the feeling: "I am not fully a man." For men and women, neurosis entailed a conflict between the psychically male and female. According to Adler, all infants viewed the feminine as inferior and tried to repress it, which meant clinicians should strive to bring to light what was psychically deemed weak, inferior, and female.

In the Vienna Society, Adler's theory was quickly challenged. Since stable notions of masculine and feminine did not exist in the psyche, had he not imported biological terms into psychology? Adler was accustomed to this kind of a reception, just as the Viennese were used to his hammering away with his own constructs. Even as he built a more flexible psychological theory, many Viennese colleagues viewed Adler as a one-trick pony, who interpreted complex phenomena in a rote manner. His ally, Wilhelm Stekel, even chided Adler for declaring that all dreams were manifestations of what he called "masculine protest."

Alfred Adler, one of the founders of the Wednesday Psychological Society. His ideas on psychological feelings of inferiority roiled the group.

But after Adler's presentation of his unified theory of neurosis, Eduard Hitschmann ominously announced that "Adler's approach is very different from ours." Paul Federn suggested that the paper afforded the opportunity to discuss the important differences between Alfred Adler and Sigmund Freud. Freud squashed such a mano a mano contest, but informed Jung that Adler was paranoid and that his theories prodded Freud "into the unwelcome role of the aging despot who prevents young men from getting ahead."

Adler's theories had been irking Freud for years, but an influx of new recruits now made the president of the Vienna Society a formidable competitor for hearts and minds. He had a coherent, unified theory while Freud still did not. Adler's ideas had some commonsense appeal, and a number of the new members who arrived between 1909 and 1910 were favorably disposed to them. Freud was concerned that Adler's denial of libido theory would undermine the theory within and give succor to doubters outside the society: "The crux of the matter—and that is what really alarms me—is that he minimizes the sexual drive and our opponents will soon be able to speak of an experienced psychoanalyst whose conclusions are radically different than ours." Carl Jung once silenced his critics by saying they were condemning a method they had never tried. What would Freud's defenders say when one of their oldest practitioners came out against libido theory?

Freud hoped to avoid such an embarrassment. In December 1910, he published a paper called " 'Wild' Psychoanalysis" that detailed the case of a doctor who had prescribed sex to an elderly matriarch with an anxiety disorder. Freud forcefully distanced himself from such wild practitioners and informed his readers that to safeguard patients, an International Psychoanalytical Association had been founded. Only those whose names were listed in this organization were really psychoanalysts.

That behind him, Freud decided to force a showdown with Adler. They were old allies, but things had changed since the days when all analysts were wild and any theory with an unconscious was good. The Professor invited Adler to give a series of lectures that outlined his new views for the Society. Adler was flattered but wary: did Freud really mean it? Was he really willing to compromise? On January 14, 1911, Adler delivered his first lecture on "Controversial Problems of Psychoanalysis." Adler went right to the heart of the matter: Freud's libido theory was wrong. Libido was artificially compounded and increased by masculine protest. A few weeks later, Adler took the podium again. His title said it all: "The Masculine Protest as the Central Problem of Neurosis." Aggression and an internalized misogyny caused the repression of all things deemed weak and feminine. Aggression was the core drive; masculine protest the result. As the coup-de-grâce, Adler concluded by saying the Oedipus complex was but another sign of masculine protest.

"He has created for himself a world system without love," Freud wrote to Oskar Pfister, "and I am in the process of carrying out on him the revenge of the offended goddess Libido." In the ensuing discussions, Freud attacked Adler as a plagiarist who stole his concepts and renamed them. In doing so, Adler created something foreign. "This is not psychoanalysis," Freud solemnly declared. While some argued psychoanalysis should allow any individual to express himself, Freud did not agree. Such a complete lack of consensus would harm the scientific standing of the field, and confirm those who believed, following Auguste Comte, that this psychology was nothing more than a personal prejudice. The only way to "guard against the subjective factor" was to encourage investigation in combination with self-analysis. Adler had failed to understand the personal biases in his work and would therefore do grave damage to the standing of psychoanalysis.

Freud's denunciation was followed by two evenings of debate in which Freud's most zealous allies unleashed their weapons. They were led by a gloomy Freud, who had written out copious notes in the service of expelling one of his oldest colleagues. But Adler was not without defenders. Carl Furtmüller and some of the newer members argued that Adler was not a danger to psychoanalysis. Stekel also supported his old friend and objected to the unscientific assumption that all of psychoanalysis was to be found in Freud. Even Paul Federn rallied to Adler's side. But near the end of the last evening, the normally quiet Maximilian Steiner, a longtime member of the group, rose to denounce Adler. The group had come together to study the unconscious and libido, and Adler had turned to consciousness and aggression. He concluded his remarks by saying Freud should be reproached for letting this go on so long.

At the following committee meeting, Dr. Alfred Adler resigned as chairman of the Vienna Society due to the "incompatibility of his scientific attitude with his position in the society." Stekel also resigned his post in the Society, and at the next meeting, Sigmund Freud was again elected to lead the Vienna Psychoanalytic Society. Eduard Hitschmann—"quite orthodox" in Freud's words—became deputy chair. But Adler's defenders were not ready to give up. Carl Furtmüller rose to counter the notion that Adler's ideas were contrary to those of the group. He argued that such a determination could not be made by fiat and suggested the Society vote on the matter. Freud tried to stop the vote, but the members promptly voted down the proposal that Adler's views were incompatible with the Society's. It was a small consolation in what had otherwise been a terrible defeat for Alfred Adler.

During this battle, Freud confessed to Oskar Pfister: "I have always made it my principle to be tolerant and not exercise authority, but it does not always work. It is like cars and pedestrians. When I began going about by car I got just as angry at the carelessness of pedestrians as I used to be at the recklessness of drivers." Afterward, Freud wrote that he was ashamed of the mess and wondered how the public could expect much from psychoanalysis if the analysts could not rise above such pettiness. For three months, Alfred Adler continued to attend the meetings of the Society, like a man awaiting his own funeral. Freud was itching for an excuse to end the charade and expel him. Freud contacted the publisher of the *Zentralblatt* and asked him to take Adler off the masthead. After that insult, Adler resigned with three other members and formed the Society for Free Psychoanalytic Investigation.

Furious at the accusation spelled out in this organization's name, Freud confronted Adler's lingering supporters who had hoped to maintain their membership in Freud's group while also attending Adler's meetings. They must choose, Freud declared, one or the other. Furtmüller denounced this "with us or against us" attitude, but when put to a vote, he and his allies lost. Six Society members resigned, which made a total of nine who left to follow Alfred Adler. Of those, seven had come to the Society in 1910; all were socialists. The "Palace Revolution" was over.

THE MARGINALIZATION AND expulsion of Alfred Adler from the Vienna Psychoanalytic Society was the result of three transformations that had taken place between 1906 and 1910: Alfred Adler changed into a pure psychologist and threw himself directly in competition with Freud; Freud changed into the leader of a movement that was intent on protecting itself; and the nature of the Freudian community changed from a ragtag group of partial adherents into a community that enforced a commitment to its core beliefs and looked fearfully about for those who might destroy it.

The process of undermining partial supporters of Freud's theories had burst into the open during the Salzburg Congress in 1908. If Carl Jung could be attacked for not being Freudian enough, who was safe? Adler's removal after the 1910 congress marked the solidification of a process by which partial support was seen as inimical to the defense of psychosexuality as the primary principle of Freudian psychoanalysis. Adler had always gone in his own direction, and Freud had long tolerated it, but as Ferenczi said in Nuremberg, *in times of war, the muses are silent*. Interesting, challenging ideas that might weaken the common defense were not interesting during wartime. Members who wished to rewrite "our Science" had to be silenced. This Nuremberg directive had Adler's name on it, but he showed little interest in modifying his views or deferring to these concerns. In the end, he was the Muse that needed to be silenced.

Adler's dismissal revealed a troubling problem for the Freudian community. In an attempt to resolve the differences between Adler and Freud, it had not been possible to demonstrate that one system of thought had more truth value than the other. Adler's theory of inferiority and aggression constituted an alternative to Freudian psychosexuality. Freud's theory itself was based on an interdisciplinary synthesis that, at its heart, defined the empirically unknowable unconscious by a series of deductions and analogies. When his academic enemies attacked libido theory, they attacked Freud's Achilles heel, not just because it was personally offensive to some, but also because it seemed to flout the rules of empiricism and scientific epistemology. When Adler presented an alternative model based on different unconscious contents, he forced those within the movement to consider how they might possibly adjudicate between Freud and Adler. In the end, Freud did not prove that Adler was wrong; rather he declared Adler was not a psychoanalyst.

If Freud thought that would solve his dilemma, he was mistaken. For now, an unrepentant Alfred Adler attempted to take possession of psychoanalysis, by setting up his own rival group. The Nuremberg proposals not only began a process of self-definition but also caused a schism that created a rival group of psychoanalysts. Suddenly there seemed to be a divide between *Freudian* and *psychoanalysis*, two words that long seemed synonymous. When Adler's followers left to pursue their agenda, it appeared that the Freudians had become just one community in an expanding psychoanalytic nation.

With Adler's resignation, Wilhelm Stekel's place in the Vienna Society was also cast into doubt. Stekel had tried to mediate the dispute between Adler and Freud, but in the end, he chose to stay with the Society. He later explained that he was financially dependent on Freud's referrals and was unsure of his own position in the theoretical controversy. With Adler gone, Stekel also became the sole editor of the *Zentralblatt*, a role he coveted.

After the Salzburg Congress, Stekel became a voice for the rights of the Viennese and the bête noire of many Zurichers, including Bleuler, Jung, and Binswanger. Freud had tolerated Stekel, even admired his uncanny clinical acumen, but was contemptuous of his theorizing. After Adler's banishment, Stekel continued his fall from favor. His long awaited book, *The Language of Dreams*, did not help. In it, the author endorsed the Adlerian view that aggressive, criminal instincts motivated dreams. To make matters worse, Stekel asked Adler to review the book for the *Zentralblatt*. Freud was enraged but felt helpless.

The time had come for another congress. The Viennese voted against another meeting in Nuremberg—they had had enough of the place—and suggested their home town, but they were rebuffed by the Zurichers. The Berliners suggested Weimar, and in September 1911, fifty-five people gathered there for the Third International Psychoanalytical Congress. They were welcomed by Carl Jung, who informed the gathering that the I.P.A. had more than doubled its numbers in one year, going from a membership of 52 to 106. Jung took credit for this expansion, saying that the new branches of psychoanalysis in Berlin, Munich, and New York had all sprung from the Zurich school. Jung urged self-restraint on these new psychoanalysts. He asked them not to give themselves over to unbridled imagination and suggested that the new groups concentrate on education and training. Further, he alluded to the defection of Adler and stressed that it was imperative to expose deviations and discredit wild psychoanalysts.

One of the wild Viennese analysts left the conference with his power enhanced. Official I.P.A. news had been carried by a short-lived publication, the *Korrespondzblatt*, but in Weimar these functions were turned over to Stekel's *Zentralblatt*, making it the official journal of the I.P.A. Freud was still unsure of Stekel's bona fides, for he seemed to be on both sides of the Adler debate. A test emerged when the Society decided to devote a number of sessions to masturbation, the subject of public disagreement between Freud and Stekel three years earlier. In these debates, Stekel did not mince words. He condemned Freud's orthodox follower Eduard Hitschmann as reactionary, and made it clear he believed masturbation did not by itself create any physical illness. Freud stepped in, seizing an opportunity to show how *his* Society was free to investigate phenomena as it saw fit. Freud reminded the audience that the questions involved were exceedingly complex, and that the task of the group was to be an open forum for all views. In 1912 as Alfred Adler was convening meetings for his Society for Free Psychoanalytic Investigation, Freud published these proceedings, telling his readers: "It is never the aim of the discussions of the Vienna Psychoanalytical Society to remove diversities or to arrive at conclusions. The different speakers, who are held together by taking a similar fundamental view of the same facts, allow themselves to give the sharpest expression to the variety of their individual opinions without any regard to the probability of converting any of their audience who may think otherwise."

Wilhelm Stekel, a founder of the Wednesday Psychological Society, was the editor of the *Zentralblatt* during Freud's battles with Adler and Jung.

Freud publicly touted his tolerance of diverse opinion, as exemplified by his barely contained capacity to endure Wilhelm Stekel, but that man's continued sympathy for Adler, combined with his position as editor of the official I.P.A. journal, made him a threat. During the fall of 1912, things came to a head. By then, Carl Jung had started to make it clear that he too could not abide by the Nuremberg rules and would soon be going his own way. Since Jung edited the *Jahrbuch*, Freud became alarmed that Jung would use that journal to promote his own views and came up with a battle plan that included making the *Zentralblatt* his own, and putting together a review board of his most loyal Viennese followers to attack Jungian work in the *Jahrbuch*. For this board, Freud elected stalwarts, including a newer member, Victor Tausk. A jurist by training, Tausk had become a psychiatrist and joined the Society in 1909. He had taken a dislike to Stekel and had frequently pointed out the older man's sloppy errors in front of his peers. Tausk had even impugned Stekel's honor by implying that the doctor fabricated his case histories. When Freud proposed that Tausk join the *Zentralblatt* board, Stekel refused.

Stekel's reaction enraged Freud and made the Professor wonder what Stekel had planned. If Stekel did not stay loyal, Freud might lose *both* the *Jahrbuch* and the *Zentralblatt* to non-Freudian psychoanalysts. He decided to force a showdown with Stekel and told Ferenczi: "(F)ew sacrifices would be too great for me to get rid of him." Freud believed Stekel had revealed himself when he announced that he intended to make his journal "independent, and open it up to *everyone*." Everyone, of course, included Alfred Adler.

In the middle of this crisis, the writer Lou Andreas-Salomé arrived in Vienna to study psychoanalysis. Frau Lou was renowned in Europe as a writer and as the soul mate of Friedrich Nietzsche and the poet Rainer Maria Rilke. Her arrival during this tumult embarrassed Freud. Despite a general prohibition that forbade Freudians from consorting with Adlerians, Freud did not dare place restrictions on Andreas-Salomé. She freely visited both camps and while visiting Adler, discovered that he had hatched a plan with Stekel to take over the *Zentralblatt*.

For Freud, this was nothing short of treason. Freud approached the *Zentralblatt*'s publisher and asked him to sack Stekel. The publisher refused. The president of the I.P.A., Jung, was in America at the time, so Freud took it upon himself to immediately launch a new official I.P.A.

journal and isolate the "Stekelblatt." An emergency meeting of all the European heads of local psychoanalytic societies was called, during which they decided to leave the *Zentralblatt* en masse and form a new official journal, the *Internationale Zeitschrift für Ärztliche Psychoanalyse*. The editors would be Freud's longtime student Otto Rank and the Hungarian loyalist, Sándor Ferenczi.

On November 3, 1912, Freud told Ferenczi that Stekel had resigned from the Society: "I am so delighted about it; you cannot realize how much I have suffered under the obligation to defend him against the whole world. He is an intolerable person." To Abraham, he insisted the split was not due to scientific matters but rather Stekel's unpleasant manner.

Wilhelm Stekel was no more obnoxious in 1912 then he was in 1907. But by 1910, he was on the list of Viennese that Jung and Freud hoped to contain. After Nuremberg, Stekel unexpectedly gained power. As editor of the *Zentralblatt*, he seemed to be plotting to have the official I.P.A. journal acknowledge a range of psychoanalyses, including that of his friend Adler. When Freud announced Stekel's departure to the Vienna Society on November 6, 1912, he minimized these matters. But Lou Andreas-Salomé wrote in her journal that Freud made it seem "as if it concerned the local Viennese only—whereas I know from Adler what Stekel's intentions are, and Freud also now recognizes them."

Stekel confronted his one-time analyst and longtime colleague, and later recalled that Freud blamed Carl Jung for the break, saying the Zuricher had bitterly opposed Stekel. As he took leave of Sigmund Freud, Stekel declared: "you have sacrificed your most faithful collaborator for an ungrateful one. Jung will not remain a Freudian for long."

IV.

WHEN HE QUIT the Burghölzli in 1909, Carl Jung left the academic culture that had transformed him from a blustery neo-Romantic into a prominent research scientist. Of course, Jung's science had served neo-Romantic ideals by seeking to establish the power of the unconscious, sexuality, and dreams. But his reputation rested on an image of the inspired lab researcher who carefully confirmed conjectures through experimentation. After leaving Bleuler's clinic and then his lab, Jung was at liberty to re-create himself. Freud hoped Jung would move away from academic psychiatry and psychopathology and dedicate himself fully to psycho-analysis. And in the beginning, he seemed to cooperate. As president of the I.P.A., Jung took pride in his role as political leader and heir apparent of the Freudians. His reluctance to accept libido theory in the psychoses appeared to dissipate. He became a full Freudian and the man Freud called the "Crown Prince" of the movement.

But while Jung immersed himself in the Freudian community, he also returned to his older interests in parapsychology and religion. In his 1909 paper, "The Significance of the Father in the Destiny of the Individual," Jung had made a connection between the study of neurosis and religion. The father was always the essential factor in the psychosexual fantasies of an individual, and this same Father complex was core to religious belief, Jung concluded. Through sublimation, the child's real father became God the Father.

Jung's linkage of neurosis and religion was not abstract. A pastor's son, Jung was painfully aware of how his own desire to compete with or submit to his father was steeped in both familial and religious history. Further complicating things, Jung discovered he had those same feelings for Freud, the "father" of psychoanalysis. Jung asked Freud to consider him a son and confided that "my old religiosity had secretly found in you a compensating factor which I had to come to terms with." In reply the godless Jew warned "a transference on a religious basis would strike me as most disastrous; it could only end in apostasy."

After the congress at Nuremberg, the Freudian movement was heavily dependent on Carl Jung. Freud and Ferenczi had overplayed their hand in creating the I.P.A. and lost leading figures like Bleuler, while concen-trating power in the hands of the esteemed academic psychiatrist, Carl Jung. And now Jung—like William James and others—was turning away from the lab to explore religious experience. "Occultism is another field we shall have to conquer," he wrote Freud, assuring the older man that he need not worry about Jung's wanderings "in these infinitudes." Armed with libido theory, Jung promised to "return with rich booty for our knowledge of the human psyche."

Around the time Jung was nominated lifetime president of the I.P.A., he threw himself into an exploration of religious symbols and myths. In these ancient stories, he gleaned the possibility for the psychological equivalent of an archaeological excavation. And other Freudians were excitedly digging alongside him. By 1909, Freud had overseen the publication of three major works on myth in his *Schriften zur angewandten Seelenkunde* book series. The first was by Jung's cousin and Zurich collaborator, Franz Riklin. In *Wish Fulfillment and Symbolism in Fairy Tales*, Riklin argued that the same Freudian mechanisms involved in dreams organized fairy tales. Soon thereafter, Jung's irksome rival, Karl Abraham, published *Dreams and Myths*, in which he argued that myths could be treated like the dreams of an infantile period of humankind. And in 1909 Freud's student Otto Rank published *The Myth of the Birth of the Hero*, which rooted the mythic accounts of heroes in the life of the heroic little boy who sought to overthrown his father. Rank nailed the problem of myth solidly to what was now considered the "core complex" of neurosis, the Oedipal complex.

Hence, when Jung entered into the psychoanalytic study of myth, the subject seemed to have already received a definitive Freudian treatment. Jung was not deterred, however, for his line of inquiry was unique. It began when Jung and his junior colleague, Johann Jakob Honegger, treated a paranoid patient whose delusions were uncannily mythic in content. This poor, uneducated man could not have acquired such extensive knowledge of myth, they believed, so what could this mean? When Honegger presented this case in Nuremberg, it created a stir. By then, Jung was feverishly studying world myth and had begun to analyze the fantasies of a young American named Miss Frank Miller.

Miss Miller had been the patient of Théodore Flournoy, the Genevan doctor, and she had published her fantasies in his *Archives de Psychologie*. Jung concluded that Miss Miller's poems and fantasies were barely disguised myths that could not have been learned, but must have been passed down and encoded in her brain. By June 1910, he finished a rough draft of a paper on the relationship of myth to individual fantasy and sent it to Freud, who voiced his approval.

In the ensuing year of turmoil with Adler, Stekel, and Bleuler, Jung continued doggedly pursuing the relationship of myth to individual fantasy, mentioning his project to Freud only in passing. Then in the summer of 1911, Jung sent Freud the latest issue of the *Jahrbuch* featuring his "Transformations and Symbols of the Libido." The long paper did not appear to be a rebel's yell, for it opened by paying homage to Sigmund Freud's dream book and took as a given that that hero of classic Greek drama, Oedipus Rex, was living inside us all. Furthermore, Jung made it clear that fantasy systems emerged from the great force of human sexuality. But more subtly, the essay found Jung turning to Flournoy, Wundt, Nietzsche, Schopenhauer, and William James. The author of "Transformations" was not simply a Freudian but also a scholar with many sources, who discovered what he believed was a new aspect of human psychology.

Carl Gustav Jung and Emma Jung in 1903.

Human thinking was characterized by two modes, Jung proposed: the verbal inner speech that made for modern science and the imagery of fantasy and dreams. This paraphrased Freud, who had recently reiterated his distinction between primary and secondary process in a paper on the two principles of mental functioning. But what organized fantasy? While Freud believed fantasies were driven by unconscious sexual drives, Jung concluded that the fantasies of any individual carried the memories of the entire race. Though Jung failed to cite any predecessors, his view would have sounded familiar to German theorists of unconscious memory like Ewald Hering and psychologists in Paris, many of whom tended toward notions of an unconscious that was phylogenetic and racial.

To argue this point, Jung turned to the fantasies of Miss Miller. Like Helly Preiswerk, Frank Miller was an adept, a woman who made the unconscious manifest. Jung claimed to have found myths in Miss Miller's unconscious that she could not have read, heard, or learned. Unbeknownst to her, Miss Miller's fantasies mirrored primitive myths of creation, and not just one set of myths, but the whole history of human myths. Her psychological struggle with the "Father Imago" (as Jung now asked readers to call the Father complex) provoked unconscious religious fantasies that traced the historical movement from the moral decadence of Roman times to the founding of Christianity and Mithraism. These religious movements arose to tame the animal urges of mankind and create brotherly love. All these memories and more teemed in Miss Miller's unconscious.

Did Jung also believe Oedipal fantasies were phylogenetic memories of prehistoric man? For the reader of Jung's 1911 paper, it was hard to tell, because in the last section of "Transformations," Jung's argument came undone. In an uninspired poem of Miss Miller's called "The Moth to the Sun," Jung discovered the myths also found in early Christian and pagan mystics, Goethe's *Faust*, Seneca, Byron, Amenhotep, the poet of Revelations, Nietzsche, and more. The list grew as did dizzying, tangential riffs on poor Miss Miller's simple poem. Jung's display of erudition was overwhelming, but his argument—as man reaches toward the Sun/Divine, he reaches inward, introverting toward the Libido—was overwhelmed as well.

One thing was clear: Jung's use of myth was unique. For the Zuricher, myth revised Freudian theory rather than simply confirming it. He hoped to write a dictionary of human unconscious fantasy that would be no less than a record of human history itself. Freud was no foreigner to such speculations, for he too was devoted to the ideas of Lamarck and Haeckel. Upon receiving Jung's article in the *Jahrbuch*, Freud wrote to congratulate the author—not for the completion of this work, but for his work as political leader "championing the cause, holding the flag high, and meting out mighty blows to our opponents."

In the next paragraph of Freud's letter, the opponents on the battlefield suddenly changed: "Since my mental powers have revived, I have been working in a field where you will be surprised to find me," Freud wrote. "I have unearthed strange and uncanny things and will almost feel obliged not to discuss them with you. But you are too shrewd not to guess what I am up to when I add that I am dying to read your 'Transformations and Symb. of the Li.'" Apparently Jung was not so shrewd and nervously asked Freud what he meant. Freud replied that now both he and Jung knew that the origin of all religion lay in the Oedipus complex.

Freud had jumped into the same forum of study as Jung. The Professor readily admitted that he was incited to creative thought by others, telling Ferenczi: "I have a decidedly obliging intellect and am very much inclined toward plagiarism." For Freud, absorbing ideas and then thinking against those same ideas had been extremely generative. As he had done with others in the past, Freud leapt on hints from Jung about myth and began to think dialectically against his Swiss colleague.

But Freud's secret insight, the one he confessed, was no secret: Jung had already floated such an interpretation in 1909. As Freud and Jung were digging in the lost fields of human prehistory, it became clear that their joint search was not amicable. The polite nature of their letters could not hide the tension. Distressed, Emma Jung wrote to Ferenczi in Budapest, asking for his advice but imploring him not to say a word to the Professor. Ferenczi immediately forwarded Emma's letter to Freud, along with a draft of his own response. After reading Ferenczi's harsh reply, Freud asked Ferenczi to strike (the German was *streichen*) all references to Freud's displeasure with Jung's paper on libido and his turn to astrology. Ferenczi did as he was told and mailed off his response to Zurich. He proudly reported to Freud that he had carried out the Professor's orders and made sure to touch on (*streifen*) Jung's turn to astrology and the libido paper. Upon rereading Freud's letter, a horrified Ferenczi realized his blunder. But the letter was already in the mail. Freud angrily castigated Ferenczi for his "false obedience," a humiliation that plunged Ferenczi into a self-analysis.

A few weeks later, Emma Jung wrote Freud. The Professor was irritated with Carl, she knew, and it had something to do with "Transformations of the Libido." She wrote again and confided that Carl was worried Freud would not accept "Transformations," and anxiously awaited Freud's opinion of it.

Freud had challenged Jung on the turf he hoped to make his own. Meanwhile Jung spoke more brazenly to Freud and others about the pre-eminence of the Zurich school and his own role as leader of the movement. Writing about his job as president of the International, Jung told Freud he was getting used to brandishing the "whip," since most people were only too happy to be tyrannized. Carl Jung was not one of them. He worked to break free and maturely face the hierarchical relation between father and son that had long occupied him. According to Jung's new theory, the Father-transference bound brothers together in reverence of the patriarch and prevented patricide. But the repressed world of Christian brotherly love had long done its work, and its strictures left men with a neurotic need to be servile. Reading these passages, Freud may well have guessed that it was Carl Jung's declaration of independence.

Before announcing that he was writing a fundamental revision of libido theory, Jung wrote Freud, saying: "You are a dangerous rival." In his reply, Freud tried to refuse the role of adversary: "I am all in favour of your attacking the libido question and I myself am expecting much light from your efforts." Freud then added innocently that his own thought moved forward when he felt "compelled to by the pressure of facts or by the influence of someone else's ideas." The latest someone else was Jung.

Meanwhile, a storm of controversy hit Zurich. A group had organized in opposition to Ernst Haeckel's scientific club, the Monist League, incensed by its forays into religion. On January 2, 1912, this Kepler League announced a special session prompted by newspaper accounts of a Zurich invention of Ludwig Frank's called psychoanalysis! Kepler League members were also up in arms after a lecture by Riklin on universal myth.

Although provoked by Frank and Riklin, the Kepler League meeting was a rousing assault on Sigmund Freud. Afterward, the *Neue Züricher Zeitung* published a flurry of letters pro and con, some of which expressed horror that their holy beliefs were being desecrated by sexual interpretations. Jung stepped in to defend the Freudians. On January 27, 1912, as president of the I.P.A., he formally denounced the "insulting and severely disparaging accusations" that had been hurled. The attacks and counterattacks only subsided when the éminence grise of Zurich psychiatry and founder of the local Monist League, Auguste Forel, stepped in. He chided critics for unfairly lumping valuable innovations like Frank's method with Freudian psychoanalysis, then seconded the condemnation of Freud's theory for its "sanctifying sexual church, its infant sexuality, its Talmudic-exegetic-theological interpretations."

In Zurich, the battle lines were drawn between Christianity and this atheistic science founded by Jews. The young Carl Jung would have had sympathy for the Kepler League, for the pastor's son had seen scientistic extrapolations as destructive to man's soul. But now he was the leader of a movement under siege for its attempts to understand the psychological foundations of religion. As Jung publicly defended psychoanalysis in Zurich, he privately quarreled with Freud. This pretext was trivial: Jung had not written to Freud as promptly as usual. Jung apologized, but Freud irritably replied that Jung's three-week neglect of their correspondence required psychoanalytic elucidation: Jung's Father complex caused him to neglect his duties to the movement. Annoyed, Jung suggested Freud's testiness came from a desire to intellectually control him. "Let Zarathustra speak for me," the Zuricher wrote, quoting Nietzsche: "One repays a teacher badly if one remains only a pupil."

Now both men were angry. If Jung hated to hear about his Father complex, Freud hated to be accused of being an intellectual bully, probably because both master psychologists knew there was some kernel of truth in these criticisms. Freud denied tyrannizing Jung and reminded his friend that he was Freud's chosen heir. Softening, Jung replied that he had no intention of abandoning Freud the way Adler had. Some relief came from the arrival of a common enemy, another German academic, who gave the two men a chance to direct their ire elsewhere. But that was short-lived. Each doctor sat in his library writing an account of the mind's prehistory. They would retell the biblical story of fathers and sons from a psychological perspective, and at the same time, they labored under the accusation that their own position as father or son had turned their work into nothing more than autobiography. Again, Auguste Comte's curse rose.

The two men also faced another conundrum. Anthropologists had concluded that incest taboos were crosscultural and nearly ubiquitous, but found it difficult to explain how a cultural prohibition could possibly be universal. How could the same social law be transmitted across time and geography? For Jung and Freud, a Lamarckian answer presented itself. In human prehistory, something happened that made incest intolerable, and that became encoded in man's biological inheritance. But what?

Freud gave Jung a hint about his solution. The Oedipal complex was the residue of a past history by which infantile sexual desires were repressed. Something happened in Darwin's primal horde that made incest forever impossible. In solving the riddle of the incest taboo, Freud also hoped to explain the origins and nature of repression itself. The question had long troubled Freud. At times, he had seen repression as the result of civilizing forces and at others, organic ones. Now he discovered a synthetic answer: in man's early history, repression had been a social force, but later it became part of man's biological endowment through unconscious memory.

Jung honed in on the same question, focusing not on the role of the tyrannical old patriarch, but on an early stage of human history when women ruled. His speculations built on the work of one of Basel's intellectual heroes, Johann Jakob Bachofen, who claimed to have established that an ancient matriarchal community predated patriarchy. Jung told Freud he believed the incest taboo emerged from this matriarchal period of history, as a proscription against incest that was not sexual, but a defense against the desire to return to the womb. Therefore, the origins of repression were not specifically sexual. Freud informed Jung that his line of thought carried a "disastrous similarity" to Adler's. Freud seemed to know that Jung had argued in Zurich seminars that Freud's sexual instinct theory did not make evolutionary sense. Pointing out that Freud postulated a number of partial instincts that united to create the mature sexual drive, Jung expressed doubt that evolution would require such a complex synthesis for its most vital function. Hunger was surely not based on the partial instincts of looking, grabbing, chewing, and swallowing. Similarly, libido was not the sum of numerous parts, but rather a holistic life energy.

While the two men imagined the establishment of the most primal of communal laws, they debated the laws of their own community. Was Freud an old tyrant, and Jung a childish rebel? Had the son eclipsed the father? Had that time come for Jung to take over the movement? What prevented Freud the father from crushing Jung, as he had other young rivals? What kept Jung and the other sons from attacking the father, and usurping his privileges? Could the Freudian community develop laws that might prevent the fragmentation caused by fratricide, womblike retreats, and incestuous inbreeding?

On May 25, 1912, Freud traveled to Kreuzlingen to pay a sympathy call to Ludwig Binswanger, who had undergone an emergency operation and was recuperating at his home near Lake Constance. By then, Binswanger had become one of Jung's arch rivals in the Zurich Society, so the visit allowed Freud to take stock of the situation in Zurich. Writing to Binswanger after the visit, Freud complained that Jung had neglected his duties as president and was again under the sway of his Father complex, not to mention the influence of a woman who was not his wife.

When Jung learned of Freud's visit, he was enraged. Freud had traveled all the way to Kreuzlingen and had not bothered to visit him a few hours away. Freud's neglect could mean only one thing: he disapproved of Jung's new theory. Despite Freud's explanations to the contrary, Jung wrote that at the next congress, he would put his presidency of the I.P.A. up for a vote to see "whether deviations are to be tolerated or not." But the next congress was over a year away. Freud and Jung had already decided to forgo a 1912 meeting, since among other things, Jung had been invited to give a series of lectures on psychoanalysis in America that September. After finishing the second half of what would now be a book called *Transformations and Symbols of the Libido*, Jung set out for the New World and had his wife send Freud the long-awaited conclusion of his studies.

When Freud opened the second installment of Jung's work, he knew in part what he would find. Jung would argue that libido theory was too narrow in its definition. While Jung did not doubt that sexual libido was at the heart of the neuroses, he would state that in the psychoses it was irrelevant. Repression and regression reanimated old childhood dramas and created neurosis but did not explain madness. The psychotic retreat from the world was too broad, and it encompassed not just sexuality but all psychic energy. Jung reiterated the earlier criticisms of Bleuler and asserted with his former chief that schizophrenic autism could not be reduced to autoerotism. Libido, Jung offered, should be defined more generally like Schopenhauer's Will; it was inner striving.

From the very beginning, Bleuler and Jung had ridden the Freudian bandwagon while maintaining certain reservations. While Karl Abraham had temporarily put Bleuler and Jung on the defensive, by 1912 both men had retrenched and publicly denied sexual libido any causal role in psychosis. These defections appeared to be contagious: Freud heard rumors that the Zurichers were fleeing psychosexuality en masse. Franz Riklin emphasized universal, nonsexual symbolism, and Alphonse Maeder was apparently preparing to assert that the theoretical differences between Viennese and Zurich schools could be explained by race. Whispers had it that Maeder would dignify the claim that the Jews in Vienna were overly concerned with sex.

In New York during the fall of 1912, Jung delivered his lectures at Fordham University. While *Transformations and Symbols of the Libido* may have been a poorly organized, speculative, and meandering work, Jung's Fordham lectures were all clarity and confidence. He spoke authoritatively on the positions of "the psychoanalytic school" and laid out the importance of childhood and family experience and the role of the unconscious. Jung openly stated what he hoped would be the new view of psychoanalysis. Libido was not only based on sexuality, and Sigmund Freud's view of infantile bodily cravings was not tenable. Jung called the earliest phase of life presexual—the same term many sexologists used before Freud's innovations.

At the same time, Jung's lectures relied heavily on Freud and offered a view of psychoanalysis in which Freud was a revered elder who has been modified and corrected by his followers in Zurich. While preserving the role of sexuality and repression for the neurosis, Jung argued that adaptation was generally a more pressing imperative. Human fantasy bore an uncanny resemblance to the symbols and myths of religion, and it was certain mythic themes like sacrifice that caused individuals to develop what the "Vienna school" saw as a castration complex. Jung referred to the origins of human history only in passing, but he promised research would soon clarify the parallels between racial and individual symbolic systems.

At this Catholic university, Jung also drew parallels between psychoanalytic treatment and confession. He praised the religious rite as a "brilliant method of guidance and education," and he noted that the psychoanalytic method went further and analyzed transference, to free the individual from the childish needs to be submissive. Perhaps the pastor's son was announcing his own liberation from infantile wishes he once harbored for his one-time father-confessor, Sigmund Freud.

Jung returned to Zurich feeling triumphant. He bluntly told Freud: "my version of psa. won over many people who until now had been put off by the problem of sexuality in the neurosis." The Harvard physician James Jackson Putnam—though he found Jung's Fordham speeches incongruent with his own beliefs—recognized that by minimizing sexuality, Jung would bring many new adherents to the field. Jung preempted Freud's standard retort by saying he regretted that Freud would attribute their disagreement to Jung's personal resistances. He had no resistances, Jung sanguinely went on, except "my refusal to be treated like a fool riddled with complexes."

Jung was in a position to take over the psychoanalytic movement. From his base, he had gained the allegiance of many in the movement who came to be Freudian through Zurich. He was renowned in America, preferred in France, and held in esteem by academics in Germany. He had put together a newly consolidated theory that was more palatable than Freudian psychosexuality, was president of the I.P.A. and editor of the *Jahrbuch*, and his major competitors in Vienna, Adler and Stekel, were imploding. Jung did not hide the fact that—as Freud put it—"he considers psa. as his own." All that stood between Jung and the psychoanalytic movement was an aging patriarch, Sigmund Freud. But as Jung well knew, Freud was a dangerous rival.

Freud understood that Jung was too powerful to confront directly, so he calmly assured his colleague that his deviations were fine. The crisis around Stekel (which Jung had no way of knowing was the result of Freud's attempt to secure a platform to attack *him*) led to a meeting of all the branch leaders of the I.P.A. At the meeting in Munich, Jung and Freud met for the first time since Jung's overt breech. Freud collapsed. It was not the first time he had fainted in Jung's presence; the same thing had happened just before their trip to America. Then Jung had been going on about dead men found in bogs, which Freud had interpreted as a death wish against him and had passed out. In Munich, Freud had little doubt about Jung's wishes and jovially reported that his latest swoon made his enemies "burn with impatience." But paraphrasing Mark Twain, he could say with confidence that the "rumors of my death are greatly exaggerated."

In Munich, Freud and Jung tried to patch things up, but the best Freud could manage to say about Jung's second paper on libido was that it *inadvertently* showed that mysticism was based on archaic complexes. It was Freud's neuroticism, Jung charged, that made him disparage and undervalue Jung's work. Jung warned that Freud should not analyze Jung's Father complex anymore and wrote: "I am forced to the painful conclusion that the majority of ΨSA.sts misuse ΨA for the purpose of devaluing others and their progress by insinuations about their complexes (as though that explained anything. A wretched theory!)."

Unbelievably, Freud confided to Jung that he too had been "disturbed for some time by the abuse of ΨA. to which you refer, that is, in polemics, especially against new ideas." But the double-talk did not prevent an explosion between the two men. All they needed was a spark, which arrived in the form of a Freudian slip. Jung wrote Freud to deny that he wanted to leave the movement, but by mistake he wrote: "Even Adler's cronies do not regard me as one of *yours*." He meant to write not one of "yours" but one of "theirs."

Freud could not resist pointing to the revealing error. Furious, Jung shot back: "your technique of treating your pupils like patients is a *blunder*. In that way you produce either slavish sons or impudent puppies." Jung insisted that he did not have a neurosis. After all, *he* had been analyzed. But the man who claimed to have analyzed himself—Sigmund Freud—had never gotten rid of his own complexes. Unlike Freud's servile followers, Jung promised to speak his mind in private, write what he believed in public, and stay in the movement.

A bitter Freud wrote Ferenczi that Jung believed he was free of neurosis because he had been analyzed by Maria Moltzer, a Burghölzli nurse who joined the Zurich Freudians and was rumored to be Jung's lover. "He is so foolish as to be proud of this work with a woman with whom he is having an affair." As a final gesture, Freud denied Jung's accusation that he treated colleagues as patients and proposed the two men break off all personal communications. On January 6, 1913, Jung agreed.

Eugen Bleuler and Carl Jung had both come to Freud for his French-derived psychology of the unconscious, and though tantalized, they could never fully endorse psychosexual theory. After a short and troubled union, both men broke their ties with Freud once it became clear that a complete acceptance of psychosexuality was being demanded. The loss of Bleuler threatened the academic standing of the Freudian movement, and the loss of Carl Jung was another grave blow. A pillar of the Freudian community and the president of the I.P.A., Jung now openly contested libido theory and echoed the popular sentiment that Freud's theories of infantile sexuality were absurd.

Freud had justified a shift of power to Zurich to further the field among scientists and protect it from anti-Semitism, but his strategy back-

fired. The Burghölzli was lost as an academic training ground and research center, and members of the Zurich Freudians were themselves
generating anti-Semitic critiques of Freud's theories. In 1912 Freud wrote
Otto Rank: "What is most regrettable about the changes in Zurich is the
certainty that I did not succeed in the union of Jews and Anti-semites
whom I hoped to unite on psychoanalytic ground." Later, A. A. Brill reported that while on his trip in America, Jung had been heard to pro-
claim: "the Jews cannot get away from the rotten sex."

When Eugen Bleuler asked Freud what kind of community he hoped
to build, Freud laid out two goals. The first was to forge an organization
larger than himself, one that defied accusations of cultish sectarianism
and was not merely Freudian but truly a science of the psyche. In this
regard, the Nuremberg congress was a disaster: a number of his most creative scientists left Freud because they found no room for themselves in
what had become a strict Freudian school. The Nuremberg proposals for
theoretical standardization and a defense organization made partial
Freudians, no matter what their value, suspect. Sándor Ferenczi's "wartime" policies legitimated a process that would spin out of control, and
soon there was no safety inside this community for freethinkers or half-
believers. A commitment to other disciplines—even those that helped
give birth to Freudian thought—became dubious. Jung attacked Bleuler
for his ties to academic psychiatry; Ferenczi attacked Phillip Stein for his
interest in academic psychopathology; both Jung and Ferenczi wanted to
keep their distance from sexologists. The only secure position became orthodox loyalty to Freud.

His second goal, Freud told Bleuler, was to define the field and control
wild analysts who had the power to sow scientific discredit and harm the
medical reputation of the field. In that regard, the founding of the I.P.A.
had achieved some success, though the price was high. After Nuremberg,
the Freudians no longer ranged from Eugen Bleuler to Otto Gross. They
would no longer have to answer for the crazed therapy or obscure speculation of some self-declared Freudian. For new adherents, the Freudian
community was a more circumscribed enterprise with clearer theoretical
and practical limits.

Thus after Nuremberg, the Freudians defined themselves: if you
wanted to be in this community, a commitment to its standard method
and theory, especially the theory of unconscious psychosexuality, was re-

quired. The I.P.A. was created to police commitment to these things. But the premature insistence on set standards made many gasp for air. The fallout was alarming, as major theoreticians, including the I.P.A. president, could not remain inside the fold. A process of attack and marginalization limited freedom of inquiry and forced those who accepted some of Freud's ideas to choose: either they fully accepted his core proposals or faced banishment. After hearing that Stekel was getting expelled, a feisty Ferenczi wrote: "I feel freer and more battle-ready since we have begun to set aside the many precautions which the 'semi-adherents' have forced us to make."

The irony was that while the Nuremberg proposals were meant to close any gap between being a psychoanalyst and being a full adherent of Freud, they did the reverse. Nuremberg changed the borders, and almost overnight, partial Freudians became non-Freudian psychoanalysts. Freud's attempt to circle the wagons had created bands of renegades who tried to make psychoanalysis their own. Alfred Adler's followers did not relinquish their identities as psychoanalysts; they refused to equate the psychoanalytic domain with the person of Sigmund Freud. Adler's Society for Free Psychoanalytic Investigation took its name to rebuke Freud; they were psychoanalysts and were free to inquire wherever their thoughts took them. After his banishment, Wilhelm Stekel continued to edit the *Zentralblatt*, which was poised to become a not particularly Freudian, psychoanalytic journal. And in Zurich, Carl Jung, though no longer a Freudian, most certainly remained a psychoanalyst. After all, he was still the president of the I.P.A.

V.

THE FREUDIAN COMMUNITY was now a tangled web of envy, jealousy, paranoia, and ambition. Amid the excitement of spreading a new psychology and therapy, the psychologists could not keep themselves from internecine conflict, and worse still, the squabbles seemed scientifically insoluble. Freud resolved his differences with Adler by forcing him aside, saying his views were inconsistent with his own, which of course did not mean that Adler's claims were wrong. But this community seemed to have no other way to arbitrate the truth value of divergent opinions regarding the nature of the unconscious. And by the summer of 1912, the appointed president of the movement, Carl Jung, had also departed from Freud's vision of unconscious psychic life.

In May 1912, with these schisms in the air, Ernest Jones arrived in Vienna. The Welshman was an unlikely savior, for he brought with him a long trail of trouble. After being forced to leave London due to allegations of sexual abuse, Jones was embroiled in scandals with women in the New World. Now he came to Vienna with a woman he called his wife, the stunning Löe Kann, who was addicted to morphine. Jones hoped Freud would take Löe into treatment.

Jones found the Freudian movement in peril. At the next congress, it seemed quite possible that Freud and his followers would lose control of the I.P.A., the two-year-old organization they had founded to protect them. The *Zentralblatt* was in Stekel's hands, while the *Jahrbuch* was controlled by Jung. Jones, along with Otto Rank and Sándor Ferenczi, began to brainstorm: What could the loyal Freudians do?

The problem, they agreed, lay in the unanalyzed analyst. Alfred Adler's theories were simply expressions of his own untreated neurosis; similarly, Jung had followed his own complexes into the stratosphere. The three men agreed that stability in the movement would come only through ceaseless self-analysis, which would purge analysts of their own neurotic reactions. Ferenczi proposed that a small group of followers be analyzed by Freud, who would rid them of subjective complexes and thereby create a purified elite to lead the Freudians of the future.

Jones shared these ruminations with Freud, who seized upon the idea. He had once envisioned the movement being carried forward by Carl Jung, and now he needed another candidate. Ferenczi was an obvious choice, but he declined, saying he was not worthy. Instead of appointing a successor, Ferenczi urged Freud to let the ideas "pave the way for themselves by means of their own specific gravity." Freud was unwilling to be so resigned.

Jones scouted around for a possible heir but was not encouraged: Rank was hindered by poverty, and Ferenczi was risky, for he like Jung was fascinated by quasiscientific matters like thought-transference and psychic powers. Jones again suggested Freud convene a small group to protect the movement. These trusted few would work in public and private to insure the interests of the movement, act as a bulwark against the irrational exuberances of the I.P.A. followers, and guarantee the purity of the science.

Freud was delighted. A "secret council" composed of the best and most trustworthy! Jones had not mentioned anything about secrecy, but Freud emphasized it: "this committee had to be strictly secret in his existence and in his actions." Once Freud envisioned such a circle formed of Jung and the presidents of the local societies, but because Jung and the local leaders in Zurich and Munich had dubious allegiances, this was impossible. Jones imagined a group like Charlemagne's paladins guarding the "kingdom and policy of their master." The Secret Committee would protect Freud and his doctrines. But who could be trusted? Mutual suspicion was intense, even among the few loyalists who came up with the idea of the committee. Ferenczi cautioned Freud against Jones, and Jones openly wondered about the reliability of Ferenczi and Rank.

As the fall of 1912 arrived, the Secret Committee was still an idea. Jones traveled to Zurich on a fact-finding mission. Carl Jung was out of town, but Jones talked at length with Riklin, Maeder, Seif, and Binswanger and came away with the view that while some of Jung's ideas were highly idiosyncratic, he still relied heavily on libido theory. Jones was hopeful that a breach could be avoided. When Jung heard about this visit, he wrote Jones to clarify his own point of view:

> The worst was, that I clearly felt, that this work was destined to destroy my friendship with Freud, because I knew, that Freud never will agree with any change in his doctrine. And this is really the case. He is convinced, that I am thinking under the domination of a father complex against him and that all is complex-nonsens [*sic*]. It would break me, if I were not prepared to it through the struggle of the past year where I liberated myself from the regard for the father. If I will go on in science, I have to go on my own path. He already ceased being my friend understanding my whole work as a personal resistance against himself and sexuality. Against this insinuation I am completely helpless.

In this same letter, Jung made it clear that he might resign his post as president of the I.P.A., though he hoped to continue editing the

Jahrbuch. He wondered if Freud would try and force him out of the
I.P.A. altogether.

When Freud and Jung broke off all relations in January 1913, Jones's
hope for reconciliation died. With the Munich Congress approaching, it
became increasingly urgent for the remaining Freudians to plan their
strategy. And so, the Secret Committee came into being. On May 25, 1913,
the committee met in Vienna for the first time. The members were Ernest
Jones, Sándor Ferenczi, Karl Abraham from Berlin, and two loyal Viennese, Otto Rank and Hanns Sachs. Distrust among the paladins was apparent when Ferenczi welcomed Rank by pointedly asking: "I suppose
you will stay loyal?" The only surprise in the group was Hanns Sachs. A
lawyer and student of literature, Sachs joined the Vienna group in 1910,
but he had impressed Freud and had served as coeditor with Rank for a
new journal, *Imago,* dedicated to psychoanalysis and the cultural sciences.

The committee maintained its secrecy with the public pretense that it
was merely a gathering of close friends. The ruse allowed them to subtly
exercise their influence. The first order of business for the committee was
Jung's theoretical divergence. Was it major or minor, a fundamental
break or a minor modification? Many were not sure. Jones had the impression that Jung was not defecting, for while he defined libido as not
just sexual, Jung nevertheless looked to psychosexuality to make sense of
the neuroses. At the first meeting of the committee, however, Sándor Ferenczi distributed a devastating critique of Jung's new theory that the
Professor had asked him to write. It left little room for doubt: Jung's deviations were radical, and his main concern lay not in psychoanalysis but
the "salvation of the Christian community." There could be no reconciliation. To help strengthen the Freudians at the upcoming I.P.A. meeting,
Ferenczi would hurriedly found a Budapest Society, so there would be
another solid voting bloc when the inevitable showdown came.

Freud had something new to distribute to the Secret Committee as
well. It was a short book on the origins of the incest taboo and the subsequent birth of law, religion, and civilization. *Totem and Taboo* would
stand in direct competition with Jung's work and (Freud hoped) provoke
a storm of protest in Zurich. It would separate the Zurich contingent
from the others the "way an acid does a salt," thereby serving to "cut us
off cleanly from all Aryan religiousness."

It is unclear if the members of the Secret Committee discussed Ferenc-zi's idea that they should all be analyzed by Freud, but after the commit-tee's meeting, a depressed Ernest Jones went to Budapest to be analyzed by Ferenczi. Jones had been caught cheating on Löe, and she had decided to leave him. His analysis would be the first by a member of the Secret Committee and the first serious analysis of any of the leading figures in the movement.

On September 7 and 8, 1913, the fourth International Psychoanalytical Congress convened at the luxurious Hotel Bayrischer Hof in Munich. The congress was the largest to date, with over eighty registered guests and a scheduled program of seventeen papers. But the atmosphere was tense. The Zurichers sat at a table opposite Freud's, and throughout the congress, pro-Jungian and pro-Freudian camps kept their distance. Jung scheduled the vote for I.P.A. officers to take place at the very beginning of the first day. When the presidency came up, Karl Abraham motioned that all those not in favor of Jung's presidency should leave their ballot blank. Of the fifty-two ballots, twenty-two came back empty. Jung would leave Munich with the presidency of the I.P.A. and a long roster of his enemies.

The Zurichers maintained that Jung's reforms would save Freudian theory. Alphonse Maeder argued that the new work from the Zurich school, as well as Alfred Adler's ideas, represented significant modifica-tions that must be open to debate. Maeder contended that dreams had been shown to have manifest mythic content, were geared toward adapta-tion to reality, and were not generated from latent sexual material. Inter-pretations must consider these problem-solving aspects of dreams, rather than rooting them in the infantile and sexual. If the Zurich and Viennese schools were to be reconciled, Maeder left no doubt about which side he believed would have to compromise.

The Freudians left Munich in a bind. Jung was still president of the I.P.A. and would be for the next two years. Freud reported to A. A. Brill that upon hearing the Zurichers' "theologizing" in Munich, he had re-nounced them, telling them that they should not masquerade as his pupils and propagandists, for their work had nothing to do with *his* psychoanal-ysis. Jung had behaved in a "brutal" way, but the Freudians were out-numbered, so he remained president. Now, the Professor told Brill, he had every intention of dissolving the I.P.A.

Six weeks after being reelected president, however, Carl Jung abruptly resigned his editorship of the *Jahrbuch*. Jung's correspondence with Jones hinted that he was ready to go his own way, but Freud and his followers were deeply suspicious. It was "too good to be true," Jones wrote. Freud was also convinced that the resignation was part of a ruse by which Jung would dissolve the journal and then reconstitute it under his sole control.

The Secret Committee debated their options. Freud, Ferenczi, Rank, and Sachs concluded that the three loyal I.P.A. groups—Vienna, Budapest, and Berlin—should propose a dissolution of the organization. If Jung did not accept their proposal, they would secede. Karl Abraham, however, worried that his Berlin group might not side with the Freudians. Furthermore, Abraham saw no room in the rules of the I.P.A. to demand this action from Jung. Freud reassured Abraham that there was a clear statement in the I.P.A. rules that the purpose of the Society was "the cultivation of Freudian Ψ." Therefore, a statement from Freud would be enough to secede. But Jones also expressed reservations. Upon returning to London after the Munich meeting, he had followed Ferenczi's example and founded a London Psychoanalytical Society with nine members. Jones argued that there was no reason to hurry, since by the next congress the Freudians—strengthened by their new groups in London and Budapest—might be in the majority. If they had the numbers to dissolve the I.P.A., Jones reasoned, they also had the numbers to hold it and force Jung out. Jones also warned that if pressed right away, the London group might not vote with the Freudians. And he warned, the same might be said for the Americans.

Ernest Jones and Karl Abraham won the day. The Freudians did not announce a disbanding of the I.P.A., and before they knew it, things took a turn in their favor. The *Jahrbuch* and its publisher stayed with Freud, and Carl Jung voluntarily resigned from the editorial board of the *Zeitschrift*. Jung seemed to be—as Freud put it—doing the Freudians' work for them. The Zuricher even let it be known that he was willing to dissolve the I.P.A. and go his own way.

Once Freud's crown prince, Carl Jung was now angry and hurt. He wrote Jones to explain that he did not demand others adopt his theories, but that his hypotheses, much like Adler's, deserved serious discussion. Instead, they were reduced to personal complexes, and scientific discourse was replaced by political intrigue, gossip, and suspicion. "I dislike advertising even in science," he grumbled.

TOTEM AND TABOO was published as a series of four essays between 1912 and 1913. Instigated by Jung's research on myth, Freud made no bones about the fact that his work was intended to root out Jung and his Christian religiosity from the Freudian world. Much like Jung's prehistory of humanity, Freud's text is rather fanciful. For men who sought to portray themselves as reserved scientists, they had both taken great leaps into speculation. But Freud's essays also function as a roman-à-clef detailing the struggles to found a psychoanalytic community.

Freud had long been interested in anthropological visions of a primal human community that first shaped the unconscious lives of modern men and women. His interest was reignited by Jung's pursuit of the psychological origins of religion. The first essay in *Totem and Taboo* was entitled "Some Points of Agreement between the Mental Lives of Savages and Neurotics," an equivalence that was old hat for Freud. He justified this essay only as a lead-in to the second, "Taboo and Emotional Ambivalence." There, Freud incorporated Bleuler's terminology to describe the similarities between obsessional rituals and the actions of aborigines regarding taboos, especially taboos surrounding enemies, rulers, and the dead. In these taboos, Freud discovered an excess of solicitude and veneration that defended against murderous feelings. Given his experience as a ruler used to what may have seemed like false veneration, Freud may have been writing from experience.

In the fall of 1912, he wrote a third essay on animism, primitive thought based on magical reasoning, and projection. Such thinking persisted among the civilized, in the form of superstitions, delusions, beliefs, philosophies, and other systems of thought, which Freud argued were defined by an overriding need for coherence. Systems of thought—could he mean post-Nuremberg psychoanalysis, as well?—were intolerant of disconnections. Hence believers fabricated meaning to fill in these gaps in the same way Maori tribesmen, neurotics, and children did.

In May 1913, as Freud's struggle with Jung boiled over, the Professor completed his last essay. In it, he imagined nothing less than the birth of civilization. A Darwinian primal horde had roamed the earth, dominated by a tyrannical father who took all the women for himself and forced the other males to look beyond the horde for mates. In the end, the father was undone when his sons stopped fighting among themselves and banded together to slay him. Guilt ridden, the sons set up a totem to honor their father and voluntarily instituted his rule of exogamy. Freud was clear: the murder of the domineering father by the sons engendered guilt, the rule of law, moral restrictions, religion, in a word, civilization.

In penning this fantasy of civilization's origins, Freud acutely described his own tragedy. As father of a movement, he had created a community in which he was repeatedly accused of being tyrannical. Now he would either have to let himself be symbolically murdered to allow the community to mature from a frightened, savage horde into a civilized brother clan or retard the civilizing process by refusing to cede his authority.

After compiling the four essays into a book, Freud added an introduction that placed his work in direct opposition to Jung's. Once the manuscript was finished, Freud grew unsure of himself and fell into a depression. He began to soften in his feelings toward Jung; he did not want to provoke a split and would rather let Jung go on his own. He wrote to Ferenczi with a note of sadness and uncharacteristic self-doubt: "Must I really always be right, always be the better one? In the long run it becomes downright improbable to one."

Freud distributed *Totem and Taboo* to the Secret Committee and asked for their comments. Ferenczi and Jones replied that Freud had "in imagination lived through the experiences he described in his book," but rather extraordinarily they suggested that the experience Freud lived through was the son killing the father. Freud's loyal sons suggested that he had become downcast because in the book he had imaginatively reexperienced his own Oedipal death wishes toward his father, not because he was guilt ridden about taking the role of the tyrant, banishing another son, and preventing a more civilized brother clan from forming. After encouragement from all five members of the Secret Committee, Freud's confidence returned. On June 30, 1913, the members of the Secret Committee convened in Vienna along with Löe Kann, and held a celebratory "totemic festival." It is not recorded what they ate, but the blood of the son, Carl Jung, was no doubt somewhere in the feast.

Freud went off to summer in Marienbad and wrote his ally in Zurich,

Ludwig Binswanger, that he was enjoying his holiday. He wearily admitted that at times he dreamt of retirement and leaving the work to others, but then he added with godlike authority: "Who among mortals could put me into retirement?"

As FREUD'S DEPRESSION lifted, he began composing another work meant to force Jung's resignation and precipitate the collapse of the I.P.A. The harshest thing Freud had ever written, he himself called it "the bomb." It purported to be an account of the psychoanalytic movement that relied on historical exegesis to expose the false paths offered by fraudulent psychoanalysts.

In prior accounts, Freud tended to generously attribute the origins of psychoanalysis to Josef Breuer and his cathartic method, but now he was in no mood to cede authority to others. Psychoanalysis was his creation. For years he alone had used it, and therefore Freud considered himself

> justified in maintaining that even today no one can know better than I do what psychoanalysis is, how it differs from other ways of investigating the life of the mind, and precisely what should be called psychoanalysis and what would be better described by some other name. In thus repudiating what seems to me a cool act of usurpation, I am indirectly informing readers of this *Jahrbuch* of the events that have led to the changes in its editorship and format.

For a master rhetorician like Freud, this was blunt and bitter talk. The rest was a brilliant hatchet job. Freud told the story of his own theoretical journey, and in so doing, organized the tale to leave no doubt that there was no psychoanalysis without psychosexuality, and that the discovery of the centrality of sexuality to psychic life was central to the entire domain. Freud created a mythic "Ah ha!" moment to explain how he replaced his theory of paternal seduction with one based on inner sexual strivings. He presented what had been a long, tenuous process as a swift empirical recognition upon which psychoanalysis was built. For those who might argue that there was an 1895 or 1900 Freudian psychoanalysis that did not require acceptance of libido theory, this was Freud's curt reply.

Freud then discussed his acolytes in Vienna and Zurich, conceding that the latter had brought him nearly all his adherents. After an initial burst of growth, however, his followers began to stray. Adler had made a genuine contribution to the study of aggression and the ego, but he demanded that the rest of psychoanalysis be overthrown as the price for this grim gift. Jung came from a devout Christian community that could not tolerate the idea that sublimated sexuality created religion and great works of culture. As a result of these differences, three distinct groups now claimed to practice psychoanalysis. Surely there was room enough on earth for all three doctrines, Freud asserted, but they should not be confused with his term, his theory, his psychoanalysis. There was no such thing as Adlerian and Jungian psychoanalysis, Sigmund Freud insisted. Psychoanalysis was his.

In between the lines, Freud's bomb carried an admission of a kind of defeat. In the beginning, he had hoped *his* psychoanalysis would be a science, tied to empirical truths rather than the personal authority of any person. But some core propositions—especially the defining of the unconscious—were very difficult to prove. Adler's claim that the core unconscious drive was aggression may have been grim, but as Freud knew, that did not make it untrue. Jung's claims might have seemed like a return to Platonic Christian beliefs, but since Freud too believed in phylogenetic inheritance, he had no way of suggesting these views were scientifically less plausible than his own. In the end, Freud appealed to historical reasoning and the reader's logic, using thought experiments to discredit his rivals. And he did precisely what academic opponents accused him of doing; he took personal ownership of the psychoanalytic field. If Freud had once hoped to create a discipline that was not predicated on his own authority but rather on the broader tenets of science, this represented a great retreat.

In February 1914, Freud finished his history and waited for the *Jahrbuch* to publish his repudiation of the journal's founding editor. In the meantime, another "salvo" in the *Zeitschrift* was fired at Zurich. In that journal, Jung's ideas were attacked in concert by four members of the Secret Committee: Ferenczi, Abraham, Eitingon, and Jones. With that, Carl Jung had had enough. On April 20, 1914, the president of the I.P.A. sent out a circular letter to all the heads of the local societies announcing his resignation. Freud and the Secret Committee were shocked. They had expected Jung to battle them for control and were stunned by his willingness to go quietly.

Freud and his secret guardians had won the battle to keep the I.P.A. Jung's nemesis, Karl Abraham, was appointed interim president by Freud; even the pro-Jungian local leaders—Seif in Munich and Maeder in Zurich—reluctantly agreed. Then in July 1914 Freud's history of the movement was published. Freud fully expected his polemic to precipitate a break with the Swiss, and he was right. On July 10, 1914, the Zurich Society resigned en masse from the I.P.A. The only person in the group who refused to give up his membership was Binswanger.

The break with Zurich was now complete. But Freud was concerned about repercussions elsewhere. The division with Jung would soon come to America with unclear results. Jones predicted the schism would cause agitation in the fledgling London Society, where he found himself outnumbered by Jung's supporters. Still, Jones and Abraham assured the Professor that new members would be streaming into their societies, so the split would not affect the growth of the movement, which would now be explicitly Freudian.

By the summer of 1914, Freud fully controlled a winnowed but more theoretically homogenous Freudian movement. The process of internal purification that began in Nuremberg had run its course. There were no partial adherents left in the movement. A boundary had been drawn, and those who sought to straddle it had been ousted. There was no Bleuler, no Jung, no Burghölzli for training and research; the entire Zurich school had gone. There was no Adler or Stekel, and the sexual revolutionaries like Fritz Wittels had been tamed or ousted. Gone too were sexologists like Magnus Hirschfeld and Iwan Bloch, who would not sign on to a purely Freudian agenda for long. Gone was any possibility of winning over mainstream academic psychopathologists. And gone were any alliances with hypnotists and other psychotherapists like Oskar Vogt and Ludwig Frank.

Sigmund Freud had cleansed his community of partial believers, competitors, and potential successors. The tribal father had run off his rebellious sons. Surrounded by little Oedipuses, King Laius had scattered them before they could rise up against him. Freud ran the movement, and his Secret Committee of loyalists was in place to see that his will be done. If it once seemed that Freud helped inaugurate a wide range of psychical and sexological research and clinical practice in Austria, Germany, Switzerland, and elsewhere, that was over. The psychoanalytic movement once welcomed all thinkers and clinicians who adopted some part of Freud's method, theory, and approach, as well as those from the disciplines that Freud had relied upon to forge his field. No more. By 1914, the Freudians would insist that to be a psychoanalyst meant an absolute commitment to Freudian psychosexuality. And while there were at least two other divergent schools that might be confused with Freud's, they were not psychoanalysis.

Sigmund Freud's battle to define the Freudian field had confirmed his critics' fears. Those who stayed in the movement lived with the ever-present knowledge that should their own clinical experience and theoretical conclusions veer from psychosexuality, they would face exile. Abraham, Ferenczi, Rank, and Jones adopted openly subservient stances to the Master. Ferenczi went so far as to chide Freud for having allowed his other collaborators too much freedom and urged him to take everything into his own hands and not rely on others.

As disturbing news regarding the assassination of the heir to the Habsburg throne reached Vienna, Sigmund Freud celebrated his victory over the Zurichers. "I cannot suppress a Hurrah," he wrote Abraham. In a few years with any luck, Freud and his loyal crew would rebuild the I.P.A. as a purely Freudian enterprise and make up for the loss of the Zurichers, the Adlerians, and the rest. As he looked forward to another summer vacation, Sigmund Freud may well have wondered, as he had a year earlier: Where was the mortal who could force him into retirement?

PART THREE

Making Psychoanalysis

EIGHT

Everything May Perish

Everything has not been lost, but everything has sensed it may perish.

—Paul Valéry, 1919

I.

BY 1914, THE Freudians were poised to become one of numerous psychotherapy schools, each of which was narrowly defined by the vision of its leader. What had happened? Freud's earliest followers came to him because his work overlapped with a broad array of their interests. The strength of Freud's theory—its stunning interdisciplinary synthesis—was an advantage for recruiting new adherents, but it became a handicap for integrating an intellectual community. Not infrequently, the Wednesday meetings stalled as different members coming from various backgrounds talked past one another. Few had expertise in psychotherapeutics, contemporary models of the brain and mind, sexual disorders, and hysteria. They could not think their way into all of that. Instead, they had to decide whether they believed Freud.

The earliest Freudians had avoided conflict on the most complex theoretical issue by adopting a loose assumption that the sexual unconscious existed. Before the Nuremberg Congress, Freud himself seemed content with this tentative, pragmatic position. "The unconscious is metaphysical, and we simply posit it as real," he declared at a typically raucous 1910 meeting of the Viennese Psychoanalytic Society. Ludwig Binswanger asked Freud to explain himself: "He says we proceed *as if* the unconscious were something real, like the conscious. On the *nature* of the unconscious, like a true natural scientist, Freud says nothing, simply because we know nothing about it for certain, but rather only deduce it from the conscious." Binswanger went on to recall that Freud affirmed that just as Kant postulated the thing-in-itself behind the phenomenal world, he had postulated an unconscious, which can never be an object of direct experience, behind consciousness.

This was a reiteration of the stance Freud had taken a decade earlier in his book on dreams. But after Nuremberg, the as-if view of the unconscious was increasingly sacrificed in the hopes of reining in partial Freudians, theoretical deviants, and wild practitioners. A process of unification transformed the Freudian community into one where the unconscious had been defined. Freud rejected challenges to libido theory, often chalking them up to the sexual repression in the scientist. Those who refused to fully embrace his unconscious were—Kant or no Kant— out. The demand for theoretical purity ripped an arm off the movement in Vienna and another in Zurich. Many left and gave weight to the accusation that the Freudians were not a scientific community open to free inquiry.

The critics had a point. Freudians demanded a commitment to the psychosexual unconscious and made it their litmus test. They believed sexuality provoked such disgust that they must safeguard their discovery in this way. The problem was that this postulate was not a fact. There was much to recommend unconscious psychosexuality as a logical inference and a theory, but scientific theories, unlike political or religious beliefs, had no power to compel acceptance without proof. Despite Freud's many efforts to build a psychology of inner life that conformed with science, the post-Nuremberg Freudians became more of a polemically driven interest group, not unlike the alcohol abstinence society Bleuler championed. To enter the group, one had to accept not just principles of evidence, but also a conclusion that could not be fully proved. After 1910, the Freudian project narrowed and libido theory hardened into an oath of loyalty. Findings that contradicted that theory—like the Zurichers' work on psychosis—were not acceptable.

When Bleuler, Adler, Stekel, and Jung left the fold, they took with them the possibility that one might be allowed to think, in part, against Freud and his theory of psychosexuality while remaining within the Freudian community. After the Zurichers resigned, loyal Freudians were assured control of the I.P.A., but what were they left with? Purified of dissent, free to champion unconscious libido, it seemed these Freudians would go on touting the value of a sexual unconscious and bring forward confirmations, while discounting possible exceptions and contradictions. The groundwork had been laid to turn Freud's great synthesis into a monotonous, closed system of thought, and the Freudians seemed destined to become a tight-knit sect unified by their belief in their leader and an unknowable entity—not God, but a different *Ding an Sich*, the sexual unconscious.

Bleuler, Jung, and Adler were convinced they knew the source of the problem. It was Sigmund Freud and his need to dominate those around him. However, in the coming years, it became apparent that Freud was not the only psychological theorist who demanded obedience from his followers. Alfred Adler and Carl Jung founded communities that became narrowly Adlerian and Jungian and were increasingly defined by the authority of a charismatic leader. The psychoanalysts were buttressing Auguste Comte's contention that psychologies were subjective in origin and fated to split into schools of opinion, rather than achieve scientific consensus. It seemed the epistemological quandary inherent in psychology would determine the fate of the Freudians and make them one minor school among an ever-expanding number of others.

Astonishingly however, that did not happen. Instead of becoming more rigid, defensive, and propagandistic, the Freudians went through a crisis and then a dramatic transformation. These surprising changes were the result of a number of factors, as we shall see, but they would be instigated by Sigmund Freud's undermining of libido theory.

It would be a stunning reversal. At fifty-eight years old, Freud was famed for his theory of psychosexuality, which was his life's greatest achievement. It would have been all too human to defend the theory to the end. And yet, though Freud could be defensive and dominating, he was also a restless intellectual who felt most alive in the heat of creation. His ideal was to be a great discoverer, a man of science who rattled the world with new truths. After Jung's departure, Freud knew the critics, who accused him of behaving like an infallible pope, appeared to have been vindicated. He was determined not to let this impression stand.

After Jung's resignation from the presidency of the I.P.A., Freud immediately tried to defend his field as science. To make his point, Freud needed a stable definition of science, and at the time in Vienna this was a matter of intense debate. The Viennese philosopher Ernst Mach had little use for psychoanalysis, but he had little use for any science that seemed to go beyond clearly observable data. Mach argued for description only in science, and he viewed synthetic explanations as unwarranted. For Mach, a theory of the unconscious could only be metaphysical and antiscientific. By the same criteria, Mach opposed Ernest Rutherford's theory of the atom, a psychological idea of a unified self, and Albert Einstein's theory of relativity. Mach's views were forcefully opposed by his former student the physicist Max Planck, who believed any effort to root out metaphysics and inferential abstractions from science was ultimately futile and would only prevent the development of a "unified world picture."

Around 1911, debates about the nature of science were a staple of conversation in Viennese intellectual circles, such as the University of Vienna Philosophical Society, a group that included Breuer, Adler, and Christian von Ehrenfels. In 1914, Freud made reference to these discussions in an attempt to shore up his view that Freudian psychoanalysis was a science. Psychoanalytic ideas, he wrote,

> are not the foundation of science, upon which everything rests: that foundation is observation alone. They are not the bottom but the top of the whole structure, and they can be replaced and discarded without damaging it. The same thing is happening in our day in the science of physics, the basic notions of which as regards matter, centres of force, attraction, etc., are scarcely less debatable than the corresponding notions in psycho-analysis.

Freud was too shrewd not to notice that this statement flew in the face of all that had just transpired in the Freudian movement. The idea of unconscious sexual libido had become Freudian bedrock after Nuremberg. Matters of close clinical observation and inference had become secondary to verifying the sexual unconscious wherever it could be found.

In 1914, when Freud wrote these words, he simply could not be taken seriously.

But as the diminished Freudians regrouped and world events took a barbarous turn, Sigmund Freud would be reminded of the responsibilities and possibilities in those same words. He would initiate a process that would do what Bleuler, Jung, Adler, and Stekel could not do. He would rip up his old notion of the unconscious, the core belief that had once defined the Freudians, and propose stunning changes. Inspired, others sought to open up their intellectual community—at times against the Professor's wishes. After 1918, instead of demanding a central ideological commitment to a man and his (now changing) theory of unthinkable thoughts, a number of Freudians tried to otherwise bound their discipline. New voices suggested that this community should be defined, not by a forced allegiance to a theory of the unconscious, but rather by more tangible means.

As we shall see, by calling into question his own theory of the unconscious, Freud helped rescue his field from becoming another closed system of thought, bent on trumpeting its confirmations, forging connections when needed, and defending itself from contradiction. After 1918, radical changes would transform Freudian studies into a broader, more diverse, more open, and ultimately more popular field. Free thinking men and women would flock to this reformed community. And the members of the movement began to view themselves differently. Many stopped calling themselves Freudians and began to see themselves as psychoanalysts. The great flowering of psychoanalysis occurred between 1918 and 1938. It was to be the best of times, and the worst.

ON JUNE 28, 1914, Archduke Franz Ferdinand, the heir to the Austro-Hungarian Empire, was assassinated. With breathtaking rapidity, much of Europe was at war and the international Freudian movement was in shambles. The I.P.A. congress in Dresden was now impossible. The readership of the psychoanalytic journals dropped precipitously, as journals published in Germany or Austria suddenly had no buyers in England, Russia, France, and then America. By 1915, paper and ink shortages threatened publication. No matter, for there was little to publish. Many

Freudians had been conscripted and were employed as emergency physicians and surgeons, not psychoanalysts. The doctors who managed to avoid conscription struggled with inflation, food rationing, flash epidemics, and the collapse of their practices.

The Battle of Verdun, which lasted from February 21 to December 18, 1916, caused over a half-million casualties.

"What Jung and Adler left intact of the movement is now perishing in the strife among nations," Freud wrote. "Our journals are headed for discontinuation; we may succeed in continuing with the *Jahrbuch*. Everything we wanted to cultivate and care for, we now have to let run wild." Freud's dire predictions were well founded; his one hope was not. The *Jahrbuch* appeared once in 1914 and then promptly expired. Soon afterward, Hugo Heller informed Freud that he could no longer publish the *Zeitschrift*, so the official journal of the I.P.A. was forced to shut down. It was cold comfort to know Stekel's *Zentralblatt* had died as well. Suddenly, the only regularly published psychoanalytic journal in the world was the recently founded *Psychoanalytic Review* in the United States.

In Vienna, the Psychoanalytic Society found it hard to bring together a weekly quorum and began to meet every two or three weeks. The Secret Committee—bowing to the difficulties of wartime censors and interdicted international mail—ceased to function. The numerous international correspondents Freud had cultivated over the last decade were now nearly impossible to contact. To make its way outside the Central Powers, a letter required elaborate plans and emissaries. Freud's pool of international patients dried up. By November 1914, he was treating only one patient.

And so, Sigmund Freud was suddenly free from the editorial, political, administrative, epistolary, and clinical labors that had so occupied him over the last decade. He returned to the private battle he had been fighting, a battle that now had two fronts. Freud would try to defend psychoanalysis from the suspicion that it was some kind of a cult, and he would labor to marginalize Carl Jung and Alfred Adler. Freud threw himself into the question of how one defined the unconscious. He was dismissive of Jung's mythic unconscious, but he found Adler's position more challenging. Adler seemed to grasp something that Freud had not adequately developed: the force of aggression and its place in willful, conscious striving. In his battle with Adler, the Professor adopted his familiar strategy of taking what he found valuable, synthesizing that with his preexisting theories, and then dismissing the rest. He would put the tastiest morsels of Adler's theories into the Freudian stew and encourage the reader to put the rest out to rot.

This reckoning became Freud's "On Narcissism: An Introduction." The term "narcissism" had been used in psychiatry to characterize a psychic retreat from reality. In 1908, Freud used the term to distinguish disorders in which there were no transferences. Two years later, he incorporated a normal stage of narcissism into his theory of development. He posited that the earliest autoerotic stage was followed by a narcissistic libidinal investment in the self, which only then led to the capacity to love others. Narcissistic neuroses, which others called psychosis, were regressions to this middle phase, a pulling back of libido into the self causing megalomania and a detachment from people. Jung rejected this argument, countering that repression of libido might make a man into a grim ascetic but not a psychotic. Freud set out to dispel Jung's seemingly convincing logic by postulating a group of "I"-drives. These drives served an individual's basic survival needs in the same way sexual drives served the survival requirements of the species. While these "I"-drives seemed asexual, Freud insisted that they were actually fueled by sexual libido. Their repression accounted for the non-sexual elements of psychosis.

After that, Freud turned to Adler's theories of inferiority and aggression, as well as his descriptions of neurotic feelings of disconnection and vanity. To explain these traits, Freud postulated that over time the "I" developed an ideal, which was the repository of all the self-love of childhood. A critical agency also emerged, which was the embodiment of parental and social criticism, and it oversaw the "I," monitoring discrepancies between it and its ideal. When one failed to live up to expectation, the critical agency leapt into action. In paranoia, these critical voices were divested of "I"-libido, and perceived as external. In a followup to these thoughts, the extraordinarily rich "Mourning and Melancholia," Freud elaborated his vision of that critical agency and its potential cruelty toward the "I," seeing this as a cause of depression.

"On Narcissism" showcased Freud's gifts as a polemicist, but more impressive was his agility as a theoretician. In his attempt to counter competing ideas, Freud soared far beyond the task and initiated entirely new theories of "I"-functions, all of which cast new light on the nature of identity, internal self-regulation, love, and disorders that involved self-loathing. By synthesizing the challenges of his rivals, Freud ended up with a much richer model that was nevertheless his own.

After "On Narcissism," Freud set out to present the only form of empirical proof he could muster in his battles with Jung and Adler; he wrote up a case history he hoped would definitively prove the existence of infantile sexuality. From dream analysis, Freud hoped to reconstruct the first four years of an adult patient's life, a rather tall order indeed. Freud presented himself as an empirical observer who had no preconceptions about what he would find in this man's unconscious, a risible claim, given his years doggedly defending libido theory. And the case proceeded in this way, displaying the Professor at his most tendentious. The patient, a twenty-three-year-old Russian aristocrat, had been ill for five years when he began an analysis with Freud. For three years, the analysis went nowhere. In the summer of 1913, Freud decided to combat the stasis by declaring that the treatment would end in one year, no matter what. Under the pressure of forced termination, the Russian's neurosis miraculously came unglued.

The patient had a dream he remembered from the age of four: in it, he saw wolves sitting in a tree outside his bedroom. (From then on, this patient would be known in the literature as the "Wolfman.") Freud interpreted the boy's castration fears and his fear of his father, but Freud went further to make the case that the dream indicated that the boy had witnessed his parents copulating when he was only eighteen months old. The reconstruction was too clever by half. Freud himself shared the concern that the reader might not find his narrative believable and hedged his bets by arguing that while the witnessing of the "primal scene" *must* have occurred, even if it didn't, Lamarckian heredity could be responsible for the child's memory of such an event. Before publishing the case, Freud doubled back and suggested that the boy may not have seen his parents having sex, but rather dogs coupling.

If this was what it took to prove that the Freudians were right and the Jungians and Adlerians were wrong, it was going to be a hard sell. The cries of the Wolfman were not only those of a frightened child waking from a nightmare, but of Freud himself, whose nightmare was a method that seemed unable to validate empirically his theory of the unconscious and prevent his hopes of a scientific community from going to pieces.

As the war thundered on in Europe, Freud put aside empirical proofs and tried to strengthen the foundation of his thought by developing his "meta-psychology." As Kant had argued that physics required metaphysics, Freud maintained that psychology required a set of assumptions that went beyond the domain of empirical study. In 1915, in a seven-week span of furious activity, Freud spun out twelve interconnected meta-psychological papers. The task, he said, was to "clarify and carry deeper the theoretical assumptions on which a psycho-analytic system could be founded." Meta-psychology, Freud argued, was necessary to organize, categorize, and clarify data. He freely admitted that these abstract concepts were not derived from observations alone. However, by comparing these ideas with empirical events, they could be honed and made more precise.

Freud published papers on drives, repression, the unconscious, and dreams. After this flurry, in the spirit of wrapping up and consolidating his life's work for its journey into an uncertain future, in 1915 Freud consented to deliver a series of basic introductory lectures at the University of Vienna. With these lectures, he created what amounted to an introductory textbook for the field.

Freud was not the first to write a textbook of psychoanalysis. In 1913, the Zurich pastor and pedagogue, Oskar Pfister, wrote *The Psychoana-*

lytic Method. The book placed a good deal of emphasis on Jung's new thought, even going so far as to let Jung write the entry on his own work. A year later, Pfister resigned from the I.P.A. with the other Zurichers. The pastor owed his primary allegiance to Jung, who had introduced him to psychoanalysis and become a friend. However, Pfister became dismayed by Jung's denunciations of Freud and his insistence that Pfister's desire to consider both Jungian and Freudian points of view was a malicious refusal to be obedient. Pfister concluded that Freud was in fact more tolerant to open scientific inquiry and switched back to the Freudian camp.

The pastor's textbook became popular among Americans like G. Stanley Hall, who lauded the book's open discussion of the various schools that had formed in psychoanalysis. An American translation was begun, and Pfister took the opportunity to make changes. While he left many of his references to Jung intact, Pfister scuttled a section in which Jung critiqued libido theory and dropped Jung's essay on character types. But Pfister left one extraordinary remnant from his first edition, which testified to the state of affairs prior to the schisms. After presenting clinical material, Pfister demonstrated how the content could be interpreted successfully by the theories of Alfred Adler or Carl Gustav Jung or Sigmund Freud. "The reader will see from our example how difficult it is, under certain circumstances, to obtain absolutely reliable interpretations or to take a position in the successive theses of the leaders of analysis," Pfister confessed. This was certainly not a view Freud wanted new students to take.

Other problems riddled the textbook written by another Zuricher, Leo Kaplan. The Russian-born Kaplan moved to Switzerland in 1897 to study mathematics, physics, and philosophy before turning to psychoanalysis around 1910. A solitary thinker, Kaplan was unknown to Freud when his *Fundamentals of Psychoanalysis* appeared in 1914. Kaplan approached psychoanalysis from a philosophical angle, in which first principles deductively led to elaborated theories. His treatise was, however, not welcome in the Freudian community. As Freud avidly defended the empirical nature of psychoanalysis, he dismissed the interpretation that his creation was a deductive philosophical system and called Kaplan a "lamentable schlemiel," an opinion that softened only after Kaplan wrote another book attacking psychoanalytic rebels.

Between the pastor's ecumenical book and the philosopher's abstract study, there was no textbook that satisfied Freud. From behind enemy lines in France, Emmanuel Régis and Angelo Hesnard had written an overview of psychoanalysis, but Ferenczi reported that their book was marked by "the ridiculous vanity of making everything essential in your teachings originate from the French." Later, Freud discovered a neurologist from The Hague named Adolph F. Meijer, who had written an introductory book in Dutch. But who read Dutch?

In the winter of 1915, Freud began his own presentation of a unified picture of psychoanalysis. Over two successive winter terms at the University of Vienna, he rose to the podium twenty-eight times. His lectures presented psychoanalysis as an empirical science that had grown naturally from the examination of certain phenomena. Freud inched his listeners forward. He assumed no prior knowledge and built his ideas from the ground up. As he proceeded, he accentuated the observable problems that his theory was built to solve. He strove to demonstrate that psychoanalysis was not a closed, speculative, philosophical system, but an accretion of different experiences that had led to a way of thinking. He initiated his audience with the most common of experiences: slips of the tongue and the little miscues of everyday life. Then for nearly three months, he lectured on dreams. A student who had completed the first year of the Professor's classes would have heard detailed descriptions of bungled acts, instances of forgetfulness, and dreams of falling and flying. They had been introduced to psychic determinism and the pleasure principle, but they hadn't heard a single word about libido or sexuality or transference.

The pacing of the lectures was masterful. There was humor and candid exposition without a trace of dogma or blind ideology. If it took Freud twenty lectures to get to libido, it was because he openly shared his hope that he would win his audience over by demonstrating the inevitability of this idea. He wanted to show, not tell. In the twenty-second lecture, Freud addressed the controversies that had recently convulsed the psychoanalytic movement. Disarming his listeners, he made light of the whole thing. Scientific controversies usually ended in sterility, he opined, because the debates become too personal. In scientific controversies: "people are very fond of selecting one portion of the truth, putting it in the place of the whole and then disputing the rest, which is no less true, in

favour of this one portion." The fierce infighter who had championed the
sole primacy of sexual libido now saw the need for both sexual and "I"
drives, the interpretation of the past and the present. Of course, Freud's
"I" was rooted in sexual libido, and his past had been given the power to
overwhelm the present.

II.

BY THE TIME Freud's lectures were over, the war in Europe had grown
steadily more horrific. Like many, Freud was initially caught up in war
fervor and had hoped the conflict would be noble, short, and swift. But a
quick victory for the Central Powers was nowhere in sight. Instead, ceaseless slaughter was the order of the day. In 1916, the battle of the Somme
claimed over half a million lives. Industrial and mechanized warfare and
the advent of deadly chemical gases, explosives, and submarines had made
for unprecedented casualties. Millions were dead. The wounded, or-
phaned, and widowed were too numerous to count.

During this new kind of war, combatants on both sides experienced
bizarre symptoms. Soldiers developed incongruous paralyses; they went
mute and deaf. They shook, blinked, trembled, and succumbed to panic
as they were overwhelmed by waves of fear. Psychiatrists were brought in
and asked to treat the men and get them back to the front. Finding a cure
for "shell shock" became urgent. But shell shock, *Nervenschock* or "war
neuroses," not to mention the many other French, German, and English
terms for these troubles, were enigmas for European psychiatry.

For the psychiatrists who solely believed in neurological causes, the
reason for such an illness was simple. The soldier's brain had been traumatized. The most famous proponent of this viewpoint was none other
than Charcot's old rival, the Berlin doctor Hermann Oppenheim. Oppenheim made his reputation arguing that traumatic neuroses were the result
of microscopic brain damage, and he extended this logic to shell shock.
But Oppenheim's theory had weaknesses. For example, what caused shell
shock in men far behind the lines, men who were nowhere near explo-
sions or gunfire?

Oppenheim believed these men had been somehow wounded and
should not be returned to combat. But military authorities were reluctant
to accept this opinion, worried it would compromise the war effort. By
late 1916, some doctors sought active strategies to get shell-shocked men

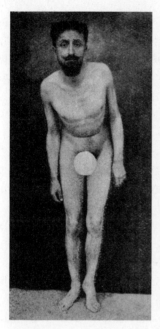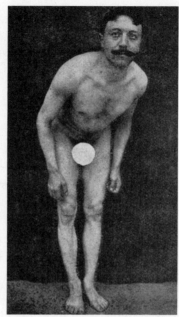

War neurosis. Two French infantrymen who were thrown by exploding shells in 1914, but deemed to have no medical reason for their subsequent inability to walk or stand naturally.

back to the front, reasoning that a psychic shock had traumatized the men. In Paris, Jules Déjerine proposed that emotional shock caused war neurosis. The French doctor Joseph Babinski declared that these illnesses were caused by suggestion from unwitting doctors; the shell-shocked should not be encouraged to continue their charade but be brusquely shipped back to the front.

Many physicians in Germany and Austria agreed. They concluded that the men were either consciously faking or deluded by suggestion. Professor Julius von Wagner-Jauregg in Vienna advocated the use of the time-honored treatment, electrical shock, in the hope that painful bursts of electricity would bring hysterical simulators to their senses. A German practitioner named Kauffman became notorious for his treatment of shell-shocked patients: without giving any warning, the doctor rushed his patient, electrodes in hand, as he executed a "surprise shock attack."

Such brutality provoked an outcry and led perplexed authorities to take a closer look at the less gruesome methods of treatment, such as hypnotism and psychotherapy. For example, a Hamburg neurologist, Max

Nonne, developed a reputation for therapeutic success with his hypnotic technique. As for the Freudians, they had been sent into military hospitals and clinics, not an easy setting to employ practices that senior doctors questioned. However, by 1916, Karl Abraham had become director of a psychiatric station in Allenstein and was able to test psychoanalytic methods on the shell-shocked. Ferenczi was transferred to Budapest to run a neurology ward, where he began to experiment with Breuer and Freud's old cathartic strategy. These Freudians were well positioned to offer treatment for traumatized men, but they were stymied by a serious problem. While Freud's psychological innovations originated to account for trauma, it was no longer 1895. Long ago, the Freudians had publicly rejected trauma theory and cathartic work, and they had embraced the sexual etiology of neurosis. And war neurosis, it would seem, could not possibly have anything to do with sex. Nonne declared that the war neuroses had proved Sigmund Freud wrong. Here were cases of hysteria with no possible connection to disturbances of libido.

But then a heretofore unknown German doctor came forward with extraordinary results. He claimed to have cured a mass of war neurotics with a modified form of psychoanalysis. His name could not be found on the roster of the International Psychoanalytical Association, but he would swiftly be embraced by the remaining Freudians. Ernst Simmel had attended the universities of Berlin and Rostock, where he wrote a thesis on psychogenesis and dementia praecox. Simmel went into private practice in 1913, only to find himself in the German Army medical corps a year later. He served as a battalion physician until 1916, when he became the medical director of a special military hospital in Posen for war neuroses. As the director of the Posen hospital, Simmel rejected the "system of tortures," the "hunger cures, dark rooms, prohibition of letters, painful electric currents, etc." He also found that the indiscriminate use of suggestion was ineffective. Instead, he turned to a combination of hypnosis and psychoanalysis. As an outsider to the Freudian movement, Simmel didn't care about the purity of his doctrine; he didn't worry about whether this was Freud or Adler or Jung. He wanted something that he could effectively use to treat a large number of patients. He opted for the cathartic method combined with analysis, dream interpretation, and hypnosis. Simmel found his patients' dreams particularly revealing and stated that he would not treat patients whose "dreams I do not know."

Simmel believed that war neurotics repressed painful experiences that acted like autosuggestions to cause their symptoms. In two or three sessions, Simmel found he could cure war neuroses, and by 1918, he claimed to have cured some two thousand men. The war neuroses were the single great issue facing psychiatry in 1918. If psychoanalysis could provide an answer, it would insure acceptance by the military, and then perhaps academic medicine and society at large. Upon hearing of Simmel's work, Freud excitedly wrote to Ferenczi: "German war medicine has taken the bait."

As the summer of 1918 approached, Austrian, German, and Hungarian Freudians pushed for a resumption of psychoanalytic congresses. There was no possibility of the congress being international, but there was a great deal to talk about, most importantly the war neuroses. Abraham's attempt to arrange a congress in Germany failed, but Ferenczi was able to set up the meeting in Budapest.

The 1918 Budapest Congress underscored the apparent centrality of that city for the future of the psychoanalytic movement. During the war, Freud had become closer to and more dependent on Ferenczi. Cut off from the Swiss, alienated from Adler's group, and unable to communicate freely with his British or Americans followers, Freud was quite isolated. His practice was weak, at times nonexistent. He leaned on Ferenczi, and Ferenczi responded generously, sending Freud patients and much-needed food. In return, Freud tried to counsel the irresolute Budapester on his warring affections for Gizella Pálos and her daughter, Elma. In an attempt to resolve his ambivalence, Ferenczi asked Freud to take him into analysis. In 1914, three weeks of analysis took place, just before Ferenczi was called up for duty. The two men met briefly two more times to take up Ferenczi's analysis.

Despite his inability to make up his mind on the romantic front, Ferenczi proved himself a searching, bold theoretician and writer. During the four years of the war, Ferenczi managed to finish thirty-eight articles despite his military obligations. And he was responsible for one of the very few encouraging wartime developments for the Freudians: he referred the wife of a wealthy Budapest brewery owner to Freud for analysis. The brewer, Anton von Freund, was a cultured man who became interested in psychoanalysis, and after being diagnosed with cancer of the testes, he became distraught and went into analysis with

Freud. Anton von Freund became a committed advocate and benefactor of the psychoanalytic cause. He donated a large sum of money to form a foundation to support the field. The combination of Anton von Freund's financial assets and Sándor Ferenczi's intellectual leadership made Budapest a city that shone with promise. In March 1918, to give further weight to Freud's expectations, the Hungarian Psychoanalytic Society was inaugurated with a very respectable number of members in attendance, nineteen.

Thanks to von Freund, the Budapest Congress was lavish. On September 28 and 29, 1918, the congress took place in the Hungarian Academy of Sciences. No one from outside the Central Powers came, save two intrepid souls from neutral Holland. Forty-two analysts and interested parties attended, including official representatives of the German, Austrian, and Hungarian governments, all of whom were there to discover what psychoanalysts could do to cure war neuroses.

The war neuroses made apparent some basic assumptions of psychoanalysis by demonstrating again the power of psychic causation, both conscious and unconscious. Against those who overstressed heredity, the war showed how the world could make a person quite crazy. And Ernst Simmel's work showed that by understanding unconscious factors, one could cure these perplexing illnesses. The psychic half of the Freudian synthesis was there for the world to see. And if ever there was a case to be made for an inner animal in man, the war had made it. However, this animal did not seem to be primarily sexual, but cruel and violent, not psychosexual but a psycho-killer.

Here Simmel could offer no help. While carefully couching his statements, he reported that his traumatized warriors had repressed feelings that were mostly not sexual. Simmel also reported that the dreams he had studied with these war neurotics did not seem to be wish fulfillment dreams. There was no pleasure to be found in these traumatic nightmares, and he concluded instead that these repetitive dreams were failed attempts at catharsis.

In Budapest, Simmel presented his stunning findings to a group of Freudians who knew little about him. Ferenczi and Abraham also presented their views on the war neuroses. Ferenczi ridiculed those who insisted on neurological disruptions in these cases, saying if Robert Gaupp had labeled these imagined brain lesions "brain mythology," he had done mythology an injustice. Ferenczi also launched a defense of the sexual etiology of the war neuroses, suggesting that while these traumatic phenomena didn't appear sexual in nature, such narcissistic regression from the world was the result of an attempt to return to the early states of development when the self was the focus of libido.

Thanks to Simmel, the congress was a roaring success. The representatives of the Prussian War Ministry and the Budapest Military Council promised to set up psychoanalytic treatment stations in the field to treat war neurotics. Freud was elated. Psychoanalysis was about to be embraced by the military and institutionalized. Anton von Freund's foundation had promised to support a press, the *Internationaler Psychoanalytischer Verlag*, a clinic for the poor, and an institute. Von Freund's money also made it possible to endow a prize for the best psychoanalytic publication, which Freud immediately gave to Ernst Simmel, as well as Karl Abraham and Otto Rank.

Then in a flash, all good fortune washed away. Before the military authorities had a chance to set up psychoanalytic clinics, the war ended in a humiliating defeat for the Central Powers. The great future for psychoanalysis in Germany and Austro-Hungary collapsed as those countries crumbled. When the armistice was declared in November 1918, the old dual monarchy in Austria and Hungary was gone. The vain, bloodstained kaiser, Wilhelm II, had fled Germany. In Budapest a socialist government took over. Freud was bitter: "No sooner does it [psychoanalysis] begin to interest the world on account of the war neuroses than the war ends, and once we find a source that affords us monetary resources, it has to dry up immediately." "Our kingdom," he added, "is indeed not of this world."

III.

SIGMUND FREUD'S DEFENSE of the great goddess Libido had been constant for fourteen years. At the war's end, he was sixty-two, an old man surrounded by true believers who were willing to push psychosexuality as far as they could. The idea had always seemed too elegant and powerful to modify. Eros and Psyche had been reunited, and a bridge had been built between body and soul, animal and man, feeling and thinking, species and individual. But for all the explanatory power of this synthesis, there were difficulties that the Great War made impossible to ignore. For religious and philosophical thinkers, the problem was evil. For biologists and psychologists, it was aggression.

Freud had never been blind to man's violent urges. In 1905, he accounted for aggression by linking it to a perverse lust, sadism, which he saw as rooted in sexuality. When the Oedipus complex came to play a central role in neurosis, Freud offered a model in which the frustration of sexual desires led to murderous competition and rivalry.

Gradually, Freud's view of hatred and aggression expanded through his study of transference. In 1908, Freud hospitalized a nobleman at the Bellevue Asylum in Kreuzlingen and complained that Herr J.v.T. was impervious to his influence. Over the next two years, this man was hospitalized at Bellevue three more times; each time he resisted the advice and influence of his doctors, which had led one, Alphonse Maeder, to inform Freud that the transference with Herr J. had "failed." In 1911, Freud revised his theory so that sexual transferences did not simply fail but were superseded by a hateful transference, though one still of "a distinctly erotic nature."

In 1912, Freud delineated a broader vision of love and hate in his short, brilliant paper "The Dynamics of Transference." When he wrote this essay, Freud was treating an obsessional woman named "Frau N.," who scorned Freud's every attempt to aid her. This was not the negativism of dementia praecox, but vengeful feelings that she had "transferred to me." Freud informed his patient that her treatment would take place on the "battlefield of transference." In his 1912 paper, Freud conceptualized that fight as one that could be understood through "Bleuler's excellent term 'ambivalence,'" a concept that had been considered a breakthrough at the Burghölzli, particularly after it had been transformed into a dynamic theory, in which one pole of ambivalent feelings could be repressed. In 1910, Franz Riklin had argued that the Oedipal complex was by its nature ambivalent, and that Oedipal transferences were similarly split. Freud now also acknowledged that transferences segregated into positive and negative poles. In more severe illnesses, Freud concluded that there was too much negative transference, and in the face of so much hatred, psychoanalysis was not possible.

The source of negative transference, for Freud, was still perverse transformations of sexuality. But Alfred Adler challenged that view, making the case that aggression was *the* essential human drive. Freud and his fol-

lowers were calling transferences sexual, he argued, and refusing to recognize that beneath a mask of fearful compliance lurked protest and rage.
Alongside Adler's privileging of aggression, Wilhelm Stekel had been
telling the Vienna Society for years that he believed anxiety was due to a
struggle between life and death forces, which he called "Eros" and
"Thanatos." Stekel later posited a criminal impulse in the unconscious
and declared that the common bond between all men was not love but
hatred.

While Adler and Stekel seemed ready to replace sexual libido with the
forces of aggression, others sought to give both a place in the unconscious.
In October 1911, a new follower, Theodor Reik, delivered his trial paper
to the Vienna Society. Reik was pursuing his doctorate at the university
by studying the psychology of Gustav Flaubert's novel *The Temptation of
St. Anthony*, and his topic was: "On Death and Sexuality." Reik argued
that the fusion of sex and death were due to guilt about sexuality, as well
as "the ring that is formed by coming into being and perishing, by Eros
and Thanatos."

In the crowd that night was another new member who was distressed
to hear her ideas being put forward by someone else. Sabina Spielrein, the
once miserable young medical student and special friend of Carl Jung,
had come to Vienna to pursue her own career in psychoanalysis. She carried with her a host of ideas she had been hatching in Zurich. Influenced
by Bleuler's concept of ambivalence and Jung's phylogenetic reasoning,
Spielrein had begun to conceive of a death drive that would exist beside
sexuality. As early as 1909, Jung had encouraged Spielrein to publish her
views, but she had not.

After Reik concluded that evening, Spielrein announced that she had
addressed many of the same problems in a paper in which she also argued
that "thoughts of death are contained in the sexual instinct itself." When
Spielrein's paper was published in 1912, it was entitled: "Destruction as a
Cause of Coming into Being." In it, she asked: Why would a reproductive
instinct harbor so much negative feeling? She posited an instinct for self-
preservation that was by its nature ambivalent. It strove to maintain
sameness and protect the individual, and was opposed by a dynamic force
that looked to resurrect the individual in a new form by destroying the
old. For Spielrein, the species imperative of creating new individuals was
at odds with the individual's desire to remain constant and live forever.

Before innovative challenges to libido theory, Freud had always followed the doubters into their strongholds, their exemplars of the nonsexual, and returned with sexual explanations. While Freud's reaction to
Spielrein's paper was not recorded, he was predictably unimpressed by
Reik's claims. In 1913, Freud formalized a theory of aggression that he
incorporated into his schema of sexual development. He created an anal-
sadistic stage of psychosexual development, to account for the power of
hatred in cases of obsessional neurosis. Sadism and masochism had
become the result of an anxious retreat from genital sexual pleasure.

If it ever seemed reasonable to contain all the manifestations of aggression in an anal-sadistic retreat from sexuality, in 1918 such an assertion
seemed untenable. After World War I, it was difficult to imagine that
human aggression came from such a limited, subsidiary place. Was the
slaughter that had bathed Europe in blood the result of an obsessional
neurosis epidemic?

Freud once saw himself as the servant of the great goddess Libido, but
during the war years he had begun to recognize the presence of another,
angrier god. Six months into the war, he wrote: "we spring from an endless series of generations of murderers, who had the lust for killing in
their blood, as perhaps, we ourselves have today." And he added: "by our
unconscious wishful impulses, we ourselves are, like primaeval man, a
gang of murderers." The "great experiment of the war"—as Ferenczi
called it—had made Freud reconsider human savagery. Soon he would
find himself reluctantly wondering if Stekel had been right. Perhaps men
were more connected to one another by hatred rather than love.

IV.

AFTER THE BUDAPEST congress, Freud collated the papers on war neuroses
for publication and supplied his own introduction. He had opened the
congress by saying: "we have never prided ourselves on the completeness
and finality of our knowledge and capacity. We are just as ready now as
we were earlier to admit the imperfections of our understanding, to learn
new things and to alter our methods in any way that can improve them."
Freud bluntly admitted the war neuroses made things look bad for libido
theory, but even if the sexual drives had not been clearly implicated in the
war neuroses, he insisted that they had not been *disproved*. War neuroses

were narcissistic neuroses, he added, though that was simply not good enough to account for the frenzy of destruction that Freud and his neighbors had witnessed. Rather implausibly, Freud also posited a "war-like 'I' " that created a lust for destruction.

The notion of a warring "I" was a halfhearted attempt to make sense of man's murderous violence. What was this warring self anyway and where did its impulses originate from? The conjecture seemed far-fetched and had the air of a man trying to make a problem go away.

Freud himself was dissatisfied with the answers he had proposed. There was something missing, something important that made humans "a gang of murderers." In January 1919, Freud informed Ferenczi that he had started a paper on the genesis of masochism. The topic was no surprise, since Freud's reflections on sadomasochism in the Wolfman case had just been revised and published. Two months later, Freud finished "A Child Is Being Beaten." Spanking fantasies were dutifully rooted in the anal-sadistic stage of development. That one of the clinical cases may have been Freud's own daughter Anna, who her father was analyzing himself, made for a disturbing footnote to this otherwise less than revolutionary paper.

But while finishing that work, Freud confided to Ferenczi that he had also begun to write a second essay with a "mysterious" title—"Beyond the Pleasure Principle." The pleasure principle was Freud's own first principle, one inherited from John Stuart Mill and other associational psychologists and generally adopted by neurologists and physiological psychologists. It was an essential premise that led Freud to psychosexuality, the lever he had pulled on time and again to make sense of so much. *Beyond* the pleasure principle? What could it mean?

Freud was not sure what he meant. He finished the paper quickly then held it back, for he felt the essay remained confused. He worked on it for another year, and when *Beyond the Pleasure Principle* finally appeared in 1920, it was earth-shattering. The man who had fought tooth and nail for libido theory, the man who had worked to explain away any challenge to the primacy of sex in human life, the man who had built a movement around the natural demands of pleasure, now radically revised himself. For those in the movement who had witnessed Freud's tenacious defense of libido theory, the paper came as a shock. By 1914, the remaining Freudians had learned psychosexuality and libido were working assumptions that could *not* be questioned. Now Freud made it clear that such questions were fair game, at least for him.

Beyond the Pleasure Principle was a vigorous shaking off of the demands of the Freudian movement and a reassertion of Freud's own intellectual freedom. The "old man" enforcing consensus made way for an unburdened freethinker. This theoretical pirouette began, appropriately enough, with a reexamination of dreams. On September 9, 1920, at the meeting of the I.P.A. in The Hague, Freud delivered an innocuous-sounding lecture called "A Supplement to the Theory of Dreams." A class of dreams, he noted, did not arise from wish fulfillment. These were traumatic dreams, which he promised to address in an upcoming work. The central problem of these dreams harkened back to Ernst Simmel's observation that shell-shocked men dreamt not of secret desires, but of terror. Their minds recreated nightmarish battle scenes again and again. The belief that the mind was driven by the pursuit of pleasure and the discharge of tension underwrote Freud's concept of libido and his theory of the unconscious. These dreams seemed to call that logic into question.

In 1911, Freud had amended his first principles to include not just a pleasure principle but also a reality principle, but the demands of the real world did not help here. There must be another force at work, a force that could override the pleasure principle, based not on reality but on something stranger. To search for clues, Freud returned to a thinker who had meant so much to him in building topographic theory. Opening *Beyond the Pleasure Principle*, the reader found that Freud had resurrected Gustav Fechner, the founder of psychophysics. For Fechner, most psychic processes were ruled by not just a search for pleasure and avoidance of pain, but also an inherent need for stability. Instability was unpleasurable in itself, and therefore a compelling need for constancy could override choices between pleasure and pain.

Within this new framework, Freud saw a host of psychological phenomena that pointed to a need for inner constancy. The recurrent dreams of soldiers were like the child's game of repeatedly saying good-bye and hello to a toy that symbolically represented his mother leaving him; it was a way to master and stabilize overwhelming stimuli. The mind's need for sameness made for a compulsion to repeat, a compulsion Freud also saw enacted in the transferences of his patients.

In this new theory, Freud returned not only to Fechner but also to a theory of trauma. Trauma represented an overwhelming of the psyche. The mind attempted to return to an inner state of constancy through repetition, no matter how painful. The war veteran replayed the shock of a shell exploding in fantasies and dreams, not out of pleasure per se, but rather in an attempt to stabilize his inner experience. The mind liked clichés, it liked conventions, it liked predictability. In the hope of regaining that equilibrium, painful traumas were repeated again and again.

Personal relationships could be scripted by this same unconscious compulsion to repeat, Freud continued. His illustration of these patterns of interaction were striking for those who had followed the Professor's strained scientific collaborations over the years:

Thus we have come across people all of whose human relationships have the same outcome: such as the benefactor who is abandoned in anger after a time by each of his protégés, however much they may differ from one another, and who thus seems doomed to taste all the bitterness of ingratitude; or the man whose friendships all end in betrayal by his friend; or the man who time after time in the course of his life raises someone else into a position of great private or public authority and then, after a certain interval, himself upsets that authority and replaces him by a new one.

To understand the full array of repetitive reenactments, Freud turned to biology, taking up Fechner's notion that an organism needed to protect itself from overwhelming external input. In the end, Freud postulated that all drives were inherently conservative, directed at the maintenance of inner peace. Peace came in its fullest form with the most massive discharge of energy: death. And so, Freud postulated a "Nirvana principle," using a term he borrowed from the British analyst Barbara Low. The idea of longing for death was an old Romantic one that had found a place in the analytic theories of Wilhelm Stekel, Theodor Reik, and Sabina Spielrein. For Freud, the "Nirvana principle" created a human desire for death. The death drive, or "Thanatos," was in conflict with "Eros." Or to put it otherwise, the desire for pleasure and the need for constancy were at odds. In this formulation, warfare had new origins. Unconsciously, human beings were driven to be at peace, which meant they unwittingly sought their own demise.

The repercussions of Freud's proposals in *Beyond the Pleasure Principle* were dramatic, but not because they immediately commanded great support. With this speculative work, Sigmund Freud shattered *the* defining characteristic of the Freudian community. For those good Freudians, the diehard believers in libido theory who had come through the schisms, the next years would be dismaying. Freud had become a heretic again. Astoundingly, he had turned against his own orthodoxy and turned against himself.

Introducing this new vision, Freud openly discussed the difficulties involved in any study of the unconscious. The pleasure principle was a hypothetical premise he had adopted because it described observed facts, he explained. However, since "this is the most obscure and inaccessible region of the mind, and, since we cannot avoid contact with it, the least rigid hypothesis, it seems to me, will be the best." Later he wrote: "in no region of psychology were we groping more in the dark." And near the end of *Beyond the Pleasure Principle*, Freud confessed that he was not so sure of the hypotheses he was putting forth, but he was developing a "line of thought and to follow it wherever it leads out of simple scientific curiosity, or, if the reader prefers, as an *advocatus diaboli*, who is not on that account himself sold to the devil."

Freud's rebellion invigorated him. He struck up a friendship with a self-styled outsider, Georg Groddeck, an eccentric Baden-Baden doctor who had delighted in shocking psychoanalysts at the 1920 congress by taking the lectern and declaring: "I am a wild analyst!" Despite the fact that members in the Secret Committee desperately tried to rein in Groddeck and his projects—especially his comic psychoanalytic novel *The Soul Seeker*—Freud stuck by his new friend.

In their correspondence, Groddeck immediately proclaimed his view that the unconscious was not knowable. The unconscious was in Groddeck's German an *Es*, an "It," inside us. Freud agreed that the unconscious was "provisional" and "indeterminate" and should not be imbued with characteristics from "secret sources of knowledge." And he confessed

I am myself a heretic who has not yet become a fanatic. I cannot stand fanatics, people who are capable of taking their narrow-mindedness seriously. By holding on to one's superiority and by knowing what one is doing, one can do a lot of things which are against the tide. The courage you intend to show I like very much, too. Perhaps the latest little work of mine that has just appeared, *Beyond the Pleasure Principle*, will change my image in your eyes a little.

Freud's image was changing in the eyes of his followers, many of whom had taken his theory of the unconscious as an immutable truth. In the midst of a flowery tribute on Freud's sixtieth birthday ("the great free spirit, the upright independent man, the unprejudiced investigator, the truly complete scholar, and the all-understanding, all-forgiving individual"), Eduard Hitschmann warned:

Beware . . . when the great God lets loose a thinker on this planet. Then all things are at risk. It is as when a conflagration has broken out in a great city, and no man knows what is safe, or where it will end. There is not a piece of science but its flank may be overturned tomorrow.

Prompted by the horrors of war, Sigmund Freud, the thinker, had thrown over Sigmund Freud, defender of the movement. In a staggering about face, he had gone back to the drawing board, changed his mind, and in the process admitted, even emphasized, the provisional nature of *any* theory of the unconscious. Perhaps no mortal could put Sigmund Freud into retirement. But by these reckonings with himself, Sigmund Freud had altered from hoary old patriarch to one of many sons and daughters of what might be a nascent science.

NINE

Searching for a New Center

Caricatures of figures attending the psychoanalytic congress in Salzburg of 1924, drawn by Robert Berény and Olga Székely-Kovacs.

I.

ONCE THE NATURE of the unconscious was thrown open to question, the old Freudian community was forced to remake itself. The bond that had tied these men and women together had vanished. While many were not pleased with Freud's revisions, it was clear that true believers in the Freudian psychosexual unconscious would have to coexist with others who contested it, especially since Freud himself was now among the others.

In the midst of this turmoil, the Freudian community found itself back in possession of a term that offered a somewhat different identity. After the schisms, different kinds of psychoanalysts existed, including Freudians, Adlerians, and Jungians, but by 1920, the Freudians were increasingly the only group allied to psychoanalysis. After the war years, Carl Jung had emerged with a panoply of new theories and terms based on unconscious archetypes, which he called Analytical Psychology. His theories were fundamentally distinct from Freudian psychoanalysis, and he disassociated himself from that endeavor. Alfred Adler had also abandoned claims to being a psychoanalyst. He changed the name of his group to the Society for Individual Psychology and founded a journal to further Individual Psychology, which was based on his theories of inferiority, the fictions of the self, and a focus on the whole personality. The only dissident who continued to claim he was a psychoanalyst was Wilhelm Stekel, but he never found many followers.

At one time the word *psychoanalysis* had been synonymous with Freudian. After 1920, *psychoanalysis* with its broader implications and impersonal nature offered new attractions, especially since being a Freudian had become so complicated. Which Freud did a Freudian follow? How could one be a Freudian when there were divergent Freuds?

The uncertainty that emerged after Freud's scrambling of drive theory was especially difficult for clinicians. Theoretical physicists are free to question the most basic assumptions of their field, but it is paralyzing for engineers to do so. Psychoanalysis did not have separate cadres of theoreticians and practitioners. Science and therapeutics, as well as the competing imperatives of lab and clinic, were all packed into the same clinical encounter. Nuremberg elevated the needs of the growing clinical community to have a stable theory to use, and Freud's reshuffling of his scientific claims could potentially throw practice into confusion. And just as there had been no clear way to adjudicate among the theories of Freud, Adler, and Jung, there seemed to be no clear way to adjudicate between Freud and Freud. Freud's multiple theories of the unconscious highlighted the provisional nature of all these claims, and their distance from empirical verification and consensus. The post-Nuremberg Freudian field had been dismantled.

Over the next years, new voices would emerge and argue that a community of psychoanalysts could be unified by other means than a commitment to a highly specific theory of the unconscious. Between 1920 and 1925, the old Freudian community changed in a way that called for new identities, new institutions, and new ways of rationalizing the exercise of authority. Such calls were not surprising, since all over Europe, identities were shifting and old authorities were in retreat.

IN 1919, THE Ottoman and Austro-Hungarian empires disintegrated. Nations and republics began fighting to be born. Communists, socialists, pacifists, anti-Semites, German nationalists, and anarchists all brandished their ideologies in the postwar chaos. In 1919, the Weimar Republic was proclaimed in Germany, while in Bavaria a Soviet-style worker's republic took over. The Czechs, Poles, Italians, and Slovenes deserted the Dual Monarchy, while in Hungary, the Red terror led to the White terror. In Vienna, the time also seemed ripe for radical change.

In 1919, Freud began a velvet revolution by throwing open the bounds of what was allowed in the Freudian community. As his followers reconstituted after the Great War, they discovered Freud entertaining speculations that previously had been heretical. The Freudian community had been forged in battles with German psychiatrists, and the likes of Fritz Wittels, Karl Kraus, Carl Jung, Eugen Bleuler, Wilhelm Stekel, and Alfred Adler. Each of these battles had drawn a line that helped define who the Freudians were. Now that old community seemed to be as upended as the rest of Central Europe.

In November 1918, the Secret Committee reconvened to discuss rebuilding psychoanalysis throughout the world. The anointed were now Abraham, Jones, Ferenczi, and Anton von Freund, as well as Otto Rank and Hanns Sachs. In 1920 to facilitate their work, the committee established a circular letter that would pass back and forth between Vienna, Budapest, Berlin, and London.

These four cities would be vital to the fate of psychoanalysis in Europe and the world. A few other psychoanalytic societies were active, such as the Dutch and the Swiss (who founded a non-Jungian society in 1919), but they remained unrepresented on the Secret Committee. There were also no Americans in the committee, and news from the New World was mixed at best. Jones informed the others: "Everything possible passes under the name psa., Not only Adlerism and Jungism, but any kind of popular or intuitive psychology. I doubt if there are six men in America who could tell the essential differences between Vienna and Zurich, at least at all clearly." From Boston, Isador Coriat confirmed Jones's impression, reporting that the Americans were in theoretical disarray. Amid all the different American brands of psychotherapy, Coriat joked that he needed to chant: "there is no psychotherapy but psychoanalysis and Sigmund Freud was its prophet."

After the war, the greatest hope centered on Budapest. Anton von Freund dedicated his wealth to psychoanalysis and encouraged Freud to announce the need for psychoanalytic clinics for the poor at the Budapest Congress. The announcement came after plans were already under way to create such a clinic as well as a psychoanalytic institute in Budapest, all funded by Freund. He had put aside the prewar equivalent of over $300,000 to establish these institutions and had secured the agreement of the Budapest mayor. After the Budapest Congress, the Hungarian Society met almost every Sunday, attracting promising young members like Sándor Radó, Imre Hermann, Géza Róheim, Melanie Klein, and István Hollós.

All of a sudden, political upheaval made von Freund's connections and money near worthless. In the spring of 1919, the liberal Károlyi government collapsed and was replaced by the Communist Revolutionary Council of Béla Kun. During the Communist takeover, Ferenczi was nominated for a university professorship in psychoanalysis and the directorship of a psychoanalytic clinic for the poor; when the Communists were overthrown and the violently anti-Communist and anti-Semitic "White Terror" commenced, Ferenczi lost his professorship and was expelled from the local Medical Society. Food was scarce, funds unavailable, and the capacity for communication outside Hungary difficult. Unable to perform his duties, Ferenczi passed his presidency of the I.P.A. to Vienna. Since things were not much better there, in the fall of 1919 the office was transferred to Ernest Jones in economically stable London. In January 1920, Anton von Freund died of cancer. The money he had set aside for the cause was frozen, and the great hopes for psychoanalysis in Budapest crashed.

However, Freund and Freud's vision of free analytic treatment spread beyond Budapest, riding a wave of egalitarian ideologies that were making their way across Europe. The idea of free clinics inspired two Berlin analysts in attendance at Budapest: Max Eitingon and Ernst Simmel. Both were returning to civilian life carrying with them their wartime experiences during which they had used modified psychoanalytic methods to treat large numbers of neurotics. Upon returning to Berlin, they hatched a plan to start a free psychoanalytic clinic supported by Eitingon's family wealth.

In 1920, the first clinic for psychoanalysis opened its doors in Berlin. In addition to providing free treatment for the poor, the clinic became a training ground for would-be psychoanalysts. That year, Abraham taught a six-week introductory course at the newly born institute. Soon this class was complemented by a practical introduction, a course for general physicians, and one on the social problems of sexuality. Hanns Sachs moved to Berlin and became the designated training analyst for students. New recruits came from various parts of Europe to study at the only formal training center for the clinical practice of psychoanalysis.

Things were not going as well in London. Early in 1921, a newspaper campaign broke out labeling all psychoanalysts charlatans. Training in psychoanalysis was not formalized in London, and the analysts there struggled to distinguish themselves from a motley horde of pretenders who declared themselves psychoanalysts as well as palm readers or spiritualists. Jones reported that some "English Psycho-analytical Publishing Co." posted announcements, reading: "would you like to earn £1,000 a year as a psycho-analyst? We can show you how to do it. Take eight postal lessons from us at four guineas a course."

Serious trouble soon followed. Legal action had been initiated against a self-proclaimed analyst who had raped a patient, and another who had seduced several women. The latter case brought a barrister to the door of Ernest Jones, who claimed not to know the culprit, a man who said he had taken a correspondence course in psychoanalysis with Jones. Lord Alfred Douglas jumped into the fray. The vitriolic former lover of Oscar Wilde had become the leader of a London-based Purity League. Douglas brought suit for indecent publication against the publisher Kegan Paul, which had put out a memoir by a theosophist who claimed to have visited Freud. Jones wrote the Secret Committee: "Douglas swears he will uproot Ps.A. in England and will make all publication impossible." Jones considered the theosophist's book rubbish but was concerned that Douglas's legal actions might set a precedent that would make all psychoanalytic works illegal. In fact, Douglas won his case and all copies of the offending publication were "burnt by the common hangman."

After disbanding the London Psychoanalytical Society in 1919 to get rid of members with ties to Carl Jung, Ernest Jones reconstituted a British Psychoanalytical Society the same year. This society had little prospect of following Berlin's example and founding a clinic or training institute. The main hope for a London-based center for training was the Brunswick Square Clinic. Founded in 1913, this psychotherapy clinic had a staff that was keenly interested in psychoanalysis, including a number of lay clinicians who looked to Berlin for formal training. Jones had little interest in laymen, but Karl Abraham and Hanns Sachs took the Brunswick Square applicants and expressed the wish that by bringing these people into the fold, they would strengthen the British Psychoanalytical Society and provide it with a home for clinical work. Jones was not so hopeful. He reported that Brunswick Square had appointed the Earl of Sandwich as chairman, and that the earl had scandalized those who assembled for the opening of the facility by prattling on about supernatural healing powers in rare individuals, such as himself. After most physicians fled the clinic, it was placed under the directorship of Miss Julia Turner, assisted by Miss Ella Sharpe and among others, a doctor named James Glover.

In Jones's view, the place was a disaster, minting a cadre of wild analysts after giving them a few weeks of so-called training. Writing to the Secret Committee, he reported that one member of the clinic who was now his patient, Dr. Ethilda Herford, acknowledged that at one time the secretary working at the clinic had freely analyzed anyone she could get her hands on. The Berliners held firm, arguing that by admitting the outsiders for more serious formal training they would be tamed. After all, members of the Berlin Society were prohibited from delivering lectures or classes without prior approval.

By the summer of 1921, the Berliners' hopes seemed warranted as the British Society and the Brunswick Square Clinic prepared to merge. But the presence of lay clinicians prevented the British General Medical Council from sanctioning the clinic, and the deal fell through. It would be five years before the British Society established a teaching clinic.

In Vienna, the Psychoanalytic Society reconstituted itself amid the upheavals of a disintegrating empire. Miraculously what remained of Austria avoided falling into complete chaos when the two major political parties—the Christian Socialists and the Social Democrats—formed a coalition government. While much was precarious in "Red Vienna," psychoanalysts saw encouraging signs. The society welcomed eleven new members between 1919 and 1920. Some of Anton von Freund's money had been transferred to Freud in Vienna before the political upheaval in Budapest, which allowed Freud to fund a psychoanalytic publishing house that took on both *Imago* and the *Zeitschrift*. A contribution of a million crowns arrived from America, which further stabilized the *Verlag*. And one of the wealthiest men in Vienna, the financier and publisher Richard Kola, was poised to purchase the publishing house, a sign that psychoanalytic publishing might be profitable or at least prestigious.

The Viennese had also been inspired by Freud's speech at Budapest and the opening of the Berlin clinic. In 1920, they began their own efforts to open a free clinic. Eduard Hitschmann suggested that a psychoanalytic department of the Vienna General Hospital be founded in an abandoned garrison hospital. Freud opposed the effort, confiding to Abraham that he doubted any of the Viennese could be entrusted with such an operation. Freud's worries were unnecessary. When the medical authorities asked the university chair of psychiatry, Julius von Wagner-Jauregg, for his opinion on the matter, the plan died. In October, Freud gave expert testimony before a commission investigating mistreatment of war neurotics by doctors who worked under Wagner-Jauregg, which exacerbated relations between the two men and made the dream of a clinic appear unrealizable.

Hitschmann and his allies persisted. They were offered abandoned space in another garrison hospital, but that was considered unsuitable.

Finally, an internist and new recruit named Felix Deutsch arranged for space at the ambulance entrance to a cardiac unit in the Vienna General Hospital. All seemed to be in order, when again, medical authorities intervened, this time to protect Vienna's physicians from new competition. Finally, after a successful appeal, the Ambulatorium opened on May 22, 1922, with a warning that non-M.D. practitioners were not to work there. Hitschmann would be the director and a new member of the Society, Dr. Wilhelm Reich, would join as his assistant. Here too the clinic provided a new forum for pragmatic learning. Classes commenced in November 1922. A year later, nine courses were being taken by more than a dozen students.

In the process of bringing psychoanalysis to workers and the poor in Weimar Germany and Austria, the Berlin and Viennese psychoanalysts found themselves with fledgling institutions for clinical teaching and training. Unlike psychoanalytic societies, which focused on novel findings and theories, these forums were practical: how to conduct an analysis, how to treat illness, and what methods to use. Before 1918, an interested person became a Society member after he or she had been nominated and presented an acceptable scientific paper before the group. Now both in Berlin and Vienna, a process of training was born.

By the time the first International Psychoanalytic Congress took place after the Great War, these new institutions had begun to function. At The Hague, 118 participants, the most ever, braved ruinous exchange rates to attend the I.P.A. meeting. The new president, Ernest Jones, welcomed eight societies to the Association. All eight—the Vienna, Berlin, Hungarian, British, New York, American, Swiss, and Dutch societies—had no formal training requirements. The question of how to formalize training was raised by the Zurich reverend Oskar Pfister, who proposed that the I.P.A. consider issuing diplomas. A number of men including Freud supported the plan, but behind the scenes, the Berliners on the Secret Committee were skeptical, and the proposal quietly died.

With new institutions for training coming into existence and a growing cadre of novices, however, leaders in the field were forced to consider what to teach, what a student had to learn, and what was required of their practitioners. The establishment of such guidelines was of tantamount importance, yet it would become a source of debate and confusion, for the answers were far from obvious.

II.

PSYCHOANALYTIC METHOD HAD become a jumble. By 1911, the published guides for conducting an analysis were either outdated or insufficient. Around Europe and America, doctors who called themselves psychoanalysts used repudiated notions of catharsis and unconscious memory retrieval or hypnosis; others latched on to the dream book and analyzed symbols and dreams; still others advocated sexual activity as a cure. These practitioners publicly claimed they were using Freudian treatments. They were also, de facto, the scientists who were supposed to be empirically testing psychoanalytic findings. Most were flying blind.

If empirical observation was to be the foundation of the field, this was a terrible problem. It meant there were no rules for amassing the evidence that led to inferences about the unconscious. It was difficult enough to achieve consensus on unconscious processes that were neither directly observable nor quantifiable. But if the object of analysis was itself hard to grasp, it was paramount that the observers themselves be stable, uniform, and of course they were not. According to their own theory, Freudians viewed all human beings as desirous, resistant, at best half-deluded. Analysts too were filled with human wants and fears and dominated by their unconscious. If an analyst in Budapest was not the same as another in Zurich, how could a science of the unconscious ever achieve a working consensus?

Followers had begged Freud to show them how to proceed, but the Professor's promises for a textbook on the practice of Freudian psychoanalysis never were fulfilled. Without Freud to guide them, clinicians turned to Stekel's book on nervous anxiety, for it contained a short account of psychoanalytic technique. These scant descriptions included the use of a couch and the observation of resistance, but little more. Ernest Jones had written two short papers in English that outlined free association, Jung's directed word-association method, and dream interpretation. These guides were spare, and for Freud at least, misleading. He did not approve of using the word-association test in treatment, and he no longer believed dream interpretation was the central act of psychoanalytic treatment. Dreams remained of special import to Freud, but he had been forced to recognize that there were dangers in an excessive reliance on them.

This lesson was brought home by Wilhelm Stekel. His forte was an uncanny ability to interpret unconscious content from dreams (like "a pig finding truffles," Freud joked). In March 1911, this dream-meister published *The Language of Dreams*, in which he argued that dream symbols were a language from man's prehistory that could be translated. A sword was always a symbol of war, a tree meant nature. Freud was amenable to the idea of phylogenetically inherited dream symbols, but he was surely displeased to find that Stekel used his universal symbols to argue that dreams were driven by an unconscious aggressive force similar to the one Alfred Adler postulated in the unconscious.

To make matters worse, Stekel's book came out just as Freud's war with Adler reached its apex. Freud detested *The Language of Dreams* and attributed it to the author's "perverse" unconscious. The Professor responded by writing a critical review saying that Stekel's universal symbolic interpretations made dream interpretation "uncertain and superficial." Freud readily admitted that dream interpretation was a difficult art that required subtlety from the practitioner and could easily go wrong. Stekel's new book was a glaring example of such a misstep.

While contemplating the instabilities of dream interpretation, Freud was persuaded to write a series of didactic articles on technique for the readers of the *Zentralblatt*. He confessed to Jung that the choice came down to writing the articles himself or leaving it to the likes of Stekel. After he finished the first paper, Freud sent it to Jung, saying "the things I write to order, without inner necessity, as has been the case with these articles, never come out right."

These written-to-order articles became critical texts for those analysts who sought out a specific psychoanalytic technique. The first paper was on dream interpretation. In it, the proud discoverer of the secret of dreams pushed aside their interpretation and subjugated it to the overall process of the analysis. Throughout Freud's text, there lurks an unnamed bad analyst who relies too heavily on rote symbolic dream interpretations. Members of the Vienna group no doubt suspected their truffle-sniffing colleague, Stekel. Sophisticated dream interpreters, Freud maintained, may see through the haze of dreams into the unconscious, but simply telling a patient the content of a dream accomplished nothing. Dream interpreters must subject themselves to the "full technical rules that govern the conduct of the analysis throughout."

What were those rules? What would take the place of dreams? The answer had long been apparent to Freud, though inexplicably he had declined to write about it. If the royal road to the unconscious was dreams, the everyday path was transference. In 1906, Freud wrote Carl Jung that transference was the key to treatment and the engine of psychoanalytic cure. Thanks to the libido they had invested in the doctor, patients might be willing to overcome their resistances and face their own hidden truths.

After 1906 a small group of cognoscenti began to follow this new idea. In 1908, Carl Jung advised Ernest Jones to pursue the mother-transference in a case the Welshman had just published. The same year, Isidor Sadger published an analysis of homosexuality in which transference was central. In 1909, Binswanger published a long analysis of a hysteric he called "Irma," in which the woman's secret love for her doctor was a key to the case. But the most significant addition to transference theory came in 1909 from Sándor Ferenczi.

Ferenczi's "Introjection and Transference" was an extensive account of transference theory, which he called one of Freud's "most significant discoveries." Ferenczi saw transference as a kind of psychic importing or "introjection," which he contrasted with the exporting of unconscious fantasies onto the world via "projection." According to Ferenczi, neurotics flew from their complexes through the introjection of objects and transference, while psychotics depleted their inner worlds by projection. Ferenczi also used transference to draw a clear line between psychoanalysis and competing therapies. Bernheim and his followers required a masochistic submission of their patients, who accepted the doctor's suggestions because of an infantile father-transference. Suggestive treatments relied on blind obedience, whereas the analysis of transference sought to free patients from these forces.

Ferenczi's paper made the discussion of transference in clinical work unavoidable. When Wilhelm Stekel prepared a second edition of his book on nervous anxiety states in 1912, he amended his discussion of technique accordingly, and argued that knowledge of different forms of transference was essential for success. Stekel tried to categorize forms of transference: those onto the physician were the most common, but transferences could also silently attach themselves to the doctor's family, household personnel, pets, even the analyst's office or apartment. Stekel made no arguments about the underlying nature of these phenomena and, neglecting the work of Riklin and Ferenczi on ambivalent Oedipal transferences, remarked that "the psychological roots of this fact have not yet been explored."

In the early months of 1912, Freud followed Stekel's paper with his own typology, in which he made the psychological origins of transference clear. "The Dynamics of Transference" was Freud's first extended discussion of this theory since his postscript to the Dora case in 1905. In Freud's view, Stekel had missed the point. The objects of a transference were fluid and changeable, but their roots were not. Love and hate were based on the templates laid down with important figures from childhood. Evidence of unconscious templates could be found in emotional ties forged in adulthood, for we come to expect love in the forms we first knew it. These "stereotypical plates" were "prototypes" projected into the world. In the clinical encounter, the physician was made over by these imprints in the patient's mind. Forget transference to the dog, the apartment, and the butler. Think: Mother, Father, Sister, Brother. These first relationships were the deep structures of transference.

Transference was mostly unconscious and did not rise into conscious memory. Unknowingly, men and women tried to live out the past in the present, reenacting a role drawn from their personal history. The longer the analysis, Freud added, the more transference becomes *the* critical source of information on the unconscious. By interpreting these transferences and bringing them to awareness, the analyst made these ghosts gradually lose their grip.

Freud had pushed aside the complex hermeneutics of dream interpretation and replaced it with the analysis of transference. From an evidentiary point of view, this move was a winner, and Freud knew it. He told Jung that transference was the "chief proof that the whole process is sexual in nature," and the only unimpeachable proof that the neuroses were due to psychosexuality. In the "Dynamics of Transference," Freud reiterated this claim, saying transferences were invaluable, for they made the patient's unconscious erotic life manifest. Transference had the great advantage of being empirically observable in the here and now. Unlike the reconstruction of past traumas, transference was not prone to falsifications of memory and the quandaries of historical retelling. Its interpretation was more straightforward than dream interpretation. Compared to dreams, transference could better satisfy the demands of both curing and knowing, medicine and science.

Of course, there was a danger in openly arguing that patients were expected to fall in love with their doctors, especially when Freud's clinicians were sometimes succumbing to their own lust. The Professor hastened to explain how important it was to work with these wild passions while avoiding erotic imbroglios. To do so, the physician must distinguish between two types of positive transferences. The first was part conscious, friendly, and affectionate and should be welcomed. It was the very "vehicle of success," because this affection allowed the patient to accept what Freud admitted were the suggestions of the analyst. (This admission incensed Ferenczi, who complained that Freud had given succor to those who would dismiss analysis as suggestion.) This positive transference was beneficial and different from an objectionable sexual one. This second form of transference was wholly unconscious, based on repressed erotic feeling, and it appeared as a demand for love. Erotic positive transferences had to be battled because they created a fierce resistance to remembering and analysis as a whole. Freud made it clear that he thought therapeutic success had more to do with the way transference was handled than with anything else.

The 1912 publication of "The Dynamics of Transference" was an event in Freudian circles. Ludwig Binswanger lectured on the paper, Jung and Ferenczi praised it, and at one of the first meetings of the London Psychoanalytical group a translation was prepared for study. But while transference seemed to solve a number of knotty problems, it also brought new ones to light. What prevented the analyst from having his own transferences? Ludwig Binswanger asked. Freud replied that the analyst should never respond with spontaneous emotion and must recognize that he has fallen in love with the patient so as to free himself from these feelings.

Binswanger's question was not an academic one. Freud had heard rumors from a patient—Frau Elfriede Hirschfeld—about transference problems with Jung and Pfister. Freud wrote Jung to warn him of the perils of such transferences, and said that an essay on countertransference was sorely needed. Jung replied that when it came to countertransference, he was "a bit refractory," but that the Reverend Pfister had been exhorting his patients "on the Christian principle: Look what I've done for you, what will you do for me?"

Countertransference was not just a problem for treatment, for during the schisms Freud was convinced that much of the infighting was caused by the countertransferences of the different theoreticians. Bleuler was puritanical, repressed, and afraid of his own sexual urges; Jung was a pastor's son, tormented by a Father complex; Adler was paranoid and discovered aggression everywhere; Stekel was thoroughly perverse. Each man looked into Freud's microscope and managed to see something different, something Freud believed had been created by the viewer's neurosis. Of course, Freud himself could not stand outside these accusations, for he too had been charged with subjectivism. Critics believed he was obsessed with sex and discovered it at every turn.

Faced again with this Pandora's box of challenges to the analyst's authority, Freud returned to the seemingly never-ending project of stabilizing the scientific observer of the psyche. Placing an emphasis on the observable phenomena of transference was a step toward minimizing the problems one man faced when trying to read the unconscious thoughts of another. In his next technical paper, Freud went further in his attempts to limit the distorting intrusions of the analyst. Just as the patient was to try to lay out his inner thoughts and feelings in free association, he encouraged the ideal psychoanalyst to follow a similar rule. Listen without a set agenda. Do not search for clues or direct associations, but wait with "evenly suspended attention." In this way, the analyst avoided the pitfalls of prejudice and bias: "In making the selection, if he follows his expectations he is in danger of never finding anything but what he already knows; and if he follows his inclinations he will certainly falsify what he may perceive."

In the hopes of containing the analyst's subjectivity, Freud created the ideal of an analyst whose desires and biases were held back. But there was a hitch. The imagined analyst floating in evenly suspended attention must be without resistances, without blind spots. "It is not enough for this that he himself should be an approximately normal person. It may be insisted, rather, that he should have undergone a psycho-analytic purification and have become aware of those complexes of his own which would be apt to interfere with his grasp of what the patient tells him." Freud asked the "purified" analyst to steel himself against excessive emotional involvement and model himself on a surgeon who puts aside his feelings. As a final hope, Freud spoke of an analyst who functioned like a mirror, faithfully reflecting what was inside the patient.

Freud's recommendations set the bar rather high. No one could claim to be so pure. Furthermore, these hopes went against the very body of knowledge on the subjectivity of perception and mentation that had been so important to Freud's theory building. Still, stabilizing the field of psychoanalytic knowledge was an urgent issue. Freud hoped the ideal of an indifferent, objective analyst would encourage analysts to keep their own desires in check. Of course, all these problems could be made easier if thoughts could be simply communicated from one mind to another without the biases of perception and cognition. Freud and Ferenczi were both intrigued by experiments in thought-transference, what F. W. H. Myers called "telepathy." If only the unconscious could be perceived without mediation. Thought-transference would remain a vain aspiration, but clinical transference became a key to analytic technique and a new conceptual center for the postwar psychoanalysts to rally around.

AFTER THE FINAL split with the Zurichers, Freud wrote his last two papers on technique. In setting up rules to keep the field on track, Freud found himself exploring new ground. If transference was truly central to analysis, then practitioners would have to deal with longstanding unconscious attempts to reenact the past. From this perspective, human beings appeared to be fundamentally nostalgic; they suffered from an unconscious "compulsion to repeat." The rekindling of old loves and hates in the transference was an attempt to live in a timeless, hallucinatory past, and a refusal to remember the past *as* past. Freud concluded that such repetitions caused the entire neurosis to be reenacted in the doctor's office; the neurosis itself could be considered "a transference neurosis." If once psychoanalysis seemed based on the belief that the truth could set one free, now it seemed that this liberation would need to occur again and again. Psychoanalytic treatment had become the untangling of unconscious repetitions.

Unfortunately, the doctors kept getting tangled up in these dramas themselves. Jung's reputed affairs now included his analyst, Maria Moltzer, and a female patient, Toni Wolff. The passions of Sándor Ferenczi included an agonized affair with Gizella Pálos and her daughter Elma, both of whom he had analyzed. Jones had fled North America after accusations of impropriety from a woman patient. Wilhelm Stekel had gained a reputation for conducting affairs with patients.

By stressing transference, Freud risked tying psychoanalysis to impropriety and sexual scandal. Undaunted, Freud boldly took up the explosive issue of transference-love, which he imagined as a female patient falling in love with a male doctor. This aspect of Freud's reasoning was quite conventional, but the rest of his argument was untouched by conventional morality. Freud did not simply deny the couple their right to love. Instead, he suggested that the problem was that the woman, having won a lover, would need another analyst, and then another. Love alone would not cure her neurosis. As for the doctor, he had been fooled into thinking her love was due to his charming person and had forgotten that for his patient he was a stand-in for much older figures. Proud of the conquest, the physician gave up his "indifference" and fell under the sway of countertransference. To counter this, Freud recommended that the patient be treated with "abstinence," the analyst's refusal to satisfy the neurotic longings that appeared in treatment, which heightened pressure on the patient to remember and work through the past.

After that, Freud stopped. It was 1915, and he had said most of what he would say on the technique of psychoanalysis. In a number of short papers, he made transference decisive for Freudian clinical practice and theory. It proved libido theory, was the focus of treatment, and the cause of clinical failures. It allowed analysts to rely less on the vagaries of dream interpretation and fostered a close focus on clinical encounters. Transference linked clinical work with a theory of neurosis and model of mind, not to mention a schema of psychosexual development that explained some of the mysteries of love. Transference was not just a technical concept, but also a central organizing notion that linked a great deal of psychoanalytic theory and practice.

In the spring of 1917, the Baden-Baden physician Georg Groddeck wrote Freud for the first time. After reading Freud's scathing history of the psychoanalytic movement, Groddeck grew nervous. He was set to publish a work he hoped to describe as psychoanalytical, though in it he stated that the unconscious was not definable. Would he be publicly repudiated too? "I began to doubt," the stranger wrote, "whether, according to your definition, I could count myself a psychoanalyst. I do not want to call myself a member of a movement as an intruder who does not belong to it." Freud replied: "I have to claim you, I have to assert that you are a splendid analyst who has understood forever the essential aspects of the matter. The discovery that transference and resistance are the most important aspects of treatment turns a person irrevocably into a member of the wild army."

Carl Jung would have been thunderstruck; his technique had not been an issue during his battle with Freud. Adler would have shared Jung's bewilderment. He too generally followed Freud's investigative method. Both had been forced out of the movement because they held aberrant views of the unconscious. Now Georg Groddeck was welcomed into the fold, based on methodological grounds? Freud's letter to Groddeck was an early sign of a realignment that would redefine what it meant to be a psychoanalyst.

III.

As THE NEED for training grew after World War I, Freud's papers on technique would become central pedagogical tools, but even with these guideposts there was much chaos and confusion. Consider, for example, the young acolytes who embraced psychoanalysis in Vienna after 1918.

The Great War created conditions for a cultural realignment that helped the Freudians in Vienna grow. The war made even the most pessimistic cultural critics look prescient. "The post-war generation emancipated itself with a violent wrench from the established order and revolted against every tradition," Stefan Zweig recalled. Neo-Romantic thought, including theosophy, occultism, and spiritualism, sprang up. Traditional medicine's condemnations of Freud and psychoanalysis were now worth their weight in gold, as Freud's community became allied with the new, the nontraditional, the outré, and the sexual.

One manifestation of the new spirit was the Viennese Youth Movement, which grew out of the *Wandervögel*, a group founded in 1901 near Berlin as a hiking organization for schoolboys. By 1911, it had grown to 1,500 members who relished the freedom that came with no adult control. Along with gymnastics and nude bathing, one observer recalled that members of the *Wandervögel* were "well instructed in sex" and took up homosexuality and lesbianism, flaunting their disregard for conventional modes of love. In Vienna, the Youth Movement united would-be reformers and rebels who lived together on the cheap in abandoned military barracks. Sexual education, free love, and contraception were espoused as a break from Wilhelmine prudery and hypocrisy.

A few members of the Youth Movement found their way to the University of Vienna Medical School, where their common interests brought them together. After a medical lecture, one recalled, "Otto Fenichel passed a slip around saying those who wanted to participate in a special, extracurricular seminar in the newer developments should stay afterward." Fenichel was an active member of the Viennese Youth Movement committed to sexual, cultural, and school reform. As a twenty-year-old medical student, he had started a library for Youth Movement literature and discovered Freud. Joining Fenichel that day were Edward Bibring, Grete Lehner, and Wilhelm Reich, all of whom would end up in the vanguard of the psychoanalytic movement.

The second wave of sexual reformers comprised these students who were attracted to Freud and psychoanalysis. In February 1919, they founded a Viennese Seminar for Sexology at the university, where they discussed sexual biology, psychoanalysis, and other topics. Fenichel was already a committed Freudian, but when the group first read Freud, several members were not impressed. Edward Bibring thought the work was ludicrous. Reich dismissed Fenichel as a fanatic who like all Freudians saw sexuality everywhere. But despite this disclaimer, Reich also believed sexuality was "the core around which all social life, as well as the inner spiritual life of the individual, revolves." It was a belief he would never give up, for he had come to it by hard experience.

Wilhelm Reich was born in Galicia in 1897, a child of privilege, and a victim of family tragedy. The son of a prosperous but autocratic Jewish landowner, Reich told the story of his calamitous childhood by disguising himself as a patient, and presenting his case to his friends in the Seminar for Sexology. "A Case of Pubertal Breaching of the Incest Taboo" told of a boy who was not degenerate or psychopathic but still had frank incestuous fantasies. The fantasies were stimulated when he witnessed his mother's affair with his childhood tutor. The young Reich imagined having sex with his mother and fantasized about blackmailing her into compliance. Finally, the boy told his father about the affair, causing the father to physically abuse and torment the mother unrelentingly until she poisoned herself to death. Not long after, the boy's father died as well, and the young Reich was left devastated.

When war broke out, he volunteered and was on the front lines three times. After the armistice, Reich moved to Vienna and spent his days in cafés, reading. He digested the standard café fare: Weininger, Schopenhauer, and Kant, and learned enough physics, chemistry, and biology to switch from law to the medical school. There, Reich hooked up with Youth Movement activists like Lehner, Fenichel, and Bibring, and joined their reform-minded, amorous group.

The Seminar for Sexology had a profound effect on Wilhelm Reich. In the spring of 1919, while completing his second year of medical school, he began to read sexologists. He was most impressed by Freud's *Three Essays* and his *Introductory Lectures*, which decided his choice of profession. Reich began to refer to himself as a sexologist, and when Otto Fenichel left for Berlin, Reich took over the seminar.

Reich took his responsibilities seriously. He tried to reorganize the seminar into biological and psychological components, and bring more order into their studies. As the chairman of the seminar, it was his responsibility to get readings for the group. Reich contacted the most famous sexual biologists and psychologists in Vienna and met with them. Wilhelm Stekel tried too hard to impress, the young man thought, and Adler disappointed Reich by launching into a tirade against Freud and displaying rigidity regarding his own interpretations. Freud impressed Reich by commending the student for his interest in sexuality and happily giving him books and reprints.

In 1919, Reich presented an overview of concepts related to the sexual instinct ranging from Forel to Freud and Jung. For Reich the idea that sexuality took hold of men at puberty was silly; he had been tormented and excited by sexual feelings from a much earlier age. Only Freud's theories explained his experiences. In September 1919, the often depressed Reich began a personal analysis with an analyst who had lectured at the sexological seminars, Isidor Sadger. That same winter, though only a medical student, Reich began practicing psychoanalysis. At the time, all that was required was an O.K. from one's own analyst and the blessings of Freud.

Reich's inauguration into clinical psychoanalysis was a catastrophe.

The young man had always struggled to control his sexual desires: it was not for nothing that his first book was entitled *The Impulsive Character*. A frequent visitor to brothels, the young Reich was rumored to have many affairs. When the novice began treatment of the beautiful nineteen-year-old Lore Kahn in the winter of 1919, the results were predictable. The young woman had lost her lover, a former leader of the Youth Movement. After entering treatment with Reich, she promptly fell in love with him. Reich had to wrestle with his own passion for Lore, as she declared that she did not want analysis but rather her analyst.

Reich resisted and reminded himself that this was transference. But after the two terminated the treatment, they met up at the sexological seminars and their affair began. In October 1920, Lore died suddenly. Her mother accused Reich of impregnating Lore and effectively killing her, through a botched abortion. The mother damned the day her daughter went into psychoanalysis and then committed suicide.

The death of Lore plagued Reich. It seemed that his sexual urges resulted inevitably in the death of the woman he loved. "Who will tell me how I should have behaved?" he wondered. "I was the cause, but could I have helped it?" The question would linger. In 1920, there was little practical guidance for the practice of analysis, except Freud's technical papers. Reich continued to practice analysis, presented a paper to the Vienna Psychoanalytic Society, and was elected to the group. But he was crushed when his paper—a reading of Ibsen's *Peer Gynt*—was dubiously received despite the fact that he had based his interpretation on incestuous infantile fantasies. Was there really an unconscious? he asked Otto Fenichel. Trying to be helpful, Fenichel replied that philosophically speaking there was no unconscious, but psychoanalytically speaking there was! An elder colleague, Eduard Hitschmann, counseled Reich to stop philosophizing and—as the young man mockingly noted in his diary—"analyze, analyze, ana-ana-anal!"

Reich looked for a more defined theory of clinical practice and technique. Few were lecturing on technique at the Society meetings, and when he went to older analysts for advice, they had little to say. Even Freud was unhelpful with his constant admonitions to be patient and keep therapeutic ambition in check. In January 1921, Reich, though still in medical school, could note with satisfaction that he had "two paying patients sent to me by Freud himself." Perhaps just by analyzing, things would work out.

In January 1921, a friend of Lore's and a member of the Youth Movement who had met Reich at the seminars, Annie Pink, entered analysis with him. Reich wrote in his diary:

> It is becoming increasingly obvious that I am analyzing Annie Pink with intentions of later winning her for myself—as was the case with Lore . . . What must I do? Terminate the analysis? No, because afterwards there would be no contact! But she—what if she remains fixated on me, as Lore did? Resolve the transference thoroughly! Yes, but is transference not love or, better said, isn't all love a transference?

As the analysis continued, the transference did not dissolve. "It is awful," Reich complained in his diary, "when a young, pretty, intelligent eighteen-year-old girl tells a twenty-four-year-old analyst that she has been entertaining the forbidden idea that she might possibly embark on an intimate friendship with him—yes, that she actually wishes it, says it would be beautiful—and the analyst has to resolve it all by pointing to her father." "Grit your teeth, maintain your facade and convey indifference!" Reich urged himself. Despite this resolve, neither Annie's transference nor his countertransference went anywhere. Six months into the analysis, Annie Pink declared her intention to terminate. Afterward, Reich wrote wistfully, "If she were decent, *she* would have to resolve *my* transference to *her*."

After terminating, Annie promptly fell ill and wrote Reich. He confessed in a letter to her that behind his "professional facade," he had struggled with his countertransference. Reich advised a cooling-down period. The two became friends. Annie went for further analysis with a more senior analyst, Hermann Nunberg. Later, their affair began. In March 1922, after they were discovered in bed by Annie's family, the two decided to marry. Together they threw themselves into the world of Viennese psychoanalysis.

In 1922, Reich graduated from medical school and took a postgraduate job in Julius von Wagner-Jauregg's psychiatric clinic, but he also was appointed assistant at the new psychoanalytic clinic in Vienna. Just a year out of medical school, Reich was asked to lecture on clinical method. This truly was a case of the blind leading the blind, for by Reich's own count, he had analyzed four women and slept with two of them. How was he to continue analyzing? Should he just treat men and older women? he wondered. Who could advise him?

In his confusion, Reich was far from alone. Take, for example, the perplexed foreign students who came to Vienna after the war. The Professor was eager for stable currency, and they were eager to study with him to become psychoanalysts. In January 1920, Ernest Jones referred Dr. John Rickman to Vienna for analysis with Freud. By the time Rickman arrived, Freud had two other Anglo-Americans in treatment. This group would be joined by the Englishman James Strachey and his American-born wife, Alix. The son of Sir Richard Strachey, James attended Cambridge, then worked as an editor before coming across Freud's work and deciding to become a psychoanalyst. In February 1920, both James and Alix were taken into analysis at the Brunswick Square Clinic. James attempted to follow Ernest Jones's recommendation to study medicine, but anatomical dissection was too much for him. He circumvented Jones by writing directly to Freud and asked to be taken into training in Vienna.

He arrived in 1920, spent time between "the Professor's dissecting table and the Opera," and soon became enchanted by Freud's method:

Each day except Sunday I spend an hour on the Prof.'s sofa (I've now spent 34 altogether),—and the "analysis" seems to provide a complete undercurrent for life. As for what it's all about, I'm vaguer than ever; but at all events it's sometimes extremely exciting; and sometimes extremely unpleasant—so I daresay there's SOMETHING to it. The Prof. himself is most affable and as an artistic performer dazzling . . . Almost every hour is made into an organic aesthetic whole. Sometimes the dramatic effect is absolutely shattering. During the early part of the hour, all is vague—a dark hint here, a mystery there—; then it gradually seems to get thicker; you feel dreadful things going on inside you, and can't make out what they could possibly be; then he begins to give you a slight lead; you suddenly get a clear glimpse of one thing; then you see another; at last a whole series of lights break in on you; he asks you one more question; you give a last reply—and as the whole truth dawns upon you the Professor rises, crosses the room to the electric bell, and shows you out the door.

In the fall of 1921, the Stracheys and Rickman were joined by five Americans, so of the nine patients Freud had in treatment, eight were students from England and America who hoped to someday practice analysis. Other than their hour a day with Freud, however, the students were at loose ends. Finally, they summoned up the courage to ask for formal classes. Freud complied; he arranged for lectures by Otto Rank, Abraham, Ferenczi, and the Hungarian analyst Géza Róheim. But the students remained mystified by the techniques they were supposed to be mastering, and Freud seemed uninterested in taking up technical questions at this point in his career. One of the Americans, the New York physician Abram Kardiner, asked Freud what he thought of himself as an analyst. Freud confessed he had "no great interest in therapeutic problems" as he had become "too much occupied with the theoretical problems all the time, so that whenever I get the occasion, I am working on my own theoretical problems, rather than paying attention to the therapeutic problems."

While Kardiner was clear that his analyst was not very interested in therapeutic matters, a number of the Brits were simply bewildered. One day, James Strachey and John Rickman asked Kardiner to tea:

> John Rickman said to me: "I understand Freud talks to you."
> I said, "Yes, he does, all the time."
> They said, "Well, how do you do it?"
> I answered, "I don't exactly know. Maybe it's the hour of the day, maybe I keep him interested, maybe I keep hopping. I don't know but he's quite garrulous. How is he with you?"
> They both said, "He never says a word."

Befuddlement spread as more and more students flocked to psychoanalysis. The 1922 I.P.A. Congress in Berlin hosted 256 people, more than double the previous record set two years earlier. The congregants heard papers elaborating complex matters in libido theory and others that followed *Beyond the Pleasure Principle* into speculative biological theorizing about the nature of cells, embryology, and organic illnesses. It was not much help for those who wondered what to say to their neurotic patients.

Before things got clearer, they were about to become more confusing. At the congress, attendees heard a declaration that heralded a new era. The bad boy from Baden-Baden, Georg Groddeck, told the audience that their conception of the unconscious was insufficient. Instead, he suggested a blind force—which he called the "It" (in German, *Es*)—was at work inside us. It was not uncommon for an eccentric member of the association to present a contrary view of the unconscious and then be shot down. But shockingly, Sigmund Freud agreed. He wrote Ferenczi saying that he was writing something that continued the explorations initiated in *Beyond the Pleasure Principle* and that it had "to do with Groddeck."

At the Berlin Congress, Freud flagged his indebtedness to Groddeck by referring to the "It" while also announcing that the "I" (*Ich*)—normally equated with consciousness—was largely unconscious. Therefore, conflicts between the "I" and the "It" were entirely out of conscious, which meant that the model of mind central to Freudian thought, which operated through the segregation of mental contents into conscious and unconscious, was of no utility when it came to describing these conflicts. Freud promised to elucidate this material in a work called *The "I" and the "It."* Implied in the words themselves was the idea that the "It" was an unrecognizable force that exerted pressure on consciousness and that the "I" at least began with every person's ordinary understanding of inner experience.

Freud's death drive had already caused perplexity and conflict among older Freudians. Even loyalists like Paul Federn and Hermann Nunberg rejected the concept before eventually coming around. Others like Sadger and Hitschmann remained vehemently opposed. But these theories of a death drive were highly speculative, as Freud himself admitted. Now, in addition, came this newer reform which was clinically based, and challenged the model of the mind Freud had constructed in *The Interpretation of Dreams*.

Amid these upheavals at the 1922 Berlin Congress, Freud announced an open competition for the best essay on "The Relation of Psycho-analytic Technique to Psycho-analytic Theory." In the last two psychoanalytic congresses, only two out of fifty-three papers had been devoted to technique. The presence of many neophytes in Berlin (including 120 from Germany alone) made it obvious that abstract speculations would not serve to orient the growing numbers of future practitioners. A cash prize of 20,000 marks would go to the paper that best explored how theory and technique were "furthering or hindering each other at the present time." By the time Freud announced the prize, he believed he knew the winner.

IV.

DURING THE WAR years, Sándor Ferenczi had established himself as the foremost thinker on psychoanalytic technique. He had addressed the challenge of the war neuroses and discovered that some patients needed active, at times aggressive psychotherapeutic measures. Inspired by Freud's arbitrary termination date with the "Wolfman," Ferenczi began to experiment with various coercive actions. He reasoned that if frustration made the neurotic ill, then limited, minor satisfactions could defuse the wish for a full recovery. Kindness and generosity could sap the desire to be cured. Therefore Ferenczi believed the analyst must energetically oppose "premature substitutive satisfactions."

This aggressive stance ran counter to the vision of the analyst as an objective observer. Despite the fact that he hoped to maintain the analyst's mirroring capacity, Freud agreed that a passive attitude would get nowhere with phobic or severely obsessional patients. Phobias must be confronted. An obsessive must not be allowed to go on and on in the name of free association. Freud embraced these modifications but cautioned: "We refused most emphatically to turn a patient . . . into our private property, to decide his fate for him, to force our own ideals upon him, and with the pride of a Creator to form him in our own image, and see that it is good."

After 1918, Ferenczi published a string of papers that emphasized enforced abstinence. He wrote about the treatment of a female patient that had been going nowhere for years, a woman who had maintained an erotic transference but never moved forward in the analysis. Ferenczi set a termination date. The woman improved a little and left treatment then returned with the same neurosis. The second try at analysis was a reenactment of the first. On his third try, Ferenczi noticed that the woman had a habit of intertwining her legs during sessions and referred to her love for her doctor as a feeling "down there." Ferenczi concluded the woman was unconsciously pressing her thighs together, stimulating herself, and discharging pent-up libido during the session. She used analysis for sexual gratification, not insight and change. Ferenczi decided to forbid her from crossing her legs during sessions. The woman became tormented, restless, shifted her body back and forth, and then lapsed into a maze of fantasies and memories of early traumas that gradually led to her improvement. Ferenczi was delighted, only to find the treatment had stagnated again. He forbade her from masturbating at home and before long reported the woman had been cured.

Ferenczi's clinical experiments impressed Freud and led him to consider new technical alterations for cases in which abstinence was vital and the analyst's neutral disengagement useless. Ferenczi instructed analysts not to answer their patients' questions but rather turn them back to their sources. With analytic abstinence as part of psychoanalytic technique, the profession also had another bulwark against the dangers of sexual countertransference. Ferenczi reported that he would force the Hungarian Sándor Feldman to resign from the local Budapest group, as well as the I.P.A., because he refused to adhere to "psa. Abstinence." Freud reported that "another execution took place more mildly," his admonishment of Wilhelm Reich for his failure to maintain abstinence.

As a proponent of active intervention, Ferenczi found himself crossing swords with those who stuck to the model of the analyst-as-mirror. One such advocate was Ernest Jones, who had fought back Pierre Janet's attack on psychoanalysis by reiterating that the analyst was not an advisor or a suggestive therapist, but an objective interpreter of intrapsychic phenomena. Ferenczi's modifications weakened this position because his active interventions could easily be seen as suggestions. Ferenczi insisted that in some cases—for example, a man who phobically avoided sexual activity—it was perfectly reasonable for his analyst to instruct him to have intercourse. The needs of psychoanalysis as an objective science of mind should not undermine its effectiveness as a treatment.

Sándor Ferenczi, the Hungarian psychoanalyst who first vigorously opposed deviations from Freud, and then espoused central revisions in technique and theory.

At the 1920 international congress, Ferenczi presented an extension of his active technique. The chief rule was to maintain an atmosphere of abstinence. To enforce that, the analyst needed to give "orders and prohibitions, always against the direction of pleasure." But these were measures to be used *only* in exceptional cases. The presenter who followed Ferenczi to the podium knew exactly what the Hungarian meant. She *was* one of those exceptions. Eugenia Sokolnicka was a Polish woman who had studied with Janet and Jung, before going into analysis with Freud in 1913. The analysis had not gone well. Freud detested Sokolnicka, whom he pronounced "a basically disgusting person" and "quite crazy." When he terminated the treatment, she tried to kill herself. In 1920, she entered analysis with Ferenczi, who suggested she refrain from masturbation. She complied for a while then began to complain and analyze Ferenczi, calling him a sadist.

Ferenczi admitted to Freud that he believed there might be some truth in her accusation, which he connected to his own marital troubles. Ferenczi's attempt to resolve his conflicted love for Gizella and Elma had become an epic of indecision. Freud vacillated between maintaining his neutrality and offering advice. Finally exasperated, he told Ferenczi that he had no views on the matter. Freud urged his colleague to stop using analysis to avoid action in the real world. In 1917, Ferenczi decided to act and, incredibly, asked the Professor to pop the question to Gizella for him. Learning a lesson from his own analysis, perhaps, Ferenczi urged analysts to intercede actively to stop "the phobia-like incapability of coming to a decision" in an analysis.

In 1922, Ferenczi asked Freud to help Sokolnicka get a job in Paris as Freud's translator. Eugenia went to Paris and four years later became a founder of the Paris Psychoanalytic Society. Ferenczi seemed to have had success treating a woman Freud had been unable to help. (He could not know that in 1934 Sokolnicka would commit suicide.) The Hungarian wrote Freud: "I now seem to be less focused on finding new things than I am on achieving better results with improvement in technique." When Sigmund Freud announced the competition for the best paper on the relationship of technique to theory, the front runner would have surely been the innovative Ferenczi.

In fact, the race had already been won. Just before the Berlin Congress, Ferenczi, along with Otto Rank, read Freud a draft of a new paper that seemed to offer a new foundation for a field in flux. It synthesized the lessons of the past two decades and consolidated a new psychoanalytic technique. When Freud announced the Berlin competition days later, the understanding was that their paper deserved the prize.

DESPITE ITS PROMISE to unify the ranks, Ferenczi and Rank's work would become ensnared in political disputes that broke out in Freud's Secret Committee. Intended to safeguard the cause, this committee had broken into opposing camps. On the one hand, there were the elected leaders of the I.P.A., the president, Ernest Jones of London, and the secretary, Karl Abraham of Berlin. On the other, there was Freud's right-hand man in Vienna, Otto Rank, the director of the psychoanalytic press. Given that Abraham was usually supported by the other Berliners, and Rank by Freud and Ferenczi, the rift pitted the leaders in London and Berlin against those in Vienna and Budapest.

Much of the early conflict was over the new publishing house Rank directed. He had returned from the war and resumed his place next to

Freud. As the scribe of missives sent from Vienna to the members of the Secret Committee, Rank assumed some of the authority usually reserved for the Professor. Increasingly, it seemed, Rank and Freud spoke in one voice. In addition, after the war, Rank became the most powerful editor in psychoanalysis. With Anton von Freund's money, he was appointed managing director of the press. He edited the *Zeitschrift* and shared editorial duties for *Imago*. When Freud could not chair a meeting of the Vienna Society, Rank did. In 1922, the Professor wondered if he had erred in dissuading Rank from attending medical school; had he become a physician, Rank would have been the obvious choice as Freud's successor.

In the summer of 1922, as Rank's power became evident, Ferenczi sought out this Viennese colleague to coauthor a call for a new psychoanalytic method. Given Rank's ties to Freud, this book would implicitly carry Vienna's stamp of approval. Written by two of Freud's most prestigious colleagues, *The Development of Psychoanalysis* was poised to define the field for the second generation of analysts, those muddled but passionate students from London, New York, Vienna, Budapest, Berlin, and elsewhere.

As Rank and Ferenczi embarked on their project, the tensions in the Secret Committee reached new heights. A meeting was called in San Cristoforo to hash out the bad blood between Jones and Rank. In the ensuing arguments over publishing matters, Jones was accused of calling Rank a "swindling Jew." Rank tried but failed to expel Jones from the committee for anti-Semitism. Jones went home thinking that in addition to problems with Rank, his former analyst, Ferenczi, would probably never speak to him again. In November 1922, Freud stepped in. Forgoing the usual procedure of sending joint letters with Rank, Freud wrote directly to the committee members to scold them and tell them what he had told Rank from the start. The attacks levied by Jones and Abraham were displacements to Rank from the Professor. Jones's reproaches, including his anti-Semitic slur, inspired Freud to recommend he finish his analysis with Ferenczi. Freud also made it clear that if Rank was assailed in the future, the Professor would understand it as an attack on himself. Freud declared he would not participate in the next meeting of the committee, so the members could work things out without the "obstacle" of his presence.

Sigmund Freud never attended a meeting of the Secret Committee again. Nor would he attend another psychoanalytic congress. As the keepers of the psychoanalytic flame feuded bitterly, Freud noticed a growth in his mouth. By the fall of 1923, it was clear that the long-time cigar smoker had oral cancer. In October, his doctors knew that the first operation in the spring had been unsuccessful and he underwent a three-hour radical surgery. "I know of course that it is the beginning of the end," Freud wrote to Groddeck. Jones and others openly admitted that Freud's tumor would take the Professor's life, not today or tomorrow, but someday in the near future. That Freud would go on to outlive Sándor Ferenczi and Karl Abraham, and precede the much younger Rank to the grave by only a month, was a trick of fate no one could have foreseen.

Freud's illness shook his followers. He had founded and defined their discipline, and he seemed to be the only one with the power to complete the task of redefining it. As for the Secret Committee running things without Freud, that now looked like a bad joke. Adding to the sense of turmoil, Freud's new work, entitled *The "I" and the "It,"* was published. Unlike the admittedly speculative *Beyond the Pleasure Principle*, this work was a force-ful correction of what Freud now saw as an inadequate model of mind.

As long as neurosis was a battle between the unconscious and con-sciousness, there was considerable value in a model that parsed mental contents into these two realms. But with *Beyond the Pleasure Principle*, Freud had suggested that neurotic conflict was all unconscious. In a battle between two unconscious forces, Eros and Thanatos, the earlier topo-graphic model had lost its utility.

On April 17, 1921, Freud wrote Groddeck:

For ages now I have been recommending in the inner circle that the unconscious and the preconscious should not be opposed, but rather the coherent "I" and the repressed material split off from this. But that does not solve the difficulty either. The "I" is deeply uncon-scious, too, in its depths . . . The more correct notion thus seems to be that the categories and hierarchies observed by us only apply to relatively superficial layers, and not to the depth for which your "It" is the right name.

In 1923, Freud altered his model to posit relatively stable mental struc-tures that were categorized not by their relationship to consciousness, but rather by their distinct *functions*. The "It" inside men and women was the unconscious pressure of libido and the death drive. Freud's "It" was similar to Groddeck's, which the Baden-Baden doctor had presented in his 1923 *The Book of the It: Psychoanalytic Letters to a Friend*. Freud congratulated Groddeck on the book, and added that he himself would be dealing with similar issues. When Groddeck read Freud's *The "I" and the "It,"* in which the author made a point of telling the reader his work was indebted to no one, Groddeck furiously accused Freud of a plot to "appropriate secretly loans made by Stekel and me," going forward in "a very sneaky way, with the help of a death instinct or destructive drive taken from Stekel and Spielrein."

While Freud's "It" took its name from Groddeck, his definition of the "I" was new. The "I" was not just the inner experience of consciousness. It began with the primitive sensations that distinguished one's body as one's own, the accumulation of sensory information on my arm, my leg, my nose. This body "I" was the essence of the "I," which over the course of childhood developed an addition of tremendous social significance.

In his work on narcissism and melancholia, Freud wrote about an "Over-I" (*Über-Ich*) that monitored the "I," a presence akin to human conscience that played a crucial role in the making of character. Freud concluded that people take in lost object ties through introjection and identification. Character was nothing more than the residue of these abandoned attachments that settled in the "I." The "Over-I" was the first and most important of these identifications: it came from a child's parents. The Oedipal phase of development was resolved when the child renounced his passionate desire for the parent of the opposite gender and identified with the parent of the same gender. This identification created the "Over-I," which now instituted parental rule in the child's mind. Religion, morality, and social conscience were all the result of this crucial acquisition. But the "Over-I" could also create neurosis, for it demanded that a boy be like his father and, at the same time, not dare to assume the prerogatives of the father.

With the publication of *The "I" and the "It,"* Freud took another shibboleth of the Freudian movement and threw it up for grabs. A structural model divided among "It," "I," and "Over-I" competed with the topographic model as the reigning model of mind. Again, the members of this community would have to choose between long-held Freudian positions or new ones. The Secret Committee members had been sanctioned as elite guardians of a set field; they were not prepared for this.

SOON THEREAFTER, SÁNDOR FERENCZI and Otto Rank published *The Development of Psychoanalysis*. When drafted in 1922, the book included a historical overview that was used as an argument for change. Ferenczi and Rank believed that the idea of a death drive had opened the door to a great deal of fatuous theorizing and that the time was ripe for a work that made the clinical and empirical demands of psychoanalysis primary. The authors wrapped themselves in Freud's statements about the provisional nature of theories of the unconscious and then laid out the problems they believed plagued the analytic community.

Psychoanalysis was intended to be both a therapeutic method and a science of mind. But over the preceding decade, the relative neglect of clinical matters had sowed confusion, they argued. During that same period, Freud—who the authors noted had long been known for his reserve on technical issues—published nothing on the subject. Now Freud's papers on technique were "antiquated," and required modification.

Ferenczi reviewed prior theories of technique, which he frankly labeled misguided. Psychoanalytic method had fossilized, he believed, and become an overly intellectualized process of educating patients about the contents of their unconscious. Analysis was not about the amassing of associations, nor was it a "fanaticism" for interpretation. Simply translating from the unconscious ignored the fact that the chief significance of unconscious phenomena derived from their place in the overall analytic situation.

The Hungarian analyst dismissed the view that analysis should focus on resolving neurotic symptoms and noted that therapists of all stripes could make neurotic symptoms go away . . . for a while. The trick was to make them not return. To do that, psychoanalysis had to focus on the whole personality. At the time, many analysts believed it was impossible to analyze patients with global character disturbances, but Ferenczi insisted that a piecemeal approach would fail unless the patient's personality was addressed.

Furthermore, libido theory had become an albatross for clinical practice, misleading analysts into simply applying this scientific knowledge in a crude and dogmatic way that neglected the true task of analysis.

Though diplomatically worded, this was a critique of the changes that had come after Nuremberg. Ferenczi argued that the need to prove, ratify, and extend libido theory had overwhelmed clinical practice. Ironically, it was Ferenczi himself who had been the fiercest of the libido warriors, but now, he called for a return to close clinical observation and thinking focused on the imperative to cure.

Such talk would have been a ticket out of the Freudian movement before 1915, but not anymore. Ferenczi took cover behind Freud, pointing out how the Professor had made a point of emphasizing the hypothetical nature of his recent work, and added that often excess speculation comes from a desire to avoid discomforting technical problems. All the worry about the scientific standing of psychoanalytic theory had been destructive and had led to theoretical rigidity, clinical paralysis, and the barren process of simply discovering what theory dictated must be discovered. The road forward was a deeper study of analytic process, especially the examination of different forms of transference and resistance. In the end, the field would stand or fall based on those things and their effectiveness for treating patients.

After Ferenczi's critique, Otto Rank delivered a new framework for the practice of psychoanalysis based on the primacy of what he called the analytic situation. After writing extensively about myth, Rank only turned to clinical matters in 1922, when he was deeply influenced by the more experienced Ferenczi. The two men shared the view that analysis had become overly scholastic. In his first clinical writings, Rank had argued that healing did not take place through the simple acquisition of new knowledge. He argued that the chief role of analytic technique was to deal actively with repetitions in the forms of resistances and transferences. Analysis allowed patients to relive their original libidinal situation with the partial satisfaction of the transference and the continual, slow abandonment of those childish wishes. Psychoanalytic technique was built to allow that past to first become manifest and then be transformed into conscious memory so as to be laid to rest. Everything relevant to the cure of neurosis happened in the analytic situation and the transference. Interpretations should focus on reactions to the analyst, for in those reactions lay infantile repetitions. A good psychoanalysis formed a "structure of two" in which the analyst played all supporting roles in the inner theater of the analysand's mind.

Writing to Groddeck, Ferenczi announced the completion of *The De-*

velopment of Psychoanalysis, which he unabashedly called a "technico-politico-scientific work" that sought to reform the discipline. While renegades like Groddeck may have been pleased by the call for change, Ferenczi and Rank's work was sure to upset more conventional analysts. The two men were asking for nothing less than the reorientation of the entire discipline around—not psychosexuality and unconscious libido—but the analytic situation. While notions like transference and resistance were highly theorized and could be taken to imply a great deal about what was transferred and what was resisted against, these theories nevertheless allowed more freedom to discover unconscious contents clinically, empirically, and inductively. In the authors' eyes, a coherent theory of technique could offer a surer foundation for the field than theoretical speculations about the nature of the unconscious. And when Ferenczi and Rank first read their work to Freud in 1922, he was wowed. It seemed this text would become required reading for the students streaming in to the field.

But as the two authors were revising their book, Otto Rank experimented with active technique, and in the process he became convinced that he had stumbled upon a great discovery. By giving a rapid termination notice to all of his patients, Rank elicited great anxiety and dreams that seemed driven by an infant's yearning to be back in the womb. He began to believe his patients were recapitulating the same primal experience in the transference, which was none other than their shocking delivery into air: the trauma of their birth. Excitedly, Rank began to dictate a new book on what he called birth trauma.

Rank announced that the resistance to ending a treatment revealed a fixation on the mother, a maternal transference. There was something new and valuable in his theory. Babies cling to their mothers: they cry when they are separated. If Freud had seen the child's discomfort as the result of a loss of pleasure, Rank suggested it was primarily due to an attachment. In naming the anxieties that came with separation, Rank also emphasized a relationship that had been given short shrift. In 1924, most analysts believed that the most important affective tie to be worked through in an analysis was between child and father. Jung, Groddeck, and others had wondered why mother transferences weren't given larger significance. Otto Rank took up that challenge and argued that in his newly structured treatment, patients repeated and worked through their separation from their mothers.

However, Rank concentrated on the separation between mother and newborn at the moment of birth, an experience no one could possibly recall. He became convinced that the analytic situation was experienced as a kind of womb, and patients were repeating a yearning for mother-libido as it was in utero. The father was relegated to the role of enemy that wrenched the child away from his mother. The fear inspired by the female genitals was not due to castration anxiety, but a reexperiencing of the scene of "birth trauma." Rank went on to reinterpret neurosis and perversions as a result of traumatic birth experiences and looked forward to the day when character typologies would be similarly constructed.

Like a number of analysts before him, Otto Rank did not understate his discoveries. He trumpeted them as something that made psychoanalytic thought "productive for our entire conception of mankind and history." Birth trauma, he boldly predicted, would be found to be a universal physical and psychical trauma with great consequences for all of mankind, for it was the very basis of psychic life. He acknowledged that his views contradicted prior ideas, but he insisted that his theories were psychoanalytic in so far as they were the result of a "consistent application of the method created by Freud."

Rank's book on birth trauma was the first major revision of the unconscious that grew directly out of a change in technique. However, if an emphasis on technique was supposed to increase empiricism and prevent excessive speculation, Rank's theory gave no evidence of it. *The Trauma of Birth* took an interesting clinical experience and then transformed speculative inferences into a grand theory that extended into every nook and cranny of human experience. Rank didn't bother to entertain countervailing arguments or possible limits and in the process found the reductio ad absurdum of the notion that early childhood experiences formed adult character.

Worse, Rank had egregiously thrown over the epistemological cautions that had been so decisive in his book with Ferenczi. He explicitly contradicted the warning not to let preexisting theory determine clinical observation. Since the entire content of the unconscious and its psychical mechanisms were known from the outset of a treatment, Rank counseled his readers to *begin* treatment by informing the patient of their birth trauma, instead of waiting for this to emerge. In December 1923, Rank sent his book on birth trauma to Freud. The book had been dedicated to the ailing Professor, and Freud accepted this dedication, writing that

Rank's productivity was deeply gratifying, for to him it meant, quoting Horace: "*Non omnis moriar*," or "Not all of me shall die."

The nearly simultaneous publication of Ferenczi and Rank's book on technique with Rank's birth theory meant the two would be measured side by side. Rank's theory was guaranteed to cause controversy, and Ferenczi and Rank's work would produce its own storm, for the authors implicitly accused the field's leaders of being lousy analysts. But the reception of these works was made worse by the fact that the two men now had powerful rivals. By 1924, Rank was at odds with Ernest Jones and Karl Abraham. In the months leading up to the two publications, Ferenczi found himself embroiled in conflict with Jones, who had cheerfully announced to the Secret Committee that a fine critique of active therapy had been accepted by him for publication in the *International Journal*. Ferenczi took offense. Jones backpedaled, saying he was not suggesting the critique was correct. When that conflict died down, there was another scuffle when Ferenczi accused Jones of plagiarizing his work.

While his disciples squabbled, Freud's illness dragged on. His cancer seemed to have provoked maneuvering among his heirs reminiscent of Shakespeare's *King Lear*. The significance of Freud's illness was particularly great for Rank, who was not just another follower. He had been essentially adopted by Freud, who supported him emotionally and financially; in return, Freud had been repaid by Rank's unswerving loyalty. Rank's substantial power in the movement came from his relationship to Freud, but now Freud was sick and perhaps dying.

However, while Freud's illness was on everybody's mind, he was still very much alive, and it became clear that his take on Rank and Ferenczi's manifesto could be decisive. Rank and Ferenczi believed they already had secured Freud's support after reading him their draft in 1922. Freud had confessed that he was unclear about some points and thought others extreme, but on the whole he had found the work very enjoyable.

Rank's book on birth trauma had also been vetted by the Professor. When Freud first heard of Rank's discoveries, the Professor declared Rank's work "the greatest advance since the discovery of Psychoanalysis, even if only 33% or 66% is true." Freud wanted to try out Rank's theory and offered his acolyte a strange compliment, saying that with ideas that good, another man would have founded his own school. What seemed to trouble Freud most was that Rank's theory was based on a thoroughly ontogenetic trauma. How could Rank have a scientific theory of the psyche that did not have a place for human inheritance? Sándor Ferenczi tried to solve this dilemma by presenting an account of human history and sexuality through the ages. In his 1924 book, *Thalassa: A Theory of Genitality*, Ferenczi fashioned a phylogenetic theory that rooted human sexuality in the history of the species and argued that birth trauma was not just an ontogenetic trauma but also the recapitulation of a phylogenetic trauma, which was nothing less than the first slithering of amphibious life out of water into air.

When *The Trauma of Birth* and *The Development of Psychoanalysis* were published in 1924, Rank and Ferenczi were immediately attacked by their rivals in London and Berlin. At the suggestion of Max Eitingon, Freud wrote a letter to the committee giving his opinion on these publications. Freud's letter supported his two close colleagues but also hinted of dissatisfaction:

> It is not without astonishment that I have heard from various sides that the recent publications of our Ferenczi and Rank, I mean their joint work and that on birth trauma, have evoked unpleasant agitation in Berlin. Apart from that, I was directly asked, by someone in our midst, to express among you my opinion of the undecided matter in which he sees the germination of a split. Thus I am complying with his wish, do not interpret it as obtrusiveness; my purpose being rather to exercise as much restraint as possible and to let each of you follow his way freely.

Freud denied the rumor that the work was divisive and suggested that he would never have accepted Rank's dedication in *The Trauma of Birth* if he had felt that way. Complete agreement could not be expected in scientific matters; only the common ground of psychoanalytical presuppositions was necessary. With these works, he insisted, that was not a question. Freud then discussed each book individually. He had little trouble accepting the book on technique as a valuable correction to his own timidity regarding therapeutic activity. It was a "refreshing and subversive intervention into our present analytic habits." The attempt to create a new mode of technique was justified and could not be judged heretical. But Freud did raise concerns about active therapy in the hands of novices

and displayed increasing pessimism about the power of abbreviated therapies. It had taken Freud six weeks to regrow his beard after his operation,
and his scars still hurt three months after surgery. Could one really expect
to change the innermost layers of the psyche in less time? he wondered. In
the end, Freud announced that he would still use classical technique himself, since his patients were mostly analysts in training who needed a
longer immersion in their own unconscious lives.

Then the Professor turned to Rank's work on birth trauma, which he
deemed far more intriguing and highly significant. Freud wondered
aloud about the fate of the Oedipus complex in Rank's model, but let the
question hang. He declared that this was not: "a coup d'état, a revolution,
a contradiction of our certain knowledge, but an interesting complement
the value of which should be recognized by us and those outside our
circle."

That should have put an end to the controversy. It did not.

In Rank's birth trauma, skeptics had before them a theory they did not
find credible. Abraham responded to Freud's letter directly, saying that
"results of whatever kind obtained in a legitimate analytic manner would
never give me cause for such grave doubts. Here we are faced with something different. I see signs of a disastrous development concerning vital
matters of Ψ." Abraham indicated that Rank's technique was not acceptable, and that his technical developments, including the mandated termination date, would be dangerous for the field. More explosively, Abraham
warned that Rank was another Carl Jung.

Freud tried to calm his Berlin stalwart. Imagine the worst case, he
wrote. Ferenczi and Rank jointly declare that difficulties with the mother
and birth trauma governed a child's capacity to negotiate the Oedipal situation, so that birth trauma dictated neurosis. Some analysts would change
their technique accordingly, but: "what further damage would ensue? We
could remain under the same roof with the greatest calmness, and after a
few years' work it would become evident whether one side had exaggerated a valuable finding or the other had underrated it."

Freud also mentioned Abraham's warning to Rank. To be compared to
Jung was to be called a traitor; Rank became incensed. By March 1923, a
frantic round-robin correspondence took place: Jones and Abraham strat-
egized while Rank and Ferenczi tried to shore up the Professor's support.
Rank assuaged Freud, saying his new theory was in "perfect harmony

with your theory of the drives." Rank had merely used a new technique, setting a termination date in advance, which caused birth experiences to appear of their own accord. Ferenczi appealed to Freud too and denounced Abraham's "boundless ambition and jealousy." He explained that while working on the joint book on technique with its intense focus on transference, Rank had come up with the concept of the analytic situation and the tactic of giving a termination notice to patients in all cases. When Ferenczi tried this out, he also discovered what Rank had found: maternal transferences and a longing for the womb. The Berliners were rejecting all these discoveries out of hand, without trying to test them. Freud agreed that Abraham was behaving badly, but the Professor's support for Rank's work also began to diminish. The idea that the trauma of birth would replace the Oedipal etiological scheme he had hammered out over twenty years made no sense to him. Rank had discovered something important, Freud granted, but he had not worked it out well.

Ferenczi and Rank anxiously passed the letters they received from Freud back and forth, trying to gauge their standing. Meanwhile Ernest Jones opened up a new front, complaining that Ferenczi and Rank had kept their innovations secret from the committee. Ferenczi explained that he and Rank had not presented their joint work to that group because it had been written for the prize competition, and the judges of the prize were on the committee. Abraham said his questions were not focused on scientific results but rather on Ferenczi and Rank's active, shortened method. The road they were taking, he opined, "lead[s] away from Ψ" Abraham lobbied for a two-and-a-half-day marathon meeting of the committee to go over Rank and Ferenczi's work. The authors refused, declaring it fruitless to convene such a meeting when none of the members could possibly have had enough time to test out their new techniques.

The bitterness among Freud's paladins was no longer restrained. Rank and the Professor agreed that the Secret Committee was dead. "Gone is gone, lost is lost," Freud wrote of his inner circle. Then Freud surprised Rank with the news that he was writing a paper in response to his birth theory titled "The Dissolution of the Oedipus Complex." Like Jung, Adler, Groddeck, and others before him, Otto Rank was startled to find the Professor poised against him. Freud read his paper to Rank, and it minced no words. Rank had replaced castration anxiety with birth trauma, but castration anxiety was at the heart of the Oedipus complex. Rank was shocked to find that, though his work had clearly instigated Freud's thinking, the father of psychoanalysis had not deigned to mention Rank in the paper. When Rank pressed Freud on this, it became obvious that the Professor had not wholly read Rank's book. The Professor had been checking out Rank's ideas by giving patients his junior colleague's book and asking them for their impressions. "Even as I write this, I still cannot believe that such a thing is possible . . . ," he lamented to Ferenczi.

As vociferous dissent in London and Berlin mounted, Ferenczi feared the worst. How could the tempest be quelled? How could these differences be resolved? The congress in Salzburg was coming. Should they ask for a plebiscite? But a scientific matter could hardly be decided by a vote, Ferenczi reasoned. So it was down to this again: "Everything depends, ultimately, upon the declarations of the Professor." The next day, Sigmund Freud announced he would not attend the congress in Salzburg.

AT THE 1924 Salzburg Congress, Karl Abraham was dutifully elected president. The rancor in the Secret Committee had grown so rapidly that Otto Rank's name was still on the I.P.A. ballot as Abraham's would-be secretary. Rank accepted Eitingon's offer to replace him, and in return, Freud offered Rank (as he had once offered a disgruntled Alfred Adler) the presidency of the Vienna Psychoanalytic Society. This congress featured the long-awaited competition for the best paper on the relationship of technique to theory. Ernest Jones presided over three papers, all written by Berliners. Rank and Ferenczi sat in the audience and were asked to make short comments. In the end, no one was awarded the prize.

During the meeting, Abraham pulled Ferenczi aside and warned him that he was heading into exile. Ferenczi entered into discussions with Abraham, and ultimately the two men came to some agreement. However, Jones and Ferenczi remained on icy terms, and Rank abruptly left the congress a day early, causing much speculation. Abraham told Freud he had heard Rank left because he couldn't stand the idea of seeing Abraham elected president.

Actually, Rank was boarding a ship to go to America. In the United States, Rank was flooded with patients who came for his new, shortened psychoanalytic treatment. He charged an astonishing twenty dollars a session at a time when most doctors charged a tenth of that. Freud's former American analysands flocked to him, and Rank reported that the word among analysts was that to get your father transference analyzed, you must go to Freud, but to get your mother transference analyzed, you must see Rank. Meanwhile, Freud informed Abraham that he was growing more skeptical of birth trauma: "I believe it will fall flat if it is not criticized too sharply," Freud advised. An American analyst wrote Freud for his opinion on Rank's theory, which the Professor called "merely an innovation in technique."

Privately, Freud was distressed. He wrote Rank in America and asked him to leave open a path of retreat from his new theory. Rank replied that he could not and would not. Freud warned Rank—much as he had Jung, Adler, and Stekel—that he was making a universal psychology of his own neurosis. Rank remained unswayed. As summer neared fall, Freud began to feel that Rank's withdrawal from the movement was imminent. The Professor was confused and angry, especially since he had first greeted Rank's theory as an important accomplishment and only later grew dubious. Now the prospect of a rebel in charge of the psychoanalytic press, two journals, and the Vienna Society had Freud scrambling.

Meanwhile, the Berliners threatened to start their own journal if Rank remained editor of the *Zeitschrift*. Berlin and London had long been abuzz about Rank deserting Freud, just as the Professor—James Strachey put it bluntly—was "supposed to be kicking the bucket." Rank's few remaining allies were falling away when he lost his last defender in the inner circle. While vigorously defending their joint work on technique, Ferenczi had argued that aspects of Rank's theory could be integrated into mainstream theory. But when Rank encouraged the financially strapped Hungarian to come to America, only to rescind the offer at the last minute, Ferenczi lost any inclination to defend Rank. And Freud was now "boiling with rage."

When Rank returned to Vienna, he met with Freud for three hours. Freud wrote Jones to inform him that all intimate relations with Rank had come to an end. Otto Rank was stripped of the directorship of the publishing house and the editorship of the *Zeitschrift*. The promised presidency of the Vienna Society was revoked. Rank fell into a deep depression, left for Paris, but then doubled back and turned up at Freud's door in need of help. Freud took Rank into treatment and a month later joyfully declared that Rank was again his old self. Rank wrote a mea culpa to the committee members in December 1924. He confessed that his behavior was a neurotic response to the Professor's illness and begged Jones, Abraham, and Sachs for forgiveness. The Berliners were not so sure they wanted to welcome Rank back and asked for more details. Max Eitingon, however, broke from the Berlin signatories and wrote a more accepting letter in which he expressed relief that the confusion Rank and Ferenczi had sowed would now be corrected.

Rank never fully repudiated his theories. After his humbling recantation, he slowly pulled himself together and moved to Paris in 1926. Two years later, he resigned from the Vienna Psychoanalytic Society and joined Breuer, Bleuler, Adler, Jung, Stekel, and others who through their hard-fought intellectual disputes helped define what was inside and what was outside the boundaries of psychoanalysis. In some aspects, Rank's expulsion resembled the schisms that took place before the Great War. Tensions between Zurich and Vienna had framed those earlier battles, and Rank's expulsion took place during a fight for supremacy between Berlin and Vienna. The knowledge that Freud had cancer incited a battle for the future of psychoanalysis. As the community of analysts burgeoned, the Berliners wanted to establish themselves as standard-bearers, and they led the charge against Rank and Ferenczi. After the British-born Edward Glover was analyzed by Abraham in Berlin, he returned to London and wrote a critique of Ferenczi's active treatment; Hanns Sachs, Berlin's designated training analyst, decimated Rank's book on birth trauma. And the first graduate of the Berlin Institute, Franz Alexander, penned a dismissive review of Ferenczi and Rank's book on technique.

Yet, while festering ill will and political jockeying may have motivated some of the scorn heaped on Otto Rank, his critics needed intellectually legitimate grounds from which to launch their attacks. And here, the case of Otto Rank differed from those of Adler and Jung. Rank was not kicked out of the movement for leaving behind the Freudian synthesis of unconscious psychosexuality, though surely he had. That prewar boundary had come undone. Freud and Abraham insisted that Rank's divergent theory did not itself place him outside psychoanalysis. Instead, Freud, Ferenczi, and Abraham focused on something else, which they highlighted as Rank's most serious failing: he had not played by the rules of science. Rank had attributed his findings not just to the technique he and

Ferenczi publicly espoused, but also to a secret modification that he refused to share with others. Therefore, no one could test his theories. A. A. Brill wrote Freud to tell him that Rank's New York analysands reported that he had declared dream analysis irrelevant, given up on sexuality, and interrupted a patient's associations to steer him directly to birth trauma. While in Berlin, one of Rank's analysands was questioned about his analyst's methods. How short were his treatments? How little exploration of an individual's inner life was necessary before Rank informed him of his birth trauma? Abraham and others were afraid that Rank's technique would lead to short indoctrinations. Sachs complained that answers to these concerns were not to be found in Rank's published work, which featured grand conclusions but none of the fundamental approaches and observations that led to these notions. Freud concluded Rank was keeping his technique secret so he could open a separate school.

Otto Rank had reason to hide his new technique, because it would provoke an outcry. Far from Vienna and Berlin, in lectures to the Pennsylvania School of Social and Health Work in 1924, Rank outlined his vision of a brief psychotherapy capable of reaching masses of neurotics who could not otherwise afford psychoanalytic treatment. Knowing the realities of birth trauma and transference, the science of psychoanalysis could dispense with the lengthy process of listening and observation and give way to the mass application of quick therapies. In Berlin and Vienna, many would have sympathized with his attempt to make psychoanalysis accessible to the populace, but Rank's solution defied the theoretical restraint and close clinical attention he and Ferenczi had advocated. Edward Glover's critique pointed out that active treatment risked distorting the field and the transference. If that was true for active treatments, it would be even more true of Rank's therapies. Facing certain attack, Rank guarded his new technique and fought Freud on theoretical grounds. But in the end, his compromising of the clinical situation brought Rank down. His expulsion demonstrated that the psychoanalytic community had moved away from strict policing of theoretical deviants toward a process of exclusion that could be based on failure to employ proper method.

Ferenczi and Rank's short, ambitious road map, *The Development of Psychoanalysis*, was also buried in an avalanche of reaction. While they had avid enemies, the two authors were also in part responsible for their failure. Their powerful critique was timely and astute. Massive and unrestrained abstract theoretical speculation had often overwhelmed technical and clinical imperatives in psychoanalysis, a fact clear to many from the Berlin power broker Max Eitingon to novices like Wilhelm Reich. This decayed state of affairs was ripe for reformers who demanded that empirical data from the clinical situation supersede theoretical speculation. But because Ferenczi and Rank's 1924 work came to light alongside Rank's *The Trauma of Birth* and Ferenczi's phylogenetic fantasy, *Thalassa*, it was hard to see the point. Both authors called for theoretical restraint and an experience-near method as the central commitment in psychoanalysis, while simultaneously publishing some of the most extravagant theories ever seen in the field. In the end, their effort to reform psychoanalysis into a more epistemologically careful, clinically rooted discipline failed in good part due to their own bad faith.

As he had with Adler and Jung, Freud evaluated Rank's work to see what part of it could be incorporated into the body of psychoanalytic thought. Freud was initially fascinated by birth trauma, but upon first hearing of it, he warned his followers: "It is not easy for me to feel my way into unfamiliar trains of thought, and I have as a rule to wait until I have found a connection with them by way of my own winding paths." Soon Freud found his way.

For more than twenty-five years, Freud never wavered from the belief that anxiety was due to pent-up sexual libido. But in 1926, after digesting Rank's work on birth trauma and separation anxiety, Freud revised that theory. "Inhibitions, Symptoms and Anxiety" appeared two months before Rank left Vienna for good. In this sprawling, important work, Freud asked what anxiety was and without any overt acknowledgment of whom he was thinking against, he developed a theory of anxiety that employed an idea of trauma similar to Rank's. Freud postulated that the "I" used "signal anxiety" to anticipate and defend against possible traumas that harked back to a child's earliest experiences of helplessness. Anxiety was a normal indicator of danger, not the result of repressed libido, and it was founded on an infantile experience of terror and weakness, the prototype of which was birth. Separation from the mother was the first terror of childhood, followed by object loss, castration, and fear of the "Over-I's" punishment. Like many others, Otto Rank was now left to protest that Sigmund Freud had co-opted his intellectual property without acknowledgment.

Meanwhile, Ferenczi positioned himself against Rank in the vain hope that he could salvage their joint reform. He argued that Rank had vulgarized their notion of the analytic situation until it was the only thing that mattered, and that he had made an individual's history meaningless save for the moment of birth. Rank's promises of quick cures—like his six-week treatment for homosexuality—discredited the close engagement that Ferenczi had supported. Ferenczi also noted the excesses committed in the name of his active technique, not losing the opportunity to blame Rank for that, too. When Rank finally went public with his new technique, Ferenczi quashed his former friend's proposals. With Rank an outcast and Ferenczi on the defensive, their reforms were dead. *The Development of Psychoanalysis* became the revolution that wasn't, and the power to remake psychoanalysis shifted to the city of Berlin.

TEN

A New Psychoanalysis

Karl Abraham trained at the Burghölzli, then established psychoanalysis in Berlin.

I.

WHEN KARL ABRAHAM moved to Berlin in 1908, he was going where no analyst dared tread. Berlin was a bastion of German academic psychiatry. There, Freud's plague did not seem so contagious. Abraham relied on his renowned relative, Dr. Hermann Oppenheim, to help build his practice;

he could also count on his old friend from university days, the Berlin sexologist Iwan Bloch, as well as one of Freud's acquaintances, the sexologist Magnus Hirschfeld. The arrival of Max Eitingon from Zurich further strengthened Abraham's tiny society, which managed to field a total of nine members when it joined the I.P.A.

In 1908, Berlin was still shaking off its reputation as a backwater, a city inferior to the great capitals of Europe. Kaiser Wilhelm II's Berlin was regarded as a stiff, formal place. Germany's crushing defeat in the Great War changed that impression. The end of the monarchy gave birth to a republic in nearby Weimar and ushered in a turbulent new democracy in which Berlin would have to accommodate Nazis, Soviet Communists, and anticapitalists. Political turbulence, strikes, beatings, and violence were common. Spiraling inflation undermined the traditional middle-class ethos of working and saving. With inflation reaching new heights daily, the most rational Berliner immediately spent his currency before it lost value. This led to a frenetic atmosphere of people living for today while desperately fending off starvation tomorrow.

In the middle of this economic lunacy, Berlin became a center for avant garde movements. Walter Gropius founded the Bauhaus school of architecture. Dadism moved from Zurich and took root among artists in Berlin who adopted its merrily absurd ways. Berlin nourished high modernism in the novels of Alfred Döblin, the theater of Bertolt Brecht, and the radical twelve-tone musical compositions of Alban Berg and Arnold Schönberg, and it fostered an extraordinary film industry exemplified by the work of F. W. Murnau, G. W. Pabst, Fritz Lang, and Josef von Sternberg. Cultural modernism and the revolt against tradition had long been prevalent in European cities like Vienna and Paris: now these values took hold in Weimar Berlin.

The collapse of the monarchy also led to a broader revolt against traditional morals. Coming from Vienna, the writer Elias Canetti felt he was a provincial by Berlin standards: "Anything went. The taboos, which there was no lack anywhere, especially in Germany, dried out here." The artist George Grosz put it bluntly: "All moral codes were abandoned." All-night dancing, pornography, and prostitution flourished in "mad, corrupt, and fantastic Berlin." Berlin became—especially for tourists with hard currency—a sexual bazaar. Nightclubs catered to a wide variety of sexual appetites. The *Weisse Maus*, for example, featured the drugged-out dancer

Anita Berber, who was known for, among other things, dousing her naked body with wine while she urinated on a patron's table. Transvestite bars like the Eldorado, lesbian hangouts like Café Dorian Gray, and gay bars like the Adonis all operated freely.

Egalitarian political ideals invigorated feminists and sexual reformers. Wilhelm had staunchly upheld patriarchy and had once revoked an award won by the artist Käthe Kollwitz simply because she was a woman. With Wilhelm gone, reformers who had long rejected conventional marriage and advocated for birth control and legalized abortion gathered new force. Magnus Hirschfeld opened his Institute for Sexual Science in 1919 and turned Berlin into a hub of sexual study and reform. And while Wilhelm's Berlin had been cool to Karl Abraham and the Freudians, Weimar Berlin swooned over them. After the war, Abraham excitedly wrote Freud: "Berlin is clamouring for psychoanalysis."

After the Budapest Congress of 1918, Max Eitingon and Ernst Simmel returned to the defeated capital of Germany intent on bringing psychoanalysis to the masses. Eitingon's family's fur business held assets in the United States, so his wealth was protected from stampeding inflation. Those funds combined with Simmel's experience treating men at the front made a psychoanalytic clinic for the poor imaginable. After getting the approval of the Society, Karl Abraham, Max Eitingon, and Ernst Simmel were appointed to run the "Poliklinik," and in February 1920, the first psychoanalytic clinic in the world opened its doors on 29 Potsdamer Strasse.

The Berlin Poliklinik was an immediate success. Consultations were done by Max Eitingon from 9 to 11:30 A.M. on every weekday except Wednesday. The treatment was either free or based on what patients thought they could afford. Each analyst in the Society was to treat one patient for free a year, or tithe 4 percent of their income to the clinic. In the first year, 193 patients came for consultations. Eitingon assigned seventy patients for psychoanalytic treatment by the permanent staff and the members of the psychoanalytic society. Even though taking on so many cases strained the clinic, they tried to maintain the same parameters of treatment. After a short experiment with a thirty-minute session, they settled on forty-five minutes instead of the traditional sixty and shortened treatments from six days a week to three or four. Artisans, clerks, and professionals came with their neuroses; the great majority were treated for less than nine months. After the first two years, the clinic could report that 94 out of 141 patients were improved or cured.

Eitingon, Simmel, and Abraham had taken psychoanalysis directly to the people. The Poliklinik was one of many public clinics that sprouted up in a Berlin that increasingly treated health care as a social obligation. Over time, institutions such as juvenile courts and youth agencies sent the analysts patients too. Then thanks to a contact Simmel had in the Ministry of Education and the Arts, the University of Berlin began to consider establishing a professorship in psychoanalysis for Karl Abraham. Despite the fact that the professorship never materialized, the very fact that such a chair was seriously considered reflected the new respectability of psychoanalysis in Berlin.

Increasingly, the problem became not the supply of patients, but having enough analysts to meet the demand. "All our plans depend on having enough new followers, and so far (we) unfortunately do not have them," Abraham told Freud in 1920. The Berliners sought to recruit young doctors and laymen by advertising courses of instruction in the newspapers. Despite a poor showing the first year, Abraham found the strategy paid off. His introductory lectures on psychoanalysis in 1921 attracted over eighty people. With its combination of a clinic and formal courses, the Poliklinik began to welcome medical students and young doctors. After losing the Burghölzli, psychoanalysis had no institutions for education and training. Berlin filled that need and became, in Freud's words, the "headquarters" of international psychoanalysis.

Early on, Abraham had decided that he would only accept students if they were well read in the field and agreed to be analyzed. Carl Jung had suggested such a training analysis long before, and Hermann Nunberg had raised it again at the Budapest Congress in 1918. Unanalyzed psychoanalysts made no sense (unless, of course, one recalled that Freud and many early Freudians were not analyzed). In Vienna, the prerequisite for becoming an analyst was simply being analyzed. In Freud's opinion, it was vital. Physician, heal thyself.

In Berlin, Abraham also decided to make a didactic analysis necessary for students who wanted to study at the clinic. Yet Abraham could not handle all the students, and he found that serving as the leader of the psychoanalytic society and treating most of the members bred trouble. He complained that his job as president was complicated by the fact that members brought their resistances from his couch to the society meetings. Abraham decided to appoint a designated *Lehranalytiker*, solely dedicated to analyzing students. Hanns Sachs, the lawyer from Vienna, came to Berlin to take this new position.

Sachs entered Freud's circle in 1910 and gained the trust of the Professor during the debates with Adler, when he proved himself an unwavering Freudian partisan. Sachs might have remained in Vienna, but near the end of the war he arrived in Budapest for the 1918 congress and began to cough up blood. Diagnosed with tuberculosis, Sachs spent the next two years recuperating in Switzerland. When the Berliners began their search for a teaching analyst, this loyal member of the Secret Committee had recovered and was at loose ends. On October 8, 1920, Sachs arrived in Berlin with a slate of student-patients waiting for him. By 1922, he had analyzed twenty-five students, thirteen of whom had begun to perform analysis themselves.

The Berliners also began to accept students for more than a passing visit. A Hungarian-born doctor who had recently completed his medical studies in Göttingen, Franz Alexander, was the first trainee to complete a set of academic requirements in a structured setting so as to become a psychoanalyst. Alexander attended classes and did a didactic analysis with Sachs that began in the fall of 1920. A year later, he was accepted as a full member of the Berlin Society.

The Berlin curriculum and its classes grew. There were not only introductory courses but also seminars and advanced classes. Max Eitingon devised a method to teach the pragmatic aspects of clinical treatment by "controlling" a novice's analysis with close one-on-one supervision. Eitingon had students take detailed notes on their sessions, which a senior analyst reviewed to detect and rectify the mistakes made by the inexperienced practitioner. As an example, Eitingon cited an "all-too-rigid attitude towards single theories and results," the very concern Ferenczi and Rank had hoped to remedy. Through supervision, Eitingon and his colleagues hoped to instruct the beginner and protect clinic patients from harm.

Before long, the Berliners were in the enviable position of taking only those applicants they thought were good bets. In 1922, six students were enrolled in the Poliklinik's training program. Three young assistant doctors from the University of Berlin's psychiatry clinic began to study psychoanalysis in 1921; however, only a year and a half later, all of the assistant doctors at that renowned clinic had entered analytic training. By then, Abraham had the luxury to complain that he and Sachs had been overwhelmed by the demand for training analyses. The exploding popularity of psychoanalysis in Berlin was never more evident than in 1922, when the I.P.A. Congress came to town. The attendance that year was by far the highest ever, 256 attendees; of those, 91 were Berliners.

The Berlin Poliklinik offered the most rigorous and structured education in psychoanalysis in the world. The best and brightest students made their way there. From London came Edward and James Glover; from Vienna Otto Fenichel and Helene Deutsch; from Budapest Melanie Klein and Sándor Radó. Some stayed and others carried the city's vision of analysis home, creating a Berlin diaspora. By the time the Vienna clinic opened in 1922, a powerhouse had already begun to emerge in Berlin.

In 1923, the leaders of the Berlin Society and the Poliklinik concluded that the time was right to establish a formal teaching institute with clear requirements. A committee headed by Max Eitingon was appointed to systematically organize their educational process. Eitingon, along with Abraham, Simmel, Sachs, and others, set to work. Some questions seemed deceptively easy: the candidates would be chosen by a committee and would be required to take a formal curriculum. But the admissions criteria for those students and the curriculum remained open to question.

It was once clear that to be a Freudian one had to accept the tenets of libido theory, but that simple requirement had vanished. If one asserted— as Freud had to Groddeck—that analysis of transference and resistance made one a psychoanalyst, that now seemed inadequate. The Berliners sought to move beyond psychoanalysis as a theoretical or methodological commitment toward its definition as a profession. The question was no longer what one had to say to be in the Freudian school, but what one had to do to be a member of the guild. In 1923, Eitingon's committee published "Directions for the Education of Psychoanalytic Therapists" and outlined three requirements: a didactic analysis, theoretical training, and practical training. With those criteria in place, the Berliners announced the formal opening of the first psychoanalytic training institute.

In this way, the Berliners ushered in a significant transformation. The earlier culture of the Freudians was now replaced by institutionalized processes of training in which ideological issues were submerged and diffused through a series of bureaucratic structures. To become a psychoanalyst meant to have passed through those structures. Max Eitingon's plan for the formalization of psychoanalytic education was presented in 1925 at the I.P.A. Congress in Bad Homburg. Training committees were founded to introduce similar guidelines at other local societies. Graduation from the Berlin Institute or one of the institutes that followed its lead was the new path toward being a psychoanalyst. The days of the wild analyst were numbered.

REQUIRING DIDACTIC ANALYSIS was the least controversial plank of Eitingon's model. By conquering his own neurosis, the analyst would protect patients from his own countertransferences and resistances. However, a required training analysis entailed the risk of indoctrination. Freud had focused heavily on the son's need to rebel against his father but had been less concerned about a vain father's Kronos-like desire to destroy his cowed sons. According to one young Berliner, Franz Alexander, Freud had been hypersensitive to the supposed parricidal wishes of his intellectual heirs. Otto Rank accused Freud of wanting to analyze him so as to transform his creative theories into neurotic pathology.

Berlin's teaching analyst, Hanns Sachs, was unmoved by these concerns. He wrote unabashedly of training analyses as akin to religious initiation and the trial period for novitiates in the church. The would-be analyst needed to see with new eyes. The didactic analysis opened him to the mysteries of the unconscious, after which the seeker would gaze upon the inner forces of the Oedipal complex, infantile sexuality, and human ambivalence.

As a counterbalance to Sachs's view, the formal educational structure in Berlin limited the training analyst's authority and decreased the dependence of the trainee on his analyst. In the apprenticeship system practiced in Vienna, a student was utterly dependent on the good graces of his analyst, who served as teacher, referring doctor, and research mentor. The analysands of Freud, Jung, and Adler were also the trainees of Freud, Jung, and Adler, and they were often dependent on these same figures for their livelihoods. The Berlin model incorporated a training analysis into a broader educational process; a student had to take classes, practice analysis in the clinic, and be analyzed to be a psychoanalyst. While being analyzed by a training analyst, he was also exposed to analysts who shared varying perspectives, including supervisors and teachers. Besides, the training analyses were not long; some analysts emerged from their treatment with Sachs obviously not parroting his views, for they were openly critical of his analyst-as-mirror style. The Berlin school had a strong perspective, but because of its educational system, it did not mint ardent followers of Karl Abraham or Hanns Sachs so much as train psychoanalysts.

The individual who had the most intellectual sway over the Berlin community was Karl Abraham. Like prewar Berlin itself, Abraham gave little hint of originality in his early writings. His plodding theoretical work pushed Freud's views forward and attacked Bleuler and Jung for their altogether reasonable assertion that psychoses could not be entirely due to psychosexual conflicts. Abraham continued to extend libido theory to make sense of manic and depressive states as well as agoraphobia. His writing was workmanlike and a bit fanatical. Freud himself could not get excited about Abraham. He was drawn to men like Wilhelm Fliess, Carl Jung, and Georg Groddeck, swashbucklers with intellectual zest, courage, and daring. Abraham was a true believer who had smelled out the heresy in Jung and Rank. In both cases, Freud had first castigated Abraham before coming around to his side. Even so, he was not very fond of his orthodox supporter in Berlin.

Abraham worked hard to link psychoanalysis to medicine, psychiatry, and science. He looked for bridges to Berlin's psychiatric clinics and tried to get psychoanalysis into the university. Importantly, he established an ethos at the institute that favored close clinical description over the reckless and exuberant theorizing that found support in other cities. For Sándor Radó, the contrast with Budapest was striking. There he recalled that society meetings were often heavily weighted toward phylogenetic speculations like those of Ferenczi in *Thalassa*. In Berlin, Abraham, while personally always assuming the truth of libido theory, nevertheless created a culture that encouraged curiosity about experience-near clinical phenomena. Wild speculation, Abraham believed, could kill the young science by discrediting it.

In this regard, Abraham shared Ferenczi and Rank's belief that more attention be paid to the analytic situation. Men and women like Sándor Radó and Helene Deutsch came to Berlin to be analyzed by Karl Abraham precisely because his writings displayed a deep engagement with the details of a patient's life. But unlike Ferenczi and Rank, who had adopted the view that analysis occurred by unraveling the transference, Abraham maintained the older view of interpreting and reconstructing the past. He was committed to using the analytic situation to understand the there and then, more than the here and now. His clinical work was sometimes seen as an old-fashioned quest for confirmations of libido theory. When Ferenczi and Rank's book appeared, it was anathema to Abraham. Ferenczi admitted that the work was a discreet dig at "the therapeutic ineptness of the Berliners (especially Abraham)."

It's difficult to know if this was a fair assessment of Abraham's technique after the war, though before then his technique did conform to Ferenczi's critical view. For example, Karen Horney entered analysis with Abraham in 1910. In her analysis, Abraham lectured her about her unconscious and interpreted in a didactic manner, often leaving his patient far behind. Exasperated, Horney related once asking Abraham "what more he really wanted to find out in me?" as if she were a lab specimen for his examination. Abraham was also clearly interested in interpreting transference. In the first entry about her analysis in her diaries, Horney described her thirst for a transference object and the way her "wildest wishes twine around" her doctor. However, Horney got the impression that transference brought up difficulties that were insoluble in the analysis.

In 1910, many analysts—Sigmund Freud not excepted—were apt to make overly intellectual interpretations and be unsure of the role of transference. But Abraham's technique—like almost everyone's—evolved over the next decade. After the war, he mocked those who lectured their patients on the structure of their neurosis. And in the 1920s, Alix Strachey felt she got more out of her five months of analysis with Abraham than her fifteen months with Freud. She also suggested that other Berliners shared her opinion that Abraham was clinically superior to the Professor.

Abraham, however, did not stress an interest in technique. He wrote only one paper on the subject, and when the time came to mount the long-awaited competition on matters of technique and theory, Abraham left it to his junior colleagues. Meanwhile, he was busy searching out the hidden life of libido. During the war, Abraham became interested in the role of aggression, which Freud still attributed to an anal phase of development. Abraham began to argue that hate and aggression were essential elements of all stages of sexual development. In 1916, Abraham wrote the paper that would mark the beginning of his mature work. "The First Pregenital Stage of Libido" proposed that in the earliest oral stage, there was not just pleasurable sucking, but also a kind of cannibalistic aggression. This was aimed not just at taking in nourishment, but also the taking in of an Other, the incorporation of a psychic object.

Over time, Abraham became the analyst of psychic hunger. Hunger for food was tied to childhood thumb sucking, hunger for a mother, hunger for an Other, desire for (or horror of) oral sex, and hunger for love. Those fixated at the oral stage were constantly hungry and always frustrated and angered by the lack of psychic nourishment they received. In their rage, they sought to devour their love objects. Developing Freud's work, Abraham argued that melancholy was a regression to the oral stage, where intense hunger and cannibalistic fantasies created guilt and a terrible feeling of loss. Like Freud and Ferenczi, Abraham linked his theory to an imaginary early prehistoric period. For Abraham, humanity's primal crime was the eating of our beloveds.

Abraham's paper demonstrated a richness of clinical detail and an explanatory power that announced him as a major theorist; Freud recognized this by awarding him the prize for best paper in 1918. Next, the methodical Abraham tried to establish a urethral stage of libido theory that also involved aggression. Premature ejaculation was an attack on woman, ultimately, the mother. Male passivity and impotence were the result of intense repressed sadism. Abraham's reputation rose further after his examination of a female castration complex, in which he described self-hating women who had been narcissistically injured by the belief that they have been mutilated. These women unconsciously believed they were men and therefore denied their own femininity, or became vengeful women who unconsciously desired to attack men.

Abraham systematically explored the kind of human relationships that emerged throughout libidinal development. In 1924, he presented his synthetic integration: six psychosexual stages correlated with six kinds of love relationships. Abraham offered the proviso that this theory was tentative and surely incomplete. Still, his synthesis became de rigueur for a whole generation of students. It accounted for melancholia, mania, manic-depression, and an inability to love. In a subtle rebuff to the founder of the field, Abraham incorporated notions of aggression without bothering to include a death drive or a separate aggressive drive. In Berlin, Abraham became the master theorist of libido. Reviewing his work, the young Otto Fenichel raved: Abraham had beautifully demonstrated new facts and incorporated them into preexisting notions, and through clinically based, empirical studies he had demonstrated the validity of libido theory and enlarged that theory.

Abraham moved forward. Sages and pedagogues had long looked for hints about the nature of human character. For over a century, psychiatrists had seen diseased character emanating from constitutionally abnormal individuals who failed to abide by societal rules. Categories such as moral insanity, psychopathy, psychopathic constitution, and psychopathic personality were developed to identify a hodgepodge of liars, criminals, eccentrics, and rule breakers who were mostly just born bad.

However, as Abraham was maturing as a theorist, a new science of character was emerging. In 1921, a German professor named Ernst Kretschmer proposed a general typology of character based on differing body types. That same year, Carl Jung published his massive and erudite *Psychological Types*, in which he reviewed the history of character typology from antiquity through Schiller, Nietzsche, and William James. In the end, Jung borrowed Otto Gross's notion of the "introvert" and "extrovert" to develop his typology. Jung's ideas were in turn adopted by the Swiss psychiatrist Hermann Rorschach, who incorporated them into a new psychodiagnostic test that would bear his name.

Abraham turned to the study of character soon after his detested rival Carl Jung had entered that forum. At the time, psychoanalysts had had little to say on the subject. Freud had written a short paper on anal erotism and character formation in 1908, and another on character types encountered in analysis. He posited that early corporeal experiences were transformed into complex, unconscious modes for interaction with the world. In 1912, soon after his expulsion from the Freudian fold, Alfred Adler published a book on the "nervous character" that extended his theories of inferiority. After that, the idea of character had been mostly dropped by analysts, since character pathology was deemed fixed and unalterable. Psychoanalysts vowed to resolve neurosis, neurotic conflicts, and symptoms. Freud warned that their methods were helpless when it came to character flaws, and in this way, he echoed the moral views of his culture: bad character was beyond redemption.

Abraham codified a new psychoanalytic characterology based on the needs of the body. Childhood sensate struggles between pleasure and pain, satisfaction and frustration, love and hate, cast the die for adult character. Abraham began by developing Freud's notion of an anal character. The submissive, crushed child forced into total obedience by his parents secretly yearned for vengeance. Unlike Freud, Abraham believed that this kind of character developed not from sublimated anal erotism, but from the child's wounded narcissism and the power of his hatred. The child became envious, brooding, and superior, a greedy possessor of others and things, addicted to the pleasure of having, not giving. All of these characteristics were vividly described and easy to picture. Abraham's argument may not have been as rhetorically clever as some of Freud's, but his character types were convincingly human.

Abraham then argued for oral sources of character. Those who have been overly gratified in this stage expected to be fed and given to at all times. Optimists, they expected the world to satisfy their yearnings, but when exposed to frustration, they become demanding, aggressive, and devouring. Finally, Abraham wrote on the healthy genital level of character, and compiled his essays into *Psychological Studies on Character Formation*. Character was the sum of instinctual reactions to one's environment, he argued. It was normal for individuals to elastically adapt their character to their surroundings. Abnormal character was the result of a failure of psychosexual maturation that led to rigidity. But character was not permanent. Since it was the result of adaptation, it could be altered. Therefore psychoanalysis could be employed not just for the treatment of neurotic symptoms, but also for character analysis.

By bringing all this together, Karl Abraham offered a newly nuanced, unified psychosexual theory. Abraham connected libido, childhood experience, character structure, love relations, and social adaptation, all of which made possible a much richer description of a human life. Radó remembered: "Abraham invented a use of Freudian language, combined with the ordinary clinical terms of psychiatry, to be able to tell the essential story of a patient."

Karl Abraham was president of the I.P.A. and the Berlin Society. He had been central to repelling Rank and Ferenczi's attempt to reshape psychoanalysis, and he had consolidated Berlin's control over the future of psychoanalysis. At the same time, in a series of much admired, closely worked out clinical papers, he had created a unified theory of character and even explained some psychiatric illnesses. Abraham traveled to Bad Homburg for the 1925 psychoanalytic congress, knowing that his Berliners would establish new training requirements for the entire field, knowing that he was about to be reelected president to the I.P.A., and knowing that he had won his struggle with Ferenczi and Rank, and had earned the right to consider himself the heir to Sigmund Freud. Karl Abraham was the most important figure in a new psychoanalysis that could no longer simply be called Freudian. Abraham must have known that he had won. Unless, of course, he realized he was dying.

Abraham had returned from the war with pulmonary problems, and in the summer of 1925, he swallowed a fish bone that caused a puncture, which led to infection, pulmonary abscesses, and a long period of ill health. Abraham made it to the Bad Homburg Congress, but a few months later, on Christmas Day, 1925, he died. He was only forty-eight years old, and his friend Ernest Jones lamented the loss as the worst blow psychoanalysis had yet faced. One of Abraham's analysands, Edward Glover, spoke to Abraham's legacy when he said that psychoanalysis was the science that Abraham "had made his own." Abraham, an acolyte who never had a truly close relationship to Freud, a man who had fanatically defended his vision of psychoanalysis even against its founder, had surprisingly brought together a new psychoanalysis. His less than reverent students would follow in his absence.

ABRAHAM'S DEATH LEFT a vacuum in Berlin. There was much consternation, and plans were hastily drawn up for Sándor Ferenczi to move there, until the usually reticent Max Eitingon stepped forward and took over the presidency of the I.P.A., the Berlin Society, and the Poliklinik. A short time later, he handed the Berlin Society to Simmel. Order returned to the flagship training center. Ferenczi and Freud were relieved; it seemed that the Berliners were capable of maintaining the center themselves.

Abraham had left behind a gaping hole, but he had also left behind many intellectual sons and daughters—a burgeoning, freethinking psychoanalytic community. As the older members took administrative charge of the organizations, the future of Berlin as a scientific center was passed on to younger members, most prominently Franz Alexander, Sándor Radó, and Karen Horney. These three theorists shared Karl Abraham's interest in character, the role of mothers, oral hunger, and aggression, but they moved beyond Abraham to develop their own ideas to help make what would be called the "new" psychoanalysis. This new vision was one Abraham himself might not have approved of had he lived, for when Freud introduced his death drive and the new notions of *The "I" and the "It,"* Abraham, like many of his generation, had hesitated. But many younger Berliners embraced these new theories, synthesized them with Abraham's work on libido and character, and in the end gave birth to something novel.

Karen Horney was a rebel and arguably the first great female psychoanalytic theoretician. Freud had put forth an Oedipal complex that in theory had the flexibility to analyze both boys and girls and their reactions to their parents, but in fact the drama Freud interpreted relentlessly was between fathers and sons. Horney burst into this patriarchal play and developed a critique of masculine biases in psychoanalytic theory. The freethinking Horney opened her important paper "On the Genesis of the Castration Complex in Women" with a challenge to Freud as well as to Abraham. Both insisted that little girls had penis envy, a deep-seated desire to be boys. Horney responded:

> In this formulation we have it assumed as an axiomatic fact that females feel at a disadvantage in this respect of their genital organs, without being regarded as constituting a problem in itself—possibly because to masculine narcissism this has seemed too self-evident to need explanation. Nevertheless, the conclusion so far drawn from the investigations—amounting as it does to an assertion that one-half of the human race is discontented with the sex assigned to it and can overcome this discontent only in favorable circumstances—is decidedly unsatisfying, not only to feminine narcissism but also to biological science.

Horney argued that penis envy was not a given but was the result of restrictions placed on girls for instinctual gratification. She also employed the new "I" psychology to make a different argument, one with large ramifications for gender identity. She brilliantly unraveled the production of a girl's penis envy as a secondary result of the frustration of any child's early sexual life, its Oedipal defeat. It was the little girl's recognition that she was not her father's primary love that led some girls to a repudiation of their own sexuality and a defensive desire to become identified with the father and be a boy. Only in this fantastic retreat did girls encounter penis envy. Similarly, Horney argued that boys can identify psychologically with their mothers and develop castration terrors. Karen Horney went on to write about the vagaries of character development in light of the Oedipal complex, for example, in papers on masculine women.

Horney's fresh, forceful writing coincided with a surge of feminism during the Weimar years, and the prominence of the so-called New Woman. Her work was also a testament to the freedom of the Berlin Institute. She openly stated that the entire edifice of psychoanalytic theory had tended to neglect female psychology, since its theoreticians were male. Horney bluntly compared the fantasies of little boys about girls with psychoanalytic theories of feminine development and concluded there was little difference. The old conundrum of the analyst's subjectivity returned, in this case, male psychoanalysts who were blind to the inner lives of women. It is hard to imagine such a frank assessment coming from within Freud's Vienna Society.

Horney took her place as one of Berlin's major thinkers beside two Hungarians who had come to Berlin. Sándor Radó had established himself as one of Ferenczi's closest associates in Budapest, before he decided to travel to the new mecca of psychoanalytic training. While in Berlin, he was tapped by Freud to assume the editorship of the *Zeitschrift*, taking Rank's place. Overnight, the young Radó gained considerable influence, which he hoped to use to make psychoanalysis into a natural science of the mind, a goal he made clear in an early paper where he linked psychoanalysis and the new physics of relativity.

After the publication of *The "I" and the "It"* in 1924, Radó enthusiastically took up Freud's new structural theory. He also borrowed Abraham's focus on the oral stage and, combining both models, attempted to explain how psychic hungers might account for drug addiction. Radó also gained recognition for his use of "I" psychology to describe the injurious effect of an overly anxious mother, who defended against her aggression by constantly warning her child of the imagined dangers she was saving him from. In a paper on melancholia, Radó further developed the dynamic interactions between a mother and her hungry child, and in the end placed emphasis on the bad mother in causing psychopathology.

Neither Radó's nor Horney's contributions were simple extensions of Karl Abraham's work or Freudian "I" psychology. Both Berliners carried out creative explorations that mined the area between Abraham and Freud. However, the Berliner who tried to build a new theoretical consensus on this middle ground was Franz Alexander.

The son of an eminent Budapest philosopher, Alexander went to medical school, where he immersed himself in experimental lab work. While there, he began to wonder about his father's criticisms regarding reductive scientific approaches to complex phenomena. Traveling to Göttingen, Alexander studied with physiologists, consorted with physicists, and became a psychiatrist who studied brain physiology. When his father asked him to review Freud's dream book for a journal, Alexander's long struggle to comprehend Freud began. His father was horrified when his son gave up his promising academic career to go to Berlin and study psychoanalysis. After being briefly analyzed by Hanns Sachs, Alexander, at the age of thirty, became the first graduate from the Berlin Institute in 1921.

In 1923, Alexander's first published paper brought him immediate recognition when Freud awarded it the prize for best paper of the year. In "The Castration Complex in the Formation of Character," Alexander expanded psychoanalytic thinking into asymptomatic characters, men and women who generally seemed fine but unraveled under pressure. Karl Abraham had described such people but had no strategy to help them. Alexander proposed that at the core of these seemingly immutable character structures was a familiar and treatable neurotic conflict. These character neuroses were driven by that old fear: castration anxiety.

Alexander massively expanded the role of castration anxiety by placing it at the core of character development. Children were filled with antisocial desires, and like animals they were broken in order to be civilized. They became men and women of good character not because of inborn nobility, but because they feared their elders. Castration anxiety was not only the progenitor of neurosis: it also forged one's character. Parents trained children, and this involved imbuing the child with the dread of punishment for their demands and desires. For Alexander, character was formed from the common capitulations that turned a child into an adult.

Alexander also described how a turbulent neurotic conflict altered into a seemingly stable "I," how the driven pleasures of the body became socially responsible adult desires, and how childhood terrors were tamed. Following Freud, he relied heavily on Gustav Fechner's principle of constancy. The human organism was driven toward pleasure as well as toward safety and sameness; neurotic characters suffered from a compulsion to repeat, and they continually replayed unresolved conflicts in search of inner constancy. As an example, Alexander cited the case of an impulsive Russian aristocrat who was asymptomatic until he lost his money in the revolution. The adaptations he once used to manage his castration anxiety vanished with his wealth, and he lapsed into a fulminant neurosis.

In 1924, Alexander shared these ideas at the Salzburg Congress. Along with Sachs and Radó, he had been selected to deliver a paper on the relationship between technique and theory for the contest Freud had proposed two years earlier. While all three speakers advertised their adoption of "I" psychology, Alexander went well beyond that. Psychoanalysis, he declared, was based on two first principles: the Freud-Breuer principle and the Freud-Fechner principle. The first dictated that mental processes function to relieve tension, and the second that these discharges occurred via a repetition-compulsion. Alexander demonstrated that the two principles supported a new way to think about neurosis as not a simple conflict, but as a mental structure. Treatment should seek to repair that structure.

In this way, Alexander focused less on the contents of the repressed unconscious, than on the nature and structure of the repressing agent. For Alexander, the agent of repression had much to do with the "Over-I" and its brutal punishments for imaginary crimes. The monitor that watched over the "I" was an introjected parental "No." Timeless and out of touch with reality, it said no to many legitimate adult pleasures. That created guilt and an array of self-destructive behaviors that indirectly expressed a patient's feeling that they deserved punishment. The job in analysis was to address and modify the brutal "Over-I." Only when it had been transferred to the doctor could it be dissipated. Alexander called this "the task of all future psycho-analytic therapy."

Back in Berlin, Alexander wrote a paper criticizing Ferenczi and Rank's work on technique. What Alexander said was simple: Ferenczi and Rank were correct that systemization was needed as psychoanalysis moved into a "new phase." But their work did not take into account "I" psychology, and therefore their book was out-of-date before the ink dried. When Ferenczi complained to Freud about the unfairness of the critique, Freud said bluntly: "I liked Alexander's critique very much."

In Berlin, Alexander lectured on the psychoanalysis of the "I" and the "Over-I" as an avenue to understanding character neuroses. Compiling his 1924 and 1925 Berlin lectures, Alexander published *Psychoanalysis of the Total Personality*. The title had an air of importance, even swagger. Radó found it pretentious and blamed the publisher, whom he facetiously addressed as the " 'Total Director' of the 'Total Press.' " If overstated, Alexander's book was nevertheless the first extensive theoretical expansion of "I" psychology since Freud introduced these concepts. Alexander's work also made all prior work look, at best, less than complete. He was frank about the division he saw in his field: there were those whose clinical practice continued with the old method of analysis, and those who had moved forward with Freud's theories on the "I" and the death drive. The old guard believed the new theoreticians had succumbed to metapsychological speculation, which Alexander compared to the contempt experimental physicists had for their theoretical brethren. He hoped his book would bridge this divide by showing the clinical merits of the new "I" psychology and rooting the new ideas in empirical descriptions. Alexander wanted to usher in "a new era in psychoanalysis."

He further implied that those who rejected "I" psychology would become irrelevant. The new psychoanalysis offered greater integration with science and medicine, and even though Freud resisted total systematization, Alexander believed the time for such a synthesis had come. "Psychoanalytic science" had to stabilize its "internal structure" and provide "an integrated picture of the whole psychic apparatus."

Despite his references to psychoanalysis as science, Alexander's style of writing was both dramatic and allegorical. The personality, he contended, was composed of three warring autonomous factions, which he described with political analogies. The "It" was the inner terrorist, the "Over-I," the corrupt police, and the "I," the long-suffering citizen. To justify his mode of presentation, Alexander argued for the necessity of such tropes. Referring to the controversies between Ernst Mach and Max Planck, Alexander asked the reader to recall that only recently the reality of the atom had not been accepted. Personifications based on synthetic abstractions were necessary for psychic forces to be seen in the mind's eye.

Alexander also added a new first principle to his metapsychology. Every pleasure, he believed, was bound to the anticipation of its cessation, hence pain. Eros and Thanatos were two wings attached to the same body. The brutal inner policeman who demanded hyperpure morality was in silent concert with the repressed criminal. The wild license of Saturday night was intimately linked to the punishment of the Sunday morning sermon. The excessive severity of the "Over-I" led to a letter-of-the-law demand for punishment that "achieves absolution for sin" as well as "justification for commiting it." Alexander cunningly demonstrated how the overly pious were prone to the most egregious lawlessness. The harshness of the punitive neurotic "Over-I" gave covert justification for raw license. Given Alexander's political allegories, it is hard not to read his theory as a premonition of what was to come, as the law-and-order fanatics who took over Germany would loose criminal aggression that went beyond the bounds of imagination.

Alexander conceived of his book as more than new clinical contributions to a growing field. He saw it as a relocation of the discipline to firmer ground. His references to physics were not happenstance but revealed Alexander's determination to think his way forward by connecting himself—if only metaphorically—to science. His reference to the first principles of Freud-Fechner and Freud-Breuer also de-emphasized Freud's persona and focused on his central scientific claims. This was consistent with a growing pride among Berliners who saw themselves as more independent of Freud, unlike the subservient Viennese.

After he finished the book, Alexander took a family vacation and asked his father, who still believed psychology was rightly a branch of philosophy, to read his book. The old man made his son very happy when he said that the book clarified for him how the totality of human subjectivity and inner life could be won back for psychology after the destruction wrought upon it by the reductive experimentalists. Alexander's psychology of the total personality seemed to address questions empirically that Alexander *père* and his heroes—Shakespeare, Kant, and Voltaire—considered central to the study of the mind.

After Alexander's book was published in 1927, the rush to Berlin turned into a stampede. Berlin had a formal training institute, with courses and a clinic. The city, already associated with the avant-garde in science and culture, became a hub for the avant-garde in psychoanalysis.

Alexander drummed home the assertion that a new day had dawned. Like the new physics that leading Berliners liked to affiliate themselves with, it seemed this new psychology would leave those who refused to change in the dust.

Although it began with speculative theories, the new psychoanalysis was brought down to earth by Alexander. He showed how these new concepts increased and broadened clinical understanding and revealed the advantage of basing a psychology on concepts that had popular referents, concepts commonly thought to exist that were easy to observe. No patient would laugh at the notion that he had an "I" or a conscience that watched over him. No phalanx of academic psychiatrists could attack the premise that a psychology was founded on character. The new character analysis included theories about the deepest animal forces, but these were organized into higher-level, experience-near psychic structures that had the great advantage of being more easily observed.

The psychology of the "I" also flattered commonsense notions of what psychology should be, and took some weight off that uncomfortable, private domain of human sexuality. Psychoanalysts could now speak in strikingly descriptive terms of not just neurotic conflicts due to sexuality, but also the psychology of characters easily recognizable in everyday life. If organized around character and identity, psychoanalysis could enter the public sphere more easily. Sexuality was so fundamentally private that psychoanalysts continually broke an unspoken rule by mentioning it. For many it was difficult to guess the sexual secrets of a close friend, much less accept that sexuality dominated their own thoughts, but it was not hard to recognize the apt description of a cheap millionaire, a guilt-ridden criminal, or a fastidious butcher next door.

Freud's analysand Abram Kardiner suddenly noticed that New York was being invaded by a wave of Berlin-trained psychoanalysts who had adopted a different approach. By 1925, Freud's illness prevented him from regularly taking students into analysis, so Kardiner's younger colleagues like Gregory Zilboorg, George Daniels, and others had gone to Germany with the hope of training with Franz Alexander, and when they returned, Kardiner was mortified:

> They all went to Alexander. And they came back and I want to tell you, I had a very bitter time. Because they brought back a technique with which I was unfamiliar. They brought the libido theory back

and were using libido theory as a scaffolding for personality analysis. Well I knew what Freud told me, and I felt completely "déclassé." Here was a great investment gone down the drain. Not only did I think I didn't know anything, but here were these wise guys coming over, and they knew everything.

Another American, Ives Hendrick, concluded that Berlin had a "far better and more cohesive organization than the Viennese." In 1928, he also went into analysis with Alexander and happily reported that he found himself in an intellectually open environment, shorn of psychoanalytic zealotry and orthodoxy. He described Alexander as a "vigorous, energetic, virile, quiet young man with the physique of a footballer. He has a good deal of common sense, always maintains a sufficiently practical view of human fallibility. He has none of the fanatical fervor." Hendrick declared that he was thrilled to be at "the best Analytic Institute in the World."

Hendrick was not exaggerating. With Eitingon's funds, the training institute had an unsurpassed faculty. And its training capacity attracted the most motivated students, some of whom stayed. In 1922, Otto Fenichel came from Vienna to study at the Berlin Institute and two years later joined the teaching staff. In 1926, Fenichel offered seminars on Abraham's contributions to "I" psychology. Fenichel had already published extensively on familiar Berlin themes, including a paper on psychoanalysis as a natural science, and a series of papers on the development of "I" psychology, including explorations of introjection and object loss, masochism, and the shaping of the "I" by identifications. In 1926, the promise of formal training lured Siegfried Bernfeld away from Vienna, despite reassurances from Freud that he did not need a didactic analysis. Karl Landauer of Frankfurt migrated to Berlin in 1925. Many did the same.

Other candidates came and went, spreading the Berlin model back to New York, Boston, Vienna, Oslo, and perhaps most importantly, London. The alliance between Ernest Jones and Karl Abraham made for close ties between the London and Berlin groups, and since the students in London had no clinic or training institute in the early 1920s, many went to Berlin for training. When they returned to London, some, like the Glover brothers—James and Edward—became eloquent spokesmen for the perspectives taught in Berlin.

The Berliners decided to expand and open the first psychoanalytic hospital. The Sanitarium at Schloss Tegel was founded in 1927 in a beautiful castle on the outskirts of Berlin. Ernst Simmel was its founder and medical director. Patients were offered psychoanalytic treatment and psychoanalytically oriented occupational therapy. They were not locked up, but the understanding was that their positive transference to their doctors would constrain them from self-destructive acts. Employing the work of Radó, Alexander, and Groddeck, the hospital took in individuals with alcoholism and drug addictions, criminal and disrupted characters, and some medical illnesses.

In 1928, the Berlin Institute moved to larger quarters. At the time, the leaders of the institute reported that 25 students were in full training and 104 analyses were being conducted in the clinic. In 1929, a subsection of the Berlin Institute opened in Frankfurt under Karl Landauer; it included Erich Fromm and Frieda Fromm-Reichmann, and had close ties with the Frankfurt Institute for Social Research, thereby giving rise to an influential mix of Marxism and psychoanalysis. The Berliners also moved into forensic medicine when Franz Alexander along with the distinguished Berlin jurist Hugo Staub began to apply psychoanalytic theories of character to criminals and delinquents.

In 1930, the Berliners celebrated their tenth anniversary as a clinic and teaching institute. They had much to be proud of. Otto Fenichel reported on the Poliklinik's 1,955 consultations and 721 analyses. Carl Müller-Braunschweig outlined the first-ever obligatory two-year academic curriculum for the training of analysts, a curriculum that included courses on dreams, sexuality, psychopathology, metapsychology, "I"-analysis, and technique. Hanns Sachs described the goals of the training analysis; Alexander the Berliners' theoretical courses; Radó the practical training; and Bernfeld the courses offered to teachers and educators.

The Berlin Institute not only weathered the death of Karl Abraham, but it had continued to blossom and establish itself as the place for professional training. A dynasty had been built in Berlin that gave it the power to shape psychoanalysis for the coming decades.

II.

SOME, HOWEVER, WERE reluctant to join the Berlin parade. If the Berliners thought they had created a new psychoanalysis, others believed they had merely grafted a new theory onto an old way of doing things. Max Eitingon was known to say there was only one technique—the correct one. But

could one really use the same technique for neurotic symptoms and solidified character structures? If Franz Alexander had shown "I" psychology
could describe many varieties of human behavior, he had not answered
that question. It would be left to a group of disgruntled Viennese.

The Viennese had opened their outpatient Ambulatorium in the
spring of 1922, and the offer of free or low-cost psychoanalysis had
brought them some two hundred patients a year. Simultaneously, the
Vienna Society began a long process of rebuilding after the schisms that
depleted its ranks, a process that returned the Society to its 1910 membership levels only in 1926. As new recruits came, the clinic offered the novices hands-on clinical instruction but at first no other training. A formal
curriculum was not offered in Vienna until 1925, when the entire I.P.A.
adopted the Berlin model of education. When the Vienna Institute was
inaugurated, it was placed under the leadership of a Berlin trainee,
Helene Deutsch. She was assisted by Freud's daughter Anna (who had
joined the Society in 1922) and the magnetic Youth Movement leader,
Siegfried Bernfeld. Rather embarrassingly, Bernfeld then quit his leadership position in the Vienna Institute to *begin* formal training in Berlin. It
was a stark reminder of how much the Berliners had eclipsed the Viennese and become the central destination for ambitious trainees.

The staff of the Vienna Ambulatorium, including in the first row the director, Eduard
Hitschmann (*fifth from right*), assistant director Wilhelm Reich (*fourth from right*), Grete
Bibring (*third from right*), Richard Sterba (*second from right*), and Annie Reich (*far right*).

While the Berliners' educational opportunities could not be rivaled, the Viennese had developed a unique teaching forum themselves. Wilhelm Reich had become fed up with the learn-as-you-go attitude of his teachers. Like Rank and Ferenczi, he longed for a more systematic method and was excited by Freud's announcement, at the Berlin Congress, of a competition on the relationship between theory and technique. After returning from the congress, Reich suggested the Viennese set up a seminar to study that subject. In 1922, the Vienna Technical Seminars were born. In Salzburg two years later, Reich listened as Alexander, Radó, and Sachs competed for the essay prize on technique. He left sorely disappointed. Later he recalled that the finalists "did not take a single practical everyday question into account and got lost in a maze of metapsychological speculations."

Reich returned to Vienna where the Technical Seminars were floundering. Under the chairmanship of Eduard Hitschmann, the seminars lacked focus. Surviving minutes show discussions that veered from sexual dysfunction to Rank's birth anxiety to ways of structuring the Ambulatorium. After two years of this, Hitschmann handed the seminars to Hermann Nunberg, who gave up after only one semester. The seminar was going nowhere when the twenty-seven-year-old Reich was handed the reins.

In some ways, Wilhelm Reich was a throwback to the older Freudians who still had a foot in sexology and emphasized the sexual in psychosexual. In 1924, Reich had proposed that patients who could not have orgasms were doomed to remain neurotic despite all the analysis in the world. Reich's emphasis on sexual dysfunction over psychological causes of illness was reflected in the fact that while he played a central role in the Ambulatorium in its first years the leading diagnosis was sexual impotence, a category not even among the list of ailments diagnosed at the Berlin Poliklinik. The Viennese also diagnosed a great deal of "frigidity," another diagnosis not in the Berlin lexicon.

Rejecting Freud's newer notion of signal anxiety, Rank's birth anxiety, and Adler's view that anxiety was due to repressed aggression, Reich stuck with the old belief that all anxiety was the result of repressed sexuality. Like some utopians in the Vienna Society before the Great War, Wilhelm Reich maintained that much human unhappiness, inhibition, fear, and cruelty could be remedied by liberating libido. If all anxiety was caused by dammed-up libido, then by all means, doctor, undam it!

Yet Reich was not among the old guard who rejected all of Freud's newer views, for he was captivated by the study of pathological character. On November 7, 1923, he gave a paper to the Technical Seminar called: "Technique of Interpreting in Character Neurosis." Many analyses of character neurosis fizzled because remembering did not cure these people, Reich argued. These patients did not suffer from reminiscences, like Freud and Breuer's hysterics. In fact, they were indifferent to their own seemingly disturbing recollections. The only hope for a cure in these people was a systematic evaluation of the way they warded off the world—a full-scale analysis of their resistances. Reich's proposal would become psychoanalytic orthodoxy, but when he first presented it, he was met with skepticism. Hermann Nunberg blandly suggested that more discussion would be profitable.

After taking charge of the technical seminars, Reich turned the group toward the study of the analytic situation. He believed that analyses often became sterile when analysts got drunk on theory, and with this in mind, he turned the Technical Seminar into a testing ground for ways to treat character neurosis. Reich quickly noticed, however, that the characters of the seminar members made this task difficult. The seminar was hampered by a reluctance to discuss clinical failures. Analysts avoided embarrassment by only recounting their grand successes.

In 1926, Reich established a bold new directive: the seminar leaders would be restricted to *only* discussing their failed cases and errors, thus encouraging students to be open about their own clinical debacles. In addition, Reich did something no one had done since Freud's Dora case: he published a botched case of his own. Most analysts lived in fear that their referrals would dry up if they publicly admitted failure. Reich bravely published an account of how his use of a passive technique—in which he focused on interpreting the unconscious content and did not interpret the transference unless it became a resistance—failed. Passive technique might be fine for less severe neuroses, he concluded, but character neuroses required an insistent focus on transference and resistance.

At the same time, Reich had come to suspect that all neurotics had an underlying character neurosis. If Alexander thought character neurosis was one kind of neurosis, Reich wondered if neurosis wasn't one kind of character structure. A man did not just become phobic without an underlying problem in his character, did he? If Reich was right, that meant analysts had been fighting the wrong battle. With the Professor's approval, Reich embarked on a book that would do what Franz Alexander had not done—adapt psychoanalytic technique for the treatment of character disorders. To prepare himself, Reich closely studied Ferenczi's work and even wrote the Hungarian a long letter (which he never sent) detailing his conviction that "true and lasting cures can be achieved only if we succeed in modifying neurotic character."

Reich reasoned that human individuality was a result of character. Animal drives were universal and invariant; people only differed in the way they managed those drives. This fact had been obscured because character could seem invisible: it had no dramatic symptoms or flagrant disruptions. Freud's dream and transference interpretations had brought attention to the hidden force of human sexuality. Rank's active termination had illuminated anxieties about human separateness. Wilhelm Reich needed a method that would bring character and its troubles into the light.

The Vienna Technical Seminar was the laboratory for this project. Following the ethos of the Youth Movement, the seminars were not run like the Society, under the imperious rule of Paul Federn. Reich encouraged the open sharing of problems and ideas, and group collaboration that offered analysts a large number of cases from which to devise their theories. Reich and his colleagues tested their ideas against the collective experience of the group. After a time, the seminar reported a striking conclusion. Reich had already demonstrated that passive technique failed in treating character neurotics, but the seminar members also concluded that Ferenczi's active technique—with its prohibition of certain behaviors— was unnecessary if the analyst focused intently on resistances.

The seminar members adopted a new mantra: no interpretation of unconscious meaning if a resistance interpretation is first needed. Clinicians were to interpret the *ways* patients held back before interpreting *what* was being held back. Reich gave an example: a pleasant, highly compliant man came to his analyst with a dream that included incestuous wishes. Reich ignored the flagrant Oedipal wishes and focused on the man's anxious compliance. Reich's interpretation of the man's defenses and his fear-ridden transference took precedence over any interpretation of unconscious psychosexual fantasies. Resistance, first!

While Reich rejected active treatment, his work resonated with Ferenczi's concern that analysts were happily offering up deep interpretations of unconscious contents that changed nothing. Treatments went dead because they had become little more than scientific validation of the Freudian unconscious. Reich proposed this corrective: "I" analysis before "It" analysis. Others had hinted at a similar approach. In 1924, the Frankfurt analyst Karl Landauer published "Passive Technique: On the Analysis of Narcissistic Illness." Arguing against Ferenczi's active privations, Landauer warned that the analyst must not "impose actively upon him [the patient] *one's own wishes, one's own associations, one's own self.*" Instead, especially in difficult cases, an insistent focus on the transference would reveal the earliest, often silent resistances that could destroy the treatment.

Reich and colleagues agreed. A majority of the failed cases were due to latent hatred for the analyst. "In speaking of 'transference,' " Reich later wrote, "the analysts meant only positive transference." He accused analysts of being too timid to open up a full exploration of the hostile, degrading, and damning thoughts patients might harbor. Classical technique encouraged one to interpret transferences only when they became resistances to free association, but in the case of negative transference, these forces were not so easy to discern. Reich and his young colleagues believed that ignoring this latent hatred would destroy treatments.

One of Reich's protégés in the Technical Seminar, Richard Sterba, mined this thought. While Reich focused on the analyst's narcissistic fragility, Sterba made it clear that analysts could simply be fooled. Negative transference was missed not just because analysts were afraid to see it, but also because hatred could be hidden behind good cheer. A twenty-five-year-old man suffering from impotence came to see Sterba, and he merrily agreed with all of his analyst's interpretations. He was warm and accommodating, but despite his analyst's efforts, he did not change. Sterba brought the case to the Technical Seminar, where discussion pointed to a latent negative transference. Before long the patient was cursing, defiantly showing up late, and expressing strong doubts about his analyst's intelligence. All this hatred had been repressed out of fear of the analyst, but now a transference of intense castration anxiety burst into the open. Facing that hatred and fear, Sterba reasoned, was the road to cure. He also implied that over time such hatred could freeze into a rock-hard character structure. For the close reader, Sterba had let the Technical

Seminar's rabbit out of the bag: character neurosis was nothing more than the consolidation of latent negative transferences. After 1927, Reich presented a rich series of expositions linking resistance analysis to these hidden hatreds.

Before the war, the answers were clear if you accepted them. Sexual libido was in the unconscious and should be made conscious. Find it, dig it out, interpret it. If you were a full-fledged Freudian, you knew the destination of the analysis, even if you did not have a map that guided you en route. But the destination in treatment was no longer clearly visible. Debates sparked over what kind of instincts existed in the unconscious, as well as the role of "Over-I" analysis alongside libido analysis and the newer "I" psychology. For Reich, the guiding compass through all this was a focus on the patient's resistance. To avoid what he called "chaotic" failures, Reich recommended that analysts avoid swift interpretations of the deepest layers of the unconscious, avoid haphazard interpretations, and instead systematically analyze the resistances, especially the latent negative transferences that appeared in seemingly obedient, trusting, conventional, all-too-correct patients.

Annie and Wilhelm Reich (*standing*), at a Danube beach in 1928, with (*left to right, sitting*) Dr. Tuchfeld (a trainee), Mrs. Tuchfeld, Anny Angel, Edith Buxbaum, Editha Sterba, Eva Reich, and Richard Sterba.

At the Tenth Congress in Innsbruck, Reich linked his method of resistance analysis to Karl Abraham's theory of character. Character was the residue of resistances: it was "character armor." Reich's phrase provided a military image for the human need to defend against dangers coming from within and without. Character armor manifested itself in a person's way of talking, walking, their affectations, their giggles, smiles, and sneers, their politesse and their rude guffaws. In shifting the analytic gaze to matters of personal style, Reich contended that it was not so much what the patient did or said, but how they did or said it. Character armor was found in the *form* of human behavior. With this insight, Reich added a new set of behaviors to the growing list of psychoanalytically meaningful phenomena and reached beyond psychopathology toward general psychology.

Reich unveiled his theory of character in a paper that included an attack on Franz Alexander. This was both an issue of turf and principle. Reich had become concerned that the new emphasis on character and "I" psychology had turned some analysts away from psychosexuality. Just as distressing for Reich was the way excitement regarding Freud's new work had led to some acceptance of the death drive. Reich lambasted Alexander and the Viennese analyst Theodor Reik for pushing the notion that neurosis came from an inner drive for self-punishment derived from the death drive. By attributing neurosis to an internal battle between Eros and Thanatos, these theoreticians had disrupted the delicate balance between outer and inner, and made any influence from the external world irrelevant. Reich assailed such reductive theorizing but neglected to name the person who had set this argument in motion. In his own defense, Franz Alexander took the liberty to remind readers that Reich's polemic was really aimed at the originator of this theory—Sigmund Freud. He left it at that.

Alexander was right. Reich had become convinced that Freud had made a disastrous error. The death instinct and the notion of a biologically driven inner need for punishment undercut thinking about the fear of punishment that came with childhood sexuality and its repression. Some analysts began to use this supposed need to suffer as a justification for their own clinical failures. A patient who did not improve in treatment was not in an inept treatment but needed to suffer. Reich paid a visit to Freud, who apparently reassured him that the death drive was just a hypothesis. Emboldened, Reich sat down and wrote his polemic against

Theodor Reik and Franz Alexander. For the time being, his battle was not with Freud.

Reich offered a version of "I" psychology divorced from speculations on the death drive and developed a way to account for failures and successes with neurotic character. By sticking to resistance-analysis, Reich avoided making great claims about the contents of the unconscious and instead drew up guidelines for clinicians to move from psychic surface to depth. He outlined the wrong turns, laid out possible confusions, took on the problem of novice analysts, and after a decade of study collated his work into a landmark study, *Character Analysis*.

Character had long been the province of novelists, dramatists, clerics, and jurists. Alfred Adler, Ernst Kretschmer, and Carl Jung had brought character into their scientific psychologies, and now Reich had brought it into both psychoanalytic theory and practice. Expanding on the work of Abraham and Alexander, he made character a primary focus of psychoanalytic treatment and advocated a turn from the shifting energies of psychoneurosis to these rigid structures. With the Vienna Technical Seminar as his proving ground and the Ambulatorium as his base, Reich's views were handed down to students. Candidates in the Vienna Institute passed through his seminar, including some whose later theories were indebted to Reich's thinking: Anna Freud, Heinz Hartmann, Robert Wälder, and Ernst Kris. Reich became an undisputed master of technique, the guru of the technical seminars, and the man Freud apparently called "the founder of modern technique." In 1934 when the British psychiatrist Aubrey Lewis visited Vienna, he was surprised to find that psychoanalysts were no longer treating psychoneurosis so much as disorders of character.

REICH'S MILITANTLY THERAPEUTIC approach was destined to move beyond the walls of the clinic into a culture that cried out for reform. Reich's diagnosis of character neurosis reverberated with the spirit of Red Vienna; he described men and women sleepwalking through their lives, aided by thick, deadening psychic armor, which was the result of their childhood capitulations to their parents and their fear of authority. The former Youth Movement activist had not only made a psychological diagnosis; he had also written a lament for youth. Reich's reformist message was clear: a change in the way young people experienced their own desires could turn serfs into citizens. Armed with that message, Wilhelm Reich took psychoanalysis to the streets, and in the process he would find himself banished from the psychoanalytic movement.

Wilhelm Reich's exile cannot be divorced from his own character. A pushy, moody, difficult man, he impressed others by the force of his intellect as well as the bruising manner with which he applied it. He made enemies easily. Reich immediately ran afoul of Freud's deputy, the acting chairman of the Vienna Society, Paul Federn. Federn had joined Freud's circle way back in 1903. With his long white beard and scathing manner, he embodied the old guard. Federn had been Reich's analyst and had developed a poor opinion of his former patient. Reich's advancement in the Vienna Society was opposed by Federn behind the scenes.

Others in Vienna treated Reich like a pariah. Hermann Nunberg had become one of Freud's trusted men in Vienna. Another of Federn's analysands, Nunberg became a training analyst in 1925. At that time, his writing displayed a tendency to interpret unconscious material far beyond a patient's capacity to understand, exactly the kind of "deep" interpretations Reich assailed. Like Paul Federn, Nunberg eventually accepted the death drive; Reich did not. In 1928, as Reich's innovations were becoming apparent, Nunberg lumped those untouchables—Jung, Adler, and Rank—with the three prominent voices for "I" psychology: Alexander, Reik, and Reich. He dismissed their findings as not new but rather as failed, one-dimensional models of psychoanalytic theory.

Despite the opposition of Federn and Nunberg, Reich's brilliance made him invulnerable, at least for a while. When Federn moved to oust Reich as director of the Technical Seminar, Freud intervened, saying Reich had "great merits for the intellectual life of the association." The Professor's goodwill, however, would not survive the political radicalization of Wilhelm Reich.

Reich was far from the only politically engaged psychoanalyst. Quite the contrary. In Vienna, many members of the psychoanalytic society were allied with the Social Democratic Party. Paul Federn was an active Social Democrat, who had held office in the district council. Hermann Nunberg, the pediatrician-analyst Karl Freidjung, August Aichhorn, Helene and Felix Deutsch, and many others were strongly affiliated with the socialist reform agenda. A few analysts like Ludwig Jekels, one of Freud's card-playing partners, were Communist.

At the time, the city of Vienna was governed by a Social Democrat mayor, Karl Seitz, but the right-wing Christian Socialist Party had gained control of much of Austria by playing on popular hopes of a return to monarchy. Ominously, both sides—left and right—had their own paid militias. Soon this volatile concoction began to burn. On January 30, 1927, in a small town called Schattendorf, several old veterans shot into a crowd of Social Democrats and killed a man and a child. Seven months later, a Viennese court acquitted the murderers. The Workers' Union in Vienna called a strike to protest the verdict; the police were called out. In the resultant melee, the police opened fire on the strikers, who set the Palace of Justice on fire. Like many Viennese, Wilhelm Reich went outside as the events began to unfold. With horror, he watched the machinelike demeanor of the police—character armor indeed—as they shot into the crowd. After it was over, eighty-nine Viennese lay dead and more than a thousand were wounded.

Reich's theory of character armor always had within it implicit political resonance. Character—that venerable term used by moralists—was in Reich's eyes based on a blind submission to authority. After four hundred years of absolute monarchy, Reich wanted to release men and women from their psychic shackles. If that was once implicit, after the July 15 riots, Wilhelm Reich's politics became explicit. He enrolled in International Worker's Relief, a group associated with the Communist Party, studied Karl Marx and Friedrich Engels, and found in them compelling convergences with his own thinking. They too viewed society and the family as oppressive structures that led to pacification and social control. A newly radicalized Reich determined to bring psychoanalysis to the downtrodden.

Hence, he stepped into a significant tradition in psychoanalysis. Freud, Gross, Adler, Rank, Simmel, Wittels, and numerous others believed that social reform or revolution would come from psychological emancipation. They believed that curing the self could cure a society, and that conversely a sick society resulted in sick men and women. In the reform-minded Weimar years, one of the most prominent attempts to use psychoanalysis as a political means was inaugurated by Federn. While personally difficult, Federn was a political pacifist (which prompted Theodor Reik to crack that Federn was "a pacifist of the most belligerent temperament"). Federn recruited one of his former analysands, Heinrich Meng, to start what became known as the Federn-Meng movement, which encouraged the use of psychoanalysis to address social and public health problems. Toward that goal they published *The Popular Book of Psychoanalysis*, which compiled essays to enlighten citizens on matters of sexual, mental, and nervous hygiene.

While Federn and Meng hoped to teach parents and educators, Reich wanted to marry Marxist principles with psychoanalytic ones, an effort that was also taking place in the Soviet Union. Reich asked for Professor Freud's approval to open sex counseling centers around Vienna. Not long before, the seventy-one-year-old Freud had demonstrated that his own capacity to scandalize had not dissipated, for he had just written a withering critique of religion called *The Future of an Illusion*. Reich proposed open clinics where psychoanalytic thinking could be used on a mass scale. Freud agreed, though he warned Reich that he was poking his nose into a "hornet's nest." Undaunted, Reich and his new lover, the kindergarten teacher Lia Laszky, traveled around Vienna, providing workers with sex counseling, sexual hygiene, child counseling, gynecological exams, and contraceptives. In January 1929, Reich helped found the Socialist Association for Sex Hygiene and Sexological Research. With four analysts and three obstetricians, Reich used the association to create clinics in Vienna for workers and help with abortions, contraceptives, and sex counseling. Despite the fact that much of this work was illegal, the small band persevered, and by 1930, Reich claimed the clinics had seen some seven hundred cases.

The now ardent Marxist had established that genitality in youth was necessary for the revolution. After centuries of political helplessness, he believed the people of Austria needed to stop being so obedient. Character armor had turned them into compliant chattel. Reich wanted to cure this at its source—the inner pursuit of pleasure, at its core sexual pleasure, must be liberated. If inner desire was freed from fear, neurosis could be thwarted and the people could advocate for their interests as citizens and comrades.

While dreaming of a happier society, Wilhelm Reich feuded with members of the Vienna Society. After lecturing on his new technique in 1928, Reich felt humiliated when Paul Federn rejected his ideas. Reich wrote to Freud and complained that the younger members of the Society, like Sterba and Bibring, refused to present their work at the Society because Federn dismissed "everything a younger analyst had to say." But it was not just the old lions that were closed-minded. Reich himself had become increasingly dogmatic, demanding that others adopt his resistance analysis without variation. His friend Richard Sterba recalled that when contradicted, Reich became belligerent and began to exhibit "increasing fanaticism" in the curative power of orgasms.

By 1929, Reich was feuding with nearly everyone. His Marxism and his eccentric focus on orgasms undermined his support from his psychoanalytic students. His social program began to lose favor with Freud, who again, spurred to think antithetically against one of his followers, took up his pen to compose a critique of Reich's utopian vision, which became the masterly tragedy *Civilization and Its Discontents*. At a meeting in December 1929, Freud informed Reich that orgasms were not the answer. When Reich refused to accept this rebuttal, Freud lost his patience, snapping: "He who wants to have the floor again and again shows that he wants to be right at any price."

In January 1930, Reich's denunciations of the Social Democratic leadership led to his expulsion from the party. His marriage to Annie Reich was falling apart in some measure due to his affairs. Reich's old friends were alarmed at the pathological character traits exhibited by the theoretician of pathologic character. They urged Reich to go into analysis with Ferenczi, generally acknowledged to be the finest psychoanalytic clinician. Reich consented to see an analyst, but not in Budapest. In 1930, he packed his bags and headed for Berlin.

By going to Berlin, Reich was boarding a sinking ship. After five years of economic stability, the Wall Street crash of 1929 had forced Berliners to recognize how much their prosperity depended on American capital. With American banks calling in loans, Berlin fell into spiraling debt and unemployment. This coincided with a deterioration of liberal democratic politics and the rising power of both the Communists and the violently anti-Semitic National Socialists. In 1930 when Heinrich Brüning's government called for elections, the residents of Germany discovered that the Nazi vote had mushroomed from 800,000 to 6.4 million in just two years. Berlin's streets were frequently the sites of beatings and violent protests. Many Jewish residents began to consider emigration.

One of them was the intellectual leader of the Berlin Psychoanalytic Institute, Franz Alexander. In 1930, Alexander left Berlin after receiving an offer to start a new psychoanalytic institute at the University of Chicago. The Berlin Psychoanalytic Institute had been weakened also by the misfortunes of Max Eitingon, their financial angel, who had lost much of his wealth in the economic implosion of the last year. But in Berlin, Reich nevertheless found a community of analysts who were more politically engaged than in Vienna. Franz Alexander and Hugo Staub had set a precedent in this area with their work on criminality and the penal system.

Some analysts in Berlin were interested in more than just the reform of the German penal system. Wilhelm Reich joined Barbara Lantos, Erich Fromm, and two old friends from the Vienna Youth Movement, Otto Fenichel and Siegfried Bernfeld, all of whom had embarked on an effort to work out an integration of psychoanalysis and Marxism. Much of these discussions occurred at a so-called *Kinderseminare*, a seminar where the younger analysts were able to talk freely about psycho-political matters. When Reich arrived, the community of young leftist analysts fell under his charismatic sway.

Reich was also welcomed into the Berlin Communist Party and began to lecture at the Marxist Workers' University. He tried to convince his often skeptical comrades that a sexual transformation was essential to political revolution. Reich organized sexual-political organizations, "Sex-Pol," to bring sexual revolution to the people. He also began his analysis with Sándor Radó, a short affair that lasted only three or four months. At the end of that time, Radó summoned Wilhelm's wife, Annie, to Berlin and grimly informed her that her husband was schizophrenic. Radó then packed his bags and left for America, where he had been recruited to bring the Berlin model of training to the New York Psychoanalytic Institute. A year later, Karen Horney and Hanns Sachs also set sail for America, furthering the Berlin brain drain.

Reich tried to step into the vacuum. He set up a technical seminar at the Berlin Society and, from this distance, risked a showdown with Freud. The confrontation was over the contents of the unconscious, the divisive issue that had separated analysts time and again. Reich wrote a paper called "The Masochistic Character," in which he argued that the death drive did not exist. As Reich saw it, psychoanalysis had once treated the conflict between inner drives and the outer world. But with the death drive, Freud, Alexander, and Reik had discounted the outer world as a source of mental conflict. For a man focused on political and social repression, this was impossible. Reich argued that the anxieties generated by the so-called death instinct were actually caused by sexual repression. Masochism had no biological source; there was no inherent drive to suffer. In line with Karl Abraham, Reich saw frustration and desire mingling to make sadism, which—when turned against the self—could beget different forms of masochism.

Reich sent this paper to the *Zeitschrift*, which was now edited by Sándor Radó. Others had argued against Freud's death drive before; some of the Professor's closest associates, including his daughter Anna, never accepted it. But when the Professor saw Reich's paper, he became enraged. He wrote Ferenczi, describing Reich's article as spewing the "nonsense that the death instinct was the activity of the capitalist system." Freud demanded the paper be preceded by a disclaimer that stated the article had been written by a loyal member of the Bolshevik Party and that party dictums demanded that the author tailor his psychoanalytic thinking to their political beliefs.

Freud's blunt interference in the journal's editorial process was met with broad disapproval. Eitingon, Bernfeld, and Jekels all said Freud had no right to interfere. Freud considered trying to suppress the article entirely and finally asked Ernest Jones and the I.P.A. leadership to make this problem go away. In the end, Reich's article was published along with an extensive rebuttal by Reich's supposed fellow traveler, Siegfried Bernfeld.

Freud was wrong to suggest that Reich rejected the death drive because of his communism, for his rejection of that theory predated his political radicalization. Yet Reich's longstanding rejection of the death drive was now articulated in an explosive political context. If there was a death drive, if suffering was predetermined and innate, then there was no reason to consider social, political, or economic systems, families, sexual mores, traumas, or bad environments. None of that would matter. If true, this theory ruined Marxist, socialist, and reformist rationales.

If Freud suspected that Reich was simply toeing the Communist Party line with this argument, he also misunderstood Wilhelm Reich's character. Reich seemed unable to toe anybody's line. After he was kicked out of the Vienna Social Democratic Party, he found himself in hot water with his Marxist comrades in Berlin. He had incensed them by arguing Marx had incorrectly predicted that economic depression would bring on revolution. In fact, Reich noted, it seemed to bring on fascism. Marx's error,

Reich pointed out, was the result of his crude psychology. He did not understand that a long repressed people were not willing to defend their own liberty. On the contrary, they were all too eager to give it back to a powerful leader. It was a sound argument but not pleasing to Reich's Marxist brethren. In 1932, as Eitingon clipped Reich's wings by excluding all students from his psychoanalytic seminars, Reich was denounced by members of the Berlin Communist Party. It was only a matter of time before he was expelled.

A year later, Reich—Austrian Jew, psychoanalyst, Marxist—would be on the run, for the golden age of Weimar Germany had turned brown. In the 1932 elections, the virulently fascist Nazi Party won 37.4 percent of the vote, making it by far the largest party in the country. Its leader, Adolf Hitler, demanded the chancellorship of Germany, and on January 30, 1933, he got his wish. With the burning of the Reichstag on February 27, 1933, Hitler blamed the deed on Communists and rounded up over four thousand of his enemies in one night. Whipped into a panic about imminent Communist revolution, Reich President Hindenburg signed a fateful emergency decree that under the Weimar constitution suspended the basic liberties and rights of the citizenry. The Communist Party was banned, and on March 23, Adolf Hitler became the dictator of Germany.

As Hitler's forces swept through Berlin after the Reichstag fire, Wilhelm Reich escaped to Vienna. With the political situation volatile there as well, Federn warned Reich not to give political speeches at the institute. Predictably, Reich refused to comply and was forbidden to participate in the Society's activities. To make matters worse, Reich discovered that his long-awaited triumph, *Character Analysis*, which was scheduled to be published by the psychoanalytic press, had now been canceled. This book was the culmination of all of Reich's work in the technical seminar, and he might have rightly expected it to become a new standard for the field. But in 1933, it was too hot to handle. Somehow Reich managed to self-publish the book along with another work, a synthesis of his psycho-political critique entitled *The Mass Psychology of Fascism*. This courageous account was quickly denounced by the Communist press for its deviations. Homeless and with few friends, Wilhelm Reich fled to Copenhagen.

The catastrophic changes that occurred in Germany cast a dangerous shadow over Europe and the psychoanalytic movement. From its inception, psychoanalysis—like psychological science in general—had been positioned as an Enlightenment enterprise pitted against the traditional authorities of church and king. In the first years of the movement, there had been fervent debate about how overtly political and radical this science of the mind should be. After the Great War, psychoanalysis in Germany and Austria rode the tide of social reformist movements and made its way into schools, clinics, and courts. It became highly identified with the Social Democratic movement. The rise of fascism in Germany threatened psychoanalysis in Berlin, Munich, and Frankfurt, as well as in nearby Austria.

The liberal dream of an objective science of the psyche was also becoming impossible in Germany. There was no way to step outside the political maelstrom and insist that psychoanalysis was only a science or a medical specialty. Given the threat of fascism, there was little capacity for disciplines to control their own borders and not be overtly politicized. After the rise of Hitler, the leaders of psychoanalysis understood that the fascists would attack them on political grounds. In an attempt to minimize this risk, Anna Freud wrote to the then president of the I.P.A., Ernest Jones, asking that he expel Wilhelm Reich. As a Marxist radical in their ranks, Reich posed too much of a risk. Besides, Anna wrote, her father found it "offensive" that Reich "has forced psychoanalysis to become political." Since he was no longer a member of the Vienna Society or Berlin Society, if Reich was forbidden from joining a new institute, he would effectively be excluded from organized psychoanalysis. Jones balked, then after a face-to-face meeting with Reich in London, replied that it was a serious matter to curtail another analyst's political activities. The matter was deferred until the next congress, which would take place in Lucerne in 1934.

ELEVEN

The Psycho-Politics of Freedom

The Lucerne Congress, 1934. *Left to right*, Erwin Stengel, Grete Bibring, and Rudolph Loewenstein talking with Wilhelm Reich. At that meeting, Reich was expelled from the I.P.A.

I.

WHEN THE INTERNATIONAL Psychoanalytical Congress met in Lucerne, its members could look back at sixteen years of extraordinary growth and change. The prewar Freudians were no more; psychoanalysts were no longer made by a simple commitment to a Freudian theory of unconscious psychosexuality. By the mid-1920s, it became possible to pit one Freudian position against another, and leverage these differences into a more open community. What was in the unconscious? Was it sexual, aggressive, both, or perhaps something else? Was psychoanalysis depth psychology or "I" psychology? These became generative questions for a profession, not oaths of loyalty to a movement.

This change from a Freudian to a psychoanalytic culture was abetted by the rise of clinics and training centers, where divisive theoretical questions were outweighed by the need to instruct students in clinical practice and build a coherent and consensually accepted method of inquiry and treatment. This emphasis on technique rose with a more experience-near "I" psychology. While it was hard to get laymen to imagine they harbored unconscious wishes to sleep with their mothers, it was easy to get them to acknowledge the voice of their own conscience or their own character quirks. And while it remained hard to claim objectivity when it came to inferring the contents of the unconscious, it was easier to defend the scientific nature of psychoanalysis with a focus on character.

By 1934, psychoanalysis had transformed into an intellectual community organized around a common technique, transmitted through universal methods of training and education. Nevertheless, much remained open to debate. Might the death drive gain greater credence and encourage analysts to search out different forms of masochism? If so, psychosexuality might take a backseat to the study of aggression. Would character analysis replace the treatment of neurosis? And with two different models of general mental functioning, "I" psychology and the topographic model, no one was quite sure if these theories might coexist or develop in ways that were contradictory.

Taking stock of her field in 1934, a British analyst, Marjorie Brierley, aptly noted that the stunning growth of the past decade had not yet been synthesized: "There has not yet been time to test all the newer hypotheses nor to effect the systematic revision of theory which they call for. Old and new concepts exist side by side and over lap each other." As the Thirteenth Congress convened in Lucerne, it was as if the speakers in these debates had suddenly been stopped in midsentence. The great Berlin school had vanished. Franz Alexander and others had immigrated to far-off America. Wilhelm Reich lived on the run. It was as if the stage had been set for a dramatic contest, only to have the principal actors hurry off before the show began.

To make matters worse, the foremost innovator of clinical technique, Sándor Ferenczi, had died in 1933. The Budapest community would never fully recover from the loss of their leader, which came as they were beginning to emerge as a potential center of their own. In the years before his death, Ferenczi had offered a new vision of psychoanalytic treatment. He retracted a number of his positions on active therapy and abstinence, which he now recognized could simply encourage a patient's masochism. This lesson was made all too clear by a patient who in taking up his doctor's methods, first forbade himself food and then considered abstaining from the satisfaction of breathing.

Around 1929, Ferenczi threw himself into clinical experimentation. Sensing distance from his "secret Grand Vizier," Freud pointedly asked Ferenczi if a "new oppositional analysis" was in the works. Ferenczi refused to be baited. The loyalist who once insisted the Muses must be silenced in times of war was determined to follow his own Muse. Ferenczi believed that analysis had become too focused on character and had forgotten the importance of childhood trauma. Ferenczi wanted to get at those traumas and heal them by a "neo-catharsis." While Reich was advocating rapid and systematic confrontations of resistance, Ferenczi moved toward a technique guided by empathy and the analyst's "elasticity." Emphasizing the analyst's responsiveness, Ferenczi offered a new "principle of indulgence" that he thought would allow patients to speak more openly and heal.

Ferenczi's new way was not well received outside Budapest, but at home his ideas held sway with followers like Imre Hermann, Alice Bálint and her husband Mihály, Vilma Kovács, and Lajos Lévy. Ferenczi's new method also quickly made trouble, when one female patient bragged that "Papa" Ferenczi allowed her to kiss him. The news soon found its way back to Papa Freud, but even after being rebuked, Ferenczi refused to stop pushing his own line of thought. He too considered the kissing a problem and began to consider whether an analyst should pretend to be indulgent when he was not feeling so inclined.

Following this query, Ferenczi began to explore the bugaboo that analysts had long left alone: countertransference. The analyst's irrationality and subjectivity were epistemological problems that shadowed every attempt to create a psychological science. Psychoanalysts had been encouraged to deal with the problem on their own, but Ferenczi understood the analyst's subjective feelings as vital to the analysis itself. In 1932, in his treatment of the American Elizabeth Severn, Ferenczi departed radically from all standards of analytic technique when he began to experiment with "mutual analysis." Severn reported being brutally raped and abused as a child. Now she was terrified that her analyst would torture her too. In an effort to get past her terror of him, Ferenczi yielded to Severn's insistence that she analyze him. Analysts from as far back as Alfred Adler would have interpreted Severn's demand as a hostile attack. Ferenczi scheduled two hours in a row to give it a try.

Ferenczi sailed into completely uncharted waters and fought with unheard-of questions. How could he keep his commitment to the confidentiality of his other patients while free-associating for Severn? What would he do if other patients demanded to analyze him, since word was leaking out? Ferenczi eventually came to think the analysis of the analyst only appropriate when the analyst's resistances were clearly in play.

Mutual analysis was the most radical challenge to the analyst's authority that psychoanalysis had ever seen. Freud had once labored to ground the psychoanalyst as a clinically distant, scientific observer. Ferenczi opened himself to the problems of the analyst's subjective engagement. By yielding to Severn's desire for mutual analysis, Ferenczi could be seen as extending his theory of indulgence, due to personal needs that he himself acknowledged. However, he also embarked on a brave effort to find a way to work with a profoundly impaired woman. In that search, Ferenczi discovered the paralyzing impact of his own latent sadism. The analyst who had recommended imposing frustrations on his patients had allowed this patient to travel through her fears by seeing him, not as an objective observer, but as a fallible subject. Through open access to her analyst's unconscious and his own angry resistances, Severn felt safe enough to explore her own howling inner world.

Reich and his colleagues had discovered that latent hatred in their patients could ruin treatment. In his mutual analysis, Ferenczi grappled with his own negative countertransference. When guides to technique were written for use in the institutes, a few courageous souls, like Ella Sharpe, Helene Deutsch, Edward Glover, and Wilhelm Reich, had touched on this sensitive subject, but no one went as far as Ferenczi.

Ferenczi never published his experiments with mutual analysis, though they were an open secret among his Budapest colleagues. He recorded his efforts in a private clinical diary and restricted himself to publicly advocating a return to the centrality of trauma theory and catharsis as a therapy. In those ways, Ferenczi distanced himself from Freud. His break was complete when he confessed to Freud that his intellectual work was not commensurate with running for president of the I.P.A., a role that was charged with "conserving and consolidating what already exists." Ferenczi was not interested in consolidation but rather exploration. Freud passed the letter on to Eitingon and bitterly prepared for Ferenczi to defect like Rank.

The Hungarian showed no inclination to leave the fold. He confidently told Freud that he hoped to present his new ideas and be understood, corrected, and amended. Mainstream leaders of the I.P.A. and the editors of its journals, like Ernest Jones, however, were seriously opposed to his views. When Ferenczi died in 1933, the hope that a Budapest-based Ferenczian theory and technique might come to rival character analysis and "I" psychology died too. When Ferenczi's widow asked the Bálints and Vilma Kovács whether she should publish her husband's clinical diary, they advised her to wait, for they worried that Ferenczi's feud with Freud and Jones would mar its reception. The clinical diary would not be published for fifty-two years. In Lucerne, Ferenczi was eulogized. Afterward the Budapest school would hang on, but without Ferenczi's leadership, they would have little influence on the central contests that would define psychoanalysis.

The congress at Lucerne was marked by the decapitation of the Budapest group and the catastrophe that had befallen Berlin. While the Berlin Psychoanalytic Institute remained open, many of the institute's members had fled. In the spring of 1933, the Nazis had demanded a purging of all Jewish doctors from Berlin hospitals, except the Jewish Hospital. Jewish members of the Berlin Psychoanalytic Institute were asked to resign. On May 6, 1933, Eitingon packed his bags for Palestine. Ernst Simmel, a leader of socialist physicians in Berlin, had already been arrested. Two Aryan institute members, Felix Boehm and Carl Müller-Braunschweig, took over the leadership positions and arrived in Lucerne to represent the institute.

These men encountered their former colleagues in Lucerne—exiled Berliners like Otto Fenichel, Siegfried Bernfeld, Georg Gerö, and Wilhelm Reich. Reich was now homeless; his papers had been revoked in Copenhagen, and then the same thing had happened in Malmö, Sweden. He arrived in Lucerne, a psychoanalyst without a country. He would soon become simply a man without a country. In Lucerne he was stripped of his membership in the I.P.A. While Reich was expelled, the uprising that Ernest Jones feared might take place against the new Aryan leaders of the "German Psychoanalytical Society" did not materialize.

Jones decided that he would do everything he could to maintain psychoanalysis in Berlin, but his support of Boehm and Müller-Braunschweig had many detractors, especially among exiled leftist analysts. A few months before the Lucerne meeting, they secretly organized under the leadership of Otto Fenichel. From Oslo in March 1934, he wrote a nineteen-page circular letter with news from Germany, Austria, France, Hungary, Holland, Switzerland, the Soviet Union, America, and elsewhere. It was as if he were writing the I.P.A. bulletin, except this bulletin was for an underground organization. "Dear Colleagues," Fenichel wrote, "we are all convinced that we have in Freud's psychoanalysis the core of the future of dialectic-materialistic psychology, and we are convinced for that reason that the care and expansion of this science is critical. If we didn't believe that we would not be psychoanalysts by profession."

For the leftist analysts, Ernest Jones's wavering held terrifying possibilities; they understood his strategy would abandon them to their own fate. Against such accommodation, Otto Fenichel summoned the ghost of Sándor Ferenczi. He asked his colleagues to remember Nuremberg and Ferenczi's insistence that the I.P.A. be founded to defend psychoanalysis against detractors and traitors from within. This was another emergency. Psychoanalysis had entered the public sphere, and powerful reactionary forces were calling for its elimination. Fenichel warned his comrades that enemies were also within the I.P.A. How far would the leadership go in hopes of appeasement?

Fenichel sent the letter to colleagues in Prague, Copenhagen, Oslo, Palestine, London, Budapest, and behind Nazi lines in Berlin. He collated the responses, then circulated them in the hope that this would knit a secret community together. But as the Lucerne Congress neared, the group began to bicker. When the congress began, tensions between Reich and Fenichel were out in the open. By the end of the congress, Reich had been expelled from the I.P.A. and Fenichel had returned to secretly writing the *Rundbriefe* to a small band of Marxist psychoanalysts. Nazism had split analysts into those who openly held political and social reform agendas and those who did not, and it was unclear if those two groups could hold together. Some worried the leaders of the I.P.A. would redefine the boundaries of the field so that the Marxist analysts would all go the way of Reich. Suddenly it seemed that political ideology might redefine psychoanalysis.

As if these crises were not enough, an unsolved bureaucratic dilemma flared at the Lucerne meeting. The Berlin Institute's training protocol had consolidated psychoanalysis as a self-policing profession with known educational and training requirements. As Eitingon fled for Palestine, he could rest assured that no matter what happened to the Berlin Institute, their rigorous educational model would survive, for it had been replicated at all the other I.P.A. institutes throughout the world. Professional standards for training were monitored and upheld by a specially appointed committee of the I.P.A., the International Training Commission (I.T.C.). Eitingon's model had won wide support, and in 1929, he himself asserted that the question of how one became a psychoanalyst had been answered.

But one deceptively simple question remained. Who was eligible for training? Did you need to be a medical doctor? The question provoked other, more difficult ones: Was psychoanalysis a part of medicine, a natural science, or something else? Was it a subdiscipline of psychology or psychiatry; was it *Naturwissenschaft* or *Geisteswissenschaft*?

From the beginning, Freudian theory was a hybrid, and the first community was similarly mixed, welcoming scientists, doctors, writers, philosophers, lawyers, and humanists. As societies for psychoanalysis opened in cities around Europe and America, much was left to the local leadership to decide who was suitable for training. But after 1920, when public psychoanalytic clinics were born, analysts lost control of these decisions.

In Vienna, medical and political authorities insisted that no lay people treat patients in the Ambulatorium. The Professional Association of Vienna Physicians delayed a psychoanalytic clinic until it was clear no lay people would be on staff. To dodge the problem, the Viennese maintained a clinic manned by medical doctors and a separate Society and Training Institute where lay members could be trained.

That finesse endured until the spring of 1926, when the lay analyst Theodor Reik was charged with quackery under Vienna's penal code. A disgruntled analysand had accused Reik of practicing medicine without a license. Freud was fond of Reik and had discouraged him from going to medical school because he saw in him the makings of a future analytic researcher. Freud had even supported Reik financially and arranged for him to be analyzed without cost by Karl Abraham in Berlin. Freud rose to Reik's defense. He tried to intervene with public officials, wrote to the press, and when all that did no good, published a polemic, "The Question of Lay Analysis."

Freud insisted medical training was not necessary for psychoanalysis. He confessed that he himself never felt like a doctor and saw no reason to exclude others who wanted to avoid medical training. Such training was often unfavorable to the development of a psychological mode of inquiry, and psychoanalysis needed a foot in the cultural sciences to progress. The field should disengage from medicine and become a self-contained discipline. Freud's position was no surprise; he had long supported efforts to bring psychoanalysis to pastors, academics, and teachers, and he held the view that anyone who had been analyzed could be an analyst. Furthermore, Freud was bitter about the travails the medical world had created for him. But his opinion broke with the Berlin leadership, who demanded codified professional standards and training. Now the Reik case gave the discipline a chance to rethink its rules. Freud threw down the gauntlet on lay analysis and demanded this field—his field—follow.

He didn't have a chance.

Sigmund Freud was no longer a monarch leading Freudian foot soldiers. While constantly paying homage to Freud, many analysts had long ago given up the idea that they were strictly Freudians. They harbored a vision of themselves as a scientific group with a discrete method of inquiry, formalized training, and a professional identity: they were psychoanalysts. Freud sensed he was swimming against the tide and complained to Ferenczi that "the inner development of psa. everywhere runs counter to my intentions, away from lay analysis to the purely medical specialty."

Most hostile to Freud were the Americans, who pleaded with their European colleagues to recognize that a swarm of hucksters had opened up shop on their shores as psychoanalysts. In the United States, the influential Flexner Report had been published in 1910, exposing the deplorable state of medical education and the numerous diploma mills that sent quacks into practice. In the wake of this report, American analysts decided that the only way to survive was to restrict analytic training to medical doctors. When the Reik affair broke in 1926, the Americans held little sway in the I.P.A., and Freud's anti-Americanism was legendary. (He once quipped: "America is gigantic, but a gigantic mistake.") He dismissed the Americans' complaints and contemplated kicking them all out. But the Americans were not standing alone.

The leader of the charge against Freud's view on lay analysis was Ernest Jones. Faux, untrained "analysts" in England had caused Jones endless grief—spawning bad press and legal suits. The timing of the Reik affair made Jones's opposition inevitable. As the controversy in Vienna took shape, the British Medical Association had just convened a commission to investigate the medical status of psychoanalysis. This provided a chance to secure the future of analysis as a legitimate medical science in England. One of the grave concerns that British physicians had raised was the practice of analysis by laymen. As Jones prepared to parry these concerns and defend psychoanalysis as a medical specialty, the founder of the field, Sigmund Freud, called for a divorce from medicine. Jones leapt into action. As editor of the *International Journal of Psychoanalysis*, he orchestrated a massive airing of views. Twenty-four leaders of the profession and two whole Societies (New York and Budapest) published their positions.

Jones wrote the first and longest piece, an acidic rebuttal of Freud that was extraordinary for its sharpness. Jones characterized the Professor's position as emotional, extreme, and absurd. Freud had argued that no public authority could prevent someone who had been analyzed from analyzing others, a position Jones thought silly, since then every patient no matter how sick could therefore be a psychoanalyst. Furthermore, psychoanalysis was in a transitional state, and in the next years, development necessitated that the field be linked to other allied sciences. To reach that

goal, psychoanalysis must resist forces that risked turning it into "an eso-
teric cult." While the field might be hurt by excluding lay analysts, the
"innumerable connections between psychoanalysis and the sciences of
biology, physiology, and clinical medicine" and its acceptance by the
scientific establishment and the public, made the cost worthwhile. Psychoanalysis, Jones bluntly concluded, was a "medical organization and
discipline." He did not want to exclude nonmedical personnel entirely but
keep them on the margins as analytic assistants supervised by physicians.

This was strong stuff. Jones openly pounded Freud and expressed the
wish that his own view would usher in unanimous consensus, which
meant unanimous except for Sigmund Freud. As the others weighed in, it
was obvious that unanimity was not the order of the day. There were
countless quibbles and minor differences, but in the end, most of the
young medical leaders lined up against Freud. A number of Berliners,
including Radó and Horney, opposed lay analysis; they were joined by
Wilhelm Reich and all the Americans. Berlin's designated teaching analyst Hanns Sachs, himself a layman, along with some Viennese and the
entire Hungarian Society, advocated for lay analysis. The most powerful
rejoinder to the medicalizers came from Theodor Reik himself, who conceded that while the clinical treatment of psychoneurosis may have been a
medical matter, the treatment of the asymptomatic character neuroses
would no longer qualify as treatment of an illness.

Seven years later at Lucerne, this rift in the I.P.A. widened, and it
became clear that years of bureaucratic diplomacy had failed. The conten-
tious issue had been initially referred to the I.T.C., which structured a
nonbinding recommendation that called on all candidates to have medical
training. The Americans insisted they would not let lay people in at any
price. A new committee was created to draw up new guidelines. Eitingon
had stacked the committee against lay analysis by choosing three members from Berlin, a society, he later admitted, that had a "definite position
on the topic." In 1929, however, the chairman, Sándor Radó reported the
committee was at an impasse. Meanwhile in London, Jones had managed
to convince the British Medical Association to recognize the legitimacy of
psychoanalysis, acknowledging that the term "psychoanalysis" referred to
those who employed a common technique first devised by Freud, as well
as a professional group, the I.P.A. By examining the practitioner's method
and his professional identity one could discern true analysts from wild

ones. Savoring his victory, Jones brokered a compromise between the Europeans and Americans on lay analysis. By Lucerne, it had all come undone. The Vienna Society was under the impression that they had negotiated a deal with the New York Psychoanalytic Society by which New York would accept lay analysts; Anna Freud complained that the New York Society was not keeping its part of the agreement. The president of the New York Society admitted as much and asked that the agreement be voided.

Left to right, René Spitz with Ernest Jones, the president of the I.P.A., in 1936.

The deal was suddenly crucial. Going to America was one of the few hopes for analysts fleeing Nazi Germany, so the credentials the Americans would accept loomed as a life-or-death issue. The Training Commission announced their issuing of the Lucerne Standing Rules, but despite the pomp, the new rules solved nothing. During the next years, the threat of an American secession hung in the air.

For the decade leading up to 1934, psychoanalytic congresses had been filled with good news of institutes springing up in places like India, Italy,

Chicago, and Japan. In Lucerne, the news of expansion in South Africa, Prague, Boston, and Palestine was undercut by widespread conflict with the American Psychoanalytic Association and the destruction of the teaching institute that had been the mother to them all, Berlin. Already uncertain of the future of psychoanalysis in German-speaking Europe, the members of the I.P.A. would not have been reassured when they read that the *Zentralblatt für Psychotherapie* had replaced Ernst Kretschmer with a new editor, Carl G. Jung of Zurich. Analysts read that Dr. Jung in his opening editorial—dated December 1933—had argued for a racial psychology that differentiated between Germans and Jews:

> In my opinion it has been a grave error in medical psychology up until now to apply Jewish categories—which are not even binding to all Jews—indiscriminately to Germanic and Slavic Christendom. Because of this the most precious secret of the Germanic peoples— their creative and intuitive depth of soul—has been explained as a morass of banal infantilisms, while my own warning voice has for decades been suspected of anti-Semitism. This suspicion emanated from Freud. He did not understand the Germanic psyche any more than did his Germanic followers. Has the formidable phenomena of National Socialism on which the whole world gazes with astonished eyes, taught them better? Where was that unparalleled tension and energy while as yet no National Socialism existed? Deep in the Germanic psyche.

Jung's embrace of racial psychology flowed from Lamarckian thought. Different races had different histories and different unconscious lives. But racial prejudice alone seemed to inform his chilling belief that the Jews had debased the Aryans, who otherwise abounded with heroic potential. Jung seemed to believe that the National Socialists confirmed his own psychological views and refuted Freud's "Jewish" psychology. But actually, Jung's diatribe seemed only to confirm Freud's view that Jung was an anti-Semite.

Readers of Jung's journal could also find a piece in the same issue from the pen of Prof. Dr. jur. Dr. med. Mattias H. Göring. In addition to his many degrees, Göring had other weighty affiliations: he was a member of the Nazi Party and the cousin of Reichsmarshall Hermann Göring. Dr.

Göring was also the leader of the reorganized German Medical Association for Psychotherapy. Rather incredibly, he declared that from that date forward it would be expected that all psychotherapists make a "serious scientific study" of Adolf Hitler's *Mein Kampf*.

For those who hoped that the madness would remain in Germany, there were more ominous developments. Germany's turn to fascism sent shock waves to Austria, which fell into a fascist dictatorship under Chancellor Engelbert Dollfuss. A member of the Christian Social Party and an anti-Semite, Dollfuss hoped to preserve independence from Germany by creating an alliance with the Italian fascist Mussolini. This unholy alliance led to the dissolution of the Austrian parliament and the abolishment of all political parties. Dollfuss offered all patriots the opportunity to join the same party, the Fatherland Front. Since most Austrian psychoanalysts were Jews and members of the Social Democratic Party, these were dangerous changes indeed. But when one considers that Dollfuss—as an anti-Nazi and anti-Communist—was seen as a middle-of-the-road alternative, one gets a sense of how perilous life in Vienna had become. Things worsened when Dollfuss was assassinated in 1934, and replaced by the anti-Semitic Kurt von Schuschnigg, who censored the press and seemed bent on restoring Austria to some fantasy of a medieval Christian empire.

The victory of the Nazis and the Aryanization of the Berlin Institute was a catastrophe not just for Germany. For those with eyes to see and ears to hear, it also announced a grave, immediate threat to Europe and more immediately, Austria. Some managed to hope that in Vienna all would be well, but many, including Ernest Jones, considered Austria's annexation by Hitler inevitable. Several Viennese analysts fled to America. Others must have asked themselves: With so much uncertain in their discipline, what would happen?

Psychoanalysis was under attack again, not by critics wielding pens, but by armed thugs who burned books, intimidated, extorted, and murdered. In Lucerne, Ernest Jones denounced all national and racial prejudice. Rather innocently, he wrote: "Politics and Science do not mix any better than oil and water." But Jones was not innocent. He wanted to delay the Nazi destruction of psychoanalysis in Germany by dissociating the field from Marxism. He warned those who had grown "impatient with social conditions" not to misuse psychoanalysis to propagate social change. Anyone who followed this course would be denounced. The expulsion of Reich was an example to the others. Later in the congress, Jones tried to arouse support for the new Aryan leadership in Berlin, defending Felix Boehm, whom Jones contended had been subject to "much criticism, some of it being of a very resentful kind." Jones even contended that irrational forces were at work in the assessment of the man who had cooperated with the Aryanization of the Berlin Institute.

After Lucerne, Wilhelm Reich went to Oslo, where he would claim that his expulsion from the I.P.A. was the result of his differences with Freud over the death drive. But many had rejected the death drive and stayed in the movement. Reich's exile was different: it was political. Freud, Jones, Eitingon, and Ferenczi had been desperate to avoid the labeling of their science as Jewish, revolutionary, or Marxist. But Wilhelm Reich had unflinchingly argued that psychoanalysis *was* revolutionary, and that the fascists who attacked psychoanalysis correctly recognized it as their enemy. A year before his death, even the maverick Sándor Ferenczi agreed that the move against Reich was "absolutely necessary to establish our political nonpartisanship." In this very real time of war, it looked as if the political Muses would be silenced.

II.

As THE ANALYSTS returned home from Lucerne, an array of unresolved debates lingered. After the events of 1933, annus horribilis, it appeared that little headway would be made on theoretical matters, not while exiled Berlin psychoanalysts struggled to relocate and the Viennese fell under dictatorship. Surprisingly, however, the next five years saw the birth of two compelling syntheses, each of which integrated many innovations of the last two decades. Even more surprisingly, those syntheses came from members of the maligned minority, the lay analyst.

Melanie Klein and Anna Freud were unlike the first wave of lay analysts. Those scholars had mingled psychoanalysis with the study of art and culture, but over time their project had lost some of its luster. In the debates over Reik and lay analysis, Professor Freud still insisted analysts bring broad humanistic learning to the field, but others openly scoffed at the idea. Reich said lay analysts in the cultural sciences had done little to advance the field. Clarence Oberndorf, an American analyst, added that the cultural sciences "have contributed little to psychoanalysis excepting in the way of corroboration and amplification." While the search for theoretical confirmation in works of literature and myth continued, there had been an important change since the days when Freud, Jung, Stekel, and Ferenczi immersed themselves in the cultural sciences. Lamarckian heredity had been discredited.

Lamarckian thinking had made it possible for the Freudians to link the unconscious memory of an individual to the history of all humankind. The study of history, archaeology, religion, anthropology, and literature became the psychologists' equivalent of an archaeological excavation. Early analysts believed that traces of early human psychic life were not just windows into the past but were also a route into the deepest strata of the unconscious, another way to know the Kantian unknowable. This logic had guided Freud's *Totem and Taboo*, Jung's *Transformations of the Libido*, Stekel's *Language of Dreams*, and Ferenczi's *Thalassa*. It was a premise so deeply held by Freud that when faced with Otto Rank's rather improbable theories of birth trauma, the Professor had been disturbed most by the fact that the theory made no place for Lamarckian inheritance.

During the first years of the twentieth century, this acceptance of Lamarckian beliefs was by no means eccentric. Not so long before, Darwin had promoted the view that acquired characteristics could be inherited, and he had advised young mothers to learn everything they could before having children in order to pass their refinement down to their brood. By 1925, however, most scientists believed Lamarckian heredity had been proved wrong. Around 1900, Gregor Mendel's experiments turned the tide, and by 1910, the gene had been conceptualized and linked to the chromosome. By the 1920s, gene theory had won the battle of ideas, aided by the failure of numerous experiments to prove acquired transmission. A general rejection of Lamarckian ideas spread through Western science, with the prominent exception of the Soviet Union, where Lamarckian notions were considered consistent with Soviet ideology and deemed sacrosanct.

Freud also refused to change his mind about phylogenetic inheritance, as he would demonstrate in one of his last works, *Moses and Monotheism*. But for psychoanalysts seeking medical and scientific legitimacy, Freud's Lamarckian positions became more and more of an embarrassment. In 1928, Ernest Jones found himself under siege by the well-known anthropologist Bronislaw Malinowski on this point. Jones had to disavow publicly Freud's theories as "both unproven and biologically improbable."

The loss of the link between ontogeny and phylogeny, the individual and the species, as well as the psychic and the biological created a crisis for analysts committed to studying the psyche through the history of human culture. Psychoanalysts lost an extraclinical avenue for understanding the unconscious, one that seemed to exist outside the complex realm of transference and countertransference. Now they had to rely more exclusively on the individual clinical encounter for clues, evidence, and confirmation of theories of unconscious mental life. History, myth, and fables no longer could lend them direct support if they hoped for credibility in scientific and medical circles.

After 1918, however, a new consortium of nonmedical analysts offered a different mode of study for extraclinical investigation and validation. Melanie Klein and Anna Freud became the leaders of a group of lay analysts who weren't interested in interpreting folktales, but rather observing children. After the war, teachers, school principals, and pedagogues in Vienna, Berlin, and Budapest turned to psychoanalysis to better understand children, and they began to use that understanding to reshape the field.

Theorizing about childhood was as old as the Freudian enterprise itself, but Freudian theories of childhood had been mostly the result of inferences gathered from the observation of adults. In a happenstance way, Sigmund Freud, Max Graf, Carl Jung, Karl Abraham, Sándor Ferenczi, Lou Andreas-Salomé, and Oskar Pfister had all tried to directly study or treat children, but the new lay analysts took up a more systematic, empirical study of children following the lead of Hermine Hug-Hellmuth.

Born into Vienna's aristocracy, Hug-Hellmuth attended the University of Vienna and received her doctorate after completing a dissertation in physics at the age of thirty-eight. By that time, Hermine was immersed in an analysis with Isidor Sadger. Trained to be a teacher, she attended meetings of the Vienna Society and published a stream of psychoanalytic inquiries, most significantly her 1913 *A Study of the Mental Life of the Child*. Hug-Hellmuth cast herself as an objective observer and chastised others who wrote without a "first-hand study of the mental life of childhood." She pieced together a general psychology of the child by highlighting Freudian sexuality, but also by adopting the work of numerous academic psychologists on cognitive, linguistic, and ethical development.

Hug-Hellmuth then published a work that embroiled her in scandal. Around 1915, she obtained a diary from someone she called a friend. It offered an intimate portrayal of female sexuality during adolescence, and the account happily coincided with Freudian theories. After the war, *Diary of an Adolescent Girl* was published with a gushing letter from Freud as its introduction. Hug-Hellmuth promoted the anonymously authored work as the "most valuable document on the subject of psychic development."

The book was racy and sold well. But critics, among them the University of Vienna psychologist Charlotte Bühler, denounced it as a fraud, saying no girl of that age could have written it. Since Hug-Hellmuth refused to name the author and could not produce the original manuscript, many began to believe she had forged the document herself. By 1922 when the third edition was published, Hug-Hellmuth came out from the shadows to identify herself as the editor of the volume. After that, the German edition was unceremoniously withdrawn from circulation.

If psychoanalysts were going to turn to the study of the child for empirical testing and observation, Hug-Hellmuth offered them a warning. Would the pursuit of child psychoanalysis bring the field fresh data, or would it be a hunt—perhaps a stacked one—for confirmations? Child observation offered the Freudians what they badly needed: an empirical forum to test, confirm, falsify, and create theory about infantile needs, wishes, and fears. But it was not possible to be part of the Freudian community before 1918 and question the legitimacy of childhood psychosexuality. Hug-Hellmuth was mirroring her community's perceived demands when she invented a firsthand account, as most critics now believe she did, to support Freud.

As suspicion about the diaries mounted, Hug-Hellmuth traveled to The Hague in the fall of 1920 to lecture on adapting psychoanalytic technique for children. Hug-Hellmuth argued that a child should be treated at home, not made to lie on a couch, and should be able to play in sessions. No child should be analyzed by a parent, she asserted, no doubt catching Anna Freud's ear, because at that moment she was in analysis with her father. But Hug-Hellmuth's major finding was that in every case the children she treated suffered from a failure of parenting, which was either too strict, too lenient, or inconsistent. For those analysts who saw animal drives as fundamentally implicated in neurosis, this made little sense. But as a pedagogue, Hug-Hellmuth naturally stressed education—good and bad. To remedy the problems of a bad upbringing, Hug-Hellmuth believed in reeducation as well as psychoanalysis.

In Red Vienna, Hug-Hellmuth's focus on destructive environments was in harmony with the politics of the times. Theories of childhood had long been a battleground for competing political ideologies; when Enlightenment figures like Jean-Jacques Rousseau argued about the nature of a child, they were also arguing about how a political or social environment might make or break an individual. Similarly, psychiatric theories about what made children insane were often politically loaded, because they fueled reformers who emphasized problems like poverty or supported conservatives who stressed a natural order. Throughout Europe in the 1920s, reformers and progressives turned to the study of deprived, poorly educated children. They engendered the Child Guidance Movement and the institutionalization of child psychiatry.

In that same spirit, the Vienna Pediatric Clinic started a department for remedial work with troubled children. Viennese Social Democrats looked to psychologists to help them create healthy future citizens. Otto Glöckel, the Austrian minister of education, recruited the academic psychologists Karl and Charlotte Bühler to found a Center for Child Study in Vienna. The authorities also summoned Alfred Adler, whose socialist credentials and emphasis on power relations made him desirable. After 1920, Adler's Individual Psychology became popular among pedagogues. Some thirty Adlerian child guidance centers were established, and Adler's devotee Carl Furtmüller took an active role in reforming Austria's school system.

The Vienna Psychoanalytic Society was swept up in this tide. Hug-Hellmuth attracted a group of young, enthusiastic, politically engaged child analysts. The dynamic Zionist activist and reformer Siegfried Bernfeld founded an institution for child analytic work in 1919, but his *Kinderheim Baumgarten*, a boarding school for Jewish war orphans, barely survived a year. In 1921, the pedagogue and schoolmaster August Aichhorn entered the Vienna Psychoanalytic Society. As the director of a residential institution north of Vienna, he had treated juvenile delinquents for a long time and had begun to employ psychoanalytic tenets in his work. When Aichhorn's book *Wayward Youth* appeared in 1925, Freud introduced it with the enthusiastic remark that "children have become the main subject of psychoanalytic research and have thus replaced in importance the neurotics on whom its studies began."

The Viennese inaugurated a Child Guidance Center at the Ambulatorium in 1922. Under the leadership of Hug-Hellmuth, the center conducted ten to twenty-five consultations a week and treated cases sent not just from families, but also from local schools and clubs. The treatment was deemed educational—Freud would call it a "pedagogic analysis"—and therefore was open to nonmedical workers.

In 1924, Hermine Hug-Hellmuth's nephew, Rudolph Hug, murdered her. An orphan, the boy was taken in by his aunt, who featured him in a number of her papers. In his legal defense, he stated that Hug-Hellmuth's constant analysis made him homicidal. The sensational murder was picked up by the press, but despite the scandal, child psychoanalysis continued to grow. After 1924, Editha Sterba and August Aichhorn continued Hug-Hellmuth's work at the Ambulatorium. As educators took up child psychoanalysis in Vienna, the late Hermine Hug-Hellmuth was eclipsed by another former schoolteacher with a formidable pedigree.

FREUD'S YOUNGEST CHILD, Anna, trained to become a schoolteacher. After the war, she became friendly with some of her father's followers who also were interested in pedagogy. As a child, she had spent many hours hanging around the Vienna Society meetings, but as an adult she began to attend these meetings and focus on the possibilities of child psychoanalysis. Anna briefly considered joining the Berlin Institute rather than face what was, at times, brutal criticism from the Society's elders, but in the end, she presented a paper to the Society on beating fantasies, and having taken her punishment, was duly elected to the Society in 1922. Around that same time, Siegfried Bernfeld, Willi Hoffer, and August Aichhorn joined Anna at Berggasse 19. While the Professor and his friends played their weekly card game, Anna and her friends held an informal study group on the place of the child in psychology, psychoanalysis, and society.

It quickly became apparent that this former schoolteacher was no ordinary Society member. Two years after her entrance into the Society, Anna Freud was inducted into the Secret Committee, and a year later she became part of the administrative leadership of the Society and the Training Institute. That year Rank's defection and her father's illness made Anna the new conduit to the Professor. When the Reik affair broke in 1926, Anna Freud and child psychoanalysis were on the rise. For Sigmund Freud, the collapse of lay analysis would mean seriously compro-

mising new research and marginalizing his own daughter. When Freud
wrote to the *Neue Freie Presse*, he made a point of defending not just Theodor Reik, but also Anna. He denied the paper's allegation that he referred adult neurotics to his daughter, who was only involved in
"pedagogic analysis of children and adolescents." The proud father could
not help but add that in the one case of neurosis that Anna had worked
on in conjunction with a doctor, she had done beautifully.

After Theodor Reik emigrated to The Hague in 1928, the strongest
voice in Vienna for lay analysts and child analysis was Anna Freud. One
of her students later remarked that six years after Hug-Hellmuth's death,
Anna could be heard insisting she had founded the field of child psychoanalysis herself. She pointedly distanced herself from Hug-Hellmuth and
challenged the only other major child analyst, Melanie Klein. This mete-
oric rise inevitably made for resentment. Before Anna leapt to the front of
the Vienna Society, Helene Deutsch had been the most prominent woman
in the group and had held the title of director of the Vienna Training Institute. Some wondered if the Black Cat club, a circle of young married
analysts surrounding Deutsch and her husband, had formed precisely to
exclude the unwed Anna.

Anna Freud in 1937. Freud's daughter emerged as the leader of a burgeoning child
psychoanalytic community in Vienna.

Nonetheless, Anna Freud's ambitions were not tempered. Arriving in Innsbruck for the 1927 congress, she delivered an exposition "On the Theory of Analysis of Children." Like Hug-Hellmuth, she stressed the value of child analysis for the scientific foundations of the field. Child psychoanalysis "gives us welcome confirmation of those conceptions of mental life of children which, in the course of years, have been deduced by psychoanalytical theory from the analysis of adults." The psychoanalytic enterprise had been snarled in problems of memory and historical reconstruction, but child psychoanalysis offered a way out. "Direct observation" in child analysis "leads us to fresh conclusions and supplementary conceptions," she declared. Like Hug-Hellmuth, Anna Freud accentuated the empiricism of child psychoanalysis. From now on, psychoanalytic theories of childhood would require coherence and support from such first hand, empirical study.

Anna Freud began to publish her views on child analysis. Her early writings are remarkable for their clarity and directness, which critics called simplemindedness. Compared to her father, whose complex rhetoric animated his texts, Anna's writing was workmanlike. But if her writing style was simple, her thinking was not. Despite Anna's intense and unswerving loyalty to Papa, she did not become a rote follower. Her father encouraged his daughter to work out her own positions, perhaps knowing that Anna would be chastised for unthinking filial devotion. Encouraged, Anna Freud navigated her way through the debates of the times. She never adopted her father's theory of a death instinct and remained for some time dubious about the value of the new "I" psychology.

In her first published lectures, Anna Freud also displayed a refreshing pragmatism. While others reached for the metapsychological heights, Anna Freud proposed nonanalytic interventions and commonsense changes in psychoanalytic doctrine. She advised an educative period of work with the child and recommended that the child analyst work hard to win the child's affection, rather than concentrate on transference interpretations. After proposing these shifts in method, Anna worried she would be called a wild analyst, a concern that would have been laughable if her father still controlled psychoanalysis. But Anna's concerns turned out to be well founded. In the course of her writing, she crossed swords with Melanie Klein, and in the battle that ensued, Sigmund Freud's daughter would be denounced as a faux analyst who recoiled from the

Oedipal complex because she'd been poorly analyzed by the man who put Oedipus at the heart of human subjectivity, her father.

MELANIE KLEIN NÉE REIZES was born in Vienna in 1882, but after marrying she found herself in Budapest in 1910, the depressed mother of two young children. She went into analysis with Sándor Ferenczi, and on her analyst's urging began to work with children, especially her own son, Erich. After the war, Klein entered the Budapest Society, but as the political situation there deteriorated, she moved to Berlin to pursue child psychoanalytic work.

Karl Abraham had become interested in the earliest developmental stages of childhood, and he looked to the first child analysts to help develop his theories. Abraham asked Hug-Hellmuth to come to Berlin to lecture, and he opened the Berlin clinic to children. He also implored Ferenczi to send Frau Klein to Berlin, since he had heard she was performing child analysis in Budapest. On February 11, 1922, Klein arrived in Berlin and was promptly taken into analysis with Abraham. In 1923, Abraham told Freud that Klein's work with a three-year-old gave "amazing insight into infantile instinctual life."

Two weeks after her arrival in Berlin, Klein gave her first paper on infantile anxiety and the development of personality. Her notes show that Klein strove to demonstrate how little Fritz's unconscious anxiety coalesced into inhibited personality traits. She also fancifully interpreted the boy's fear of ice-skating as castration anxiety ("It's boring to go back and forth with one's legs"). The reception was mixed. Abraham's support was strong, but other Berliners like Radó and Alexander did not share his enthusiasm. When the two Hungarians attacked Klein at future Society meetings, Abraham came to her defense. After Karl Abraham's death, however, Melanie Klein found herself in a hostile environment. According to her friend Alix Strachey, many in the Society thought Miss Klein's rambling commentary at Society meetings proof she was an "imbecile." Alix agreed that Klein was garrulous and at times let her thoughts wander, but she also recognized her colleague's creativity. She knew Melanie had no future in Berlin or Vienna, where she would face serious opposition from "those hopeless pedagogues, and I fear, by Anna Freud, that open or secret sentimentalist."

Alix enlisted her husband in a scheme to get Klein out of Berlin. In 1925, they got Ernest Jones to invite Klein to deliver six lectures at the British Society, where several members were already engrossed in child analysis, including women from the now defunct Brunswick Square Clinic like Ella Sharpe and Nina Searl, and the principal of an experimental school in Cambridge, Susan Isaacs. In London, Klein did not waste her opportunity. According to Jones, she made "an extraordinarily deep impression." She was asked to join the British Psychoanalytical Society and by the fall of 1926 had relocated to London. The fact that Ernest Jones entrusted the analyses of his wife and two children to the newcomer was a sign of the enthusiasm with which she was received in the new city. When Ferenczi came to visit London in the summer of 1927, he noted that many had been won over to Klein's views.

Over the next years, numerous theoretical differences appeared between Melanie Klein and Anna Freud, but most centrally they differed on the age-old problem of nature and nurture. Anna Freud, like Hermine Hug-Hellmuth and many left-leaning Viennese child analysts, identified deprivations of the environment, in the form of bad, cruel, or incompetent parents, starvation and trauma as sources of a child's suffering. Child analysts created remedies that were a mixture of psychoanalysis and corrective pedagogy.

Melanie Klein was not a teacher or social reformer. Her theory was not interdisciplinary but was born exclusively from psychoanalysis. Klein rejected the emphasis on education and pedagogic analysis. As time went on, she would pay less and less attention to an unloving or cruel environment and argue that a child's troubles were his own psychic making. At first, her work gave no hint of this perspective: In 1921, Klein delivered a rote work on the deleterious effects of an unhealthy home environment and argued that a psychoanalytic upbringing minimized the child's swings between resistance and submission. She even called for the founding of kindergartens with women analysts as teachers.

But two years later, Klein turned to the inner fears and anxieties that consume children regardless of their environment. In school, a child's castration fears led to inhibition. While acknowledging that a kind teacher might help a bit, Klein concluded that ultimately the teacher had little effect. She had discovered children in the best of circumstances who were terribly anxious; conversely, cruel teachers didn't seem to engender more inhibition. This conclusion encouraged Klein's move from the complexities of inner and outer to just the inner. She simplified her field of study and pursued the intrapsychic causes of childhood fear. However, Melanie Klein did not bracket potential outer causes so much as eliminate them. The developmental, traumatic, and environmental became increasingly irrelevant to her. Klein later amended her theory to acknowledge the complex interaction between reality and fantasy, but by then her reduction had done its work and helped her build a fundamentally different theory of the child's inner life.

One of the critical innovations guiding Klein's work was a novel form of play therapy. Many nineteenth-century psychologists wrote about the symbolic meaning of children's play, and Hermine Hug-Hellmuth had discussed its uses and meanings. But playing with Melanie Klein was different. For instance after a girl bit a doll, she was told this represented a desire to rip out her mother's innards. Using her play technique, Klein interpreted fantasies and transferences of children as young as two and three, something no one had dared do before. She argued that the analysis of infants should be the same as that of adults and be founded on analyzing the infant's resistances and transferences in the analytic situation. When she announced her new method could cure infantile neurosis, Klein forced those around her to take notice.

Klein also challenged the standard Freudian developmental scheme. While studying night terrors in children, she found that these episodes of fear could erupt as early as two years old, which meant a punitive "Over-I" had to be in place by that time. Freud was therefore wrong when he said the "Over-I" appeared around the age of five, after the Oedipal phase. In London, she took her idea further and claimed she detected an Oedipal complex by the second year of life. According to Klein, by that time, children already suffered under a fierce and punitive conscience.

This contradicted Freud's theory, as well as the work of Hug-Hellmuth, Bernfeld, Aichhorn, and Anna Freud in Vienna. The British Society's enthusiastic reception of Klein's ideas was bound to result in a direct confrontation with the Viennese. In 1925, Sigmund Freud laconically told Jones that Klein's views were met with skepticism in Vienna, though he himself had no opinion on the matter. Yet, after Anna Freud published her lectures on psychoanalysis in 1927, in which she criticized Klein's work, the ill will could no longer be contained. Jones wrote to

Freud and told him he could not agree with Anna's work, which he surmised was due to her own "imperfectly analyzed resistances," a neurotic dimension to her personality that, Jones cryptically added, he could prove "in detail."

Freud was incensed, but really he had no right to be. He had let the ad hominem genie out of the bottle long ago, and now in a moment of supreme irony, the president of the I.P.A. reduced the theoretical positions of Freud's own daughter to her neurosis. Freud shot back:

> I will make only one comment on the polemical part of your letter. When two analysts have differing opinions on some point, one may be fully justified in ever so many cases, in assuming that the mistaken view of one of them stems from his having been insufficiently analyzed, and he therefore allows himself to be influenced by his complexes to the detriment of science. But in practical polemics such an argument is not permissible, for it is at the disposal of each party, and does not reveal on whose side the error lies. We are generally agreed to renounce arguments of this sort and, in the case of differences of opinion, to leave resolutions to the advancements in empirical knowledge.

In reply, Jones reassured Freud that his comments were private and would never be used in public debate. The comment by Jones was a veiled confession, for Freud had never mentioned anything about Jones making such claims *public*. In London, Jones and his colleagues had already held an open debate in which it was easy to discern the accusation that Anna Freud's unanalyzed neurosis had compromised her thinking. On May 4, 1927, the British Society had convened a two-day symposium to demolish Anna's theories. Melanie Klein opened the proceedings with a no-holds-barred attack that ran thirty-two pages in length when it appeared in print. After that, Joan Riviere, Nina Searl, Ella Sharpe, Edward Glover, and Ernest Jones piled on. Three months after the British symposium was held, the proceedings appeared in toto in the pages of the *International Journal of Psychoanalysis*. The editor of that journal, Ernest Jones himself, had printed over fifty pages of attacks on Anna's little book.

At the International Congress in Innsbruck a few weeks later, the acrimony between Melanie Klein and Anna Freud was intense. Klein rejected Anna Freud's major views on child analysis point by point, saying her views were erroneous and illogical. Anna had argued that child analysis must not disturb the child's relationship with its parents. Klein read this as a refusal to analyze the Oedipal complex. Anna reasoned that a transference neurosis could hardly be achieved in young children; Klein insisted it could. Anna and the Viennese felt that work with children should include a pedagogic element. Klein argued pedagogy canceled out effective analytic work. Anna Freud's advice to work hard to win the child's affection and trust was, in Klein's view, a manipulation that made the child submissive and avoided negative transference. Reanalyzing one of Anna's cases, Klein argued that Anna had stopped just at the point where she would have met with the child's hatred of her mother and her Oedipal complex. In the end, Melanie Klein accused Anna Freud of simply not analyzing the unconscious.

Freud *père* was furious. He wrote a scathing letter to Jones asking whether he was again "yielding to your inclination to make yourself unpleasant?" Freud accused Jones of organizing a public campaign against his daughter, accusing her of being insufficiently analyzed. "I can assure you," Freud seethed, "that Anna has been analyzed longer and more thoroughly than, for example, you yourself." Jones replied with a long denial.

Melanie Klein and Anna Freud were both deeply immersed in psychoanalytic precepts, and yet they were miles apart. If the question: "Who was more fully analyzed?" was bankrupt and led nowhere, how could these differences be resolved? An appeal to Professor Freud's authority had no validity anymore. Melanie Klein pointed to Freud's *Beyond the Pleasure Principle* as a primary inspiration of her work on the internal determinants of anxiety, while Anna Freud pointed to her father's *Inhibitions, Symptoms and Anxieties* as ballast for her focus on real danger situations that traumatize and overwhelm the "I." Sigmund Freud was a brilliant synthetic thinker, but he was by his own admission, not a coherent system builder. He did not tie up the loose ends or repudiate the former theories he later seemed to contradict. When he wrote *Beyond the Pleasure Principle*, he "put dynamite in the house" of Freudian psychoanalysis, as Max Eitingon said to Melanie Klein. But that dynamite hadn't demolished all prior Freuds. By 1930, a series of compelling Freuds existed that were not reconcilable. And so, it had become meaningless to appeal to Freud's authority. Like Wilhelm Reich and Franz Alexander, Melanie Klein and

Anna Freud each had her own Freud. Their differences would have to be resolved some other way.

Each woman returned home intent on developing her own theories. After 1927, Melanie Klein dropped polemics and over the next five years churned out one original idea after another describing the inner world of the child. Klein reconceptualized the Oedipal complex as the result, not of castration fears, but of earlier oral sadism that came about in part from the frustration of weaning. By the tender age of two, the child was riddled with sadistic urges toward its mother and had turned to the father, entering the complex world of competing, triangular relationships. From this new perspective, Klein challenged Freud's *Inhibition, Symptoms and Anxieties* and defined her own danger situations, which were not external but entirely fantastic. Children were terrorized by monstrous enemies they conjured up by projecting their aggression onto others.

Klein's stature was solidified by her first book, *The Psychoanalysis of Children*, where she laid out her positions on the Oedipal complex and included several additions that made the book less a summa and more a transitional document. Klein had changed her views on the cause of a young child's sadism. It was no longer the frustration of weaning, but an inborn death drive which, when projected outward, manifested itself as human aggression.

Klein's thinking led to a new theory of technique. Dismissing Wilhelm Reich, she believed technique had not progressed since Ferenczi and Rank's book on the analytic situation. Her play technique, however, allowed for direct interpretations of a child's unconscious contents. She gave extensive illustrations of her method. A child who bumped some cars together might hear Klein announce: "You wanted to bump your thingummy along with Daddy and Mummy's thingummy." Klein, like the earliest Freudians, valued deep interpretation, in which putative sexual content far from consciousness was nonetheless interpreted. Her method also led to the same epistemological worries the early Freudians faced about the power of suggestion and self-delusion.

Melanie Klein proposed compelling new transference paradigms based on the idea that young children split inner representations of objects into all-good and all-bad caricatures as a defense against their fear of a persecuting mother. Children held separate representations of a witch-Mother and an all-giving-Mother, a devilish-Father and a kind, heroic one. Klein showed how splitting was a device that allowed a young child to manage its aggression and not be abandoned to a terrifying world inhabited only by enemies. She encouraged analysts to interpret these cartoonish, all-good or all-bad parental transferences.

Klein also entered the debates on the nature of femininity that had been initiated by Karen Horney and carried on by a number of women who joined the psychoanalytic movement after 1918. As their numbers grew, female analysts reexamined female sexuality in light of Horney's accusation—seconded by Ernest Jones—that Freud saw girls as nothing more than damaged little boys. In 1931 and 1933, Sigmund Freud answered his critics and supported his theory of penis envy, but first he had to address the accusation of his own male bias. Aware that his account would be seen as perpetuating male prejudice, Freud nonetheless insisted that a girl's development turned on her belief that she had been castrated. Prior to that, the girl had a "pre-Oedipal" attachment to her mother, which was transformed by the recognition of her and her mother's damaged state. Dismissing Horney and Jones, Freud looked for support from female psychoanalysts like Ruth Mack Brunswick, Jeanne Lampl-de Groot, and Helene Deutsch, implying (rather unpsychologically) that they could not be under the sway of male bias. In the end, Freud was forced to admit that before those who considered his ideas the result of his male subjectivity, he stood "defenceless."

While addressing the pre-Oedipal attachment between mother and daughter, Freud also took up the work of Melanie Klein. Her dating of the Oedipus complex, he wrote, was not compatible with his observation, based on adult analyses, of a long pre-Oedipus attachment of a girl to her mother. Melanie Klein had a powerful retort to this: in 1932, she devoted a chapter of her book to the sexual development of girls as demonstrated, not by reconstructions from women, but by analyses of girls. Among these young patients, their deepest fear was not castration, but rather bodily invasion and destruction. Klein elaborated a complex scheme in which the girl turned from the breast to the father's penis and generated fantasies of her mother allowing her father's penis into her body. The little girl developed sadistic fantasies of destroying her mother's insides and through projection, was terrified the same would happen to her. Although it began with direct clinical observation, the theory was, to say the least, highly ornate.

The Psychoanalysis of Children received mixed reviews. Franz Alexander described the book with barely disguised contempt as filled with "many illuminating, even though often improbable statements," which showed great intuition but suffered from clumsy and illogical inferences. Alexander noted that Klein took the hypothesis of the death drive for granted, and chastised the author for confusing theory with observational data. But Klein also found supporters. Edward Glover, the scientific secretary of the British Society and the secretary of the International Association, was a tough-minded medical man. He had been deeply impressed by Klein when she arrived on British shores and now heralded her book as a major new work, sections of which were equal to Freud's classic contributions. Glover supported Klein's predating of the dates of the Oedipus complex and "Over-I," and noted that Klein was the first to bring into clear perspective the contributions of the mother to "Over-I" formation. He also sided with Klein on the fantastic nature of the most primitive fears and anxieties.

In London, Klein's work proved to be persuasive among a number of influential analysts such as James Strachey. After finishing his analysis with Freud, James had returned to London and enthusiastically followed Klein's development as reported to him by his wife in Berlin. The erudite Strachey was part of the Bloomsbury circle, and despite his retiring nature, a powerful ally in London. He consorted with the likes of Clive and Vanessa Bell, John Maynard Keynes, Virginia and Leonard Woolf, and his brother Lytton, a figure of literary stature. Though James published little on psychoanalysis himself during those years, the little he did write sparkled with wit and wry intelligence. Reviewing a book, he mischievously wrote: "This volume is chiefly remarkable for its dust-cover, and we therefore propose in this instance to review the dust-cover instead of the book which it contains."

After Klein came to London, in good part thanks to the Stracheys, James eagerly followed her work. Though he preferred to remain behind the scenes, in 1933 Strachey began to lecture at the British Society. A year later, he published a paper on psychoanalytic theory that quickly became a classic. If Melanie Klein was at the time theoretically sloppy, a sprinting intelligence that sometimes ran far ahead of itself, this paper showed that James Strachey was a thinker of calm reserve and sure-footed logic, someone who could package Melanie's theories for a skeptical world. "The Nature of Therapeutic Action in Psycho-Analysis" was not just a promotion of Melanie Klein, but also an employment of Klein's theory in the service of a full-fledged theory of psychoanalytic cure. To avoid misjudging Klein's deep interpretation, Strachey argued that a shared definition of "deep" as well as "interpretation" was needed. His logic moved incrementally from there. If Klein tended toward a dogmatic and declarative style, Strachey was the opposite, carefully anticipating and attending to his reader's doubts. Surveying theories of cure, Strachey concluded that Klein offered more than Anna Freud, Alexander, or Radó. Klein best described the "neurotic vicious cycle" that psychoanalysis had to break. The analyst must focus on the projection of sadistic imagos and through transference interpretations modify the "Over-I," or as it was now referred to in English, the "super-ego," over time. These "mutative" interpretations were the only effective agent psychoanalysts offered. Rote interpretations were useless; mutative interpretations were specific and concrete. Deep interpretation was safe and possible only if it fit those criteria. While other means like suggestion and reassurance were part of any analysis, only on-the-mark transference interpretations made therapy effective.

Strachey's eminently reasonable extension of Klein was just what she needed to gain respect outside London. She could not have wished for a more eloquent spokesman. However, Strachey never got a real chance to be her emissary, for Melanie Klein was not done building her theory, and her further developments undercut her own credibility in a way that would be difficult to repair.

Left to right, Princess Marie Bonaparte, Melanie Klein, and Anna Freud with Ernest Jones in Paris, 1938.

Before her acceptance of the death drive, it could have been said of Klein that, like Alfred Adler, she had fashioned a world without love. Klein's child was filled with sadism and riddled by twisted aggressively distorted forms of libido. In her shift to the grim notion of Thanatos, Klein appeared to opt for an even more loveless view of human nature, but in fact, Klein's acceptance of Thanatos led her also to embrace Freud's theory of Eros. With that, Melanie Klein made room in her theory for a richly elaborated, tragic theory of love.

Psychoanalysts already had a tragic vision of love. For Freud, the child longed for the unattainable parent and lived the rest of his life in search of someone that might dimly resemble that lost and impossible love. Looking for love was a search for ghosts. In the Freudian model, the child's aggression was primarily toward his rival for that love, but in the work of Karl Abraham, a prehistory of struggle between mother and child was added to the story line. Conflict marred the Edenic world of mother and child.

To this already unhappy scenario, Melanie Klein added a tragic twist. She presented her ideas during the tumult of the Lucerne Congress in a paper entitled "The Psychogenesis of Manic Depressive States," which postulated a broad conception of what would be called Object Relations theory. The mind was organized by the relation of the self to its imaginary inner others or objects. Early on, the child experienced its objects as partial, split into the all-good and the all-bad. In her description of a child at nursing age, Klein contended that part-objects were the good breast that nourished and fed or the bad breast that frustrated the hungry child. The bad breast was not so much a result of normal weaning, but a creation of the child's projected aggression. The child lived in paranoid fear of this imagined monster and took solace by identifying with the good breast.

While worthy of Hieronymus Bosch, the schema was simply a consolidation of Klein's earlier work. She then argued that at some point the child shifted into a saner mode of relating to others. The child eventually perceived a whole mother, not simply breasts that fed or frustrated. With the experience of wholeness, the object was experienced as both good and bad, and the emotional life of the child was radically altered. In a moment of self-recognition fit for classical tragedy, the child was horrified by its raging attacks on its beloved mother. Afterward, remorse, guilt, and depression set in. Klein argued that these two developmental positions—

paranoid and depressive—existed in a dynamic. The primitive paranoid position defended against the more mature, but unhappy depressive one. A retreat to a paranoid world where one felt accosted by monstrous bad objects was a common human smokescreen erected to avoid experiencing crushing guilt for one's own destructiveness.

Besides its clinical usefulness, Klein's new theory had the added advantage of making sense of the world around her. Across Europe, ideologues justified murder by hiding their bloodthirstiness behind whipped-up grievances. Unfortunately, Klein was not content just to describe the shift from the paranoid to the depressive mode. She also dated it. And if many had scoffed at her pushing up the Oedipal complex so early, the date she assigned to this new phase led to dismay. According to Melanie Klein, the shift from part to whole objects occurred between the fourth and fifth month of life.

The difficulties of gaining access to another person's inner world had always made scientific psychology a problematic task. The positing of an unconscious by which the inner world was not even accessible to the subject himself made psychological claims even more perilous. Melanie Klein's claim for such precise knowledge of a preverbal child's unconscious recklessly disregarded this quandary. To purport to know with certainty what goes on inside the nearly mute four-month-old was bound to produce incredulity.

Klein asserted her dating of events was based on clinical observations of children who around five months shifted toward depression. But those observations could not support the complex fantasies she assigned to the preverbal child's mind. Melanie was already deeply distrusted outside London, where she was criticized for an inability to distinguish between empirical observations and her theories. These provocative ideas alienated her further, not only from those in Vienna and the scattered Berliners, but more ominously from some colleagues in London.

Edward Glover had brought his substantial scientific gravitas to Klein's endeavors. He had once vigorously defended her, but he found this new model untenable. With Jones on his way to retirement and Glover in line to become president of the British Society, this was a serious problem for Klein. A Viennese analyst recalled that Glover's break meant Klein was "no longer the Goddess" of London. To make matters worse, Glover's ally in his upcoming battle against Melanie Klein and her theories of sadistic children was Melanie's own daughter, Melitta Schmideberg. After completing medical training, Melitta had joined the British Society and gone into analysis with Glover. The British Society's meetings soon devolved into a bizarre family feud. The outspoken Melitta condemned her mother and her mother's theory of mothers, while Melanie defended her theory of sadistic children to, among others, her own daughter.

Melanie Klein had benefited from the wave of interest in child study, but in creating Object Relations theory, she had squandered her empirical credentials as a child observer by confidently making assertions that were—in the registers of science—highly dubious. While insisting on the priority of facts and observations, Klein—like Otto Rank a few years before her—undermined her own authority. All of human development and psychopathology had been reduced to this earliest oral stage of life. Her claim that disturbances in this realm led to serious mental illness like manic depression—something with which she had had little experience—furthered the impression that Klein was so enamored of her own theories that she had forced the world to conform to them.

It was by now a familiar story in the history of psychoanalysis. The brilliance of many of Klein's innovations and observations might be ignored or lost because she was unable to restrain her claims, which swelled far beyond the observations that first gave them life. As Klein's support began to crack in London, the psychiatrist Aubrey Lewis reported that as he ventured across Europe in 1937, he heard little good said about Melanie Klein.

AFTER BEING SAVAGED at the London Symposium, Anna Freud did not publish much. As Klein published paper after paper and made striking contributions, Anna Freud busied herself with building a robust community of child analysts in Vienna. She had started lecturing on child analysis at the institute in 1926. A year later, she founded an experimental school with Dorothy Burlingham and Eva Rosenfeld, which attracted others to psychoanalytic pedagogy, including the teacher Peter Blos and an artist, Erik Homburger Erikson.

In 1929, Anna started a technical seminar for child analysis, and the same year August Aichhorn, who Anna introduced to the Vienna Society, began to lecture on pedagogy. By 1931, The Vienna Society offered six

courses for pedagogues on the winter curriculum alone, which matched
the number of courses offered for the regular analytic students. Despite
the closing of the Burlingham-Rosenfeld school, the Vienna Society's formidable program in child psychoanalysis continued to grow. In 1932, the
clinic thrived under the direction of Aichhorn and included Editha
Sterba, Willi Hoffer, and one of Aichhorn's protégés, Kurt Eissler.

After the rise of the Nazis, many German analysts came to Vienna to
participate in the stimulating analytic scene, despite the risk that Hitler's
shadow loomed over Austria. Much of the excitement came from child
analysts. In 1934, Willi Hoffer became editor of the new *Zeitschrift für
psychoanalytische Pädagogik*. A year later, when Anna was asked to edit a
special issue of the American journal *Psychoanalytic Quarterly* devoted to
child analysis, she did not have to look far for contributions, tapping Bernfeld, Erikson, Dorothy Burlingham, Bertha and Steffi Bornstein, Anny
Angel, Editha Sterba, and Edith Buxbaum, all Viennese.

By 1935, Anna Freud had become the leader of the Viennese child analysts, but she had shown little promise as a cutting-edge theoretician. Her
published papers were mostly lectures given to nonanalytic audiences,
and they had the light touch of a popularizer. In 1930, Anna published
four lectures Edward Glover considered "tame and uninspired." The
same year she began to plan a book that would change her reputation.
Developing Wilhelm Reich's emphasis on resistances and their structures,
and using conclusions that had been hashed out in the child technical
seminars, Anna focused on the way the "I" defended itself. The project
might have seemed tame in comparison to the daring Klein, but when
Anna Freud published *The "I" and the Mechanisms of Defense*, it profoundly altered the psychoanalytic community.

By 1936, psychoanalysis had an impressive list of high-flying theoreticians who after genuine innovation became so enamored with their own
speculations that they discredited themselves and tumbled back to earth.
As if unaware of the treacherous waters surrounding the unconscious,
they had one by one drowned in excess and empty assertion. They had
coined new terms, new concepts, and new theories, following the dangerous example set by Freud's speculative work on the death drive. Worse,
few had mastered Freud's cautious dialectic. Instead, subtle clinical findings had led to exorbitant claims about the mind, the unconscious, and
human nature.

Melanie Klein was the latest in a distinguished line of analysts to drive off this cliff. Hypothesis became conviction, which then infiltrated observation and became confirmation. Were there no constraints on psychoanalytic theorizing? After the departures of Jung, Adler, and Rank, Freud groused that the dissidents in his movement all succumbed to the same temptation: they took one facet of mental life and tried to make the part into the whole. Between the "Over-I" analysts and character analysts, between an unconscious driven by Thanatos and Eros or libido or orgasms, it seemed psychoanalysis had separated into irreconcilable worldviews. Now, perhaps, there would be a new round of schisms. Alexander, Horney, Reik, Reich, Ferenczi, and Klein would all go their own way, fragmenting psychoanalysis into many small schools. It seemed no one could bring unity to this raucous chorus.

The problem had been observed firsthand by a young Viennese philosopher named Karl Popper. While working with children in Vienna after the war, Popper met both Adlerians and Freudians, and he was amazed that members of each group gave totally dissimilar but seemingly irrefutable accounts on the very same issue. Popper went on to construct a theory of science that insisted scientific theories must be potentially falsifiable. For Popper, psychoanalytic theories were not falsifiable.

Daunting epistemological problems lay at the heart of any psychological science that entertained inner subjective states, including unconscious ones. It was hard to imagine that Anna Freud would address these weighty philosophical matters. Her early forays in scientific debate were not up to the task, Her simple words might have come from a grade school teacher lecturing her students. But Anna Freud had been seriously underestimated. After her drubbing at the hands of the Londoners, she had been quietly learning, discussing, debating, and working with her father and a circle of bright Viennese analysts. Moreover, she had become the heir apparent in Vienna and spoke with authority on the political issues of the day. A new maturity became evident in her prose. Once simplistic, Anna Freud's writing was now clear and forceful.

In the very first page of her new book on the "I" and its defenses, Anna Freud gave a definition of psychoanalysis that included some startling assertions. Seeking to quickly undermine the field's unhappy history of orthodoxy and schism, and put an end to ideological accusations about who was a real analyst, Anna Freud presented herself as a voice for the dissidents who would be drummed out of the field by some narrow, parochial definition of psychoanalysis. Psychoanalysis was more than the study of unconscious psychic life, she insisted. Those like herself who saw analysis as a discipline that encompassed the study of the "adjustment of children or adults to the outside world" were also psychoanalysts. "The odium of analytic unorthodoxy" should no longer apply to the study of the "I," she wrote.

Anna's plea for tolerance can be read as a counterpoint to her father's polemical 1914 history of the psychoanalytic movement. Instead of ideological litmus tests and a politics of exclusion, Anna offered a vision of a community of researchers who traveled beyond unconscious psychosexuality to arenas like "health and disease, virtue or vice." In many ways, she described a change that had been slowly taking place. The new generation of analysts were democrats who espoused both individual freedom and social equality, and the youthful Anna Freud embodied this more open community. Still, if the question was no longer whether you accepted Freud's theory of the unconscious or not, no one had successfully redefined psychoanalysis in a way that grasped its theoretical multiplicity. Who could hold together such a diverse field without simply resorting to personal authority? Was there a fair, rule-bound path between dogma and anarchy?

Anna Freud insisted that her call for tolerance did not neglect the need for standards and minimal commitments. Those commitments would be the same for psychoanalysis as for any science. You did not have to side with Anna Freud or Melanie Klein or Sándor Ferenczi or Wilhelm Reich or Sigmund Freud. You did not have to accept any one definition of the drives in the unconscious. But psychoanalysts should be characterized by a commitment to empiricism. Of course, their object of study, the psyche, had multiple domains that varied greatly in their "accessibility to observation." The "It" was known only indirectly through its effect on consciousness. The "Over-I" had conscious elements that were accessible through self-observation, but also had unconscious elements only known when in conflict with the "I." Anna Freud concluded that the psychoanalyst's field for observation began in all cases with the "I."

Anna Freud had softly but firmly redefined psychoanalysis for many of her generation. The "I" observed the internal processes of the mind, and only through it could one glean information regarding the other forces. In their haste to discuss libido or the death drive or the "Over-I," analysts had neglected the fact that their data had come through a mediating "I." That more empirically observable "I" had to be the central object of psychoanalytic inquiry, she insisted. Psychoanalysis demanded a primary commitment to this object of observation.

Anna then moved on to establish a common method of exploration, and again she demonstrated her gift for making complex problems clear. Sketching out the range of analytic techniques, she surveyed the traditional modes of garnering inferences about the unconscious and argued against all "one-sided" techniques. Without naming names, she mentioned a too-singular focus on resistance (Reich), transference (Strachey and Klein), and dreams and symbols (Hitschmann). Instead of any one of these tasks, she concluded: "It is the task of the analyst to bring into consciousness that which is unconscious, no matter to which psychic institutions. To put it another way, when he sets about the work of enlightenment, he takes his stand at a point equidistant" from the "It," the "I," and the "Over-I." This was Anna Freud's solution to the debates between sex and death, depth analysis and resistance analysis, fantastic dangers and real threats. All were partially right and in their partiality, wrong. After fifteen years of debates, this now seemed obvious to many of her readers, but like many good ideas, it was only obvious after someone had put it into words.

One of Anna Freud's close colleagues, Robert Wälder, buttressed Anna's position. Wälder was a prodigy who began his career in physics, received his doctorate in Vienna in 1921, and then turned to psychoanalysis to conquer his own inhibitions. He and his wife, the child analyst Jenny Wälder, were close to the Freuds and played a leading role in the Viennese scene. In 1930, Wälder published a paper with a curiously abstract title. "The Principle of Multiple Function: Observations on Over-Determination" could have been a paper in theoretical physics, but it was Wälder's attempt to grasp the opportunities offered by "I" psychology to reposition the foundations of psychoanalytic observation and inference. Wälder challenged clinicians to use all available theories in a lighter way to develop a highly specific and nuanced description of a particular individual. The "I" faced at least four "groups" of problems, Wälder asserted—from the drives, the world, the "Over-I," and the forces of the repetition compulsion. Mental life was guided by "the principle of multiple function" whereby each psychic act was an attempted solution to all of these problems. This was only possible because each psychic act had multiple meanings. The full meaning of any psychic act was only exhausted if it had been interpreted as an attempted solution to problems that arose in each of these four domains.

Many had seen neurosis as the result of two conflicting trends and had therefore fastened on a single psychic facet, he argued. Psychoanalysis offered greater possibilities to grasp the "enormous many-sidedness of motivation and meaning." Wälder's many-sided geometry of the mind centered around the "I," which refracted psychic meaning differently in different facets. Psychoanalytic theoreticians had falsely positioned themselves against one another, when in fact they were describing different aspects of the same whole.

Anna Freud adopted the same assumption in a watered-down version. Her notion of an analyst who strove for equidistance from the various psychic agencies opened up the same possibility for greater understanding and theoretical flexibility (if not in Wälder's overwhelming number of axes). Influenced by Reich's attempt to categorize resistances systematically, Anna Freud then described different "I" defenses and linked them to specific kinds of anxiety. With examples from children and adolescents, she delineated the ways people managed threats from outside and inside. She made no divisive claims as to the nature of the unconscious and simply spoke of aggression, without tackling its source. To say sexual and aggressive drives were part of human inner life was not polemical in a Darwinian world. To speak generally about unconscious sex and aggression avoided unnecessarily elaborate and divisive speculation and helped build common ground. Similarly, Anna refused to organize the "I" defenses in a lockstep schema. Even one of her harshest critics, Ernest Jones, had to admit: "There are many places where the author could have carried her arguments and analyses further than she has, but she has preferred not to pass beyond the sphere of what she considers definitely ascertainable knowledge." While chiding Anna for only touching on the Kleinian ideas of defense, Jones recognized the strength of Anna's method:

A more ambitious author would doubtless have proceeded light-heartedly to embark on classifications of the ten varieties of defense

and would have become entangled in all manner of positive assertions regarding the chronological order of their appearance and
their inter-relationships. Anna Freud has clearly and wisely perceived the limitations affecting the present state of our knowledge
and shown the factors and circumstances which must first be clarified before any such attempts can hope to achieve success.

Through her careful observance of the limits of psychoanalytic knowledge, Anna Freud outlined a way to consolidate gains from the many
streams of psychoanalytic inquiry that flowed after 1918. While quietly
arguing against Klein, Anna made a place for Alexander and Reik and
Reich and Nunberg and Hitschmann and many others. She accommo-
dated their views but limited their theories to partial truths that were
pieces in a more complex whole. She had proposed a new common de-
nominator for psychoanalysis, the "I." If you accepted this, you could
pursue many different numerators without losing your place as a psychoanalyst.

Even Ernest Jones admitted that the book became a classic overnight,
and brought "order and clarity into the prevailing confusion." The Viennese lay analyst Ernst Kris applauded Anna's emphasis on empiricism
and her working from the surface downward. Kris was especially enthusiastic about the way Anna suggested a research program that he hoped
might create a fertile alliance between the psychoanalysts and academic
psychologists. Child analysts had hoped for such a marriage, but needed
to prove their empirical, scientific value to win over psychologists. In
Vienna, such a union looked plausible when the university's department
of psychology began to allow more of its students to study at the analytic
society.

With tact and restraint, Anna Freud provided a proposal that would
realign and consolidate much of psychoanalysis. Firmly, she spoke of the
requirements of empirical science, not allegiance to leaders and teachers.
One had to start with observables, with the "I," and the known. And one
must do this, not because of Anna Freud's personal authority, but rather
out of a commitment to the rules of scientific observation. By tending
carefully to what Glover called her "scientific reserve," Anna Freud undercut concerns about her own lack of scientific training, much as Melanie Klein had heightened those concerns with her excessive theorizing.

Marianne Kris remembered that in Vienna, Anna's book had an enormous impact. She further recalled Heinz Hartmann and Robert Wälder warmly praising the work to Sigmund Freud, who testily replied that Anna's book was what he had been saying all along.

By 1936, eighteen years of extraordinary theoretical expansion coupled with the destruction of the Berlin Institute resulted in the consolidation of two rival psychoanalytic perspectives, both of which grew out of the study and analysis of children. The English were led by Melanie Klein along with Joan Riviere, Susan Isaacs, James Strachey, and more importantly, the most powerful patron in organized psychoanalysis, the president of the I.P.A. and editor of its journal, Ernest Jones. Klein's theories developed into a model of Object Relations and focused on how projection and introjection led to a fantastically peopled inner world. It was a theory that was highly attuned to the manifestations of human sadism and aggression, and the constitutive role of fantasy.

The Viennese more loosely organized themselves around the new "I" psychology, as framed by Anna Freud. Anna Freud began this integration with the argument that analysts should pay primary attention to the "I" as defined within what would be called a structural model of the mind. Instead of concentrating on whether mental contents were conscious or unconscious, this model hoped to understand the conflicts among internal mental structures—the unconscious drives, the "I," and the "Over-I." Anna Freud's approach kept the interpretation of the unconscious primary, but it left open what was to be interpreted. She made no specific claims about the nature of the unconscious.

Unlike Klein, Anna Freud placed experience at the heart of psychic development. For her, inner instincts and fantasy were powerful, but experience in the world could not be underestimated. The child "fears the instincts because it fears the outside world," she reasoned. In this, Anna was speaking from her own experience. The aftermath of the Great War exposed many Viennese child analysts to starving and abandoned war orphans. For these analysts, a general psychology that had no prominent place for really losing one's parents or starving was unimaginable. When anxiety was objective and came from the world, it weakened the child's defenses and only made the child's suffering worse. In situations like that, one did not attack the defenses, Anna reasoned, one changed the environment.

After 1936, the British and the Viennese vied for the loyalty of psychoanalysts. In the contest of ideas that followed, each theory would be scrutinized for its clinical effectiveness, theoretical subtlety, and epistemological soundness. Each school would also have to negotiate amid political disaster; in the harsh years ahead, defining psychoanalysis's political implications became a matter of survival.

III.

HITLER'S RISE TO power initially caused no great alarm in the major capitals of Europe. He wasn't the first thug to take power, and many reasoned he would be sated by the pleasures of victory and tamed by the requirements of governance. As Hitler consolidated his hold on Germany, however, a feeling of crisis deepened throughout Europe. Under Hitler, the rule of law and the freedom of the press were rapidly undermined. Declaring that the nation was imperiled by a Communist takeover, Hitler began to remilitarize Germany. Bizarrely, he equated Bolsheviks with Jews and stoked the populace's anti-Semitism. In the fall of 1933, the Nazis burned books by prominent Jewish authors, including those written by Sigmund Freud. In 1935, the Nuremberg Laws deprived German Jews of their rights as citizens. By then, most leftist and Jewish analysts had fled Berlin; the few that remained were forced to resign from the psychoanalytic society. In July 1936, the Psychoanalytic Institute in Berlin was dissolved and incorporated into the Nazi Mattias Göring's German Institute for Psychological Research and Psychotherapy. As Joseph Roth put it, Germany had embarked on an "Auto-da-fé of the Mind."

In Austria, Sigmund Freud watched the growing menace with helplessness and dismay. In February 1934, he wrote Arnold Zweig, predicting: "something is bound to happen. Whether the Nazis will come, or whether our own home-made Fascism will be ready in time, or whether Otto von Habsburg will step in, as people think." Either way, the future looked dark. Freud believed psychoanalysis could "flourish no better under Fascism than under Bolshevism and National Socialism."

In these perilous times, even Freud, a man who proudly saw himself as a disturber of mankind's sleep, contemplated censoring himself. He had turned his attention to anti-Semitism, but he was worried that publishing a work on the subject might result in the banning of psychoanalysis by the

Austrian authorities. Freud had witnessed the way Austrian clerical forces had worked with the Vatican and Italian fascists to shut down Edoardo Weiss's psychoanalytic journal in Italy. If he and his colleagues were accustomed to public hostility, they operated under the assumption that if their work was sanctioned as medical and scientific, it would not be outlawed. "I am a liberal of the old school," Freud wrote wistfully to Zweig. The old liberal had mostly tried to buffer his science from direct political engagements, fearing they would harm the objective status of the field, and in return, he expected the reigning political forces to let his science be. But with Dollfuss in Austria, Hitler in Germany, Mussolini in Italy, and Stalin in the Soviet Union, it had become impossible to shield rational scientific inquiry from politics in Europe. Theories once evaluated for their inventiveness, reliability, and clinical usefulness now were scrutinized for their politics.

By 1933, the political meanings to be found in psychoanalytic theory had diverged and multiplied. In the early years of the Freudian movement, the topographic theory of mind encouraged liberationist ideals, but with the death instinct, some analysts began to press analytic theories that had socially conservative implications. These differences became apparent when psychoanalysts entered forensic debates. In 1916, Freud had described criminals who acted out of guilt, those "pale" ones who broke the law in order to be punished. In Berlin, thirteen years later, Franz Alexander and Hugo Staub proposed that a great majority of criminals were neurotics who succumbed to a masochistic need for punishment. This perspective suggested that social and political reform would do nothing to prevent crime. In London, Kleinian analysts like Edward Glover and John Rickman contended that criminals lacked the ability to temper their innate aggressiveness. By 1934, Melanie Klein believed criminal impulses were inborn and only marginally affected by matters such as degrading surroundings.

These positions enraged leftists like Wilhelm Reich and Otto Fenichel, who were convinced that social repression, poverty, and desperation had some relationship to aggression, law breaking, and crime. The socialist analyst Ernst Simmel countered these conservative perspectives by declaring in 1932 that the "hellfire of collapsing capitalism" produced aggression, delinquency, and crime.

By 1933, it was legitimate to ask whether psychoanalysis was intrinsically a liberationist social and political theory or whether it was a drive-based psychology that viewed social problems as inherently psychological, absolving social structures of all responsibility. This question took on urgency as the survival of psychoanalysis appeared to hinge increasingly not on its broad cultural appeal or scientific standing, but rather on its political identity.

As the threat mounted, psychoanalysts tried to fathom the rise of dictatorial rule. Why would men and women willingly abdicate their own freedom to tyrants? In 1921, Freud had explored such willful enslavement in his kaleidoscopic *Mass Psychology and "I" Analysis*. He argued that the freedom of an unruly and riotous mob was not caused by a simple loosening of self-restraint, but was dictated by the intense interpersonal bonds that constituted a community. Members of an army shared a common "I-ideal" based on their love of their leader. They followed their leader as if he were their own idea of perfection, obeying his orders as if these commands came from themselves. Mass movements were driven by a human need for identity and could easily result in the abdication of individual freedom. Liberal ideals could be swept away by attachments to nation, church, or army.

In the late 1920s with crisis hanging in the air, more European analysts began to practice psychoanalytic sociology. The most prominent center was in Frankfurt, where the Psychoanalytic Institute operated alongside that university's Institute for Social Research. The Institute for Social Research teemed with brilliant minds that hoped to integrate Marx and Freud. Run by Max Horkheimer, an analysand of Karl Landauer's, the "Frankfurt School" included Herbert Marcuse, Theodor Adorno, Karl Mannheim, and Erich Fromm, who headed the social psychology unit. At the 1929 opening ceremonies for the Psychoanalytic Institute, Fromm gave his vision of psychoanalytic sociology, and he went on to study the fear of freedom that encouraged an individual to lose himself in a mass.

As conflict loomed, the League of Nations' Institute for Intellectual Cooperation encouraged psychoanalysts to take on social problems, especially those related to warfare. They sponsored the famous exchange on the nature of war between Albert Einstein and Sigmund Freud, as well as a series of lectures in London by Edward Glover. Glover's lectures on war, pacifism, and sadism offered a simple but powerful Kleinian analysis. He contended that introjection created the melancholic whose aggression was directed at himself, and projection resulted in the paranoiac, who attacked others. The first group tended to be suicidal pacifists, and the second group homicidal militants. In this way, men and women did not simply fight for their own conscious interests, but more sinisterly fought out of irrational, unconscious need. Glover reserved his harshest words for the pacifists. In 1933, pacifism was a powerful political sentiment in England, despite the growing belligerence of Adolf Hitler. Glover spoke witheringly about the suicidality of those who would avert war in the service of their own unconscious self-destructiveness.

Glover highlighted how elegantly the Kleinian model linked individual psychology to politics and war, but at the same time he exposed a weakness in the Kleinian position. With its radical emphasis on drives, the Kleinian model had no place for historical specificity. By this account, men and women should always be at war. Why war now? Why war here? Those were questions the Kleinians simply could not address. If war was fundamentally drive-based and neurotic, then social injustice, nationalism, militarism, racism, economic turmoil, corruption, and a host of other factors were all irrelevant. If Glover was right, the Treaty of Versailles had nothing to do with making Nazi Germany. Glover's commandment: "Know thine own (unconscious) sadism" might have been a wake-up call to the pacifists in his own country and a prescient attack on those who would support Neville Chamberlain's policy of appeasement, but it was hollow for those in Europe who looked at Nazi Germany and wondered what had happened to the land of Goethe and Schiller.

Viennese analysts were quick to make this point. Robert Wälder had joined the League of Nations' Institute for Intellectual Cooperation, hoping to explain the growing irrationality that had gripped Europe. In 1935, he traveled to London to speak to the Royal Institute of International Affairs, but took time out to deliver a withering critique at the British Psychoanalytical Society. Wälder went directly for Melanie Klein's Achilles heel—the evidentiary vulnerability of a theory predicated on claims about the inner life of a four-month-old—but he also stressed that no child's inner life could be understood without examining its social situation. In Wälder's opinion, the exiled Marxist analyst Wilhelm Reich had taken an extreme position when he said all neuroses were caused by social phenomena, but Melanie Klein had veered to the opposite extreme, giving far too much emphasis to biology and fantasy.

In Vienna, Wälder was part of a close-knit group that longed for a psychoanalytic model that did not succumb to the "temptations" of Reich or Klein. The man who would become most associated with this effort was Heinz Hartmann. Born in 1894, Hartmann came from an impressive Austrian family. His grandfather had been a liberal legislator, who was forced to flee to Switzerland after the failed revolutions of 1848; his father was a respected professor of history who, during the Weimar years, served as Austria's ambassador to Germany. Hartmann's mother was the urbane daughter of a famous Viennese physician, Rudolf Chrobak. As a boy, Hartmann discovered Freud's book on wit on his father's bookshelf; after reading it, he decided to become a psychoanalyst.

In 1920, Hartmann studied social science with Max Weber before completing his medical studies at the University of Vienna. He then secured a position as a Second Assistant at the Neurological-Psychiatric Clinic under Julius von Wagner-Jauregg. For the next fourteen years, Hartmann worked at the clinic alongside the prolific psychiatrist and psychoanalyst, Paul Schilder, who was busily exploring the boundaries between neurology, psychiatry, and psychoanalysis. Hartmann joined the Vienna Psychoanalytic Society in 1925; by then, he had conducted research on hallucinations, depersonalization, body image, parapraxes, and obsessive-compulsive disorders.

In 1926, the ambitious Hartmann decided to address the scientific standing of psychoanalysis. Armed with a Rockefeller grant, he headed to Berlin on a six-month sabbatical, undertook a personal analysis with Sándor Radó, worked in the psychology lab of the Gestalt psychologist Kurt Lewin, and wrote *The Foundation of Psychoanalysis*. Psychoanalysis, Hartmann argued, was a unique natural science because it had a special methodology. It was distinct from the humanities insofar as it searched for general laws, and distinct too from competing psychologies, especially those that relied on understanding. Psychoanalysis, on the other hand, offered explanations.

Returning to Vienna, Hartmann entered the elite circles of the Vienna Society. Already close to Anna Freud, Hartmann and his wife also became part of Felix and Helene Deutsch's Black Cat Club, joining Robert and Jenny Wälder, Ernst and Marianne Kris, Edward and Grete Bibring, and Willi and Hedwig Hoffer. In 1928, when Paul Schilder left Vienna to join Adolph Meyer at Johns Hopkins, Hartmann became the most academically well positioned of the Viennese analysts. Despite his success, Hartmann and his Jewish wife, Dora, considered immigration after the events of 1933. Distressed at the prospect of losing yet another potential Viennese leader, Sigmund Freud offered to take Hartmann into a personal analysis if he stayed in Austria. Hartmann had already been analyzed, but it was an offer he could not refuse. In 1933, Hartmann went into analysis with Sigmund Freud, abandoned his academic career at the psychiatric clinic, and opened a full-time psychoanalytic practice.

That same year, Hartmann returned to the questions that preoccupied him during his sabbatical. What was the essence of psychoanalysis? The question had changed shape since 1926. The challenges posed by competing social sciences, neurobiologies, and psychologies had been supplanted by pressures exerted by political ideologies. Hartmann had found himself in the middle of one such political tug-of-war when Freud asked him to be one of the replacements for Otto Fenichel, who the Professor had sacked as editor of the *Zeitschrift* after Fenichel sided with "Reich's Bolshevism." By 1933, Wilhelm Reich, Otto Fenichel, and others wanted to secure a Marxist psychoanalysis. Carl Jung seemed to throw his weight behind a racial psychology, and in a dystopian vision of what the future could hold, Mattias Göring had called for a *Mein Kampf*–based mind science. Could there be several psychoanalyses: one revolutionary, one racist, another fascist? By what criteria could these politicized studies of the psyche be rejected? Was psychoanalysis a natural science that could rightly exclude such intrusions, or was it simply another "Weltanschauung"?

For over a century scholars had debated the meaning of *Weltanschauung*, that densely textured German word for "worldview." During the years of the Weimar Republic, the word was broadly used to discuss what appeared to be missing. There was a communal need for a system of meaning in a world where science had dismissed the self and Nietzsche had declared God dead. Ideologues on the extreme right and left agreed that a new worldview was required. As dictators stepped in to fill this need, psychoanalysts wondered if their science could also fill the gap.

In 1929, Robert Wälder argued psychoanalysis could be a new Enlightenment Weltanschauung. It was a rationalism that "attempted to rescue and make accessible to its scrutiny the truths and ethical values inherent in irrationality." Psychoanalysis was an amendment and corrective to the Enlightenment's narrow insistence on reason. In 1933, Freud rejected this assertion. Psychoanalysis had no particular Weltanschauung, but was simply part of the scientific worldview. Psychoanalysis was ordered by the rules of science, and stood in opposition to religious and Marxist worldviews.

In 1933, Heinz Hartmann also weighed in on whether psychoanalysis was a Weltanschauung. Unlike Freud, Hartmann refused to erect a wall between science and politics. It was futile for psychoanalysis to insist that it had no moral or political values embedded in it, impossible to speak of psychoanalysis as removed from all social and political aims. Instead, it behooved analysts to clearly state their assumptions.

Psychoanalysis was based on rationalism, empiricism, and naturalism, and its moral code was the increase of happiness and the relief of suffering. Psychoanalysts might offer society technical means for reaching its aims, but psychoanalytic knowledge in itself could not and should not be twisted to conform to the ethics of a community. For those who sought to transform psychoanalysis into a political ideology, this was Hartmann's curt reply: Reich was wrong, Jung was wrong, Göring was wrong. There was no Marxist psychoanalysis, no Jewish psychoanalysis, there could be no Nazi psychoanalysis. There was only psychoanalysis. Or rather, a rationalist, empiricist, and naturalist psychoanalysis.

How could such an endeavor survive when all around Europe, political radicals demanded that science and art serve the revolution? In Nazi Germany, Einstein's work was dismissed as Jewish physics. In the Soviet Union, Marxist-Leninist theory had become the objective science of human action, and by 1931 Pavlov's reflexology had been crowned the only true psychology, true that is to the revolution. Between revolution and counterrevolution, was it possible not to take sides?

This dilemma was skillfully treated by one of the Frankfurt School sociologists, Karl Mannheim. In *Ideology and Utopia*, Mannheim argued that since human consciousness was permeated with social life, the social context for any truth claims were essential. Knowledge was always situation-bound. Each age produced its own truths, its own Weltanschauung. These could be categorized as ideologies that used excessive idealizations of the past and overemphasized stability to justify the status quo, or utopias that encouraged fantasies of a future that justified radical change. The only hope for a path between ideology and utopia came from a band of unattached intellectuals who could free themselves from their personal beliefs and more clearly comprehend their world.

Heinz Hartmann was profoundly influenced by Mannheim. Like his grandfather and father before him, Hartmann was a liberal. He railed against scientists who compromised their work to support their politics, a corruption Hartmann had already detected in psychiatric debates on heredity. Psychiatry and psychoanalysis, he believed, should not be a disguised form of politics. But in a polarizing world preparing for war, where was that path?

Hartmann thought Anna Freud's work was a good place to start. He had befriended Freud's daughter in the days when she went on educational visits to Wagner-Jauregg's clinic and was excited by her description of the "I" as mental ringmaster. This approach, he believed, could restore wholeness to an otherwise atomized self. Anna's demand that theoreticians recommit themselves to the authority of empiricism and therefore closely study childhood development, also tallied with Hartmann's hopes. Hartmann later recalled that he strove to build conceptual bridges that would link child observers from academic psychology with psychoanalysts.

In addition, Hartmann sought out common ground that placed individuals in their societies, and biological organisms in an environment. On November 17, 1937, at the Vienna Society, Hartmann previewed his thinking; two years later he published a monograph called *"I" Psychology and the Problem of Adaptation*.

Hartmann's notion of adaptation was neither Darwinian nor Lamarckian; he was not concerned with species adaptation but rather individual adaptation. To develop such a theory, Hartmann referred back to his own research on identical twins. In the 1920s, scientists realized that identical twins raised in different environments offered an extraordinary opportunity to tease apart the contributions of nature and nurture. Hartmann took up twin study; he found that differing environments did not affect the intelligence of the twins much, but it did have impact on their character. Hartmann concluded that while intellect seemed biologically endowed, character structure was created through adaptation to a given social world.

Like Freud's notion of psychosexuality three decades before, Hartmann's concept of adaptation was a synthetic one that integrated outer and inner forces, drives and social demands, biology and politics. Such a perspective seemed to expand psychoanalysis into a more full-fledged psychology, throwing light on mental functions like memory, cognition, rational action, will, and perception. Analysts previously attended to these things only when they seemed compromised by unconscious forces. Building on Wälder's work, Hartmann insisted that only psychoanalysis could contend with the way these mental activities normally served multiple functions, including healthy adaptive ones. Intellectualization could act as a defense against instinctual dangers and be a force for maintaining the status quo, but it was also an essential requirement for comprehending reality. Similarly, a defensive retreat into fantasy could not be reduced to a denial of reality alone; it might also hold the seeds of imagining new possibilities and serve as an adaptive kind of problem solving.

In this description, Hartmann mirrored Karl Mannheim's framework. But between psychological defenses and wishes, between social ideologies and utopias, where was the way to a clearer apprehension of the world? Like Mannheim's isolated band of dispassionate men and women, Hartmann made a limited place in the mind, a "conflict-free" sphere of the "I," uncompromised by drives and defenses. In this realm of the mind, moral and political action could—at least in theory—be based on reason and reality.

Adaptation was the compass Hartmann used to steer "I" psychology between the more extreme positions of Melanie Klein and Wilhelm Reich. Adaptation required that some psychic agency respond to both Klein's inner demons and Reich's social privations. Autonomous "I" functions served this need and were united under a common organizing goal: individual adaptation. Man molded his environment and was molded by it, which created a seamless continuum between biology, psychology, and sociology. Defending against dangerous impulses, helplessly clinging to mother, learning to walk and count, surviving in wartime: all these were adaptations.

Hartmann's approach broadened psychoanalysis so that it took in cognitive processes normally reserved for academic psychology. By adding a nonconflictual autonomous sphere to the "I," Hartmann redefined the boundaries of psychoanalysis so as to make it fit with developmental biology, psychology, and sociology. The conflict-free "I" also carried a clear political message. If some Marxist analysts saw men controlled by social forces, if Melanie Klein put them at the mercy of their drives, Heinz Hartmann sought to restore the dialectic between inner and outer and in the process, posit a conflict-free domain that allowed for the possibility that men and women, in some limited way, could be free. Men and women were not mechanical dolls nor were they just animals. They were potentially more, thanks to a restricted but nevertheless real capacity for creativity and adaptation.

Hartmann's work was itself an adaptation to the problems created by the collapse of liberalism in Europe. He insisted there was a place in the mind in which reason and free will might reign, following the less successful efforts of his friend Robert Wälder, who in 1934 tried to define psychoanalysis itself in terms of human freedom. While Freud had declared that the goal of analysis was: "where It was, there I shall become," Wälder, the theoretician of multiple functions, could not help but admit that freedom in one domain might mean something different in another. If health was an elaborate compromise between many needs, where was freedom in that? To get out of this quandary, Wälder defined human freedom as a function that came with a functional "Over-I." This self-scrutinizing and self-regulating part of the mind allowed man to rise above being an instinct-driven animal.

This was an old answer dressed up in new clothes. Philosophers had long believed that moral and ethical self-reflection distinguished human beings. And there was something immediately problematic about Wälder's view for his colleagues. Following Alexander, many thought of the "Over-I" as charged with unconscious aggression. For them, it operated irrationally, even brutally. In Wälder's hands, it had become the source of human goodness. Wälder's strained attempt to account for human freedom highlighted the soundness of Hartmann's decision to posit a new autonomous sphere in the "I," rather than succumb to a similar embarrassment.

Heinz Hartmann took political and philosophical notions like autonomy and freedom and redefined them for "I" psychology. It was not at all clear how much clinical leverage these formulations offered, but that was not the point. With this repositioning, the Viennese psychoanalyst presented an image of his field that did not flirt with "that malady of the times whose nature it is to worship instinct and pour scorn on reason," a trend, Hartmann noted, that had "assumed a highly aggressive and political complexion."

In 1936, at the Fourteenth Congress in Marienbad, President Ernest Jones welcomed the 198 attendees to democratic Czechoslovakia, this "island of freedom" without dictators where free inquiry made scientific work possible. Heinz Hartmann mapped out a small pocket of freedom inside the psyche and saw psychoanalysis defending and expanding that place. He championed the potentials of human liberty while islands of freedom in Europe were being devoured. For Hartmann, Wälder, and their followers, psychoanalysis was a way to help individuals not conform but find their own adaptations in complex, changing environments. Their theory would soon be put to the test, for much was soon to change in Austria, Czechoslovakia, and the rest of Europe.

WHILE PROMINENT BRITISH and Viennese psychoanalysts were delivering their psycho-political views to august bodies like the League of Nations, the Marxist analysts were whispering in the dark. The Soviets had condemned them, and the Nazis had banned their political parties. The I.P.A. had denounced them. Many lived in exile and needed to exercise great caution if they did not want to be left wandering the unsafe borders of Europe.

They too were trying to redefine psychoanalysis so as to include at its core the human values of, not liberalism, but Marxism. From a new home in Oslo, Otto Fenichel worked to keep the scattered Marxist opposition alive. Fenichel had always been an organizer, a consensus builder who brought people together, and his commitment to Marxism was strong. In his steadfastness, Fenichel had insisted on Reich's right to publish his paper critiquing the death drive and was rewarded with the loss of his job as editor of the *Zeitschrift*. Freud's autocratic reaction made it easy for Fenichel to take the high ground and appear dignified. In addition, he had many allies who held his encyclopedic knowledge and prodigious work in high esteem. After the Nazis took over Germany, Fenichel fled to Oslo; from there he pulled together his epistolary community of Marxist analysts. He had become convinced that all of Europe would be swept into fascism, and like it or not, the future for psychoanalysis was in America. For Fenichel, this was another kind of disaster, for he believed that the state of psychoanalytic knowledge in the United States was abysmal.

Otto Fenichel with Bertha Bornstein (*on the left*) and an unidentified woman at the 1936 Marienbad Congress. After fleeing Berlin in 1933, Fenichel was at the center of an underground community of leftist psychoanalysts.

In exile, Fenichel tried to preserve the field he loved from the twin perils of Nazi fascists and American know-nothings. The result of his labors was *Outline of Clinical Psychoanalysis*, which Fenichel sent off to be published in the U.S. In this encyclopedic work, Fenichel managed to integrate a huge number of clinical papers by a vast array of authors from Freud, Abraham, Rank, Jung, Jones, Ferenczi, Alexander, Radó, Horney,

Reik, Reich, Schilder, Aichhorn, Klein, Federn, Nunberg . . . the list went on and on. When the nearly five-hundred-page book appeared, it was as if some modern-day Noah had tried to herd every psychoanalyst onto an ark in preparation for the Flood.

At the same time, Fenichel's small band of leftist analysts feuded, especially about Reich. After 1934, Reich placed ever more weight on the primacy of orgasm as well as muscular and physical bodily energy. Reich's aggressive technique with patients had also begun to worry Fenichel. He had seen patients so emotionally bruised by Reich's relentless attacks that they had fallen into depressions. Brushing aside such concerns, the supremely confident Reich pressured followers to become Reichian analysts. Georg Gëro refused to give such an oath of loyalty. Reich—Gëro wrote to the members of the circular letter—demanded true believers, and attacked those who wished to engage in more, open scientific inquiry and discussion. In the spring of 1935, there was a total break between the Reichians in Norway and those who maintained membership in the I.P.A., like Fenichel.

Otto Fenichel moved to Prague and began to work more closely with the Viennese. Even if most were not politically radical like his Berlin colleagues, Fenichel believed a united front could be made with the Viennese against common enemies, in particular the reactionary theories of Melanie Klein. In February 1935, Fenichel reported to his secret correspondents that Anna Freud had taken a decisive stand against the English school. Fenichel himself entered this battle and accused Edward Glover of glibly discussing war in a way that was divorced from reality. The Viennese began to court Fenichel. Between Reich's demands for intellectual submission and the reactionary Kleinians, Otto Fenichel had nowhere else to turn, despite his view that the Viennese were mostly bourgeois liberals and not at all revolutionary. In private, Fenichel continued his attempt to fuse Marx and Freud, but publicly he strove to preserve psychoanalysis for the future.

Intrigued by Anna Freud's synthesis, Fenichel asked the Vienna Society if he might come from Prague to lecture on these ideas and lead a class on weekends at the Society. In October 1936, Otto began teaching psychoanalytic technique. A number of analysts had taken a stab at pulling together a textbook on the subject since Ferenczi and Rank's book appeared. For example, in the summer of 1932, Edward Glover developed an elaborate questionnaire to poll analysts on their views on technique, including active method, neutrality, to what extent theory drove interpretation, and whether it was permissible to smoke during sessions. He hoped to collate the responses and produce a textbook that reflected actual technique in practice. However, Glover's project became obsolete before he completed it. His questionnaire made no mention of Reich's resistance analysis, Anna Freud's work on defenses, or Melanie Klein's transformation. When his work was published eight years later, it was useless.

In 1936, Otto Fenichel began to form a coherent psychoanalytic technique. In his class at the Vienna Institute, he sorted through others' positions and moved his way forward in a dialectical manner. Quoting Ferenczi, Fenichel told his colleagues that the theory of technique would not give iron-clad rules but rather advice. He also firmly rejected Theodor Reik's argument against the very possibility of a formalized technique. Somewhere between prescriptive commandments and no guidance at all, a method could be found. Echoing Ferenczi and Rank, he argued against theories that focused on talking too much, but had little sympathy for those that insisted on feeling too much. He laid out arguments for some forms of abstinence but did not advise Ferenczi's active treatment. He emphasized transference and resistance analysis, but rejected those who wanted to restrict analysis to that. He praised the pioneering work of Reich but criticized his method of shattering character defenses.

To organize these many positions, Fenichel turned to Anna Freud's idea of equidistance, asking that analysts not align themselves with any single sphere of the mind. In seeking such neutrality, he cautioned analysts not to become wooden and lifeless. They should not try to be mirrors. "The patient should always be able to rely on the 'humanness' of the analyst," Fenichel cautioned, in a way that made it clear he believed some patients were not always able to do so.

The Marxist also weighed in on the place of adaptation in psychoanalytic technique. Without mentioning Hartmann, Fenichel cautioned that some analysts saw their patients as cured when they had come to regard their circumstances as unchangeable. This profoundly conservative vision of adaptation—which could not be fairly ascribed to Hartmann's work— was wrong, Fenichel cautioned. Adaptation should be measured not by capitulation to social circumstance, but rather by the individual's capacity to judge reality and act accordingly. Later Fenichel would review Hartmann's work and appreciate his view of adaptation as in part dependent on "the social structure of the particular society."

In under 150 pages, Fenichel produced an unprecedented synthesis of psychoanalytic technique that reached beyond earlier, more narrow technical treatises by the likes of Glover and Ella Sharpe. Otto Fenichel was a man politically and personally committed to the commune, the common purpose. As such, he saw himself as a orchestra conductor, not a prophet. After he finished his manuscript, Fenichel sent it to numerous correspondents and fielded complaints from many, including Reik. While some in Budapest said he had slighted their innovations, the most glaring omission was the absence of a discussion of Melanie Klein's latest work. Like Anna Freud, Fenichel had given voice to many positions, but for the followers of Melanie Klein, there had been only what the Viennese called death by silence.

Otto Fenichel worried that with the fall of Europe, psychoanalysis would disappear into a dark age. He sent his little book to the New World, and in 1938 *Problems of Psychoanalytic Technique* was published there. It became a classic guide for those Americans who became psychoanalysts over the next fifty years. The book would not be published in Europe, for psychoanalysis on that continent, as Fenichel feared, was dying.

Fleeing Nazi Vienna, Sigmund and Anna Freud arrive in Paris in 1938.

IV.

On March 11, 1938, Austria fell to Adolf Hitler. As Nazi tanks rolled into Vienna for the long-awaited *Anschluss*, they were welcomed by roaring crowds. Shockingly, the rest of the world did nothing. After several tense weeks, on June 4, 1938, Sigmund Freud boarded the Orient Express and left Vienna for good. The interpreter of King Oedipus's tragedy was led into exile by his faithful daughter, "Anna-Antigone" as he once called her. What the elderly Freud left behind was an Austria that had shown the German Nazis a thing or two about how to terrorize Jews. Impressed by the shameless manner with which Austrians beat and robbed their Jewish neighbors, the authorities in Berlin would soon pass expropriation laws to steal whatever they could from their own Jewish population. Hitler and his leadership also began to execute a plan to systematically eliminate Jews from Europe. Freud and other fleeing psychoanalysts were leaving behind a holocaust.

Overnight, few remnants of psychoanalysis remained in its birthplace. The Vienna Society had disbanded. One of the few non-Jewish members of the Society, Richard Sterba, had been recruited by Müller-Braunschweig of Berlin to follow in his footsteps and become leader of an Aryanized Vienna Society. Sterba was also informed that the Society, once rid of its Jewish members, would be incorporated into Göring's Institute. Ernest Jones himself seemed eager for Sterba to collaborate and was enraged to find that by the time he arrived in Vienna to help Freud and other Jewish analysts, Sterba had fled to Switzerland. Müller-Braunschweig sent an appeal to Sterba, who again refused to participate in the scheme. In the spring of 1938, organized psychoanalysis was no more in Vienna.

On the train from Vienna to Paris, Sigmund Freud retraced the journey he had taken over half a century earlier when he traveled to France in search of a future. He was no longer a young, failed scientist, but "Freud," a world-famous thinker and physician. He would not arrive in Paris unknown and alone, but would be accompanied by family and members of the discipline he had done so much to create. In Paris, he was greeted by the press and Princess Marie Bonaparte. Freud owed his freedom in part to Princess Marie, whose ransom, along with the diligent work of Ernest Jones and the intervention of powerful friends like William Bullitt, the American ambassador to France, and the British home secretary, Sir Samuel Hoare, secured Freud's freedom.

And so, Freud was back in Paris, the home of Ribot, Charcot, and Janet, the place that had once given him inspiration. He found himself in the bizarre position of knowing there was no psychoanalysis in Vienna, while long recalcitrant Paris had a growing psychoanalytic group built around medical analysts and a new generation of aesthetic radicals called Surrealists. After a short stop in Paris, Freud and his entourage continued on, for to the dismay of Melanie Klein and her followers, the Professor and his daughter had decided to settle in London.

As Freud sailed from the European continent, he followed many European analysts who had already fled, and he preceded a mass of psychoanalysts who would soon flee Austria, Czechoslovakia, Poland, Denmark, France, and Hungary. Psychoanalysts began to stream onto foreign docks. In London, Anna Freud arrived with Willi and Hedwig Hoffer, Robert and Jenny Wälder, Edward and Grete Bibring, Ernst and Marianne Kris, Otto and Salomea Isakower, and a group of other members of the Society, including Freud's physician, Max Schur.

In New York, after failing to secure a hospital position in Chicago, Detroit, or Toledo, the Freudian stalwart Paul Federn joined the New York Psychoanalytic Society. Freud's cherished student Heinz Hartmann made his way to New York, where he was given a visa because a position had been created for him at the Bank Street School for teachers. Edith Jacobssohn now "Jacobson," the Berlin Communist analyst and colleague of Fenichel's whose anti-Nazi activities had resulted in over two years of imprisonment, escaped and made her way from Prague to New York. Rudolph Loewenstein, one of the founders of the Paris Psychoanalytic Society, joined the New York group, as did the Hungarian analyst and anthropologist Géza Róheim. After 1938, a number of Viennese child analysts also joined the New York Society like Berta Bornstein, Marianne Kris, not to mention the Hungarian Margarethe Mahler-Schönberger, and René Spitz, who came by way of Paris. The leader of Polish psychoanalysis, Gustav Bychowski arrived in New York in 1939 along with the leader of Italian psychoanalysis, Edoardo Weiss. The Communist Ludwig Jekels, Freud's card-playing partner, came via Australia, as did the analyst who had provoked the battle over lay analysis, Theodor Reik. Between 1938 and 1941, this astounding collection of analytic talent joined the New York Psychoanalytic Society.

Others made their way inland and up the coast: Grete and Edward Bibring settled in Boston, along with Eduard Hitschmann and Robert

and Jenny Wälder (though they would soon move to Philadelphia). Richard and Editha Sterba went to Detroit in 1939. Siegfried Bernfeld went to
San Francisco, and Otto Fenichel joined his Berlin friends Ernst Simmel
and Hugo Staub in Los Angeles.

As analysts escaped the European continent, they found themselves
joined by old allies and enemies. Wilhelm Stekel fled to England in 1938
and committed suicide two years later. In 1935, both Alfred Adler and
Otto Rank came to the United States; Adler died in 1937, and two years
later, Rank, Eugen Bleuler and Havelock Ellis were dead. Wilhelm Reich
managed to escape to the United States. When he disembarked at New
York's harbor, he was greeted by an old analysand who was shocked by
the change in his analyst. Reich, he recalled, looked broken and depressed.
Wilhelm Reich would set up a lab in his basement in Forest Hills, Queens,
and come to believe that he had discovered the energetic source of all life,
which he named Orgone. His new Orgone-based analysis would move
Reich far from psychoanalysis and win him new followers as well as the
attention of the American authorities, who imprisoned him for fraud. He
died in jail in 1957.

By 1940, the psychoanalytic community in continental Europe had
been wiped out. The great network of individuals and organizations that
sustained the training of psychoanalysts, their scientific publications, and
their clinical practices was gone. Psychoanalysis had lost its birthplace.
The mass exodus marked the end of a thriving psychoanalytic community that had come of age after the Great War and blossomed in Berlin,
Vienna, and Budapest as well as many smaller centers. Of the great centers, only London remained.

These exiles lost their birthplace and their mother tongue. Psychoanalysis had always paid great attention to linguistic nuance, in the belief that
these subtleties of expression held clues to the psyche. Now, immigrants
with strong accents worked in foreign lands. Furthermore, the language
that analysts had created to describe and name the inner world also began
a new life in translation. In 1941, no German-language psychoanalytic
publications existed. The *Zeitschrift*, *Imago*, the *Almanach der Psychoanalyse*,
Psychoanalytische Bewegung, and *Zeitschrift für psychoanalytische Pädagogik*
died. As Hitler's victories piled up, the language of *Psychoanalyse* would
have to survive in other tongues.

The translation of psychoanalytic terms and concepts had a long history in France, America, and England, but before 1938, these translations were auxiliary and did not threaten to fundamentally alter the semantics of the field. The collapse of Austria changed that balance. When émigrés poured into Britain and America, it became critical to consider what English translations had been adopted.

Of the many difficult choices that Freud's English translators faced, the ones that proved most fateful were little words. Freud had used everyday words to usher in his "I" psychology. He had used the German *Es* and *Ich*, which were simply "It" and "I." And he had explicitly defended this choice:

> You will probably protest at our having chosen simple pronouns to describe our two agencies or provinces instead of giving them orotund Greek names. In psychoanalysis, however, we like to keep in contact with the popular mode of thinking and prefer to make its concepts scientifically serviceable rather than to reject them. There is no merit in this: we are obliged to take this line; for our theories must be understood by our patients, who are often very intelligent but not always learned.

At the time that Freud wrote these words, Ernest Jones was fighting for the acceptance of psychoanalysis by British medical authorities. In the winter of 1924, Jones enlisted his former analysand, Joan Riviere, to translate Freud's *Das Ich und das Es* into English. They consulted their glossary committee, which included James Strachey, who reported on the meeting to his wife, Alix:

> I had a very tiresome hour with Jones and Mrs. Riviere from 2 to 3 today. The little beast (if I may venture so to describe him) is really most irritating . . . Our names will be ousted from the title page all right, Mark my words. They want to call "das Es" "the Id."

Jones and Riviere's decision to use the medical-sounding Latin word "id" for Groddeck and Freud's "It" was a running joke for the Stracheys, as was Jones and Riviere's Latinate choices for Freud's common *Ich*, which they decided to translate as "ego," so that *Über-Ich* would become "superego." This translation of *Ich* was not novel, for there was a tradition of employing the term "ego" in English for philosophical usages of this German term, but the translation defeated Freud's purpose of using common words that could easily be understood. In 1926, John Rickman wrote to the Hogarth Press publisher, Leonard Woolf, to complain: "I am not quite happy about the title. I feel myself that a literal translation into *The I and the It* might be better." Nonetheless, Riviere's translation *The Ego and the Id* appeared in 1927. From then on, in James Strachey's greatly influential *Standard Edition* of Freud's works and in the English-speaking world, "I" psychology would be known as ego psychology.

Super-ego, ego, and id were obviously different from Over-I, I, and It. The Latinate terms stripped the words of any connection to ordinary internal experience. A part of "I" psychology had been lost in translation. There was no easy bridge from the technical meanings of ego to everyday diction. There was no obvious link between Super-ego and the common part of mental life when we feel and hear a presence that watches over us. These philosophical terms divorced psychoanalysis from child observers and psychologists. The very name, ego psychology, suggested a theoretical discipline, not one predicated on empirical experience. Much hinged on this fateful choice.

As for the French, they stuck closer to a more literal translation, by rendering the German *Ich* as *le moi*, or "the me." After the war ended, the French, led by the charismatic Jacques Lacan, would develop a strong critique of ego psychology in America.

Psychoanalysts lost their birthplace, they lost their mother tongue, and then the organization that had provided some stability during turbulent times seemed on the verge of collapse. During the years of political turmoil, the I.P.A. had been a steadying, if not uncontroversial, influence. Since 1932, Ernest Jones had presided over the group from London. As each European capital of psychoanalysis imploded, the I.P.A. had tried to secure the interests of the field. In 1938, the import of the I.P.A.'s capacity to work along with the American Psychoanalytic Association had never been more apparent. Together the two organizations saved many lives by getting Jewish analysts out of Europe.

On August 1, 1938, the I.P.A. held its fifteenth congress in Paris. A reception was held at the Hôtel Salmon de Rothschild, where a representative of the French minister of education welcomed the group. Unbeknownst to its members, it would be the last time the psychoanalysts would congregate for over a decade. Ernest Jones opened with the news that the "mother of all psycho-analytical societies" no longer existed, but he bravely declared that psychoanalysis had grown out of its infancy and could withstand any opposition. Further, he announced the welcome news that all but 6 of the 102 analysts and candidates in Vienna had been rescued. Given the terrible complexities of negotiating with the Nazis, this was a tremendous feat.

Jones regarded fascism as something psychoanalysis could wait out. He had worked hard to maintain psychoanalysis in Berlin under the Nazi rule and openly ascribed his accommodations to the hope that analysis would outlast these disastrous times. Jones had supported the surviving Aryan analysts in Berlin, but to his credit, he did not count on the Nazis becoming good citizens, and he had busily prepared behind the scenes for more grim possibilities. As president of the I.P.A., Jones had made plans to evacuate Jewish analysts if the worst should come in Austria. When that day came, a number of analysts owed their lives to Ernest Jones and the I.P.A.

Despite the collapse of psychoanalysis in Germany and Austria, there was no reason to think that the I.P.A. would be in jeopardy, but at the business meeting in Paris, Jones revealed a shocking development. Just before the congress, he had received a dossier from the president of the American Psychoanalytic Association. The thirty-seven-page document suggested that the I.P.A., in Jones's words, should "cease to exist as an administrative and executive body and should resolve itself entirely into a Congress for scientific purposes only." The Americans were especially eager to disband the I.T.C. and end the long-standing dispute over the question of lay analysis. At this moment of intense instability, the Americans had decided to declare the I.P.A. null and void. They made it clear that they would not abide by its authority.

Ignoring the Americans was no longer an option. In 1938, 30 percent of the I.P.A. membership lived in America, and that percentage was about to grow exponentially. Of the nearly one hundred Viennese analysts that needed to be settled, all but four would go to English-speaking countries. Since England could only accommodate a few, most of them would end up in the United States, driving the American stake in the I.P.A. up to nearly 50 percent. If the immigrants were lay analysts and had no protection from the I.P.A., they could not practice in America. With the I.P.A.

weakened and psychoanalysis on the European continent in ruins, Jones was helpless. He needed the Americans' to help settle the exiled Europeans. He convened a committee to negotiate between the two groups, but the future of the I.P.A. looked dismal.

Their birthplace, their language, and any semblance of a unified community had vanished. Then they lost Sigmund Freud. Having settled in Maresfield Gardens in London, the eighty-three-year-old Freud had a recurrence of his malignancy on the jaw. After another operation, he was in constant pain, his sores ulcerated and became fetid. Freud's doctor, Max Schur, had long ago agreed to spare Freud any unnecessary pain when his time had come. When Freud made his last request, he did not forget to tell Schur to consult with Anna beforehand. On September 23, 1939, Dr. Schur gave Sigmund Freud a lethal dose of morphine.

Freud had lived to see the beginning of what would be the bloodiest struggle in human history. On the first day of September in 1939, Adolf Hitler invaded Poland. With this, the British and French governments did what they had long resisted doing: they declared war. The ensuing conflagration spread over the globe and killed a staggering number of people, over thirty-five million in Europe alone. Germany, France, Austria, Poland, Denmark, the Netherlands, Hungary, Czechoslovakia, Norway, Belgium, and other nations would all fall under Nazi rule.

Psychoanalysis was born in Europe. It was the child of European cultures, nurtured by *Geisteswissenschaft* and *Naturwissenschaft*, post-Kantian philosophy, neo-Romanticism, and sexual reform. It found inspiration in the creative tensions between Germany and France, and grew up in Europe's medical institutions and the liberal urban centers that produced a radical modern aesthetic—such as the literary, philosophical, and artistic movements in Paris, Vienna, Zurich, Berlin, Frankfurt, Munich, Prague, and Budapest. It was a discipline founded by the brilliant synthetic work of a Viennese physician who named the complex workings of the psyche in German. The world that nourished psychoanalysis had disappeared.

Yet, Ernest Jones was right. Psychoanalysis would not die, even with the destruction of the Viennese community. Its books would be burned, its followers hounded, exiled, and even murdered, but its theories and methods had long flown from the fires of Europe and found fertile soil far away. After 1939 the question was not whether psychoanalysis would live, but rather what form it would take after losing so much.

Epilogue

I.

In 1949, the first postwar meeting of the International Psychoanalytical Association took place in Zurich. Ernest Jones, that dogged warrior for the movement, surveyed a transformed landscape. So much had been lost, so much destroyed. The Nazis had perpetrated a genocide of European Jews and massacred leftists, sexual "deviants," and those deemed degenerate or mentally unfit. During the Nazi terror, only the Dutch managed to keep a clandestine psychoanalytic group together. The French, Italian, Hungarian, and German psychoanalytic communities were either destroyed, dispersed, disbanded, or Nazified. Many European cities had been bombed into rubble, while others, Prague, Budapest, and half of Berlin, survived the war only to be pulled into the totalitarian Soviet orbit. Jones's hope of waiting out the war and returning to the places that once nurtured psychoanalysis was for most analysts impossible.

Accustomed to eulogizing a few members of the I.P.A. who had died since the prior meeting, Ernest Jones now confronted a list far too long for individual memorials. The dead included at least fifteen members murdered by the Nazis, as well as the prewar founders and leaders of the movement: Sigmund Freud, Max Eitingon, A. A. Brill, Ernst Simmel, Otto Fenichel, Hanns Sachs, Paul Schilder, Susan Isaacs, and Hugo Staub. Paul Federn was sick and could not travel. Of the original group that met in Salzburg in 1908, only Ernest Jones and Eduard Hitschmann remained.

Despite the devastation in Europe, the I.P.A. had grown to an astonishing eight hundred members, over half of whom were Americans. Nearly three-quarters of the association now hailed from English-speaking countries. Ernest Jones proudly announced that he had received applications for twenty new societies. The applications from Italy, Vienna, Belgium, Argentina, and Chile were approved, while Greece, Ireland, Montreal, Prague, Brazil, and Johannesburg were given more provisional status. Five new American institutes were sanctioned, in Detroit, Los Angeles, Topeka, San Francisco, as well as a second institute in New York. Questions were raised about the application by Müller-Braunschweig and Felix Boehm to reinstate the German Society. To the disappointment of Jones and the Germans, this effort was rebuffed and sent to an investigatory committee for deliberation.

As Jones continued his survey, it was clear that the mass exodus from continental Europe had changed the topography of the field. Of the five largest psychoanalytic communities in the world, four were American. And though Jones only obliquely referred to it, the two largest capitals of psychoanalysis, New York and London, had been thrown into turmoil by the arrival of so many European immigrants. For among the immigrants were some of the discipline's major theoreticians and power brokers. They arrived in London and New York and ignited struggles for control that reopened old questions and proved fateful as psychoanalysis expanded into the dominant psychology of subjective experience in the Western world.

BETWEEN 1934 AND 1938, the major centers for analysis were in Vienna and London. The psychoanalytic world had made room for their theoretical differences, a freedom that was supported by geographic distance and considerable local autonomy. Then in 1938, the Viennese and Londoners found themselves staring at one another across the aisle as members of the same institute.

Melanie Klein was understandably unnerved by the arrival of the Viennese; suddenly, they made up a substantial minority in the British Society. While Jones planned to help many of these émigrés relocate to the United States over the next few years, Anna Freud stayed in London, and her presence helped ignite a fierce debate.

Shortly before he died, Freud sent a congratulatory note to the British Society, declaring that he counted London as the "main site and center of the psychoanalytical movement." The gesture was not devoid of pathos; over the years Freud had expressed similar feelings to leaders in Zurich, then Budapest, then Berlin, only to watch each of those hubs either desert him or collapse. Still, there was no doubt that the future of the field seemed heavily predicated on what would happen in London. In London, Anna Freud's open-minded vision of ego psychology was of little help, for she was stuck with the one faction she had refused to make a place for under her big tent. To make matters more divisive, Melanie Klein outlined a further extension of her theory in 1938, which expanded her idea of the depressive position in a way that rendered the Oedipal complex irrelevant. Klein's newer theories had already estranged former supporters in her own society, who whispered that her work was not Freudian.

When the blitz of London began, Melanie Klein and a number of her loyal followers fled to the countryside. Ernest Jones also hurried to his house in Sussex. However, as alien immigrants, the Viennese did not have freedom of movement. They stepped into the vacuum and attended British Society meetings, where bizarrely, they were at times a majority. When Melanie Klein and her followers returned to London near the end of 1941, the emboldened Viennese were ready.

James Strachey anticipated a clash and wrote one of the most vociferous of Melanie Klein opponents, Edward Glover:

> I should rather like you to know (for your personal information) that—if it comes to a showdown—I'm very strongly in favour of compromise at all costs. The trouble seems to me to be with extremism on both sides. My own view is that Mrs. K has made some highly important contributions to PA, but that it's absurd to make out that {a} that they cover the whole subject or {b} that their validity is axiomatic. On the other hand I think it's equally ludicrous for Miss F. to maintain that PA is a Game Reserve belonging to the F. family and that Mrs. K's ideas are totally subversive.

Strachey's temperate approach did not carry the day. By the beginning of 1942, both sides prepared for a showdown, which would come to be known with British reserve as the Controversial Discussions.

Operating under the threat of German air raids, the two camps gathered to determine the future of their society. Things began innocuously but quickly evolved into a debate about whether the Kleinians were part of the psychoanalytic community. Of course the question required the participants to define exactly what they meant by the psychoanalytic community. Klein's son-in-law and enemy, Walter Schmideberg, asked the British Society to affirm that its very aim was to further "Freudian psychoanalysis" and deal with deviations from it. He declared that any writings that ran counter to Freudian psychoanalysis should be censored by the society's publications, which happened to include the I.P.A.'s journal, the *International Journal for Psychoanalysis.*

Schmideberg's call to arms may have worked in 1910, but thirty-two years later his appeal was ineffective. Schmideberg did not bother to characterize what "Freudian" meant, which was a problem, since there were different Freuds, and one could be claimed by Melanie Klein as well as by Anna Freud. In addition, Sigmund Freud himself was dead and could no longer adjudicate these warring claims. There was another dangerous aspect to Schmideberg's appeal. If psychoanalysis was to return to being a strictly Freudian field, the death of Sigmund Freud would mean that in some fundamental way the field itself must come to an end too.

Schmideberg's diatribe fell on deaf ears. Since 1918, a more pluralistic psychoanalytic community had changed the parameter of the discipline, nowhere more so than in London, where leading analysts long made it clear that their allegiance was to psychoanalysis not just Freud. Ernest Jones had publicly dissented from Freud's views on matter like lay analysis, Lamarckian heredity, the nature of femininity, and the theories of Melanie Klein. This independent community was not easily swayed by Schmideberg's call for a return to pure Freud. Susan Isaacs, a Klein ally, rose to reject the appeal to the authority of Sigmund Freud. Isaacs insisted that the society was there to further not so much Freud, as "psychoanalytic science." Another of Klein's followers, Joan Riviere, was careful to put the science before the Freudian, arguing that the society was for "cultivating and furthering the science of psychoanalysis, founded by Freud." Not Freudian psychoanalysis, but the science of psychoanalysis, founded by Freud. There was a world of difference in those two phrases.

The Kleinians' call for scientific freedom was potent. As Sylvia Payne pointed out, British boys were fighting for freedom at that very moment. However, she cautioned that all freedoms can be abused; democracies and scientific societies must provide safeguards to ensure against such exploitation. Walter Schmideberg exclaimed that he was acting in the spirit of "my teacher, Professor Sigmund Freud." As for his mother-in-law, she was a plagiarist and possibly worse. Kleinian theory, Walter fumed, made him feel as if he were "following Alice through the looking-glass." "We must," he exhorted his colleagues, "draw the line somewhere." A talented member of Klein's group named Donald Winnicott protested that the Professor would never have wanted to "limit our search for truth." He too asked the society to adopt language that put the aim of the group as the furthering of "the psychoanalytical branch of science founded by Freud."

All this semantic jousting was not without irony. The Kleinians had taken the high ground of science, despite the fact that their leader had been accused of dramatically departing from basic scientific principles. Like the old Freudians, the Kleinians had become defenders of an empirically unknowable belief regarding unconscious mental life. Nonetheless, the Kleinians draped themselves in the principles of free inquiry. Like others before them, they seemed to want the freedom of scientific pursuit without accepting the responsibilities that came with it.

The ironies did not end there. The Viennese analysts who followed Anna Freud had labored to make psychoanalytic theory more empirical and had tried to foster a more open psychoanalytic community. They were now defended by British allies who demanded a return to orthodoxy in which no dissent was permitted. If Walter Schmideberg thought he was in Wonderland, Anna Freud must have felt she had walked into an upside-down world. Once attacked as a wild analyst by Jones and Klein, Anna had responded by decrying analytic orthodoxy and censorship, and now, she was zealously being defended with an appeal for orthodoxy, censorship included.

Anna Freud disassociated herself from Walter Schmideberg's call for censorship, but she stated that scientific discussions could reveal who was legitimately within the bounds of psychoanalysis. Scientific differences in the British Society must be worked out before administrative changes are adopted, otherwise it would be like "renovating a house before we know who wants to live in it."

The British Society embarked on an extended discussion of the scientific differences between the two schools. A series of ten discussions took place between January 27, 1943, to May 3, 1944. The substantive discussions on the nature of fantasy and early childhood changed few minds. There were those who believed in Klein's proposals and those who didn't.

Again, James Strachey attempted to rescue his colleagues from their predicament. Any psychoanalytic society must tolerate a range of ideas, but clearly there was a limit, which if trespassed qualified as outside the field. Strachey called for an assessment of this range of theories and unification around a common psychoanalytic method of investigation. His emphasis on a set technique allowed for robust scientific differences over theory without fostering anarchy.

Unfortunately, the bookish Strachey had little clout in the society. His comments were brushed aside by Anna Freud, who perhaps had miscalculated the popularity of her position. Technique and theory were bound together, Anna insisted. The Kleinians put too much emphasis on transference to the detriment of analyzing dreams, associations, and memories. Besides, Anna added, the British idea of an open forum was one that was open to Klein, not other major European psychoanalysts.

Control of the British Society was up for grabs. In 1943, the Freudians were unable to prevail and the British Society seemed headed for schism. Edward Glover resigned from the society on February 2, 1944. Anna Freud resigned from the Training Committee and withdrew from society activities. The British Training Committee had no Viennese members on it. Melanie Klein and her supporters had won.

Then a fascinating twist took place. There had always been a healthy number of analysts like Sylvia Payne who were committed to neither Anna Freud nor Melanie Klein, but saw themselves as dedicated to a science of psychoanalysis. Payne took over the presidency of the society in 1944, and she, along with other moderate members, brokered a compromise that guaranteed the Kleinians survival, while also giving Anna Freud and her followers a home. The compromise allowed both Anna Freud and Melanie Klein to train candidates on separate educational tracks. The society maintained Kleinian and Anna Freudian forms of training, along with a third group of independents that were not attached to either group. To keep any candidate from too much one-sided influence, it was mandated that the second supervisor of each student, no matter what track, be from the independent, middle group.

In effect, the compromise re-created the lost landscape of European psychoanalysis. In London, the Viennese School had a home with some autonomy and security, as did the English School. Between the two schools there was plenty of room for independent analysts. The compromise stabilized the British Society as it grew and prospered in the postwar years. It would provide a home for a number of creative theorists such as John Bowlby, D.W. Winnicott, Michael Bálint, Wilfred Bion, and Paula Heimann.

UNLIKE THE BRITISH Society, the American analytic community had no major school of theory to defend. Instead, the Americans slowly made an identity for themselves by exploiting the opportunities created by the mental hygiene movement, academic psychiatry, and medicine. As they grew in number, the Americans increasingly imagined the future of the field was theirs. In 1938, Franz Alexander delivered his presidential address, "Psychoanalysis Comes of Age," to the members of the American Psychoanalytic Association. The Chicago-based analyst announced that a new American direction would make psychoanalysis a discipline with one foot in medicine and another in the social sciences. Psychoanalysis would rid itself of obscure theoretical superstructures and be given a new foundation in observation. It would no longer be a Weltanschauung but would assume a "more scientific character."

As their confidence grew, the Americans became less willing to bend to the will of the I.P.A. In January of 1938, the president of the New York Society, Lawrence Kubie, and its educational director, Sándor Radó, met with four other analysts appointed by the American Psychoanalytic Association. Kubie's committee had been asked to deal with the problems between the Americans and the I.P.A. In January 1938, they decided to resolve these conflicts once and for all: they would break from the I.P.A. with its training requirements. And eight months later, they did exactly that in Paris. However, if that was the committee's main problem when it met in January, it was not two months later, as the annexation of Austria unfolded before their eyes. Kubie hurriedly reconvened the group so they could take up the dire issue of saving their European colleagues from the Nazis. On March 19 the committee formed an Emergency Committee on Relief and Immigration to face this urgent task.

The problems the committee faced were daunting. In the decade leading up to 1938, a total of some 1,400 German, Austrian, and Italian doctors had immigrated to the United States, but after the annexation of Austria, the number of refugees who landed in America surged to over

1,400 in 1938 alone. While America was a vast space, forty-two of its states prohibited foreign medical doctors from practicing until they became citizens, a five- or six-year process that effectively limited immigrants to a small number of places, most often their port of entry, New York. Kubie recognized that the concentration of specialists in one place could be economically disastrous and started to plan to seed analysts throughout the country.

Further complicating matters, the U.S. State Department demanded affidavits and security deposits of $5,000 for an immigrant family of three or four. Kubie pleaded with American analysts: money had to be pumped in if they were to save the European analysts. Many analysts contributed five, ten, or twenty dollars a month, but the need outstripped these resources. As the Nazis moved into other parts of Europe, desperate requests came from Holland and the vestiges of free France. In the end, the Emergency Commission secured over $47,000 dollars. Working with Jones, Kubie and the Americans gave financial support to some 68 individuals, provided affidavits for 82, and were in contact with another 136. By 1943, 149 exiled psychoanalysts and psychiatrists had been relocated somewhere in the United States.

The exodus brought some of the most successful psychoanalysts to America. They had few possessions and little money, but they carried with them a proud identity. They were the ambassadors of a great civilization. Now a dark age had engulfed their homes and threatened the civilization that made their work possible. Many like Otto Fenichel believed that it was their task to make sure psychoanalysis did not die in America, that money-crazed foreign place. For the Americans, who believed they were the future of the field, the Europeans seemed to have arrogated to themselves the rights of a dispossessed royalty. There was bound to be trouble.

American psychoanalysts had long suffered from the contempt of their European colleagues, but the Americans had done little to controvert Freud's prejudice that the New World was a hopeless place for science. In 1925, reviewing his own nation's contributions to psychoanalysis, Clarence Oberndorf admitted that there had been little original work. After 1930, that view began to change with the arrival of the Berlin elite—Alexander, Radó, Horney, Sachs, and Simmel—as well as Viennese like Helene Deutsch and Paul Schilder. In 1937, Radó said that a break from the I.P.A. posed no danger, since the strength of psychoanalysis had been "concentrated in America for many years."

The main concentration of talent in America was at the New York Psychoanalytic Society. The New York Society had opened its training institute in 1931. Two years later, Sándor Radó, who had been brought from Berlin to be education director, took over the teaching of theory. Radó's lectures were supplemented by the work of Americans like Abram Kardiner, David Levy, George Daniels, and Bertram Lewin. While these men had varied interests, they, along with Kubie, Schilder, and Alexander, shared a desire to make psychoanalysis more scientific. Kubie was interested in merging Pavlov's reflexology with analysis; Kardiner hoped to use anthropology to enhance psychoanalytic theories; David Levy had conducted extensive research on young children; and Radó himself had become interested in Walter B. Cannon's physiology of emotion, the so-called fight or flight reactions, and had used it to model how the ego defended itself against anxiety.

Like others on the American scene, Radó spoke with great enthusiasm about the future of psychoanalysis as a natural science and worked to advance this perspective in the classroom. He was determined to reform psychoanalysis and avoid the fractures that had ripped it apart in the past. Why did schisms take place in psychoanalysis and not in physics and chemistry? Radó demanded. When theory outweighed facts, then science became a matter of opinion, and differences were transformed into a battle of wills. "Competing authorities do not like to dwell under the same roof," he continued. "One of them has to get out. And when he does he founds a school of his own in which he can be just as authoritarian as is his opponent in the school which he left." Radó's words demonstrate that he had learned a lesson from the breaks between Freud, Jung, and Adler, but the same words could have been written almost a century earlier by Auguste Comte when he described the peril facing any science of inner life.

In America, Radó hoped to alleviate the situation by underplaying theory and stressing the critical, empirical, and scientific. Only a few years after he arrived, his European colleagues raised their eyebrows when Radó identified himself as an American and seemed a little too eager to cast off his Old World identity. Then the Europeans started unpacking their bags in New York. In 1939, Kubie and Radó's New York Institute listed a teaching faculty of twenty-seven; a year later the faculty ḥad added nineteen new members. The arrival of the traumatized, displaced Europeans would take a few years before its full impact was felt, but an early sign of strife came in the person of Paul Federn.

Federn's acidic manner was legendary in Vienna, and it had earned him the animosity of a generation of students. He prided himself on his absolute loyalty to Freud and was in that sense an unreconstructed old Freudian. After arriving in New York, Federn joined the New York Society, and the strain between him and Radó began almost immediately.

Radó had been invited to deliver a presentation to the New York Academy of Medicine and the New York Neurological Society on May 2, 1939. Given the audience, Radó spoke on a broad subject: the analytic treatment of neuroses. Before this medical crowd, he did not mince words: Freud's turn to signal anxiety was vital but further clarification had been impeded by his death drive, "the highly speculative hypothesis" that was "so vague and remote as to be of questionable value." Radó suggested that more recently analysts had tried to tease out the metaphysics from the facts in their work on the ego, and illustrated this with his own theory of emergency control, based on the work of Walter B. Cannon.

Radó's lecture was cheered by his allies from the New York Institute. George Daniels congratulated him and suggested analysts in general hoped to remake psychoanalysis in a similar way. They looked to biology to clear away "vestiges of outworn theories of instincts." *Outworn theory of instincts?* David Levy commented that the last ten years of American psychoanalysis pointed in the direction that Radó was going, as a reaction "against authoritarianism in psychoanalysis." "The stage of obeisance to Freud" had given way to a freer utilization of his doctrine. *Obeisance to Freud?* As exemplars, Levy proudly pointed to research done by Franz Alexander, Abram Kardiner, and Karen Horney.

There was only one discussant left. Paul Federn, the bearded patriarch and standard-bearer who had come to America carrying the minutes of the Vienna Psychoanalytic Society in his bags. Federn was outraged by what he had heard. What happened next was hotly disputed, but for those who had seen Federn preside over meetings of the Vienna Society, it was not surprising. Federn took the floor as he had so many times in Vienna to denounce supposedly new findings. His accent was so thick that the stenographer could not follow much of what he said. Her notes indicate that

Federn scolded Radó for departing from drive theory, and accused him of leaving both Freud and the unconscious behind. Radó replied that Federn had completely misunderstood him, and the idea that he was ignoring the unconscious was astounding, but Radó did not reject the accusation that he was moving beyond Freud. Instead, he defended the legitimacy of replacing psychoanalytic theories that "can be neither verified nor refuted because they are beyond the available means of investigation" with theories that fell into the range of scientific method.

Months later Radó received the page proofs from the soon to be published transcript of the meeting. To his dismay, he found that Federn's discussion now included a full-scale denunciation of Radó as a heretic. Federn was to have said:

> A few years ago, Radó broadened the meaning and importance of masochism in a rather infantile way. Now he thinks he has discovered that there is no sadomasochistic phase and that there are no different components to the sexual drive. It would not be of intrinsic merit to watch the details of his desertion of Freud. I have watched similar desertions in the cases of Adler, Jung and Rank. Radó is the chief teacher in the New York Psychoanalytical [*sic*] Institute, however and since he is trying to draw that school behind him, he must be opposed.

In the transcript no reply to this diatribe appeared from the accused because in Radó's recollection, Federn had never uttered those words at the meeting. Certain that Federn had doctored the transcript, Radó accused his adversary of behaving unethically. Lawrence Kubie tried to intercede, at first denying that any alterations were made. David Levy jumped into the fray on Radó's side. Finally, it came out that the befuddled stenographer, unable to make heads or tails of her notes, had asked "Dr. Fedor" to write up his response. Federn had freely obliged.

It was a nasty little quarrel and a harbinger of things to come. Radó, Levy, Kardiner, and Karen Horney wanted to sustain an open, pluralistic psychoanalysis that was built on pragmatism and a suspicion of metaphysical European theory. As Federn discovered, they all supported discarding Freudian drive theory and hoped to resituate psychoanalysis by dissolving its ties to unconscious libido. Some Americans like Kardiner,

Horney, and Harry Stack Sullivan would look to the social environment to make a new sociology of character. Others like Radó and Alexander turned toward new physiological findings to ground their efforts. Then suddenly while they were busily reframing their own views of psychoanalysis, they were confronted by a group of Viennese led by Paul Federn who begged to differ.

The tension in the New York Psychoanalytic Society became electric. David Levy feared that the American commitment to a scientific psychoanalysis would be compromised by the newcomers. He wrote to Kubie:

> The old-timers from Vienna are a good example of the danger of starting with holy writ . . . they never ask themselves the question, "What is the truth?" They ask rather, "Does he agree with Freud?" implying "There is no truth but Freud." And that is why the discussions are so little grounded in empiricism, so thoroughly dialectic.

Kubie asked Levy to be patient and went on to suggest that this attitude was in part loyalty to Freud, but was also characteristic of "the refugee who doesn't want to feel that everything that he has had and everything he knows had been swept away and superseded by something better."

Patience wore thin. Kubie began to act in a more highhanded manner, prompting one analyst to suggest he had become a "little Hitler." As it became clear that Kubie had misrepresented Federn's actions in the conflict with Radó, David Levy became furious. Normally dapper and unruffled, Kubie replied by accusing Levy of "vilifying Federn." Levy had already accused Kubie of stacking teaching appointments with those deemed orthodox; soon he found that his own course on child analysis had been handed over to the Viennese analyst Bertha Bornstein.

Meanwhile, the New York Psychoanalytic Society had added not just Federn, but also Ludwig Jekels, Van Ophuijsen, Annie Reich, and a group of guest lecturers like Edith Jacobson, Ernst and Marianne Kris, and Heinz Hartmann. Hermann Nunberg attended meetings at the society, though he had been denied membership for refusing to pledge that he would not endorse lay analysis. As one member recalled, in 1939 one heard far more German spoken at the New York Society meetings than English. The society that had once recruited Radó to bring weight to their fledgling institute, now had at its disposal a depth of experience. Kubie decided that it was no longer necessary to follow Radó into a fight with the Viennese. In December of 1940, he orchestrated the removal of Sándor Radó as the educational director, an action justified by Kubie's claim that "we have come to disagree with him on certain theoretical issues."

Kubie knew that Radó was now fuming "like Achilles in his tent." Kardiner and Levy were also bitter about the way Kubie had allowed the newcomers to move in. Exacerbating the tension, Karen Horney began to push for radical revisions that rocked the society. She had forged her identity in a Berlin psychoanalytic community where freedom from a slavish fawning before Freud had been a matter of pride. After following Alexander to Chicago, she came to New York in 1934 and joined the society there. In 1937, she published *The Neurotic Personality of Our Time*, in which she argued that cultural forces gave the form of a neurosis. She knew that some would ask whether this cultural emphasis was still psychoanalysis:

> If one believes that it [psychoanalysis] is constituted entirely by the sum total of theories propounded by Freud, then what is presented here is not psychoanalysis. If, however, one believes that the essentials of psychoanalysis lie in certain basic trends of thought concerning the role of unconscious processes and the way they find expression, and in a form of therapeutic treatment that brings these processes to awareness, then what I present is psychoanalysis.

Like the transplanted Berliners, many Americans would have signed on to that declaration in 1937. Then two years later, Horney published *New Ways in Psychoanalysis*, a book-length critique of psychoanalytic fundamentals. The book was unabashedly radical: Horney rejected libido theory, the Oedipal complex, the childhood origins of neurosis, the notion of a repetition-compulsion, and transference; she did not accept the superego, ego, and id, not to mention a host of other minor theories. For Horney, personality and neurosis were due to environmental influences that disturbed a child's relation to self and other. The New York Society asked Horney to present her new theories before a scientific meeting. During the meeting, Horney was accused of hiding the extremity of her ideas and was vigorously upbraided. An enraged Abram Kardiner wrote Kubie to protest the attacks that he had allowed during the meeting, including repeated claims that Horney's theories were due to her own neurosis. "Is this science or a racket?" Kardiner fumed. "I was under the illusion that I was a member of a scientific society and not a club."

As reviews came in, Horney found little support elsewhere. One damning critique was written by Otto Fenichel. Though he shared Horney's distaste for orthodox Freudians, Fenichel could not countenance Horney's rejection of inner drives, which he believed eviscerated one of the central assertions of psychoanalytic theory. In times like these, he counseled his *Rundbriefe* colleagues, the task was to save psychoanalysis from dissolution. Fenichel assailed Horney's book. After all that she rejected, he asked, what was left? Another crushing review came from Horney's former Berlin ally, Franz Alexander, who spent over thirty pages picking apart her arguments. Horney created a cartoonish, one-dimensional Freud only to then present her own one dimensional antithesis, he wrote. Between libido and culture, one empty word replaced another. Psychoanalysis had to attend to both biology and environment, the present and the past, family and culture. In arguing against a straw man, Horney had created a theory riddled with the opposite of errors, not truths but different errors.

At the New York Society, the once fiery radical and now fiery conservative Fritz Wittels emerged as the most vocal critic of Karen Horney. After resigning from the Vienna Society in 1910, Wittels had made his way back into Freud's good graces in 1927. Following his immigration to New York, he joined the society and took up the fight against heretics like Horney. In 1940, he wrote Kubie saying: "The issue is Freud or no Freud." Horney's students made "fools of us by insisting on what they call democratic methods in a scientific body." What if some internist claimed microbes did not cause malaria and his colleagues voted on this view? Horney, Wittels demanded, should recant or resign.

Kubie was in a difficult position. Before the arrival of the Europeans, he had fashioned himself as a standard-bearer of science in psychoanalysis. He had supported Radó, Horney, Kardiner, Levy, and others as they quietly worked to reinvent psychoanalytic theory for America. He told Wittels: "I am basically opposed to any form of a 'purge' in a scientific organization." Yet Kubie agreed that intellectual freedom did not bring with it the right to indoctrinate students. Horney's followers complained of intimidation in the classrooms. Some of them signed a petition stating that the advancement of the science of psychoanalysis had been grossly impeded by the committee.

In June 1941, Horney was demoted to lecturer, and her training privileges were revoked. Having in essence been fired, she quit. Horney, four other faculty members, and fourteen students resigned from the society to form the Association for the Advancement of Psychoanalysis and the American Academy of Psychoanalysis. Horney left with Sándor Ferenczi's former student Clara Thompson, and the two found allies elsewhere, including the homegrown American analyst Harry Stack Sullivan of Washington. Under the mentorship of Adolph Meyer and William Alanson White, Sullivan had developed an interpersonal theory that left sexual drives and the technique of transference interpretation behind. Instead, he focused on the pathologies of human relationships. With the former Frankfurt School analyst Eric Fromm, Sullivan supported Horney's new group. They all hoped to advance psychoanalysis by creating a pragmatic, unesoteric field stabilized by social science. However, while these analysts were for lay analysis, Horney was firmly against it, creating a schism within the schism—again replete with rhetoric on academic freedom. In the end, Thompson, Fromm, and Sullivan split from Horney and established the William Alanson White Institute in 1942.

The departure of Karen Horney and her allies did not end the accusations of dogmatism and heresy in the New York Society. Disgruntled analysts like Radó, Levy, Daniels, and Kardiner, who had once been the core faculty of the New York Institute, plotted their own secession. Radó and Levy looked to Alexander for assistance to set up a psychoanalytic center within a medical school, thereby concretely establishing a link between medicine and psychoanalysis. With the aid of Adolph Meyer, Radó and his associates opened a new analytic center at Columbia University, which began to train candidates in 1945 and was accepted by the I.P.A. in 1949. Before long accusations arose that the leader of the institute, Sándor Radó, had formed a school around his own teachings, in the name of scientific freedom, of course.

After 1945, the stage was set. The Viennese émigrés became "orthodox Freudians" opposed to the former Berliners and Americans, who came to be known as "neo-Freudians." That the latter were called neo-Freudian was in itself a defeat. Horney, Radó, and Kardiner wanted to be part of a psychoanalytic science and did not want to be forever shadowed by the ghost of the man they sought to move beyond. Decades later, after delivering a paper, Sándor Radó was asked a question from the audience about Freud's views. Wearily he replied: "for thirty years Radó gives lectures by Radó. For thirty years Radó gets questions about Freud."

In New York, Washington, Philadelphia, Chicago, and elsewhere, orthodox Freudians were pitted against neo-Freudians. In New York, Lawrence Kubie would be cast off. The new leader of the New York Society had impeccable Viennese credentials and could trace his analytic lineage to Freud, for he had been one of the Professor's last training cases. Heinz Hartmann served as director of education, then president of the New York Society, and president of the I.P.A. With his old colleagues from Vienna and Paris, Ernst Kris and Rudolph Loewenstein, Hartmann set the theoretical agenda for American ego psychology for the following three decades. His notions of adaptation cohered nicely with American values of self-reliance, and his hope to link psychoanalysis to academic psychology would be taken up by allies, such as David Rapaport. He could also count on the support of the heir to the Freud legacy, Anna Freud in London, whose emphasis on defense analysis became central to ego psychologists.

Against the neo-Freudians, Hartmann also had the support of Otto Fenichel and the underground community of leftist analysts. Without openly pushing for a Marxist psychoanalytic sociology in America, Fenichel, who was privately dismissive of Hartmann's ego psychology, concluded that the greater peril lay in the reactionary interpersonal and cultural work of Horney and Kardiner, and the driveless model of Radó, all of which abandoned the radical propositions of Freudian sexuality. He advised his *Rundbriefe* colleagues that the only choice was siding with the orthodox for now. As his epistolary community began to flag in 1945, Fenichel terminated the circular letter and began a grueling medical internship, so as to get an American medical license. He died before he finished it.

A coalition of ego psychologists and orthodox Freudians controlled the American Psychoanalytic Association by 1946. Four years earlier, the New York Society first proposed a resolution that gave authority to the association to grant certification and diplomas for psychoanalysts. The New York group also tried to pass an amendment that banned any secessions without prior approval from the association, an amendment that seemed to misunderstand the nature of a secession. After World War II, spurred by the refugees who now dominated the New York Society, the American Psychoanalytic Association transformed itself from a loosely knit federation into a central power that policed standards throughout the country. The association enforced its standards on teaching and training, and no longer yielded authority on these matters to local societies. The Hartmann era in American psychoanalysis had begun.

AFTER MORE THAN half a century of work, psychoanalysis emerged with the richest systematic description of inner experience that the Western world had produced. Its theories spanned such fundamental matters as sex, love, and death; childhood, parenting, and family; cruelty, fear, jealousy, envy, and hate; identity, conscience, and character; desire and mourning. In addition, a new social space had been created for the in-depth examination of mental life through use of methods that might also mitigate psychic suffering. And yet, these ideas were not timeless truths, immune to social flux. Psychoanalysis took root in western and central Europe, and that culture permeated its logic and assumptions. Once transplanted to foreign lands, it was inevitable that this body of knowledge would in part be remade.

After World War II, the psychoanalytic community that had existed in Europe became a memory as a small army of immigrants adapted to their new homes. In London, stalemate and political compromise allowed the Anna Freudians and Kleinians to pursue their own work and share the same house, but sleep in different bedrooms. In America, ego psychologists wrapped themselves in Freud's coat, attempting to collapse the difference between a Freudian, an ego psychologist, and a psychoanalyst. "I" psychology with its connection to lived experience mutated into an abstract, impersonal ego psychology in Hartmann's hands. In support of the new Freudian orthodoxy, self-justifying myths about Sigmund Freud were propounded by a troop of Viennese refugees like Anna Freud, Ernst Kris, Ernst Federn, and Kurt Eissler, along with the Londoner who had so often opposed Freud, Ernest Jones. They would create a legend of Freud as a solitary genius, who created psychoanalysis in splendid isolation, unaided by contemporaries and attacked by prigs and rebellious followers who often suffered from grave mental illness.

Against the self-anointed Freudians stood a small number of the leftist analysts whose political beliefs remained secret, and the outmaneuvered neo-Freudians, who had been relegated to the margins. Some, like Franz Alexander and Sándor Radó, remained inside the International Psychoanalytical Associations, while others, like Karen Horney, would not. All these groups retreated from one another and published in their own forums. While not exactly a return to the pre-1914 Freudian world, this was not the pluralistic field of the inter-war years. If there was not outright war among different camps, nor was there peaceful, vigorous competition. Like much of the Western world, psychoanalysis fell into something like a cold war.

These different communities could not agree on the same question that had dogged this field since its inception: What defined psychoanalysis? The question arose almost as soon as Sigmund Freud began to pull together his grand, interdisciplinary synthesis. It led to confusion that had engendered feuds, rivalries, and schisms. The emphasis on a common method and the creation of educational institutions after 1920 had moderated and managed these charged questions, but when the exiles of Europe were thrown together and the battles became pitched, the analysts again struggled to find rules for resolving theoretical differences and building consensus. How could they stabilize their field? Charismatic authority? An enforced commitment to as yet unprovable "truths" that were so disturbing that they must be protected? Common modes of investigation? A strict commitment to empiricism? Furthermore, should analysis organize around the imperatives of knowing or doing, discovery or cure? Should it strive to be a natural science, a human science, or a new Weltanschauung?

By 1945, any single answer to this array of questions was impossible. The extraordinary complexity and paradoxes of a science of human subjectivity were overwhelming and could only be contained by preliminary, pragmatic, or political means. Still, the answers individual psychoanalysts proposed won them followings. Despite the fault lines in their community, psychoanalysis thrived thanks to the explanatory power of its ideas. From its centers in London and New York, an outpouring of interest lifted all boats and made it possible for rival groups to pursue their separate visions. Following World War II, the Kleinians found fertile ground in much of South America, and Heinz Hartmann and his allies pressed their theories across North America. In Paris, Jacques Lacan provided a new psychoanalytic amalgam that increased in popularity despite his banishment from the I.P.A.

Psychoanalysis emerged from the rubble of postwar Europe as the leading modern theory of the mind. Its model of unconscious passions, its notions of defense and inner conflict, and its methods of unraveling self-deception, encroached upon traditional sources of self-understanding like religion. In the U.S., psychoanalysis made its way into the courts, schools, and hospitals, and informed literature, cinema, television, journalism, theater, and art. Its ideas spread into popular discourse as adages, clichés, and jokes. And all the while, as psychoanalysts fanned out, wittingly or unwittingly, they brought with them the culture of Kant; the assumptions of *Geisteswissenschaft* and a European classical education; they brought evolutionary biology, positivism, and Newtonian physics along with the thought of Ribot, Charcot, Bernheim, Breuer, Brentano, Krafft-Ebing, Fliess, Brücke, Helmholtz, Mach, Schelling, Fechner, Hering, Haeckel, Ehrenfels, Forel, Bleuler, Jung, Gross, Adler, Stekel, Sadger, Rank, Ferenczi, Abraham, Horney, Alexander, Fenichel, and many more. However, most of these ancestors would be increasingly diminished, forgotten, or dismissed. Instead, a ghostly presence would carry all that had been inherited and destroyed, all the possibility and all the loss. The culture that had given birth to psychoanalysis had become a graveyard. It was no more. Exiled survivors and followers in new lands fell into the vastness of their future accompanied by a word, a name, a talisman: Freud. A man had come to represent a history, and as a symbol he would live on, haunting his sons and daughters, his enemies and friends.

Acknowledgments

My deepest thanks go to my parents, Jack and Odette Makari, whose passion for intellectual inquiry and whose perseverance have been my inspiration. I am delighted to express my gratitude to those teachers and friends who helped me on this long climb, including my erudite initial guide, Sander L. Gilman, the late Eric T. Carlson, Norman Dain, Theodore Shapiro, Arnold Cooper, Robert Michels, William Frosch, Lawrence Friedman, Gerald Fogel, Roy Schafer, Elizabeth Auchincloss, Nathan Kravis, the inimitable Michael S. Harper, Rob Seidenberg, David Levin, Adam Bresnick, Leora Kahn, Leonard Groopman, Nicola Khuri, Sam Messer, Eleanor Gaver, Constance Herndon, *il miglio fabbro* Anthony Walton, Cathy Frankel, and Michael Beldoch.

During my years of research and writing, I was sustained by supportive colleagues at Rockefeller University, the Center for Psychoanalytic Research and Training at Columbia University, and especially Cornell's Department of Psychiatry and its Institute for the History of Psychiatry. I was enriched by longstanding discussions with the exceptional faculty of the Institute, as well as dialogue with other scholars including the late Roy Porter, Sonu Shamdasani, Brett Kahr, Peter Swales, Richard Skues, Elke Mühlleitner, Patrick Mahoney, German Berrios, Ernst Falzeder, Eric Engstrom, Cheryce Kramer, Nathan Hale, and John Burnham.

Parts of this work were presented at the Muriel Gardner Lecture at Yale University; the Samuel Perry Lecture hosted by Weill Medical College of Cornell University; the Sandor Radó Lecture at the Psychoanalytic Center of Columbia University; Rockefeller University; University College London; and Emory University. I thank the sponsors of these events, and the audiences for their feedback.

This book would not have been possible without the kindness and wisdom of many librarians and archivists from Zurich to Los Angeles. Of these, I must mention Erica Davies, J. Keith Davies, Jill Duncan, Nellie Thompson, Matthew von Unwerth, Thomas Roberts, Marvin Krantz, Leonard Bruno, Andreas Jung, Ulrich Hoerni, Tina Joos-Bleuler, Marie-Renée Cazabon, Verena Michels, Giselle Sharaf, Sanford Gifford, and the remarkable Diane Richardson, Special Collections Librarian of the Oskar Diethelm Library in New York. The late Doro Belz, Heidi Ziegler, and Astrid von Chamier helped greatly with German materials, and over the years, my assistants at the Institute—Charles Gross, Tanya Uhlmann, and Siovahn Walker—kept my mountain of work from crushing me.

Mitchell Feigenbaum offered me sanctuary at Rockefeller University where this book was written; more so, he and Gunilla Feigenbaum provided me with a steady stream of encouragement. *Revolution in Mind* benefited from the insightful readings of Robert Michels, Sander L. Gilman, Nathan Kravis, Lawrence Friedman, Sonu Shamdasani, Anthony Walton, Arabella Ogilvie, the Feigenbaums, and most of all, the generous and gifted Siri Hustvedt, whose devotion to this book was unsurpassed.

I count myself among the blessed for having the good fortune to work with Sarah Chalfant, as well as Edward Orloff and the rest of the team at the Wylie Agency. Tim Duggan at HarperCollins was a wonderful editor, offering me an abundance of sure-footed advice; his assistant, Allison Lorentzen, tolerated my foibles with patience and good humor. This work would have been impossible without a timely research grant from the International Psychoanalytical Association, and the ongoing support of the DeWitt Wallace Reader's Digest Program of the New York Community Trust. My final thanks are reserved for my exemplary chairman, Jack D. Barchas, who rode shotgun on this project and urged me on from start to finish.

Abbreviations for Frequently Cited Sources

For multivolume works, the volume number will be listed after the abbreviation. Full citations will only be given the first time a work is referenced.

Bleuler-Freud Letters—Eugen Bleuler-Sigmund Freud Correspondence, Sigmund Freud Collection, Manuscript Division, Library of Congress, Washington, D.C.

C.W.—Jung, C. G. (1957–79), *The Collected Works of C. G. Jung*, 20 vols., ed. H. Read, M. Fordham, G. Adler, W. McGuire, trans. R. F. C. Hull, Princeton, N.J.: Princeton University Press.

David Levy Papers—David Levy Archive, Oskar Diethelm Library, Weill Medical College of Cornell University.

Fenichel Rundbriefe—Reichmayr, J., and E. Mühlleitner, eds. (1998), *Otto Fenichel, 119 Rundbriefe: Band I, Europa (1934–1938)*, Frankfurt am Main and Basel: Stroemfeld Verlag; and Reichmayr, J., and E. Mühlleitner, eds. (1998), *Otto Fenichel, 119 Rundbriefe: Band II, Amerika (1938–1945)*, Frankfurt am Main and Basel: Stroemfeld Verlag.

Ferenczi's Clinical Diary—Dupont, J., ed. (1988), *The Clinical Diary of Sándor Ferenczi*, trans. M. Balint and N. Z. Jackson, Cambridge, Mass.: Harvard University Press.

Ferenczi-Groddeck Letters—Ferenczi, S., and G. Groddeck (1982), *Correspondance (1921–1933)*, trans. J. Dupont et al., Paris: Payot.

Freud-Abraham Letters—Freud, S., and K. Abraham (2002), *The Complete Correspondence of Sigmund Freud and Karl Abraham, 1907–1925*, ed. and trans. E. Falzeder, London and New York: Karnac.

Freud-Binswanger Letters—Freud, S., and L. Binswanger (2003), *The Sigmund Freud-Ludwig Binswanger Correspondence, 1908–1938*, ed. G. Fichtner, trans. A. Pomerans, Tom Roberts, New York: Other Press.

Freud-Brill Letters—Sigmund Freud–A. A. Brill Correspondence, Sigmund Freud Collection, Manuscript Division, Library of Congress, Washington, D.C.

Freud-Ferenczi Letters—Freud, S., and S. Ferenczi (1993), *The Correspondence of Sigmund Freud and Sándor Ferenczi: vol. 1, 1908–1914*, ed. E. Brabant, E. Falzeder, P.

Giampieri-Deutsch, trans. P. Hoffer, Cambridge, Mass.: Harvard University Press. Freud, S., and S. Ferenczi (1996), *The Correspondence of Sigmund Freud and Sándor Ferenczi, vol. 2, 1914–1919*, eds. E. Falzeder and E. Brabant, trans. P. Hoffer, Cambridge, Mass.: Harvard University Press. Freud, S., and S. Ferenczi (2000), *The Correspondence of Sigmund Freud and Sándor Ferenczi, vol. 3, 1920–1933*, eds. E. Falzeder and E. Brabant, trans. P. Hoffer, Cambridge, Mass.: Harvard University Press.

Freud-Fliess Letters—Freud, S. and W. Fliess (1985), *The Complete Letters of Sigmund Freud to Wilhelm Fliess, 1887–1904*, trans. and ed., J. Masson, Cambridge, Mass.: Harvard University Press.

Freud-Groddeck Letters—Groddeck, G. (1970), *The Meaning of Illness*, ed. L. Schacht, trans. G. Mander, London: Maresfield Library.

Freud-Jones Letters—Freud, S., and E. Jones (1993), *The Complete Correspondence of Sigmund Freud and Ernest Jones, 1908–1939*, ed. R. Andrew Paskauskas, Cambridge, Mass.: Harvard University Press.

Freud-Jung Letters—Freud, S., and C. G. Jung (1988), *The Freud/Jung Letters: The Correspondence Between Sigmund Freud and C. G. Jung*, ed. W. McGuire, trans. R. Manheim and R. F. C. Hull, Cambridge, Mass.: Harvard University Press.

Freud Letters—Freud, S. (1960), *Letters of Sigmund Freud*, ed. E. L. Freud, trans. T. and S. Stern, New York: Basic Books.

Freud-Pfister Letters—Freud, S. and O. Pfister (1963), *Psychoanalysis and Faith: The Letters of Sigmund Freud and Oskar Pfister*, eds. H. Meng, E. Freud, trans. E. Mosbacher, New York: Basic Books.

Freud Reviews—Keill, N., ed. (1988), *Freud Without Hindsight: Reviews of His Work (1893–1939)*, Madison, Conn.: International Universities Press, Inc.

Freud-Rank Letters—*The Otto Rank Papers*, Rare Book and Manuscript Library, Butler Library, Columbia University, New York.

Freud-Silberstein Letters—Freud, S., and E. Silberstein (1990), *The Letters of Sigmund Freud to Eduard Silberstein, 1871–1881*, ed. W. Boehlich, trans. A. Pomerans, Cambridge, Mass.: Harvard University Press.

G.W.—Freud, S. (1940–1952), *Gesammelte Werke*, 17 vols., eds. A. Freud, et al., London: Imago Pub.

I.J.P.—*International Journal of Psychoanalysis*.

I.R.P.—*International Review of Psychoanalysis*.

I.Z.P.—*Internationale Zeitschrift für Psychoanalyse*.

Rank Diaries—Rank, O. (1903–1905), "Opinions and Thoughts about Men and Things in the Form of a Daybook," *Otto Rank Papers*, Rare Book and Manuscript Library, Columbia University, New York.

S.E.—Freud, S. (1953–1974), *The Standard Edition of the Complete Psychological Works of Sigmund Freud*, 24 vols., trans. James Strachey with H. Freud, A. Strachey, and A. Tyson, London: Hogarth Press.

Secret Committee RB—Wittenberger, G., and C. Tögel, eds. (1999), *Die Rundbriefe des "Geheimen Komitees," Band I, 1913–1920*, Tübingen: Editíon Diskord; Wittenberger, G., and C. Tögel, eds. (2001), *Die Rundbriefe des "Geheimen Komitees," Band II, 1921*, Tübingen: Edition Diskord; Wittenberger, G., and C. Tögel, eds. (2003), *Die Rundbriefe des "Geheimen Komitees," Band III, 1922*, Tübingen: Edition Diskord; Wittenberger, G., and C. Tögel, eds. (2006), *Die Rundbriefe des "Geheimen Komitees," Band IV, 1923–1927*, Tübingen: edition diskord.

Vienna Minutes—Nunberg, H., and E. Federn, eds. (1962–1975), *Minutes of the Vienna Psychoanalytic Society*, 4 vols., trans. M. Nunberg, New York: International Universities Press, Inc.

Notes

PROLOGUE

2 **"a whole climate of opinion":** Auden, W. H. (1976), "In Memory of Sigmund Freud," *Collected Poems*, London: Faber and Faber; 217.

2 **in its birthplace: western and central Europe:** Recent histories of psychoanalysis have not filled this need. They include Joseph Schwartz's (1999) *Cassandra's Daughter: A History of Psychoanalysis*, New York: Viking; and Eli Zaetsky's (2004) *Secrets of the Soul: A Social and Cultural History of Psychoanalysis*, New York: Knopf.

3 **"Freud Is Not Dead":** See "Is Freud Dead?" *Time*, Nov. 29, 1993; and "Freud Is Not Dead," *Newsweek*, March 27, 2006.

4–5 **nineteenth-century intellectual communities:** On intellectual communities, see Fleck, L. (1935), *Genesis and Development of a Scientific Fact*, Chicago: The University of Chicago Press, 1979; and Kuhn, T. S. (1962), *The Structure of Scientific Revolutions*, Chicago, The University of Chicago Press.

PART ONE: MAKING FREUDIAN THEORY

1. A MIND FOR SCIENCE

9 **"*I* is an *other*":** Rimbaud, A. (1976), *Complete Works*, trans. P. Schmidt, New York: Harper and Row, 100.

9 **described by Church fathers:** Descartes, R. (1637), *Discours de la méthode*, Paris: Bordas, 1984, 100.

10 **set out to change that:** Nicolas, S., and D. J. Murray (1999), "Théodule Ribot (1839–1916), Founder of French Psychology: A Biographical Introduction," *History of Psychology* 2:277–301.

10 *Contemporary English Psychology (The Experimental School):* Ribot, T. (1870), *La Psychologie anglaise contemporaine (école expérimentale)*, Paris: Librairie Philosophique de Ladrange.

10 **methods of natural science:** Ibid., 21–2.

11 **prophet of science, Auguste Comte:** Guillin, V. (2004), "Théodule Ribot's Ambiguous Positivism: Philosophical and Epistemological Strategies in the Founding of French Scientific Psychology," *Journal of the History of the Behavioral Sciences* 40: 165–181.

11 **program was dubbed positivism:** See Comte, A. (1855), *The Positive Philosophy of Auguste Comte*, trans. H. Martineau, New York: AMS Press, Inc., 1974.

11 **as there are observers:** Ibid., 33.

12 causes, inference, reasoning by analogy": Ribot T. (1870), *La Psychologie anglaise con-temporaine (école cxpérimentale)*, 23.

12 scientific psychology required both: Ibid., 30.

13 madness of "most men": Locke, J. (1690), *An Essay Concerning Human Understanding*, New York: Dover Pub., 1959, 528.

13 "affective phenomena in general": Ribot, T. (1870), *La Psychologie anglaise contempo-raine (école expérimentale)*, 72–3.

13 good deal of psychological functioning: Ribot, T. (1873) *L'héredité: étude psychologique: sur ses phénomènes, ses lois, ses causes, ses conséquences*, Paris: Librairie Philosophique de Ladrange; also Ribot, T. (1873), *Heredity: A Psychological Study of Its Pheno-mena, Laws, Causes, and Consequences*, New York: D. Appleton-Century Company, 1895.

13 experimentation is more rare": Ribot, T. (1870), *La Psychologie anglaise contemporaine (école expérimentale)*, 31.

14 "What a cerebral orgy!": Ribot, T. (1957), "Lettres de Théodule Ribot à Espinas, ed. by Raymond Lenoir," *Revue philosophique de la France et de l'étranger* 147:13.

14 thirty-six editions in France alone: The English translations of these works are Ribot, T. (1882), *Diseases of Memory: An Essay in the Positive Psychology*, New York: D. Ap-pleton-Century Company, 1896. Ribot, T. (1883), *The Diseases of the Will*, trans. M. Snell, Chicago: The Open Court Publishing Company, 1894. Ribot, T. (1885), *The Diseases of Personality*, Chicago: The Open Court Publishing Company, 1891. On the success of these works, see Nicolas, S., and D. J. Murray (2000), "Le Fondateur de la psychologie 'scientifique' française: Théodule Ribot (1839–1916)," *Psychologie et His-toire* 1: 1–42.

14 highly original, rich orientation: Janet, P. (1902) "Psychologie expérimentale et com-parée," *Annuaire du Collège de France*, Paris: E. Leroux, 27. This can be found in the Archive of the Collège de France, Paris.

14 I have a good foothold there": Lenoir, R. (1975), "Lettres de Théodule Ribot à Alfred Espinas (1876–1893)," *Revue philosophique de la France et de l'étranger* 165: 157.

14 stirred up by the sea: Fouille, A. (1891), "Le Physique et le Mental à propos de l'hypnotisme," *Revue des deux mondes* 105: 429–61.

14 destitute, or deemed incurable: On Charcot, see Goetz, C. G., M. Bonduelle, T. Gel-fand (1995), *Charcot: Constructing Neurology*, New York: Oxford University Press, 55–67.

15 what was hysteria?: On the history of this elusive disorder, see Veith, I. (1965), *Hyste-ria: The History of a Disease*, Chicago: The University of Chicago Press; and Micale, M. (1995), *Approaching Hysteria*, Princeton, N.J.: Princeton University Press.

15 disorder in these newer terms: Briquet, P. (1859), *Traité clinique et thérapeutique de l'hystérie*, Paris: Masson.

16 objectively observable outward signs alone: Charcot, J. M. (1892), *Leçons du Mardi à la Salpêtrière: Policlinique 1887–1888*, Paris: Aux Bureaux du Progrès Médical, V. A Delahaye et Lecrosnier, 103–05

16 pathologist's microscopic slide: Bourneville and P. Regnard, eds. (1879–1880), *Iconog-raphie photographique de la Salpêtrière*, Paris: Bureaux de Progrès Médical.

17 frequent anticlerical jokes: Goldstein, J. (1987), *Console and Classify: The French Psy-*

chiatric Profession in the Nineteenth Century, Cambridge, U.K.: Cambridge University Press, 360.

17 **exposed as simply hysterical:** Goetz, C. G., M. Bonduelle, T. Gelfand (1995), *Charcot: Constructing Neurology*, 183.

17 **for decades to come:** See Gauld, A. (1992), *A History of Hypnotism*, Cambridge, U.K.: Cambridge University Press; and Crabtree, A. (1993), *From Mesmer to Freud: Magnetic Sleep and the Roots of Psychological Healing*, New Haven: Yale University Press.

17 **not some spooky mesmeric power:** Charcot, J. M. (1883), *Exposé des titres scientifiques*, Paris: Goupil et Jordan, 149.

17 **strict emphasis on bodily symptoms:** Binet, A., and C. Féré (1888), *Animal Magnetism*, New York: Appleton and Co., 85.

17 **abnormal nervous state:** Charcot, J. M. (1892), *Clinique des maladies du système nerveux*, Paris: Bureaux du Progrès Médical, 95–100.

18 **on the body freely and automatically:** Charcot, J. M. (1887), *Lecture on the Diseases of the Nervous System*, vol. III. London: New Sydenham Society, 1889, 290. One of the unresolved historical questions is how much Charcot was influenced by his young protégé Pierre Janet, who began to publish on unconscious psychological action and dissociation around the same time as Charcot.

19 **paralyses of the imagination, Charcot concluded:** Ibid., 289.

19 **impunity of a hypnotic command:** Ibid., 304–05.

19 **paralysis became real:** See Charcot, J. M., and P. Marie (1892), "Hysteria, mainly Hystero-epilepsy," in *Dictionary of Psychological Medicine*, vol. I., ed. D. H. Tuke. London: Churchill, 627–41.

19 **that new monster, the locomotive:** On railway spine see Caplan, E. (1995), "Trains, brains and sprains: Railway Spine and the Origins of Psychoneurosis," *Bulletin of the History of Medicine* 69:387–419.

19 **outside the confines of consciousness?:** Binet, A., and C. Féré (1888), *Animal Magnetism*, New York: Appleton and Co., 182.

20 **dissociation and self-suggestion:** Charcot, J. M. (1889), *Lecture on the Diseases of the Nervous System*, vol. III. London: New Sydenham Society. 383–85.

20 **so good as to aid us:** Ibid., 308; also see 293.

20 **no talk could remedy:** See, for example, Charcot, J. M. (1877), "De l'influence des lésions traumatiques sur le développement des phénomènes d'hystérie locale," in *Oeuvres complètes de J.-M. Charcot, leçons sur les maladies du système nerveux*. Paris: A. Delahaye and E. Lecrosnier, vol. 1., 1886, 449.

21 **none had quite fit:** Of the many biographies of Freud, the most comprehensive, though flawed, account remains Jones, E. (1953–1957), *The Life and Work of Sigmund Freud*, 3 vols., New York: Basic Books. Also see Peter Gay (1988), *Freud: A Life for Our Time*, New York: W. W. Norton.

22 **against the "dirty Jew":** Freud Letters, 78–79.

22 **botany, physiology, and physics:** See Bernfeld, S. (1951), "Sigmund Freud, M.D., 1882–1885," *I.J.P* 32: 204–17.

23 **weird psychological occurrences, he advised:** Brentano, F. (1874), *Psychology from an Empirical Standpoint*, trans. A. Rancurello, D. Terrell, et al., London: Routledge, 1995, 28–43.

23 **Freud became one of his students:** See Fancher, R. (1977), "Brentano's *Psychology*

from an Empirical Standpoint and Freud's Early Metapsychology," *Journal of the History of the Behavioral Sciences* 13: 207–27. Over the next two years, Freud took five courses with Brentano, including courses on logic and Aristotle. During his first eight semesters of university study, Freud's only courses outside the sciences were with Brentano. See Bernfeld, S. (1951), "Sigmund Freud, M.D., 1882–1885," 204–17.

23 **wrote the young man:** Freud-Silberstein Letters, 95.

23 **undeveloped for any such union:** Freud-Silberstein Letters, 102–03.

24 **Darwinian struggle for existence:** Ibid., 97.

24 **publications from his work:** For a partial bibliography of Freud's extensive neuroscientific writings, see Freud, S. (1974), "Freud Bibliography," *S.E* 24:47–51.

24 **like the emancipation of woman:** See Molnar, M. (1999), "John Stuart Mill Translated by Sigmund Freud," *Psychoanalysis and History* 1: 195–205; and Bernfeld, S. (1949), "Freud's Scientific Beginnings," *The American Imago* 6: 163–96.

25 **in the winter of 1877:** Freud Letters, 30.

25 **prove critical to mavericks like Freud:** See Lesky, E. (1976), *The Vienna Medical School of the 19th Century*, trans. L. Williams and I. S. Levij, Baltimore: The Johns Hopkins University Press, 340–41. On the tension between the asylum doctors and those in the university-based clinics, see Engstrom, E. J. (2003), *Clinical Psychiatry in Imperial Germany: A History of Psychiatric Practice*, Ithaca, N.Y.: Cornell University Press.

25 **left much of an impression:** Freud's case reports can be found in Hirschmüller, A. (1991), *Freuds Begegnung mit der Psychiatrie: Von der Hirnmythologie zur Neurosenlehre*, Tübingen: edition diskord, 490.

25 **searching for new breakthrough treatments:** Freud Letters, 107.

25 **he wrote Martha, his fiancée:** Ibid., 108.

26 **help end his addiction:** Ibid., 107, 195, 200–01.

26 ***Centralblatt für die gesammte Therapie:*** Freud, S. (1884), "Über Coca," in *Cocaine Papers*, ed. R. Byck, New York: Stonehill, 1974, 49–73.

26 **effective and harmless:** See Bernfeld, S. (1953), "Freud's Studies on Cocaine, 1884–1887," *Journal of the American Psychoanalytic Association* 1: 581–613; and Swales, P. J. (1989), "Freud, Cocaine and Sexual Chemistry," in *Sigmund Freud: Critical Assessments*, vol. 1, ed. L. Spurling, London: Routledge, 273–301.

26 **in the nervous disorders:** Freud Letters, 88–9.

27 **go on kissing you:** Ibid., 154.

27 **in the Latin Quarter:** Jones, E. (1953), *The Life and Work of Sigmund Freud*, vol. I, 183.

27 **autopsies at the Paris Morgue:** Freud, S. (1886), "Report on My Studies in Paris and Berlin," *S.E.* 1:8.

27 **cases of Pin and Porez:** Freud, S. (1886), "Report on My Studies in Paris and Berlin," *S.E.* 1:12.

27 **Hermann Oppenheim of Berlin:** On this debate, see Lerner, P. (2003), *Hysterical Men: War, Psychiatry, and the Politics of Trauma in Germany, 1890–1930*, Ithaca, N.Y.: Cornell University Press, 27–39.

27 **on the other from fraud":** Freud, S. (1886), "Report on My Studies," 11–13.

27 **acknowledgement of a miracle":** Freud-Silberstein Letters, 177.

28 **new idea about perfection:** Freud Letters, 184–85.

28 **patients in Germany," Freud gushed:** Ibid., 188–89.

28　**created by the imagination:** After a number of years, he would publish a paper on the subject; Freud, S. (1893), "Some points for a comparative study of organic and hysterical motor paralyses," *S.E.* 1:157–74.

29　**traumatic paralysis in Vienna:** See Freud, Letters, 218. His October 15, 1885, presentation to the Gesellschaft der Aerzte in Wien was reported in Freud, S. (1886) "Ueber männliche Hysterie," *Wiener Medizinische Wochenschrift*, 43:1445–57. The responses of four Professors—Rosenthal, Meynert, Bamberger, and Leidesdorf—are also recorded there.

29　**presented such a case to the group:** Freud, S. (1886), "Observation of a Severe Case of Hemi-Anaesthesia in a Hysterical Male," *S.E.* 1:23–31. Also see Freud, S. (1886) "Beiträge zur Kasuistik der Hysterie," *Wiener Medizinische Wochenschrift*, 49:1633–38, 1674–76.

29　**ruined his former pupil:** Freud, S. (1889), "Review of August Forel's *Hypnotism*," *S.E.* 1:95–96.

29　**forced into the Opposition":** Freud, S. (1925), "An Autobiographical Study," *S.E.* 20:15–6; also see Jones, E. (1953), *The Life and Work of Sigmund Freud*, vol. 1, 231–32.

29　**"the wickedness of Paris":** Freud, S. (1889), "Review of August Forel's *Hypnotism*," *S.E.* 1:96.

29　**power of prejudice to blind:** See Freud's discussion of his status as an outsider with the Frenchman Gilles de la Tourette reported in Freud Letters, 203.

30　**sleep brought on by suggestion:** Liébeault, A. A. (1866), *Du Sommeil et des états analogues*, Paris: Masson, 293–353.

30　**psychological explanation of hypnosis:** Bernheim, H. (1886), *De la Suggestion et des ses applications à la thérapeutique*, Paris: Octave Doin. The English translation appeared a year later; Bernheim, H. (1887), *Suggestive Therapeutics*, trans. C. Herter, New York: Putnam and Sons, 1889.

30　**men and women of all temperaments:** Bernheim, H. (1887), *Suggestive Therapeutics*, vii.

30　**essential to normal psychological life:** Ibid., 132.

31　**he insisted, were wholly imaginary:** Ibid., 91.

31　**hypnotism could be found in Nancy:** Bernheim, H. (1891), *Hypnotisme, Suggestion, Psychothérapie*, Paris: Octave Doin, 167–69. Bernheim, H. (1887), *Suggestive Therapeutics*, 88–89.

31　**but also "anti-psychologic":** Janet, P. (1901), *The Mental State of Hystericals*, trans. C. Corson, New York: Putnam, 238.

31–32　**under the influence of suggestion":** Babinski, J. (1891), *Hypnotisme et Hystérie*, Paris: G. Masson, 21.

32　**hysteria hit the French press:** See Hillman, R. (1965), "A scientific study of mystery: the role of the medical and popular press in the Nancy-Salpêtrière controversy on hypnotism," *Bulletin of the History of Medicine* 39: 163–83; and Harris, R. (1989), *Murders and Madness: Medicine, Law, and Society in the Fin de Siècle*, Oxford, U.K.: Clarendon Press.

32　**one evil hypnotist away:** Forel, A. (1889), *Hypnotism and Psychotherapy*, trans. H. Armit, New York: Rebman Co., viii. Also see S. Freud, *Letters*, 1960, 188.

32　**potentially devastating critique:** Freud-Fliess Letters, 24. On Freud's navigating of the suggestion debates, also see Andersson, O. (1962); *Studies in the Prehistory of Psychoanalysis*, Stockholm: Svenska Bokförlaget, 47–79, and Makari G. (1992) "A History of Freud's First Concept of Transference," *I.R.P.* 19:415–32.

32 **"beyond scientific understanding":** Freud, S. (1888), "Preface to the Translation of Bernheim's *Suggestion*," *S.E.* 1:78.

32 **never saw Charcot's stages:** Moll, A. (1890), *Hypnotism*, London: Walter Scott, 77.

32 **519 could not be hypnotized:** Bramwell, J. M. (1930), *Hypnotism: Its History, Practice and Theory*, London: Rider and Co., 57.

33 **what those laws might be:** Freud, S. (1888), "Preface to the Translation of Bernheim's *Suggestion*," *S.E.* 1:79–86.

33 **brought on by autosuggestion:** Ibid., 83.

34 **study of the inner world:** Ibid., 83–84.

34 **based on unconscious psychology:** In 1889, Freud traveled to Nancy to consult with Bernheim on a difficult case. See Freud-Fliess Letters, 24; also Freud Letters, 394. On the case of Anna von Lieben, see Swales, P. (1986), "Freud, his teacher and the birth of psychoanalysis," in *Freud: Appraisals and Reappraisals*, vol. 1, ed. P. Stepansky, New York: Analytic Press, 3–82.

34 **plagued men since Adam:** Morel, B. A. (1860), *Traité des maladies mentales*, Paris: Masson. On the role of heredity and degeneration in French psychiatry, see the trenchant study, Dowbiggin, I. (1991), *Inheriting Madness: Professionalization and Psychiatric Knowledge in Nineteenth-Century France*, Berkeley: University of California Press.

35 **could be found among "Israelites.":** Charcot, J. M. (1892), *Leçons du Mardi à la Salpêtrière: Policlinique 1887–1888*, Paris: Aux Bureaux du Progrès Médical Louis Battaille, 477–78.

35 **minds sadly ill-suited for science:** Ribot, T. (1873), *Heredity: A Psychological Study of Its Phenomena, Laws, Causes, and Consequences*, 112–14.

35 **common in Jewish families":** Freud Letters, 210.

36 **prior infection of syphilis:** Goetz, C. G., M. Bonduelle, T. Gelfand (1995), *Charcot: Constructing Neurology*, New York: Oxford University Press, 262.

36 **with the other 10%:** Gelfand, T. (1988), " 'Mon Cher Docteur Freud': Charcot's Unpublished Correspondence to Freud, 1888–1893," *Bulletin of the History of Medicine* 62:573–74. © The Johns Hopkins University Press. Reprinted with permission of Johns Hopkins University Press. General Paresis was also known as "Paralyse générale progressive" or P.G.P., as Gelfand notes.

36 **could not be defended, he wrote:** Freud, S. (1892–1894), "Extracts from Freud's Footnotes to His Translation of Charcot's Tuesday Lectures," *S.E* 1: 142. On Freud's Judaism and his relation to French degeneration theory, see the excellent studies, Gilman, S. (1985), *Difference and Pathology*, Ithaca, N.Y.: Cornell University Press, and Gelfand, T. (1987), "Réflexions sur Charcot et la famille névropathique," *Hist. des Sciences Medicales* 21: 245–50.

37 **must apply to psychology":** Freud, S. (1893), "Some Points for a Comparative Study of Organic and Hysterical Motor Paralyses," *S.E* 1:171.

38 **hypnosis in Vienna prior to 1886:** Benedikt, M. (1894), *Hypnotismus und Suggestion: Eine klinisch-psychologische Studie*, Leipzig: Verlag der Buchhandlung M. Breitenstein, 17–28. Other Viennese advocates for hypnotism included Johann Schnitzler and Heinrich Obersteiner; see Hauser, R. I. (1989), *Sexuality, Neurasthenia and the*

Law: Richard von Krafft-Ebing (1940–1902), Ph.D diss., University College, University of London. Also see Borch-Jacobsen, M. (1996), *Remembering Anna O.: A Century of Mystification*, New York and London: Routledge, 111–18.

38   **other than a piece of absurdity":** Freud, S. (1889), "Review of August Forel's *Hypnotism*," *S.E.* 1:91–95. Also see Freud's "Preface to the Translation of Bernheim's *Suggestion*," *S.E.* 1:75, and his (1914), "On the History of the Psychoanalytical Movement," *S.E.* 14:9.

39   **not just any private practitioner:** Anti-Semitism often imposed a glass ceiling on academic physicians in Germany and Austria, but Breuer's biographer has argued that it was not central in this case. See Hirschmüller, A. (1978), *The Life and Work of Josef Breuer: Physiology and Psychoanalysis,* New York: New York University, 1989, 20–29.

39   **know her as "Anna O":** Freud Letters, 40–41.

39   **make sense of this striking case:** A transcript of this report is reproduced in Hirschmüller, A. (1978), *The Life and Work of Josef Breuer*, 276–311.

39   **real one and an evil one":** Ibid., 281.

39   **similarly altered mental condition:** Ibid., 282.

40   **lost their power to harm her:** Ibid., 288.

40   **Bertha's illness was not faked:** Ibid., 290–91, 295.

41   **new cases to his junior colleague:** Cranefield, P. F. (1958), "Josef Breuer's Evaluation of His Contribution to Psycho-Analysis," *I.J.P.*, 39: 319–22.

41   **Breuer adopted Charcot's terms:** Ibid., 320. See Breuer, J., and Freud, S. (1895), *Studies on Hysteria, S.E.* 2:237.

41   **advertise this new twist:** Freud, S. (1888), "Hysteria," *S.E.* 1:56.

41   **disorder could be cured:** Freud, S. (1889), "Review of August Forel's *Hypnotism*," *S.E.* 1:100.

41   **patient cured by hypnosis:** Freud, S. (1892–93), "A Case of Successful Treatment by Hypnotism," *S.E.* 1:117–28.

42   **promptly lost its power:** Ibid., 126.

42   *produced* **by "laborious" suppression:** Ibid. Also see *G.W.* 14. I have followed Strachey's translation of *Unterdrückung* as suppression while reserving Freud's *Verdrängung* for the term "repression."

42   **doctor furiously exclaimed:** Freud, S. (1921), "Group Psychology and the Analysis of the Ego," *S.E.* 18:89.

43   **in the process create illness:** Freud, S. (1892–93), "A Case of Successful Treatment by Hypnotism," *S.E.* 1:128.

43   **priority in a hot field:** Breuer, J., and S. Freud (1893), "On the Psychical Mechanism of Hysterical Phenomena: Preliminary Communication," *S.E* 2:3–17. Also see Freud's letter dated June 29, 1892, in Freud, S. (1892), "Letter to Joseph Breuer," *S.E.* 1:147.

43   **Alfred Binet, Pierre Janet, and Joseph Delboeuf:** Breuer, J., and S. Freud (1893), "On the Psychical Mechanism of Hysterical Phenomena," *S.E.* 2:7.

43   **notably Paul Möbius:** On this under-appreciated figure, see Schiller, F. (1982), *A Möbius Strip: Fin-de-Siècle Neuropsychiatry and Paul Möbius*, Berkeley: University of California Press. Hirschmüller lists Möbius, Benedikt, Adolf von Strumpel, and Oppenheim as innovators in this. Hirschmüller, A. (1978), *The Life and Work of Josef Breuer*, 147.

43 **authors jointly declared:** Breuer, J., and S. Freud (1893), "On the Psychical Mechanism of Hysterical Phenomena," *S.E.* 2:7.

43 **feeling of relief—a "cathartic effect.":** Ibid., 8.

44 **it was the cathartic method:** Bernays, J. (1880), *Zwei Abhandlungen über die Aristotelische Theorie des Dramas*, Berlin: Wilhelm Hertz. For a discussion of catharsis, see Lear, J. (1998), *Open-Minded: Working Out the Logic of the Soul*, Cambridge, Mass: Harvard University Press, 191–218.

44 **one aspect of that legacy:** Breuer, J., and S. Freud (1895), *Studies in Hysteria, S.E.* 2:3–305.

44 **cured by her treatment:** Ibid., 47.

44 **helped the patient more:** On the fate of "Anna O." see Hirschmüller, A. (1978), *The Life and Work of Josef Breuer*, 115. There is considerable debate surrounding Bertha's illness and the subsequent course of her life. For opposing views, see Borch-Jacobsen, M. (1996), *Remembering Anna O.: A Century of Mystification*, New York: Routledge; and Skues, R. (2006), *Sigmund Freud and the History of Anna O.—Reopening a Closed Case*, New York: Palgrave Macmillan.

44 **Fräulein Elisabeth von R.:** On the dating of the case of Emmy, see Tögel, C. (1999), " 'My Bad Diagnostic Error': Once More about Freud and Emmy v. N (Fanny Moser)," *I.J.P.* 80:1165–1173. On "Katarina," see the extraordinary historical reconstruction, Swales, P. (1988), "Freud, Katharina, and the First 'Wild Analysis,' " in *Freud Appraisals and Reappraisals: Contributions to Freud Studies*, vol. 3, ed. P. Stepansky, Hillsdale: Analytic Press, 80–164.

44 **shut their eyes, and concentrate:** Breuer, J., and S. Freud (1895), *Studies on Hysteria, S.E.* 2:109; on his later "pressure" technique, see 270.

44 **work for dissociated memories. He found it did:** Ibid., 110.

45 **it cannot be otherwise":** Ibid., 185.

45 **patients and co-exist with conscious mental life":** Ibid., 221.

45 **ideas alone caused all erections:** Ibid., 187.

45 **developed in those with pathological inheritances:** Ibid., 214–22, 234, 245.

45 **not a question of heredity:** Freud, S. (1894), "The Neuro-Psychoses of Defence," *S.E.* 3:46–47.

46 **links that Freud called "false connections":** Ibid., 52.

46 **same process as autosuggestion:** In 1889, Forel published an influential paper in which he defined "suggestion" as the automatic and unconscious process of "dissociating that which was associated and associating that which was not associated before." Forel, A. (1889), *Der Hypnotismus: Seine Bedeutung und seine Handhabung*, Stuttgart: Ferdinand Enke, 159.

46 **theory of autosuggestion:** Forel wrote: "Dr. Freud in Vienna has built a whole doctrine and method of treatment based on the fact of such auto-suggestions and the way they arouse emotions." Forel, A. (1903), *The Hygiene of Nerves and Mind in Health and Disease*, trans. H. Aikens, New York: Putnam & Sons, 221. Freud's review of Forel found him noting that " 'auto-suggestion'—[is] a term which incidentally, only appears to be an enrichment of the concept of 'suggestion' but is, strictly speaking, an abrogation of it." Freud, S. (1889), "Review of August Forel's *Hypnotism*," *S.E.* 1:97.

46 **process he called "psychical analysis":** Freud first used this phrase—*psychische Anal-*

yse—in 1894; see Freud, S. (1894), "The Neuro-Psychoses of Defence," *S.E.* 3:47, also see *G.W.* 1:61.

46 **but rather psychic conflict:** Breuer, J., and S. Freud (1895), *Studies on Hysteria, S.E* 2:266.

46 **brains of *dégénérés* and *déséquilibrés":* Ibid., 294.

47 **of memories or the connection of events":** Ibid., 295.

47 **It will do no harm":** Ibid.

48 **mistakenly tied to him:** Ibid., 302–03.

48 **"false connection have taken place":** Ibid., 303.

49 **altered states of consciousness:** See Janet, P. (1886), "Les acts inconscients et le dédoublement de la personnalité pendant le somnambulisme provoqué," *Revue philosophique* 22:577–92; Janet, P. (1887), "L'anesthésie systematisée et la dissociation des phénomènes psychologiques," *Revue philosophique* 23:449–72; and Janet, P. (1888), "Les acts inconscients et la mémoire pendant le somnambulisme," *Revue philosophique* 25:238–79.

49 **philosophers, animal magnetists, and alienists:** Janet, P. (1889), *L'Automatisme psychologique: essai de psychologie expérimentale sur les formes inférieures de l'activité humaine*, Paris: F. Alcan, 1973.

49 **solid ground for an objective psychology:** Ibid., 27.

49 **result of psychological dissociation:** Ibid., 415.

49 **thank them for their amiable citation":** Janet, P. (1893), "Quelques définitions récentes de l'hystérie," *Archives de Neurologie* 25:437.

50 **reinforced existing French theories:** Ibid., 438.

50 **form of psychological excess:** Breuer, J., and S. Freud (1895), *Studies on Hysteria, S.E.* 2:240.

50 **together broken associations in his patients:** See cases of Marie and Lucie, in Janet, P. (1889), *L'Automatisme psychologique: essai de psychologie expérimentale sur les formes inférieures de l'activité humaine*, Paris: F. Alcan, 1973. 458.

50 **Janet's understanding of neurosis:** On Janet and Freud see MacMillan, M. (1979), "Delbeouf and Janet as influences in Freud's treatment of Emmy von N.," *Journal of the History of the Behavioral Sciences* 15:299–309; Decker, H. (1986), "The lure of non-materialism in materialist Europe: investigations of dissociative phenomena, 1880–1915," in *Split Minds/Split Brains*, ed. J. Quen, New York: New York University Press: 31–62; and MacMillan, M. (1990), "Freud and Janet on organic and hysterical paralyses: a mystery solved?" *I.R.P.*, 17:189–203.

50 **dominant role in hysteria:** Janet, P. (1893), "Quelques définitions récentes de l'hystérie," *Archives de Neurologie* 26:27.

51 **hopefully referred to as "Frend":** On Janet's fall out of history, see Ellenberger, H. (1970), *The Discovery of the Unconscious*, New York: Basic Books, 407.

51 **in some ways they were right:** See, for example, the complaints registered in Anonymous (1913), Review of 'La Doctrine de Freud' by Régis and Hesnard," *Archives Internationales de Neurologie* 2:128–30, where the authors claim that most of Freud's concepts were derived from French psychopathology.

51 **to the Viennese doctor:** Janet, P. (1914), "Psychoanalysis," *The Journal of Abnormal Psychology* 9:1–187. The "Freudian" was Carl Jung.

52 **course of that affection":** Breuer, J., and S. Freud (1895), *Studies on Hysteria, S.E.* 2:160–61.

2. CITY OF MIRRORS, CITY OF DREAMS

53 **"No more strange than us," I said:** Plato (2000), *The Republic*, ed. G. R. F. Ferrari, trans. T. Griffith, Cambridge, U.K.: Cambridge University Press, 220.

54 **limit on such knowledge:** Kant, I. (1781), *Critique of Pure Reason*, trans. N. K. Smith, New York: St. Martin's Press, 1929; also Kant, I. (1783), *Prolegomena to Any Future Metaphysics*, trans. P. Carus, Indianapolis: Hackett Pub., 1977.

55 **could never be a science:** See Friedman, M. (1992), *Kant and the Exact Sciences*, Cambridge, Mass.: Harvard University Press; and Gary Hatfield (1992), "Empirical, rational and transcendental psychology: Psychology as science and as philosophy," in *The Cambridge Companion to Kant*, ed. P. Guyer, Cambridge, U.K.: Cambridge University Press, 200–27.

55 **read, misread, cited, and appropriated:** See Sassen, B. (1997), "Critical Idealism in the Eyes of Kant's Contemporaries," *Journal of the History of Philosophy* 35: 421–55.

55 **two aspects of the same world:** Schelling, F. W. J. (1802), "On the Relationship of the Philosophy of Nature to Philosophy in General," in *Between Kant and Hegel: Texts in the Development of Post-Kantian Idealism*, eds. G. di Giovani and H. S. Harris, Albany: State University of New York Press, 1985: 363–82. Also Robinson, H. (1994), "Two Perspectives on Kant's Appearances and Things in Themselves," *Journal of the History of Philosophy* 32: 411–41.

55 **known as dual aspect monism:** Leary, D. E. (1982), "Immanuel Kant and the Development of Modern Psychology," in *The Problematic Science: Psychology in Nineteenth-Century Thought*, eds. W. R. Woodward and M. G. Ash, New York: Praeger, 17–41.

55 **inner lives and the natural world:** Pinkard, T. (2002), *German Philosophy, 1760–1860: The Legacy of Idealism*, Cambridge, U.K.: Cambridge University Press, 141.

55 **electricity and magnetism with striking results:** Kuhn, T. S. (1977), "Energy Conservation as an Example of Simultaneous Discovery," in *The Essential Tension: Selected Studies in Scientific Tradition and Change*, Chicago: University of Chicago Press, 66–104.

55 **life in a grand evolutionary sweep:** See Schelling, F. W. (1803), "Ideas on a Philosophy of Nature as an Introduction to the Study of This Science," in *Philosophy of German Idealism*, ed. E. Behler, New York: Continuum, 1987, 167–02.

55 **nature into the consciousness of man:** For Carus, unconscious psychic life was unified with and the basis for consciousness. See Carus, C. G. (1851), *Psyche: On the Development of the Soul*. Dallas: Spring Publications, 1970.

55 **mind might somehow cure the body:** On Romantic medicine see the fine discussion in Ellenberger, H. (1970), *The Discovery of the Unconscious*, 202–15.

56 **crucial tasks for philosophy:** Fichte, J. G. (1794–1802), *The Science of Knowledge*, Cambridge, U.K.: Cambridge University Press, 1982. Also Breazeale, D. (2001), "Fichte's Conception of Philosophy as a 'Pragmatic History of the Human Mind' and the Contributions of Kant, Platner, and Maimon," *Journal of the History of Ideas* 62:685–703; and Neuhouser, F. (1990), *Fichte's Theory of Subjectivity*, Cambridge, U.K.: Cambridge University Press.

56 **perceived them as existing out there:** Schopenhauer, A. (1815), *Textes sur la vue et sur les couleurs*, Paris: J. Vrin, 1986, 50–51. The philosophical debates regarding vision were examined in Crary, J. (1990), *Techniques of the Observer*, Cambridge, Mass.: MIT Press. Also see Lauxtermann, P. F. H. (1987), "Five decisive years: Schopen-

hauer's epistemology as reflected in his theory of colour," *Stud. Hist. Philos. Sci.* 18:271–91.

56 *The World as Will and Representation*: Schopenhauer, A. (1819), *The World as Will and Representation*, vol. I, New York: Dover Publications, Inc., 1969; Schopenhauer, A. (1844), *The World as Will and Representation*, vol. II, New York: Dover Publications, Inc., 1969.

56 **distorting powers in mental life:** Mandelbaum, M. (1971), *History, Man, & Reason: A Study in Nineteenth-Century Thought*, Baltimore and London: The Johns Hopkins University Press, 315–25, Crary, J. (1990), *Techniques of the Observer*, 73–85.

56–57 **"fantastic apparitions of vision":** Müller, J. (1826), *Ueber Die Phantastischen Gesichtserscheinungen*, Coblenz: Jacob Hölscher.

57 **scientific proof of Schopenhauer's contentions:** Müller, J. (1833), *Elements of Physiology*, trans. W. Baly, London: Taylor and Walton, 1838, 1059–1065.

59 **not representative of the whole:** The confusing misnomer—the "Helmholtz School"—seems to have originated from Bernfeld, S. (1944), "Freud's earliest theories and the school of Helmholtz," *Psychoanalytic Quarterly* 13:341–62. I follow Cranefield in referring to this emerging community as the "biophysics" movement. See Cranefield, P. F. (1966), "The Philosophical and Cultural Interests of the Biophysics Movement of 1847," *Journal of the History of Medicine* 21:1–4; Cranefield, P. F. (1966), "Freud and the 'School of Helmholtz,' " *Gesnerus* 23:35–39.

59 **teacher and avid follower of Fichte:** Helmholtz, H. (1891), "Autobiographical Sketch," *Popular Lectures on Scientific Subjects*, vol. 2, trans. E. Alkinson, London, New York: Longmans, Green and Co., 1903, 266–91.

59 **amount of energy remained unchanged:** Helmholtz, H. (1847), "The Conservation of Force: A Personal Memoir," in *Selected Writings of Hermann von Helmholtz*, ed. R. Kahl, Middletown, Conn.: Wesleyan U. Press, 1971; also see Helmholtz, H. (1862–63), "On the Conservation of Force," in *Science and Culture*, ed. D. Cahan, Chicago: University of Chicago, 1995, 96–126.

59 **tracked different transformations of energy:** Helmholtz, H. (1890), "Interaction of Natural Forces (1854)," in *The Correlation and Conservation of Forces: A Series of Expositions*, ed. E. Youmans, New York: D. Appleton and Co., 211–50. On the origins of this theory, see Heimann, P. M (1974), "Helmholtz and Kant: The Metaphysical Foundation of Uber die Erhaltung der Kraft," *Studies in the History and Philosophy of Science*, 5:205–38; and Kuhn, T. S. (1977), "Energy Conservation as an Example of Simultaneous Discovery," in *The Essential Tension*, 66–104.

59 **human sciences (*Geisteswissenschaft*):** Veit-Brause, I. (2001), "Scientists and the cultural politics of academic disciplines in late 19th century Germany: Emil Du Bois-Reymond and the controversy over the role of cultural sciences," *History of the Human Sciences* 14:50.

59 **education of the German elite:** Berrios, G. (2002), "Introduction to Robert Gaupp's 'About the Limits of Psychiatric Knowledge,' " *History of Psychiatry* 13:327–31.

59 **psychology of the knower:** Helmholtz, H. (1862), "On the Relation of Natural Science to Science in General," in *Science and Culture*, ed. D. Cahan, Chicago: University of Chicago, 1995, 76–95.

60 **reflex-driven, churning machine:** See Lesky, E. (1976), *The Vienna Medical School of the 19th Century*, Baltimore: The Johns Hopkins University Press, 228–37; and Am-

acher, P. (1965), *Freud's Neurological Education and Its Influence on Psychoanalytic Theory*, New York: International Universities Press, 18–19.

60 **education and criminal law:** Exner, S. (1894), *Entwurf zu einer physiologischen Erklärung der psychischen Erscheinungen*, Leipzig: Franz Deuticke.

61 **was published in 1884:** Meynert, T. (1884), *Psychiatry: A Clinical Treatise on Diseases of the Fore-Brain Based upon a Study of Its Structure, Functions, and Nutrition*, trans. B. Sachs, New York: Putnam & Sons, 1885.

61 **this set of assumptions:** Ibid., viii–ix. Also see Meynert, T. (1890), *Klinische Vorlesungen über Psychiatrie auf Wissenschaftlichen Grundlagen für Studirende und Aerzte, Juristen und Psychologen*, Wien: Wilhelm Braumüller.

61 **flickering light with sharp pain:** Meynert, T. (1884), *Psychiatry*, 272.

62 **particular regions of the brain:** Ibid., 153.

63 **human beings into mechanical dolls:** Hartmann, E. (1884), *Philosophy of the Unconscious: Speculative Results According to the Inductive Method of Physical Science*, trans. W. C. Coupland, New York: Harcourt, Brace and Company, 1931, xxx.

63 **inner experience from internal templates:** In lectures, Meynert cited Schopenhauer while putting forward the view that our perceptions and associations were clouded, colored, directed, and transformed by an inner bodily force, a "Will-impulse." Meynert, T. (1890), *Klinische Vorlesungen über Psychiatrie*, 119. Also see Meynert, T. (1884), *Psychiatry*, 183.

63 **away from the self by defenses:** The term Meynert used, "Abwehr," was the same used by Freud. See Meynert, T. (1890), *Klinische Vorlesungen über Psychiatrie*, 227–78. On this matter, I am indebted to the excellent work of William McGrath, who pointed this out; see his (1986), *Freud's Discovery of Psychoanalysis: The Politics of Hysteria*, Ithaca, N.Y.: Cornell University Press, 142–49.

63 **conflict with the primary body "I":** See Meynert, T. (1884), *Psychiatry*, 175–76; Meynert, T. (1892), *Klinische Vorlesungen über Psychiatrie*, 228.

63 **dubious anatomy and physiology":** James, W. (1890), *The Principles of Psychology*, New York: Henry Holt and Company, 80.

64 **return beaten and bloodied:** On the reception of Exner and Meynert, see Hirschmüller, A. (1978), *The Life and Work of Josef Breuer*, 341; and Lesky, E. (1976), *The Vienna Medical School of the 19th Century*, Baltimore: The Johns Hopkins University Press, 338, 494.

64 **energies toward other projects:** See Harman, P. M. (1982), *Energy, Force, and Matter: The Conceptual Development of Nineteenth-Century Physics*, Cambridge, U.K.: Cambridge University Press.

64 **beyond the reach of science:** Du Bois-Reymond, E. (1872), "Ueber die Grenzen des Naturkennens," in *Reden von Emil Du Bois-Reymond*, Leipzig: Verlag Von Veit, 1886, 105–30. For an English translation, see Du Bois-Reymond, E. (1874), "The Limits of Our Knowledge of Nature," *The Popular Science Monthly* 5:17–32.

64 **issues vital to any psychology:** Du Bois-Reymond, E. (1880), "Die sieben Welträthsel," in *Reden von Emil Du Bois-Reymond*, 1886, 381–411. In English, see Du Bois-Reymond, E. (1882), "The Seven World-Problems," *The Popular Science Monthly* 20:432–47.

65 **manifest in madness and clairvoyance:** Fechner, G. T. (1836), *The Little Book of Life After Death*, trans. M. Wadsworth, Boston: Little Brown, & Company, 1912.

65 **governed rocks and air:** Fechner, G. T. (1848), "Ueber das Lustprinzip des Han-

delns," *Zeitschrift für Philosophie und philosophische Kritik*, 19:1–30, 163–94. Also, Marshall, M. (1982), "Physics, Metaphysics and Fechner's Psychophysics," in *The Problematic Science: Psychology in Nineteenth Century Thought*, eds. W. Woodward and M. Ash, New York: Praeger, 65–87; and the classic text, Boring, E. G. (1957), *A History of Experimental Psychology*, New York: Appleton-Century-Crofts, Inc., 275–96.

65 **omnipresent spirit force in Nature:** Fechner, G. T. (1848), *Nanna; oder über das Seelenleben der Pflanzen*, Leipzig: Leopold Voss.

65 **outer stimulus and inner sensation:** If the difference between two stimuli was proportionally constant, he claimed, the perception of that difference would be the same. So, the human perception of the difference between 1 and 100 units of touch was supposed to be the same as the difference between 2 and 200, 3 and 300, etc. In this way, Weber hoped to give inner perception a precise mathematical relation to external sensory stimulation, and create a fundament for psychological investigation.

65 **physical stimuli to qualitative mental experience:** Fechner, G. T. (1860), *Elements of Psychophysics*, trans. H. Adler, New York: Holt, Rinehart and Winston, Inc., 1966.

66 **poor recommendation if they were":** Ibid., xxviii.

66 **crossed a line into consciousness:** Ibid., 199.

67 **not be consciously seen or heard:** Ibid., 205–06, 208.

67 **young physicist Ernst Mach:** Mach, E. (1863), "Lectures on Psychophysics—Conclusion (1863)," in *Ernst Mach—A Deeper Look: Documents and New Perspectives*, ed. J. Blackmore, Dordrecht: Kluwer Academic Pub., 1992, 111–14.

67 **to adopt Fechner's approach:** Hirschmüller, A. (1978), *The Life and Work of Josef Breuer: Physiology and Psychoanalysis*, New York: New York University Press, 39–40. Also see Breuer's letter to Brentano, Ibid., 247–55.

67 **nor spirit without matter":** Haeckel, E. (1900), *The Riddle of the Universe at the Close of the Nineteenth Century*, trans. J. McCabe, New York: Harper & Brothers Publishers. 20–21.

67 **process Helmholtz called "unconscious inference":** Helmholtz, H. (1868), "The recent progress in the theory on vision," in *Selected Writings of Hermann von Helmholtz*, 1971, 168–69; Helmholtz, H. (1878), "The facts of perception," in Ibid., 369. The phrase for unconscious inference in German is *unbewusster Schluss*.

67 **from the past with a foreign present:** Helmholtz, H. (1867), *Handbuch der Physiologischen Optik*, Leipzig: Leopold Voss, 430, 447–49. In English, see Helmholtz, H. (1910), *Helmholtz's Treatise on Physiological Optics*, ed. J. P. C. Southhall, The Optical Society of America, 1925, 10–12; and Helmholtz, H. (1868), "The recent progress in the theory on vision," in *Selected Writings of Hermann von Helmholtz*, 213–17.

68 **nonphysical mind forces:** Others sought to explain away Helmholtz's work as simply a new term for the passive association of ideas by habit. And indeed, Helmholtz at times could make it seem as if this was so. See Helmholtz, H. (1910), *Helmholtz's Treatise on Physiological Optics*, 26. On Helmholtz's attempt to resolve unconscious inference into a process of association see Hatfield, G. (1990), *The Natural and the Normative:Theories of Spatial Perception from Kant to Helmholtz*, Cambridge, Mass.: The MIT Press, 204.

68 **subjective idealists were correct:** Helmholtz, H. (1878), "The facts of perception," *Selected Writings of Hermann von Helmholtz*, 384–85.

68 **being subsumed into the other:** Wundt, W. (1873–74), *Principles of Physiological Psy-*

chology, trans. E. B. Titchner, London: Swan Sonnenschein & Co., vol. 1, 1904, 102–03. For the German, see Wundt, W. (1873–74), *Grundzüge der Physiologischen Psychologie*, Leipzig: Wilhelm Engelmann. While Wundt initially treated psychophysics as synonymous with physiological psychology, a terminological confusion would soon arise, as "physiological psychology" was used to mean both psychophysics and its opposite, that is the attempt to reduce psychology to physiology.

68 **constantly come and go":** Stumpf, C. (1895), "Hermann von Helmholtz and the New Psychology," *The Psychological Review* 2:1–12.

69 **inner experiences of sensory perception:** Freud, S. (1888), "Gehirn," in *A Moment of Transition: Two Neuroscientific Articles by Sigmund Freud*, ed. M. Solms and M. Saling, London: Karnac, 1990, 41, 62.

69 **lifting a thought into consciousness:** Ibid., 62–63. Scholars agree that this passage marks a rejection of epiphenomenalism and reductive monism. After that, they disagree. Silverstein saw Freud adopting a mind-brain interactionalism; Silverstein, B. (1985), "Freud's psychology and its organic foundation: sexuality and mind-body interactionism," *Psychoanalytic Review*, 72:23–28. Solms and Saling argue Freud adopted a purely psychological, dynamic model, based on psychophysical parallelism, which I believe is more than can be concluded from this text. See their essay, "Significance of 'Gehirn' for Psychoanalysis," in *A Moment of Transition: Two Neuroscientific Articles by Sigmund Freud*, eds. M. Solms and M. Saling, London: Karnac, 1990, 94.

69 **psychic intentions might act in competition:** Freud, S. (1888), "Gehirn," in *A Moment of Transition*, 63–65.

69 **made it into our conscious awareness:** Herbart, Johann Friedrich (1834), *A Textbook in Psychology: An Attempt to Found the Science of Psychology on Experience, Metaphysics, and Mathematics*, trans. M. K. Smith, New York: D. Appleton and Co., 1891, 12–13.

69 **while in the Gymnasium:** Siegfried and Suzanne Bernfeld discovered this, as reported in Jones, E. (1953), *The Life and Work of Sigmund Freud*, vol. 1, 374. At the University of Vienna, however, he had gotten a very unfavorable appraisal of Herbart's work from Franz Brentano. The extent of Herbart's influence on Freud is controversial. Those who emphasize the connection include Maria Dorer in her (1932), *Historische Grundlagen der Psychoanalyse*, Leipzig: Hirschfeld Verlag. It was picked up by others like Ernest Jones, Didier Anzieu, and most recently Rosemarie Sand in her (2004), "The Unconscious: What Freud Learned in High School," presented to the Institute for the History of Psychiatry, Weill Medical College of Cornell, October 6, 2004. Against these enthusiasts stand Peter Amacher and William McGrath, who I believe are more convincing.

69 **calling it a "masterpiece":** On Freud and Hering, see Strachey, J. (1957), "Freud and Ewald Hering," *S.E* 14:205. On Breuer and Hering, see Hirschmüller, A. (1978), *The Life and Work of Josef Breuer: Physiology and Psychoanalysis*.

69 **took up unconscious psychology:** Freud-Silberstein Letters, 49, 84, 118–20. In 1883, Freud told Martha that Helmholtz was one of his "idols." Cited in Jones, E. (1953), *The Life and Work of Sigmund Freud*, vol. 1, 41.

69 **quantitative sensations alongside qualitative perceptions:** This was one of three courses Freud took with Exner. See Bernfeld, S. (1951), "Sigmund Freud, M.D.,

1882–1885," *I.J.P.* 32:204–17. On Exner's repudiation of this endeavor, see Green-berg, V. (1997), *Freud and His Aphasia Book: Language and the Sources of Psychoanalysis*, Ithaca, N.Y.: Cornell University Press, 131–34.

70 **speech or the comprehension of language:** Freud, S. (1888), "Aphasie," in *A Moment of Transition*, 31–37.

70 ***On Aphasia: A Critical Study:*** Freud, S. (1891), *On Aphasia: A Critical Study*, New York: International Universities Press, 1953.

70 **anatomical approach to this illness:** The aphasias were a daunting arena for Freud to make this case. After the landmark work of Carl Wernicke and Paul Broca, it seemed that a strictly anatomical model had yielded precise localizations for speech and language. To parry this line of thought, Freud turned to the work of his friends Sigmund Exner and Josef Paneth, who had shown that cutting the surrounding cortex of a cerebral center has the same effect as removing it entirely. So were these brain regions really centers, or just connections for more global activity? Freud voted for the latter.

70 **anatomical language only fostered confusion:** Freud, S. (1891), *On Aphasia*, 54–55.

70 **different approach to mind and brain:** Wundt, W. (1873–74), *Principles of Physiological Psychology*, 308.

70 **who championed "psychophysical parallelism":** Freud, S. (1891), *On Aphasia*, 56, 61.

70 **how the psychic and physical interacted:** Hughlings Jackson, J. (1887), "Remarks on Evolution and Dissolution of the Nervous System," in *Selected Writings of John Hughlings Jackson*, vol. 2, ed. J. Taylor. New York: Basic Books, 1958, 76–118.

71 **called his private "tyrant":** Freud-Fliess Letters, 129.

71 **in short all of psychology":** Ibid., 136.

71 **specifiable material particles":** Freud, S. (1895), "Project for a Scientific Psychology," *S.E.* 1:295.

71 **Kant is apparent in the "Project":** Trosman, H., and R. D. Simmons (1973), "The Freud Library," *Journal of the American Psychoanalytic Association* 11:646–87.

71 **motion and the conservation of energy:** Breuer wrote Auguste Forel in 1907 and stated that the "idea of the etiological significance of affective ideas" was his own, while the notion of a "conversion of affective excitation" was Freud's, see Cranefield, P. F. (1958), "Josef Breuer's Evaluation of His Contribution to Psycho-Analysis," *I.J.P.* 39:319–22.

71 **association or motor action:** Freud, S. (1892), "Letter to Josef Breuer," *S.E.* 1:147; also Breuer, J., and Freud, S. (1895), *Studies on Hysteria, S.E.* 2:86, 201, and Freud, S. (1894), "The Neuropsychoses of Defence," *S.E.* 3:60–61.

72 **inner experience, consciousness, and perception:** Freud, S. (1895), "Project for a Scientific Psychology," *S.E.* 1:309.

72 **Consciousness thereby resulted:** Ibid., 311. This hunch probably came out of Freud's impassioned dialogues with Wilhelm Fliess, a physician who firmly believed in the periodicity of all natural phenomena.

72 **in only one place: dreams:** Ibid., 322–27, 340, 360–61.

72 **would function on its own":** Freud-Fliess Letters, 146.

73 **a "kind of madness":** Ibid., 152.

73 **tacitly reinstated the 'idea' ":** Breuer, J., and Freud, S. (1895) *Studies on Hysteria*, S.E. 2:185.

74 **"ideal and woebegone child—metapsychology":** Freud-Fliess Letters, 172, 216. In this passage Freud referred to Ribot's friend Hippolyte Taine, whose work *L'Intelligence* suited Freud "extraordinarily well." On Taine and Freud, see Makari, G. J. (1994), "In the eye of the beholder: Helmholtzian perception and the origins of Freud's 1900 theory of transference," *Journal of the American Psychoanalytic Association* 42:549–80.

74 **described a new psychology":** Freud-Fliess Letters, 180, 208.

74 **neuroses and the new psychology":** Ibid., 219.

74 **of a man's life":** Freud, S. (1900), *The Interpretation of Dreams, S.E.* 4:xxvi.

74 **unconscious, hypnotic hallucinations:** On the continuity between Freud's self-analysis and the self-exploration of hypnotists like Forel, Vogt, and others, see Mayer, A. (2006), "La Spécificité de l'auto-analyse freudienne: un Expérimentalisme sans laboratoire," *Psychiatrie, Sciences Humaines, Neurosciences*, 16:23–33.

75 **self-reproach for neglecting his father:** Freud-Fliess Letters, 202.

75 **he felt "so very certain":** Ibid., 243.

75 **point to areas of conflict:** See, for example, Freud, S. (1900), *The Interpretation of Dreams, S.E.* 4:317; I am indebted to Patrick Mahoney for this insight; see his (1994), "Psychoanalysis—The Writing Cure," in *100 Years of Psychoanalysis*, eds. A. Haynal and E. Falzeder, Geneva: Cahiers Psychiatriques Genevois, 101–20.

75 **told Wilhelm Fliess on October 15, 1897:** Freud-Fliess Letters, 270.

76 **here transplanted into reality":** Freud-Fliess Letters, 272. Note that at this time, Freud was making an observation about normal psychology, not neurosis. On Freud and his culture's fascination with antiquity, see Armstrong, R. (2005), *A Compulsion for Antiquity, Freud and the Ancient World*, Ithaca, N.Y.: Cornell University Press; for Freud's engagement with Sophocles' great drama, see Rudnytsky, P. (1987), *Freud and Oedipus*, New York: Columbia University Press.

76 **younger brother Julius died:** Freud-Fliess Letters, 268.

76 **than I had imagined":** Ibid., 300.

76 **hypnotists and psychophysicists:** See, for example, the scientific publications on this subject listed in Warren, H., and L. Farrand, eds. (1894) *The Psychological Index*, 1:40–41; Ibid. (1895) 2:43–44; and Ibid. (1896), 3:69–71.

77 **madness a long dream":** Freud, S. (1900), *The Interpretation of Dreams, S.E.* 4:90.

77 **provoked illusions, he insisted:** Ibid., 29, 40–41, 222. Also Wundt, W. (1897), *Outlines of Psychology*, trans. C. H. Judd, Leipzig: W. Engelmann, 272–74.

77 **indigestion might catalyze dreams:** Freud, S. (1900), *The Interpretation of Dreams, S.E.* 4:36.

77 **hide from the problem:** Ibid., 41–42.

78 **revealed itself to Dr. Sigm. Freud":** Freud-Fliess Letters, 417.

78 **fantasy, thought, and feeling:** Freud, S. (1900), *The Interpretation of Dreams, S.E.* 4:101.

79 **dignitaries like Alphonse Daudet:** Ibid., 107–18, 121, 126.

79 **elude the psychic censors:** Ibid., 142.

79 **sneak past the censor:** Ibid., 283, 305–07.

79 **model of the mental life:** Ibid., 145.

80 **by reading Gustav Fechner:** In his *Autobiography*, Freud would openly admit "I was always open to the ideas of G. T. Fechner and have followed that thinker upon many

important points." Freud, S. (1925), "An Autobiographical Study," *S.E.* 20:59. Freud's interest in Fechner dated to his youth. In 1874, he made inquiries regarding Fechner's teachings, and in 1879 he reported reading the essays of this "great philosopher and wit." Freud-Silberstein Letters, 66, 71, 175.

80 **crude map of that territory":** Freud-Fliess Letters, 1985, 299.

80 **peculiarities of dream-life intelligible":** The italics are the Viennese doctor's. See Freud, S. (1900), *The Interpretation of Dreams, S.E.* 5:536.

80 **neuron or brain region:** Freud would never abandon this idea. As late as 1925, a student of Freud's was disappointed to find the Professor attacking his theoretical proposals for not correlating with "laws operative within the physical universe," see Burrow, T. (1958), *A Search for Man's Sanity: The Selected Letters of Trigant Burrow*, ed. W. Galt et. al., New York: Oxford University Press, 95.

80 **remain upon psychological ground:** Freud, S. (1900), *The Interpretation of Dreams, S.E.* 5:536; also see *S.E.* 4:48.

80 **rather as organized, psychological structures:** Ibid. On the influence of Fechner on Freud, see the Freud-Binswanger Letters, 176–79.

80 **prevent disruptive elements from invading consciousness:** Freud, S. (1900), *The Interpretation of Dreams, S.E.* 5:540–42.

81 **its effects on consciousness:** Freud cited Kant's work seven times in *The Interpretation of Dreams*. Years later to Binswanger, Freud would equate his unconscious with Kant's *Ding an Sich*. See Binswanger, L. (1957), *Sigmund Freud: Reminiscences of a Friendship*, New York and London: Grune and Stratton, 9.

81 **without passing the defenses:** Freud, S. (1900), *The Interpretation of Dreams, S.E.* 4:157. Freud's need to make his rule universal led him to explain counter-wish dreams as motivated by the *wish* to thwart the analyst. This is both an early notion of transference and an early indication of a capacity for tendentiousness in Freud's reasoning. Ellenberger pointed out that Fechner, using the pleasure principle, employed the same logic; Ellenberger, H. (1956), "Fechner and Freud," *Bulletin of the Menninger Clinic* 20:201–14.

81 **Freud called this "regression":** This echoed John Hughlings Jackson's "doctrine of 'functional retrogression.' " Hughlings Jackson, J. (1887), "Remarks on Evolution and Dissolution of the Nervous System," in *Selected Writings of John Hughlings Jackson*, vol. 2, ed. J. Taylor, New York: Basic Books, 1958, 87.

81 **won free passage was during sleep:** Freud, S. (1900), *The Interpretation of Dreams, S.E.* 5:565–66.

82 **"consciousness, quality, and so forth":** Freud-Fliess Letters, 325, 329.

82 **Cs. system and the perceptual systems":** Freud, S. (1900), *The Interpretation of Dreams, S.E.* 5:616.

82 **communications of our sense organs":** Ibid., 613.

82 **described hysterical false connections:** On the early history of the concept of transference, see Kravis, N. (1992), "The prehistory of the idea of transference," *I.R.P.* 19:9–22, and Makari, G. J. (1992), "A history of Freud's first concept of transference," *I.R.P.* 19:415–32.

82–83 **attachment for a transference":** Freud, S. (1900), *The Interpretation of Dreams, S.E.* 5:562–64.

83 **Helmholtz and Theodor Lipps:** Lipps, T. (1883), *Grundtatsachen des Seelenlebens*,

Bonn: Cohen, 87–95. On Lipps's notion of transference, see Makari, G. J. (1994), "In the eye of the beholder: Helmholtzian perception and the origins of Freud's 1900 theory of transference," *Journal of the American Psychoanalytic Association* 42:549–80.

83   **"royal road to the unconscious":** Freud, S. (1900), *The Interpretation of Dreams, S.E.* 5:608.

3. THE UNHAPPY MARRIAGE OF PSYCHE AND EROS

85   **serpent she must wed . . .":** Apuleius (1960), *The Golden Ass*, trans. J. Lindsay, Bloomington: Indiana University Press, 108.

86   **"a sort of insult":** Breuer, J., and S. Freud (1895), *Studies on Hysteria, S.E.* 2:259–60.

86   **could cause infantile hysteria:** See Baginsky, A. (1877), *Handbuch der Schulhygiene zum Gebrauche für Ärzte, Sanitätsbeamte, Lehrer, Schulvorstände und Techniker*, Berlin: Denicke. On Baginsky, Kassowitz, and Freud, see Bonomi, C. (1994), "Why Have We Ignored Freud the 'Paediatrician'? The relevance of Freud's paediatric training for the origins of psychoanalysis," in *100 Years of Psychoanalysis*, eds. A. Haynal and E. Falzeder, Geneva: Cahiers Psychiatriques Genevois, 55–99.

86   **traumas were sexual in nature:** There are serious debates about the extent to which Freud's notions of sexual trauma emerged from his experiences with patients and how much they were suggestions foisted upon patients by Freud.

87   **due to "sexual abuse":** Fliess, W. (1893), "Les réflexes d'origine nasale," *Archives Internationales de Laryngologie, Otologie, Rhinologie et de Broncho-Oesophagascopie* 6:266–69. He wrote of *les abus sexuels* on p. 268. At times Freud referred to the sexual etiology as "our etiological formula," indicating the intimate cooperation with Fliess; see Freud-Fliess Letters, 45.

87   **potential causes of this disorder:** Beard, G. (1884), *Sexual Neurasthenia; Its Hygiene, Causes, Symptoms, and Treatment*, New York: E. B. Treat, 127.

87   **neurasthenia, homosexuality, and hysteria:** See Makari, G. (1998), "Between Seduction and Libido: Sigmund Freud's Masturbation Hypotheses and the Realignment of His Etiologic Thinking, 1897–1905," *Bulletin of the History of Medicine* 72:638–62.

87   **would have overlooked it":** Freud-Fliess Letters, 66.

88   **great secrets of nature":** Ibid., 74.

88   **led to repression and psychoneurosis:** Ibid., 76–78.

88   **neurotic, hysterical, or simply sexually unruly:** See Merkel, F. (1887), *Beitrag zur Casuistik der Castration bei Neurosen*, Nürnberg: Kaiser Wilhelms-Universität Strassburg.

88   **neurasthenia was caused by masturbation:** Freud, S. (1895), "On the Grounds for Detaching a Particular Syndrome from Neurasthenia under the Description 'Anxiety Neurosis,'" *S.E.* 3:87–115. On Breuer's approval see his letters to Fliess of Oct. 16 and 24, 1895, reprinted in Hirschmüller, A. (1978), *The Life and Work of Josef Breuer*, 316–18.

89   **refraining from any sexual activity:** On Freud's sexual abstinence, see Freud-Fliess Letters, 54; on his withdrawal from cigars, see 84–87. It seems that Sigmund and Martha Freud stopped having a sexual relationship after the birth of their sixth child in 1895; in 1911, Freud alluded to the sexual death of his marriage, telling Emma and Carl Jung that his marriage had "long been amortized"; see Freud-Jung Letters, 456. However there is reason to believe, following Peter Swales, that Freud conducted an affair with his sister-in-law, Minna Bernays, starting around 1898. See Maciejewski, F. (2006), "Freud, His Wife, and his 'Wife,'" *American Imago* 63:497–506.

89 called his "thesis of specificity": Freud-Fliess Letters, 77.
89 disease it was believed to cause: In 1882, Koch discovered the tubercle bacillus and a year later he discovered the cholera bacillus. These extraordinary breakthroughs led to sweeping public health measures, so that while 106,441 Austrians died of cholera in 1873, only 1,288 died of the disease in 1886. See Lesky, E. (1976), *The Vienna Medical School of the 19th Century*, Baltimore: The Johns Hopkins University Press, 259–60.
89 General Paralysis of the Insane: See Quétal, C. (1990), *History of Syphilis*, Baltimore: Johns Hopkins University Press, and Crissey, J., and Parish, L., *The Dermatology and Syphilogy of the Nineteenth Century*, New York: Praeger, 1981.
89 alcoholism, trauma, and sexual assault: See Harris, R. (1989), *Murders and Madness: Medicine, Law, and Society in the Fin de Siècle*, Oxford, U.K.: Clarendon Press, 103–04, 139–40.
90 cause that must be prevented: On the power of germ theory on Freud's thought, see Carter, K. C. (1980), "Germ theory, hysteria, and Freud's early work in psychopathology," *Medical History* 24:259–74.
90 fall victim to incurable neuroses": Freud-Fliess Letters, 44.
90 objection to Freud's thinking: Löwenfeld, L. (1895), "Über die Verknüpfung neurasthenischer und hysterischer Symptome in Anfallsform nebst Bemerkungen über die Freudsche Angstneurose," *Münchener Medizinische Wochenschrift*, 42:121–39. He believed a neuropathic disposition along with a host of environmental influences—including sexual ones—caused both neurasthenia and hysteria.
90 from the patient's sexual life: Freud, S. (1895), "A Reply to Criticisms of My Paper on Anxiety Neurosis," *S.E.* 3:123–39.
90 consequence of a presexual *sexual* shock": Freud-Fliess Letters, 144. Also see Freud's (1895), "Project for a Scientific Psychology," *S.E.* 1:356–57 for an earlier seduction hypothesis.
90 children under the age of twelve: Casper, J. L. (1861–1865), *A Handbook of the Practice of Forensic Medicine, Based upon Personal Experience*, trans. G. W. Balfour, London: New Sydenham Society, 283.
90 documenting the rape of children: Tardieu, A. (1860), "Étude médico-légale sur les sévices et mauvais traitements exercés sur des enfants," *Annales d'hygiéne publique et de médicine légale*, 13:361–98. See especially the case of Adelina Defert, 377–89. This article was discussed in Masson, J. M. (1984), *The Assault on Truth: Freud's Suppression of the Seduction Theory*, New York: Farrar, Straus and Giroux, 14–27.
90 murdered a four-year-old: See Harris, R. (1989), *Murders and Madness: Medicine, Law, and Society in the Fin de Siècle*, Oxford, U.K.: Clarendon Press, 95.
91 what Freud called "deferred action": On this concept, see Freud, S. (1895), "Project for a Scientific Psychology," *S.E.* 1:356; also Freud-Fliess Letters, 187–90. Strachey's translation of the German term *nachträglich*, or *Nachträglichkeit*, has been a source of controversy. See Laplanche, J., and J.-B. Pontalis (1967), *The Language of Psycho-Analysis*, New York and London: W. W. Norton & Company, 111–14.
91 riddle of hysteria had been solved: Freud, S. (1896), "Heredity and the Aetiology of the Neuroses," *S.E.* 3:143–56; Freud, S. (1896), "The Aetiology of Hysteria," *S.E.* 3:191–221; and Freud, S. (1896), "Further Remarks on the Neuropsychoses of Defence," *S.E.* 3:162–85.
91 specific kind of sexual abuse: See Sulloway, F. (1992), *Freud: Biologist of the Mind*, Cambridge, Mass.: Harvard University Press, 87–88.

91 **Freud's theory as a great advance:** See two newspaper reports reproduced in Sulloway, F. (1992), *Freud: Biologist of the Mind*, 507–09. As indicated here, the notion that Breuer was horrified by the idea of sexual causation is surely a myth. Also see Breuer's letter to Forel reprinted in Cranefield, P. F. (1958), "Josef Breuer's Evaluation of His Contribution to Psycho-Analysis," *I.J.P.* 39:319–22.

92 **I do not believe it!":** Freud-Fliess Letters, 151.

92 **he complained in 1907:** While Breuer insisted Freud's views on sexuality were rooted in clinical experience, he added that Freud had "a psychical need which, in my opinion, leads to excessive generalization. There may in addition be a desire d'épâter le bourgeois." Cranefield, P. F. (1958), "Josef Breuer's Evaluation of his Contribution to Psycho-Analysis," *I.J.P.* 39:319.

92 **"scientific fairy tale":** Freud-Fliess Letters, 184.

92 **and paranoia (ages eight to fourteen):** Freud-Fliess Letters, 187–90.

93 **sexuality in and of itself:** On the history of sexology see Bullough, V. (1994), *Science in the Bedroom: A History of Sex Research*, New York: Basic Books; and Lanteri-Laura, G. (1979), *Lecture des perversions. Histoire de leur appropriation médicale*, Paris: Masson.

93 **residing in a male body:** Ulrichs, C. (1865), '*Formatrix*' *Anthropologische Studien über mannmännliche Liebe*, Leipzig: Heinrich Matthes.

94 **psychiatrist named Richard von Krafft-Ebing:** See Westphal, C. (1870), "Die conträre Sexualempfindung, Symptom eines neuropathischen (psychopathischen) Zustandes," in *Gesammelte Abhandlungen*. Berlin: August Hirschwald, 1892. On Krafft-Ebing see Oosterhuis, H. (2000), *Stepchildren of Nature: Krafft-Ebing, Psychiatry, and the Making of Sexual Identity*. Chicago: The University of Chicago Press. Also Hauser, R. I. (1989), *Sexuality, Neurasthenia and the Law: Richard von Krafft-Ebing (1840–1902)*.

95 **sexual studies for years to come:** Krafft-Ebing, R. (1886), *Psychopathia Sexualis: eine klinisch-forensische Studie*, Stuttgart: Ferdinand Enke.

95 **incest, bestiality, and necrophilia:** Compare the first edition with Krafft-Ebing, R. (1903), *Psychopathia Sexualis with Especial Reference to the Antipathic Sexual Instinct*, trans. F. Klaf, New York: Stein & Day, 1965.

95 **sexual differences through degeneration:** Charcot, J. M., and V. Magnan (1882), "Inversion du sens génital," *Archives de Neurologie* 3:53–60 and 4:296–322.

96 **most often a natural variation:** Krafft-Ebing, R. (1901), "Neue Studien auf dem Gebiete der Homosexualität," *Jahrbuch für sexuelle Zwischenstufen*, 3:1–36.

96 **Hungarian woman named "Ilma":** On the "Ilma" case, see Lafferton, E. (2002), "Hypnosis and Hysteria as Ongoing Processes of Negotiation: Ilma's Case from the Austro-Hungarian Monarchy," *History of Psychiatry*, 3:177–97, 4:305–27.

96 **Reportedly, the treatment worked:** Krafft-Ebing, R. (1888), *An Experimental Study in the Domain of Hypnotism*, trans. C. G. Chaddock, New York: De Capo, 1982, 15.

96 ***Suggestive Therapeutics in Psychopathia Sexualis:*** Schrenck-Notzing, A. (1892), *Die Suggestions-Therapie bei krankhaften Erscheinungen des Geschlechtssinnes mit besonderer Berücksichtigung der conträren Sexualempfindung*, Stuttgart: Ferdinand Enke. Also see Schrenck-Notzing, A. (1889), "Un cas d'inversion sexuelle amélioré par la suggestion hypnotique," *Premier congrès international de l'hypnotisme expérimental et thérapeutique*. E. Bérillon. Paris: Octave Doin: 319–22.

96 **perversion had been greatly overstated:** Schrenck-Notzing, A. (1892), *Therapeutic*

Suggestion in Psychopathia Sexualis, trans. C. G. Chaddock, Philadelphia, F. A. Davis. 1895, v–xi.

97 **may have been formative:** Ibid., x–xi, 168–64, 193.

97 **improvement and even cure" in such cases:** Kraepelin, E. (1899), *Psychiatry: A Textbook for Students and Physicians*, ed. J. Quen, Canton, Mass.: Science History Publications, 421–23.

97 **to be sexual molestation:** See Makari, G. (1997), "Towards defining the Freudian unconscious: seduction, sexology and the negative of perversion (1896–1905)," *History of Psychiatry*, 8:459–85.

97 **ancient Greece and boys' schools:** Schrenck-Notzing, A. (1892), *Therapeutic Suggestion in Psychopathia Sexualis*, 126–35, 153, 175–78.

97 **servant girls, governesses, and nursemaids:** Krafft-Ebing, R. (1892), *Psychopathia Sexualis, with Especial Reference to the Antipathic Sex Instinct*, 369–71.

97 **maids, governesses, and relatives:** Freud, S. (1896), "The Aetiology of Hysteria," *S.E.* 3:208–09; also Freud-Fliess Letters, 37.

97 **valuable reality confirmation," he wrote:** Freud-Fliess Letters, 219.

98 **psychoneurosis simply had no precedent:** While father-child incest was discussed by forensic experts such as Casper, it was rarely mentioned as an etiology of perversion. There is a massive literature on Freud's seduction hypothesis or theory; see for example Israëls, H., and M. Schatzman (1993), "The Seduction Theory," *History of Psychiatry*: 4:23–59. Blass, R., and B. Simon (1992), "Freud on his own mistake(s): the role of seduction in the etiology of neurosis," *Psychiatry and the Humanities* 12:160–83, and Makari, G. (1998), "The Seductions of History: Sexual Trauma in Freud's Theory and Historiography," *I.J.P.* 79: 857–69.

98 **Fliess in December of 1896:** Freud-Fliess Letters, 210, 212. On trauma masquerading as "pseudoheredity" also see Freud (1896), "The Aetiology of Hysteria," *S.E.* 3:209.

99 **Freud appealed to an omnipresent bisexuality:** Freud-Fliess Letters, 212.

99 **female in men and the male in women:** Ibid., 213.

99 **hysteria was the negative of perversion:** Ibid., 227.

99 **unknowable region, the unconscious:** This equation was expanded to link perversions with all neurotics. See, for example, Freud, S. (1905), *Fragment of a Case of Hysteria, S.E.* 7:50. Also see Makari, G. (1997), "Towards Defining the Freudian Unconscious: Seduction, Sexology and the Negative of Perversion (1896–1905)," *History of Psychiatry* 8:459–85.

99 **testing his teacher's theories:** On the interesting role of Gattel, see Hermanns, L., and M. Schröter (1992), "Felix Gattel (1870–1904): Freud's First Pupil. Part I," *I.R.P.* 19:91–104; also Sulloway, F. (1992), *Freud: Biologist of the Mind*, 513–15.

99 **sexual abstinence and anxiety neurosis:** Gattel, F. (1898), *Ueber die sexuellen Ursachen der Neurasthenie und Angstneurose*, Berlin: August Hirschwald, 47–48.

99 **by the patient's father:** Ibid., 25.

99 **hysteria was very high?:** Sulloway, F. (1992), *Freud: Biologist of the Mind*, 513–15. Sulloway's argument is important, but it should be noted that Freud could have discounted the incidence of hysteria coming from Gattel's sample because his was a sample of psychiatric patients, not one of the general population.

100 **account for all cases of hysteria:** Freud, S. (1896), "The Aetiology of Hysteria," *S.E.* 3:207–209.

100　**childhood abuse did not emerge:** Freud-Fliess Letters, 264–66.

100　**back into the picture:** Ibid., 265.

100　**into such lewd acts:** For example, Mauriac, C. (1879), "Onanisme et excès vénéri-
ens"," *Nouveau dictionnaire de médecine et de chirurgie pratiques*, ed. Dr. Jaccoud,
Paris: J. B. Baillière, 518; Beard, G. (1884), *Sexual Neurasthenia; Its Hygiene, Causes,
Symptoms, and Treatment*, New York: E. B. Treat, 120; Schrenck-Notzing, A. (1892),
Therapeutic Suggestion in Psychopathia Sexualis, 150–53.

100　**incorporated this logic into his theory:** Freud-Fliess Letters, 274. Also see S. Freud
and W. Fliess (1986), *Sigmund Freud Briefe an Wilhelm Fliess 1887–1904*, ed. J. M.
Masson and M. Schröter, Frankfurt am Main: S. Fischer, 1986, 296. In this letter
Freud seems to hold on to his notion of a pseudoheredity, by claiming early sexual
experiences and later longing form degenerate character.

100　**already overstimulated nervous system:** Freud-Fliess Letters, 275; for the German, see
S. Freud and W. Fliess (1986), *Sigmund Freud Briefe an Wilhelm Fliess 1887–1904*, 296.

101　**causal role in illness:** Some sexologists argued that masturbation was pathological
while intercourse was not, due to the intense fantasies associated with the former. See,
for example, Rohleder, H. (1899), *Die Masturbation*, Berlin: H. Kornfeld, 35, 170–71.

101　**"embellishments of them":** Freud-Fliess Letters, 239.

101　**daydreams and neural irritation:** Ibid., 279–80. The meaning of this passage is obscured in Masson's English translation. Freud wrote: *"Daher etwa die Anästhesie der
Frauen, die Rolle der Masturbation bei den zur Hysterie bestimmten Kindern und das
Aufhören der Masturbation, wenn eine Hysterie daraus wird."*; S. Freud and W. Fliess
(1986), *Sigmund Freud Briefe an Wilhelm Fliess, 1887–1904*, 304. Hence Freud wrote
not of children "predisposed" to hysteria, as the present translation would have it,
which would imply a hereditarian disposition, but rather children "destined" or
"fated" to become hysterics.

101　**"in hysteria is enormous":** Freud-Fliess Letters, 287.

101　**"in a very sensible manner":** Freud-Fliess Letters, 338. Lewis and Landis lists a copy
of this Havelock Ellis 1898 reprint in the holdings purchased by Columbia; Nolan
D. C. Lewis and Carney Landis (1957), "Freud's Library," *Psychoanalytic Review*
44:327–354, #256. However, this reprint seems to be lost. On Freud's library, see
J. K. Davies and G. Fichtner, eds. (2006), *Freud's Library, A Comprehensive Catalogue*,
Tübingen: edition diskord.

101　**"a transformation of autoerotism":** Ellis, H. (1898), "Hysteria in Relation to the
Sexual Emotions," *The Alienist and Neurologist* 19:614.

102　**autoerotic lives created conflict:** Ellis, H. (1898), "Auto-erotism: A Psychological
Study," *The Alienist and Neurologist* 19:260, 273–74, 280–94.

102　**sexual impulses and fantasies:** I argue the reprint of this article in Columbia University's Freud Collection was in fact annotated by Freud. See Makari, G. (1998), "Between Seduction and Libido: Sigmund Freud's Masturbation Hypotheses and the
Realignment of His Etiologic Thinking, 1897–1905," *Bulletin of the History of Medicine* 72:638–62.

102　**inching closer to Ellis:** Freud attributed this insight to his self-analysis. I side with
Macmillan, who argues that it is improbable Freud could have gleaned such knowl-
edge from the analysis of his own dreams alone; see Macmillan, M. (1991), *Freud
Evaluated: The Completed Arc*, Amsterdam: North-Holland, 286–87.

102　**"alloerotism (homo- or heteroerotism)":** Freud-Fliess Letters, 390–91.

103 successor to the dream book": Ibid., 379.

103 man of this sort": Ibid., 398.

103 contours of Lucifer-Amor": Ibid., 421.

103 unconscious causes of Ida's misery: Ibid., 427.

104 political figure in Austrian politics): On Ida Bauer's life and milieu, see Decker, H. (1991), *Freud, Dora, and Vienna 1900*. New York: The Free Press.

104 disruption of the girl's sexual life: Freud, S. (1905), "Fragment of an Analysis of a Case of Hysteria," *S.E.* 7:24.

104 "entirely and completely hysterical": Ibid., 28.

105 analogous to that of a trauma": Ibid., 27.

105 far superior to their own: Ibid., 50.

105 deemed perverts by society: Ibid., 51.

105 the *negative* of perversion": Ibid., 50.

105 desires lurked in Dora's unconscious: Ibid., 51. Thumb sucking had been associated in medical discourse with both masturbation and infantile hysteria. See Carter, K. C. (1983), "Infantile hysteria and infantile sexuality," 186–96.

106 veiled accusation of masturbation: Freud, S. (1905), "Fragment of an Analysis of a Case of Hysteria," *S.E.* 7:78–80.

106 because of her Oedipal love: See Blass, R. (1992), "Did Dora Have an Oedipus Complex?" *Psychoanalytic Study of the Child* 47:159–87.

106 love for her father: In a footnote Freud added that such pathogenic reinforcement of Oedipal love could hypothetically come in three ways: from the spontaneous early appearance of genital sensations, "or as a result of seduction or masturbation." Freud, S. (1905), "Fragment of an Analysis of a Case of Hysteria," *S.E.* 7:57.

106 and then shutting it: Ibid., 76–78.

106 I have written so far.": Freud-Fliess Letters, 433–34.

106 frank recognition of bisexuality: Ibid., 434.

107 date of death is menstrual": Fliess, W. (1897), *Die Beziehungen zwischen Nase und weiblichen Geschlechtsorganen*, Leipzig and Vienna: Franz Deuticke, 213. This can be found in French as Fliess, W. (1897), *Les Relations entre le nez et les organes génitaux de la femme: Présentées selon leurs significations biologiques*, Paris: Éditions du Seuil, 1977, 255.

107 made use of Fliess's theory: Freud-Fliess Letters, 212.

107 masculine and feminine periods: Ibid., 465.

107 four individuals are involved": Ibid., 364.

108 to play on your flute": Ibid., 229.

108 not accept his grand theory: Ibid., 450.

108 own thoughts into other people": Ibid., 447.

108 wanted to kill him: Swales, P. J. (1989), "Freud, Fliess, and Fratricide: The role of Fliess in Freud's conception of paranoia," in *Sigmund Freud: Critical Assessments*, vol. 1, ed. L. Spurling, New York: Routledge, 302–29. Swales puts forward the fascinating hypothesis that Fliess's father's secret suicide was the topic that led the Berlin doctor to make his stinging rejoinder.

108 dismissing him as a mind reader: Freud-Fliess Letters, 447.

108 as worthless as the others do": Ibid., 450.

109 eruptions from the unconscious: Freud, S. (1901), *The Psychopathology of Everyday Life, S.E.* 6:1–279.

109   **explanation of the neurotic:** Freud-Fliess Letters, 448.

110   **lack of mathematical skill:** Ibid., 450.

110   **inner accident is invariably intentional (unconsciously)":** Ibid., 424.

110   **central cause of repression:** See for instance Freud, S. (1916–17), "Introductory Lectures on Psychoanalysis," *S.E.* 16, 370.

110   **with a misleadingly modest title:** Freud, S. (1905), *Three Essays on the Theory of Sexuality, S.E.* 7:130–243. I will also refer to an alternate, at times less euphemistic translation, Freud, S. (1905), *Three Contributions to the Sexual Theory*, trans. A. A. Brill, New York: The Journal of Nervous and Mental Disease Publishing Company, 1910.

111   **laws of the natural world:** Darwin, C. (1859), *On the Origin of Species by Means of Natural Selection*, London: Watts and Co., and Darwin, C. (1871), *The Descent of Man and Selection in Relation to Sex*, 2 vols., London: John Murray.

111   **entranced by Darwin's writings:** See Ritvo, L. (1990), *Darwin's Influence on Freud*, New Haven: Yale University Press.

111   **Albert von Schrenck-Notzing, and others:** Freud, S. (1905), *Three Essays on the Theory of Sexuality, S.E.* 7:135.

112   **affected some but not all:** Ibid., 139–140.

113   **distort cooperation into an antithesis":** Freud Letters, 284.

113   **met environmental stimuli "half-way":** Freud, S. (1905), *Three Contributions to the Sexual Theory*, New York: The Journal of Nervous and Mental Disease Publishing Company, 1910, 7; Strachey translates this as "cooperation of," Freud, S. (1905), *Three Essays on the Theory of Sexuality, S.E.* 7:141.

113   **female brain in a male body:** Freud, S. (1905), *Three Essays on the Theory of Sexuality, S.E.* 7:142.

113   **use of a fetish:** Ibid., 148–52.

113   **influences of actual life":** Ibid., 171–72.

114   **all a bit perverse too:** Ibid., 167.

114   **nor a biological variant:** Ibid., 171–72.

114   **adolescence and sexual life:** Mantegazza, P. (1894), *The Physiology of Love*, New York: Cleveland Pub. Co., 55–57.

114   **made for mature sexuality:** Moll, A. (1897), *Untersuchungen über die Libido sexualis*, 2 vols., Berlin: Fischer's Medicin Buchhandlung, H. Kornfeld. Also see Freud-Fliess Letters, 297.

115   **highly arousable: erotogenic zones:** Krafft-Ebing, R. (1892), *Psychopathia Sexualis, with Especial Reference to the Antipathic Sex Instinct*, 31; Moll, A. (1897), *Untersuchungen über die Libido sexualis*, Berlin: Fischer's Medicin Buchhandlung, H. Kornfeld; and Bloch, I. (1902–03), *Anthropological Studies in the Strange Sexual Practices in All Races of the World*, trans. K. Wallis, New York: Anthropological Press, 1933.

115   **they were in childhood:** See Freud-Fliess Letters, 297. On the history of the concept of erotogenic zones, see Kern, S. (1970), *Freud and the Emergence of Child Psychology: 1880–1910*, Ph.D diss., Columbia University, New York.

115   **manually stimulate themselves:** In his thinking on anal stimulation, Freud followed Krafft-Ebing, Moll, Bloch, and others who had focused on the excitements of this region to explain the desire for anal sex. Freud, S. (1905), *Three Essays on the Theory of Sexuality, S.E.* 7:179.

116   **unconscious incestuous fantasies:** Ibid., 222–25.

116   **innate, polymorphously perverse disposition:** The German is "polymorph perverse

Anlage." See Freud, S. (1905), "Drei Abhandlungen zur Sexualtheorie," *G.W.* 5:91. To environmentally oriented sexologists, the concept would have seemed close to their idea of an innate "undifferentiated" sexual disposition. See Bloch, I. (1902–03), *Anthropological Studies in the Strange Sexual Practices in All Races of the World*, 207.

117 **recognize the origins of man:** Freud wrote "zu allen Perversionen nicht das allgemein Menschliche und Ursprüngliche zu erkennen." Freud, S. (1905), "Drei Abhandlungen zur Sexualtheorie," *G.W.* 5:92. Strachey translated this to read: "a general and universal human characteristic"; Freud, S. (1905), *Three Essays on the Theory of Sexuality, S.E.* 7:191. Brill renders this last phrase as "the universal and primitive human"; Freud, S. (1905), *Three Contributions to the Sexual Theory*. New York: The Journal of Nervous and Mental Disease Publishing Company, 1910, 50.

117 **storage in the unconscious:** Hering, E. (1913), *Memory: Lectures on the Specific Energies of the Nervous System*, Chicago: The Open Court Publishing Company, 6, 20–23.

118 **in those deemed primitive:** Freud, S. (1905), *Three Essays on the Theory of Sexuality, S.E.* 7:139.

118 **earlier experience of the species.":** Ibid., 131. Also see, Freud Letters, 284. On Freud's evolutionary logic, see Sulloway, F. (1992), *Freud. Biologist of the Mind*, Cambridge, Mass.: Harvard University Press.

118 **mystery of the unconscious:** See, for example, Freud, S. (1914), "On the history of the psychoanalytic movement," *S.E.* 14:75.

119 **(3) the metapsychological":** Freud-Fliess Letters, 362.

119 **of the psychoneuroses in general":** Freud, S. (1905), *Fragment of a Case of Hysteria, S.E.* 7:113.

121 **transference," he told Fliess:** Freud-Fliess Letters, 409.

121 **entering in the present paper":** Freud, S. (1905), *Fragment of a Case of Hysteria, S.E.* 7:74.

121 **physician at the present moment:** Ibid., 116.

122 **him and his young patient:** Ibid., 118–19.

123 **yet breathtakingly simple:** *psychosexuality:* Freud had used the terms "Psychosexuelle" and "Psychosexuellen" a few times before 1905—see, for example, Freud, S. (1900), *The Interpretation of Dreams, S.E.* 4:236—but it was in the *Three Essays on Sexuality* that he showcased this as a unifying concept. See his use of that word in *G.W.* 5:46, 99, 116, 118, 128, 129, or *S.E* 7:136, 147, 199, 215, 217, 227.

124 **heads had been chopped off:** Wittels, F. (1924), *Sigmund Freud: His Personality, His Teaching, & His School*, trans. E. and C. Paul, New York: Dodd, Mead & Company Publishers, 130.

PART TWO: MAKING FREUDIANS

4. VIENNA

129 **new physicians of the soul?:** Nietzsche, F. (1881), *Daybreak: Thoughts on the Prejudices of Morality*, trans. R. J. Hollingdale, Cambridge, U.K.: Cambridge University Press, 1982, 33.

129 **drew but a scattered few:** In 1906, Freud's lectures brought him but seven students; see Jones, E. (1953), *The Life and Work of Sigmund Freud*, vol. I, 14.

130 **sexual problem by Freud:** Stekel's account of reading the *Interpretation of Dreams*, becoming an advocate and then going into treatment with Freud, is contradicted by

Ernest Jones, who supposedly was told by Freud that Stekel's treatment came first. See Stekel, W. (1950), *The Autobiography of Wilhelm Stekel: The Life Story of a Pioneer Psychoanalyst*, ed. E. Gutheil, New York: Liveright Publishing Corporation, 107; and Jones, E. (1955), *The Life and Work of Sigmund Freud*, vol. II, 8.

130   **"a new era in psychology":** Freud Reviews, 150.

130   **translation for a second edition:** On Kahane, see Mühlleitner, E. (1992), *Biographisches Lexikon der Psychoanalyse: Die Mitglieder der Psychologischen Mittwoch-Gesellschaft und der Wiener Psychoanalytischen Vereinigung 1902–1938*, Tübingen: edition diskord, 176.

130   **Freud's new psychological treatment:** Ibid., 176–77. On Kahane, also see the fine monograph, Rose, L. (1998), *The Freudian Calling: Early Viennese Psychoanalysis and the Pursuit of Cultural Science*, Detroit: Wayne State University Press, 52.

131   **poor hygiene, poverty, and ignorance:** See the best account of Adler to date; Stepansky, P. E. (1983), *In Freud's Shadow: Adler in Context*, Hillsdale, New Jersey, and London: The Analytic Press, 14–31.

131   **on one of his cases:** This was the discovery of the scholar Ernst Falzeder, who is credited by Mühlleitner, E. (1992), *Biographisches Lexikon der Psychoanalyse*, 17, as well as Handlbauer, B. (1998), *The Freud-Adler Controversy*, Oxford, U.K.: Oneworld, 172.

131   **envy of the Western world:** See Lesky, E. (1976), *The Vienna Medical School of the 19th Century*, trans. L. Williams and I. S. Levij, Baltimore: The Johns Hopkins University Press; Bynum, W. F. (1994), *Science and the Practice of Medicine in the Nineteenth Century*, Cambridge, U.K.: Cambridge University Press; and Bonner, T. N. (1995), *Becoming a Physician: Medical Education in Britain, France, Germany, and the United States, 1750–1945*, New York: Oxford University Press.

132   **not much had changed:** Johnston, W. M. (1972), *The Austrian Mind: An Intellectual and Social History 1848–1938*, Berkeley: University of California Press, 227–28.

132   **nature alone could heal:** Lesky, E. (1976), *The Vienna Medical School of the 19th Century*, 156–57.

133   **that he knows nothing":** Cited in Bonner, T. N. (1995), *Becoming a Physician*, 237.

133   **give up on his patients:** On the unjustly neglected Oppolzer, see Lesky, E. (1976), *The Vienna Medical School of the 19th Century*, 117–28.

134   **influence in medical school:** Furtmüller, C. (1973), "Alfred Adler: A Biographical Essay," in *Alfred Adler, Superiority and Social Interest*, ed. H. Ansbacher and R. Ansbacher, New York: W. W. Norton, 332–33.

134   **suggestive cure he called psychotherapy:** Oosterhuis, H. (2000), *Stepchildren of Nature*, 124.

135   **and sometimes themselves:** Paul Federn was a depressive who conducted a self-analysis under Freud's guidance; Eduard Hitschmann went into analysis with Freud. See Weiss, E. (1966), "Paul Federn, 1871–1950," in Alexander, F., S. Eisenstein, M. Grotjahn, *Psychoanalytic Pioneers*, New York: Basic Books, 145–46, and Becker, P. (1966), "Edward Hitschmann, 1871–1957," Ibid., 167.

136   **social, and political world:** Helmholtz, H. (1995), *Science and Culture: Popular and Philosophical Essays*, 76–79.

136   **vain, untalented, and not German:** Billroth, T. (1924), *The Medical Sciences in the German Universities: A Study in the History of Civilization*, New York: The Macmillan Company, 105–08.

136   **elected mayor of Vienna in 1897:** On the collapse of liberalism in Austria, see Mosse, G. L. (1964), *The Crisis of German Ideology: Intellectual Origins of the Third Reich*, New York: Howard Fertig, 130, 135, 195; Janik, A., and S. Toulmin (1973),

Wittgenstein's Vienna, New York: Simon & Schuster; and Schorske, C. (1980), *Fin-De-Siècle Vienna: Politics and Culture*, New York: Vintage Books. For a broader account of changes in Europe, see the illuminating study by J. W. Burrow (2003), *The Crisis of Reason: European Thought, 1848–1914*, New Haven: Yale University Press.

137 his 1892 book, *Degeneration*: Nordau, M. (1892), *Degeneration*, Lincoln: University of Nebraska, 1993.

137 and illiteracy throughout Europe: Ibid., 559–60.

138 sickness carried by Jews: See Hamann, B. (1999), *Hitler's Vienna: A Dictator's Apprenticeship*, trans. T. Thornton, New York: Oxford University Press, and Mosse, G. (1964), *The Crisis of German Ideology*, 143.

138 collapse of European liberalism: Gilman, S. (1985), "Sexology, Psychoanalysis and Degeneration," in *Degeneration: The Dark Side of Progress*, eds. S. L. Gilman and J. E. Chamberlain. New York: Columbia University Press, 72–96.

138 resembled the Tower of Babel: Hamman, B. (1999), *Hitler's Vienna*, 117–18.

139 these trickling streams": Musil, R. (1930), *The Man Without Qualities*, vol. I, trans. E. Wilkins and E. Kaiser, London: Picador, 1979, 34. On questions of identity in 1900 Vienna, see Le Rider, J. (1990), *Modernité Viennoise et Crises de l'Identité*, Paris: Presses Universitaires de France.

139 physician to a diseased culture: Nietzsche, F. (1979), *Philosophy and Truth: Selections from Nietzsche's Notebooks of the Early 1870's*, ed. and trans. D. Breazeale, New Jersey: Humanities Press International.

139 health, future, growth, power, life: Nietzsche, F. (1887), *The Gay Science: With a Prelude in Rhymes and an Appendix of Songs*, trans. W. Kaufmann, New York: Vintage Books, 1974, 35.

140 smallest part of all this": Ibid., 298–99.

140 morals of the church: Thomas, R. H. (1983), *Nietzsche in German Politics and Society, 1890–1918*, La Salle, Illinois: Open Court.

140 even the psychopathological self: Bahr, H. (1891), "The Overcoming of Naturalism," in *The Vienna Coffeehouse Wits: 1890–1938*, ed. H. B. Segal, West Lafayette, Indiana: Purdue University Press, 1993, 50.

140 what Bahr called a "nervous romanticism": Ibid., 49.

140 physical thrills of his smuttier episodes": Mahler-Werfel, A. (1999), *Alma Mahler-Werfel: Diaries 1898–1902*, trans. A. Beaumont, Ithaca, N.Y.: Cornell University Press, 189.

141 "One cannot wear too little!": Altenberg, P. (1906), "From 'Pròdrômôs,'" in *The Vienna Coffeehouse Wits: 1890–1938*, ed. H. B. Segal, West Lafayette, Indiana: Purdue University Press, 1993, 155.

141 against the old way of life": Musil, R. (1930), *The Man Without Qualities*, vol. I, 59.

141 insincerity in this matter of sex": Zweig, S. (1943), *The World of Yesterday*, Lincoln: University of Nebraska Press, 1964, 67.

142 "gigantic army" of prostitutes: Ibid., 85.

142 sexual life by a prostitute: Cited in Hamann, B. (1999), *Hitler's Vienna*, 365.

142 Whoever it might be": Mahler-Werfel, A. (1999), *Alma Mahler-Werfel: Diaries 1898–1902*, 420–21.

142 to view men naked: Hamann, B. (1999), *Hitler's Vienna*, 371.

142 Berliner Helene Stöcker: Anderson, H. (1992), "Psychoanalysis and Feminism: An Ambivalent Alliance, Viennese Feminist Responses to Freud, 1900–30," in *Psycho-*

analysis in Its Cultural Context, eds. E. Timms and R. Robertson, Edinburgh: Edinburgh University Press, 71–80.

142 **making men and women ill:** Meisel-Hess, G. (1909), *Die sexuelle Krise*, Jena: Eugen Diederichs.

143 **one of Freud's early patients:** Stocker's group was the *Bund für Mutterschutz* and Pappenheim's was the *Jüdische Frauenbund*. On Emma Eckstein and her catastrophic treatment with Freud and Wilhelm Fliess, see Masson, J. M. (1984), *The Assault on Truth: Freud's Suppression of the Seduction Theory*. New York: Farrar, Straus and Giroux, 55–106.

143 **"earth in the Cretaceous Age":** Altenberg, P. (1896), "The Primitive," in *The Vienna Coffeehouse Wits: 1890–1938*, ed. H. B. Segal, West Lafayette, Indiana, Purdue University Press, 1993, 149.

143 **'discovered' the primitive!":** Ibid., 152.

143 **Freud's friend Christian von Ehrenfels:** See, for example, Ehrenfels, C. V. (1903), "Die aufsteigende Entwicklung des Menschen," *Politisch-anthropoligische Revue* 2: 45–59; Ehrenfels, C. V. (1903), "Sexuales Ober- und Unterbewusstsein," *Politisch-anthropoligische Revue* 2:456–76; Ehrenfels, C. V. (1903), "Monogamische Entwicklungsaussichten," *Politisch-anthropoligische Revue* 2:706–18; and Ehrenfels, C. V. (1903), "Die sexuale Reform," *Politisch-anthropoligische Revue* 2: 970–93.

143 **socially sanctioned and the sexual:** Ehrenfels, C. V. (1903), "Sexuales Ober- und Unterbewusstein," 457.

143 **could be done through sexual reform:** Ibid., 476.

144 **serial monogamous relations or simultaneously:** Ehrenfels, C. V. (1907), *Sexualethik*, Wiesbaden: Verlag Von J. F. Bergmann.

144 **who annotated the work carefully:** The book can be found in Freud's library in London. He annotated it and noted the place (p. 13) in the text where Ehrenfels called polygyny the only healthy sexual ethic. Also see Freud-Pfister Letters, 19.

144 **widespread dangers of repression:** Freud, S. (1908), "Civilized Sexual Morality and Modern Nervous Illness," *S.E.* 9:181–204.

144 **traumas of a courtesan:** Salten, F. (1906), *The Memories of Josephine Mutzenbacher*, North Hollywood, Cal.: Brandon House, 1967. The authorship of this book has also at times been attributed to Arthur Schnitzler.

145 **the startling work of Otto Weininger:** On Otto Weininger, see Sengoopta, C. (2000), *Otto Weininger: Sex, Science, and Self in Imperial Vienna*, Chicago: The University of Chicago Press.

146 **Weininger became a literary sensation:** Weininger, O. (1903), *Sex and Character*, London and New York: William Heinemann and G. P. Putnam's Sons, 1906.

146 **capacity for motherhood":** Ibid., 217.

146 **for secretly being lusty:** Ibid., 265–67. Weininger's recourse to female self-hatred is truly a case of the pot calling the kettle black, for Weininger was the archetype of the self-hating Jew and famously derided Jews for having no capacity for morality.

146 **took up the Hervay case:** Kraus, K. (1904), "Der Fall Hervay," *Die Fackel* 165:2–12. Kraus used the phrase "die polygame Frauennatur" earlier; see Kraus, K. (1903), "Untitled," *Die Fackel*, 142:17.

147 **larger implications of a small affair":** Cited in Timms, E. (1986), *Karl Kraus Apocalyptic Satirist; Culture and Catastrophe in Habsburg Vienna*, New Haven: Yale University Press, 94.

147 **Freud felt isolated:** Freud, S. (1925), "An Autobiographical Study," *S.E.* 20:28. Numerous scholars have sought to show that Freud exaggerated his isolation; see for example, Decker, H. S. (1977), *Freud in Germany: Revolution and Reaction in Science, 1893–1907*, New York: International Universities Press, 172–73.

147 **to sell its 600 copies:** Freud Reviews, 12.

147 **two reviews in professional journals":** Freud-Fliess Letters, 454.

147 **to pursue the masses directly:** Freud, S. (1904), "Freud's Psychoanalytic Procedure," *S.E.* 7:249–54; Freud, S. (1905), "On Psychotherapy," *S.E.* 7:257–68; Freud, S. (1905), "Psychical (or Mental) Treatment," *S.E.* 7:283–304.

147 **stoop to such measures:** Freud-Fliess Letters, 407–08.

147 **one obituary over two years:** Freud, S. (1903–04), "Contributions to the *Neue Freie Presse*," *S.E.* 9, 253–56; Also see Solms, M. (1989), "Three Previously Untranslated Reviews by Freud from the Neue Freie Presse," *I.J.P.* 70:397–400.

147–48 **nonsensical or self-evident":** Freud, S. (1903–04), "Contributions to the *Neue Freie Presse*," *S.E.* 9:253–54.

148 **savaging of prominent figures:** Mahler-Werfel, A. (1999), *Alma Mahler-Werfel: Diaries 1898–1902*, 284.

148 **It uses psychiatry instead":** Kraus, K. (1904), "Irrenhaus Österreich," *Die Fackel*, 166:1. The English translation is from Kraus, K. (1904), "The Case of Louis von Coburg," in *Karl Kraus and the Soul-Doctors*, ed. T. Szasz, Baton Rouge: Louisiana State University Press, 1976, 133.

148 **conduct a parallel investigation:** The group was called the *Kulturpolitische Gesellschaft*. On this meeting, see *Neue Freie Presse*, February 2, 1905, 7–8. Also see the fine essay by John Boyer (1978), "Freud, Marriage, and Late Viennese Liberalism: A Commentary from 1905," *Journal of Modern History* 50: 72–102, which includes a transcript of Freud's testimony.

149 **female representatives in all marriage courts:** See Boyer, J. (1978), "Freud, Marriage, and Late Viennese Liberalism," 91.

149 **scientific justification for "women's emancipation":** Ibid., 93. Also see *Vienna Minutes* I, 211.

149 **satisfying normal sexual instincts":** Boyer, J. (1978), "Freud, Marriage, and Late Viennese Liberalism," 93.

149 **his sister-in-law, Minna Bernays:** This affair was rumored to exist by Carl Jung, denied by Ernest Jones, and resuscitated in the 1980s by Peter Swales. See Swales, P. (1982), "Freud, Minna Bernays, and the Conquest of Rome," *New American Review*, 1:1–23. Most recently a German scholar, Franz Maciejewski, discovered an entry from 1898 in an inn in the Swiss alps, in which Freud wrote "Dr. Sigm Freud u. Frau," or "Dr. Sigm. Freud and wife." See Maciejewski, F. (2006), "Freud, His Wife, and 'His Wife.' " *American Imago*, 63:497–506.

149 **nor in the madhouse," he wrote:** Kraus, K. (1905), "Die Kinderfreunde," *Die Fackel*, 187:21.

149 **partially coincide with mine":** Freud Letters, 249–50.

150 **rid themselves of sexual thoughts:** Soyka, O. (1905), "Zwei Bücher," *Die Fackel*, 191:6–11. For the English, see Freud Reviews, 300–02.

150 **Freud has left it far behind":** Soyka, O. (1905), "Zwei Bücher," *Die Fackel*, 191:9.

150 **sexual morality for *Die Fackel*:** Soyka, O. (1905), "Psychiatrie," *Die Fackel*, 186:20–22. Also see Soyka, O. (1906), "Sexuelle Ethik," *Die Fackel*, 206:10–13.

150 **foundation for a new sexual ethic:** Soyka, O. (1906), *Jenseits der Sittlichkeits-Grenze. Ein Beitrag zur Kritik der Moral*, Wien and Leipzig: Akademischer Verlag.

150 **perversions were not diseases:** Ibid., 43.

150 **acceptance of masochism and sodomy:** Ibid., 85–57.

150 **published his own treatise,** *The Battle of the Sexes:* Frey, P. (1904), *Der Kampf der Geschlechter*, Wien: Wiener Verlag.

150 **"solid fundamentals for a novel theory":** *Freud Reviews*, 305.

150 *Towards a Critique of Femininity:* Mayreder, R. (1905), *Zur Kritik der Weiblichkeit.* Jena: Eugen Diederichs. In English, see Mayreder, R. (1905), *A Survey of the Woman Problem.* Westport, Ct.: Hyperion Press, 1982.

150 **woman as sexual fetish object:** Mayreder, R. (1905), *A Survey of the Woman Problem*, 224–41.

150 **true scientific value," Mayreder commented:** *Freud Reviews*, 305.

150 **repression characterized female psychology:** Ibid., 307. Mayreder would lose her enthusiasm for Freud later in her life, which may have been the result of the consultation Mayreder's husband had with Freud in 1915; see Anderson, H. (1992), "Psychoanalysis and Feminism: An Ambivalent Alliance, Viennese Feminist Responses to Freud, 1900–30," in *Psychoanalysis in Its Cultural Context*, eds. E. Timms and R. Robertson. Edinburgh: Edinburgh University Press, 71–80.

151 **with the author's theories," wrote one:** *Freud Reviews*, 308.

151 **brought forth a "rich harvest":** Ibid., 310.

151 **called the thin book essential:** Ibid., 311.

151 *Jung Wien* **writers, Hermann Bahr:** Graf, M. (1942), "Reminiscences of Professor Sigmund Freud," *Psychoanalytic Quarterly* 11:469.

151 **derived inspiration from my writings":** Freud Letters, 251. Schnitzler and Freud never met, and later, Freud concluded that he had been held back by a "kind of reluctance to meet my double." See Freud Letters, 339.

152 **Freud have done about that?:** Freud-Fliess Letters, 464.

152 **Sigmund Freud and Wilhelm Fliess ended:** Freud-Fliess Letters, 459–68.

152 **Freud's self-incriminating, final letter:** Fliess, W. (1906), *In eigener Sache: Gegen Otto Weininger und Hermann Swoboda*, Berlin: Emil Goldschmidt.

152 **Kraus for years to come:** Vienna Minutes, II, 391. Kraus would become a fervent detractor of psychoanalysis after 1910, claiming this "cure" was a disease itself.

152 **cigar smoking and sexual matters:** Reprinted in Handlbauer, B. (1998), *The Freud-Adler Controversy*, Oxford, U.K.: Oneworld, 18–21.

153 **work smoothly together:** Freud, S. (1914), "On the History of the Psychoanalytic Moment," *S.E.* 14:25.

154 **not to contradict their professors:** While doing research with Meynert, Forel wrote: "I had been as cautious and considerate as possible in putting forward opinions which differed from his own." Forel, A. (1937), *Out of My Life and Work*, trans. B. Miall, New York: W. W. Norton, 81–82.

154 **powerful aspects of Freud's theory:** Vienna, Minutes, I:92–102.

154 **maintained by the Zurich School":** Ibid., I:99.

155 **like a psychic water cure:** Ibid., I:101–02.

155 **curbed by tested rules":** Ibid., I:237.

155 **but also his father:** Freud, S. (1909), "Analysis of a Phobia of a Five-Year-Old Boy," *S.E.* 10:5–151.

155　doctor and his daughter: Freud, S. (1909), "Notes upon a Case of Obsessional Neurosis," *S.E.* 10:209, 284. A fuller set of Freud's notes can be found in Freud, S. (2000), *L'Homme aux rats: Journal d'une analyse,* trans. E. Ribeiro-Hawelka, Paris, Presses Universitaires.

156　contradiction of the Professor's beliefs: Vienna Minutes, I:70.

156　had succumbed to repression: Ibid., I:71.

157　of preserving the species": Ibid., I:83.

158　similar cures in two other cases: See Stekel, W. (1908), *Nervöse Angstzustände und ihre Behandlung,* Berlin and Vienna: Urban & Schwarzenberg.

158　could not help but be convinced: Forel, A. (1937), *Out of My Life and Work,* 101.

158　not in accordance with his theory": Stekel, W. (1950), *The Autobiography of Wilhelm Stekel: The Life Story of a Pioneer Psychoanalyst,* ed. E. Gutheil, New York: Liveright Publishing Corporation, 118.

158　underlying these cases," he remembered: Ibid., 137.

158　to Berggasse 19, manuscripts in hand: See Stekel's 1950 account as well as an earlier account, Stekel, W. (1926), "On the History of the Analytical Movement," trans. Jaap Bos, *Psychoanalysis and History,* 7:99–130, 2005.

159　Stekel's cases were all misdiagnosed: Vienna Minutes, I:175–82. Stekel would later complain that Freud stole the idea of Eros and Thanatos from him; Stekel, W. (1950), *The Autobiography of Wilhelm Stekel,* 138.

159　his presentation went poorly: Ibid., 204–11.

159　essence of a case: Ibid., 242–47.

159　psychoanalytic theory and method: Stekel, W. (1908), *Nervöse Angstzustände und ihre Behandlung,* Berlin and Vienna: Urban & Schwarzenberg.

159　wrote those pages for him: Stekel W. (1950), *The Autobiography of Wilhelm Stekel,* 118.

160　"neurotic anxiety-states," Freud wrote: Freud, S. (1909), "Analysis of a Phobia in a Five-Year-Old Boy," *S.E.* 10:115–56.

161　nobles that frequented Breuer and Freud: On the purported difference in patient populations between Freud and Adler, see Wasserman, I. (1958), "Letter to the Editor," *American Journal of Psychiatry,* 12:623–27.

161　first referred to Freud's work: Adler, A. (1904), "Der Arzt als Erzieher," in *Heilen und Bilden,* Germany: Fischer Taschenbuch, 1973.

161　key word was "organic": The first title can be found in Ibid I:14, the second on p. 36; the discrepancy was pointed out by Handlbauer, B. (1998), *The Freud-Adler Controversy,* Oxford, U.K.: Oneworld, 41.

161　organic cause of neurosis: Adler, A. (1907), *Studie über Minderwertigkeit von Organen,* Berlin and Wien: Urban and Schwarzenberg. For the English, see Adler, A. (1907), *Study of Organ Inferiority and Its Psychical Compensation: A Contribution to Clinical Medicine,* trans. S. E. Jelliffe, New York: The Nervous and Mental Disease Pub., 1917.

161　fundamentally rewritten Freud's theory: Adler, A. (1907), *Study of Organ Inferiority and Its Psychical Compensation,* 65.

162　called cerebral "over-compensation": Vienna Minutes, I:37.

162　painters had optical peculiarities: Paralleling Marxist theory, Adler's system was built on a materialist determinism that generated an ideogenic superstructure. Freud later complained that socialism inflected Adler's psychology.

162 **psychoneuroses is cleared up:** Adler, A. (1907), *Study of Organ Inferiority and Its Psychical Compensation*, 65.

162 **refused to join in the praise:** Vienna Minutes, I:42–47.

163 **cohere with Adler's own theory:** Ibid., 142.

163 **symbols and inferior organs:** Ibid., 171, 179, 252, 322.

163 **praised Adler's work as physiology:** Furtmüller took this to mean that Freud wanted Adler to "stick to physiology." See Furtmüller, C. (1973), "A Biographical Essay," *Alfred Adler: Superiority and Social Interest*, 340.

164 **no pressing need to decide:** Adler made his theory more enticing by acknowledging the empirical presence of infantile sexuality and allowing for the *seeming* clinical relevance of sexuality, but he also clearly undermined those notions. On this, I am in agreement with Stepansky, P. E. (1983), *In Freud's Shadow: Adler in Context*, 54.

164 **his religion to unaffiliated:** On Rank, see Lieberman, E. J. (1985), *Acts of Will: The Life and Work of Otto Rank*, New York: The Free Press, 4.

164 **he wrote in his diaries:** Rank Diaries, October 26, 1903, 27.

164 **"nourisher" of his generation:** Ibid., October, 1903, 18.

164 **no different from prostitution:** Ibid., January 6, 1904, 36.

165 **despise what they do," he wrote:** Ibid., March 19, 1904, 5.

165 **psychological works into German:** Lieberman, E. J. (1985), *Acts of Will*, 20–21.

165 **self, and psychology:** See the influential work of Schorske, C. (1980), *Fin-De-Siècle Vienna: Politics and Culture*, New York: Vintage Books, as well as Spector, S. (1998), "Beyond the Aesthetic Garden: Politics and Culture on the Margins of Fin-de-Siècle Vienna," *Journal of the History of Ideas* 59:691–10.

165 **"Meistersinger (Freud–Interpretation of Dreams) dreamsong":** Rank Diaries, October 17, 1904, 23.

165 **power that drives everything":** Ibid., May 27, 1905, 20.

165 ***The Artist: Towards a Sexual Psychology:*** Rank, O. (1907), *Der Künstler: Ansätze zu einer Sexual-Psychologie*, Leipzig und Wien: Hugo Heller.

165 **conscious of this projection, my birth":** Rank Diaries, Book III, January 14, 1905, 2.

165 **sexual drives with clarity and control:** Rank, O. (1907), *Der Künstler: Ansätze zu einer Sexual-Psychologie*, 56.

166 **incestuous themes in art:** There are complications with using the notes made by Rank as a source, since these notes were often modified by other members who borrowed them; see Handlbauer, B. (1998), *The Freud-Adler Controversy*, 35.

166 **one member complained:** Vienna Minutes, I:7.

166 **"far-fetched angle":** Ibid., 9.

166 **"over-extending an elastic band":** Ibid., 12.

166 **without going beyond Freud":** Ibid., 25.

167 **medical and psychiatric evaluation:** Schiller, F. (1982), *A Möbius Strip: Fin-de-Siècle Neuropsychiatry and Paul Möbius*, Berkeley, University of California Press, 80.

167 **Berlioz, Guy de Maupassant, and Tolstoy:** The series was called *Grenzfragen des Nerven-und Seelenlebens*. Journals dedicated to such studies included the *Schriften zur angewandten Seelenkunde*.

167 **loose on the public:** Freud, S. (1942), "Psychopathic Characters on the Stage," *S.E.* 7:305–10.

167 **hero of the German people, Goethe:** For a bibliography of Sadger's writings, see

Mühlleitner, E. (1992), *Biographisches Lexikon der Psychoanalyse: Die Mitglieder der Psychologischen Mittwoch-Gesellschaft und der Wiener Psychoanalytischen Vereinigung 1902–1938*, Tübingen: edition diskord, 284–85.

168 **university lectures between 1895 and 1904:** Max Kahane had encouraged Sadger to attend; see Sadger, I. (2005), *Recollecting Freud*, ed. A. Dundes, Madison: University of Wisconsin Press, 7–13.

168 **explained by psychosexual arguments":** Vienna Minutes, I:68.

168 **about the writer's sexual development:** Ibid., 256.

168 **nothing more than personal insults:** Ibid., 257–58.

168 **appearing at the Professor's lectures:** See Wittels, F. (1995), *Freud and the Child Woman: The Memoirs of Fritz Wittels*, ed. E. Timms, New Haven and London: Yale University Press, 25.

169 **more believers to fill the pews:** Avicenna, (1907), "Das größte Verbrechen des Strafgesetzes (Das Verbot der Fruchtabtreibung)," *Die Fackel*, 219–220:1–22. Reprinted in Wittels, F. (1909), *Die Sexuelle Not*, Wien and Leipzig: C. W. Stern.

169 **I subscribe to every word of it":** Wittels, F. (1995), *Freud and the Child Woman*, 48.

169 **"his personal distaste":** Vienna Minutes, I:162.

169 **harmless class of people":** Ibid.

170 **marked by the author's neurosis:** Ibid., I:160–65.

170 **understand a man's psychology:** Ibid., I:196.

170 **not reach a man's cultural achievements:** Ibid., I:199–200.

170 **scandalous for the Society to record:** See Wittels, F. (1995), *Freud and the Child Woman*, 169–70.

171 **Wittels cited Sigmund Freud:** Avicenna (1907), "Das Kindweib," *Die Fackel* 230–31:14–33.

171 **one is abstinent under protest:** Vienna Minutes, I:200.

171 **he was rendered speechless:** Ibid., 195–201.

171 **jumped on his bandwagon:** On this tension, see Rose, L. (1998), *The Freudian Calling: Early Viennese Psychoanalysis and the Pursuit of Cultural Science*, 25–90.

171–72 **then published his essay:** Avicenna (1907), "Das Kindweib," *Die Fackel* 230–31:14–33; see Wittels, F. (1995), *Freud and the Child Woman*, 60–62.

172 **"running wild would not help either":** Wittels, F. (1995), *Freud and the Child Woman*, 63, also see 50–51.

172 **so syphilis behind Christianity":** Avicenna (1907), "Die Lustseuche," *Die Fackel* 238:10. Also see Wittels, F. (1995), *Freud and the Child Woman*, 76.

172 **female medical students, and venereal disease:** Vienna Minutes, I:238–41.

173 **interested only in theoretical-psychological aspects":** Ibid., 241.

173 **he was ostracized and denounced:** Vienna Minutes, I:350–51. Alfred Adler seconded that view and then launched into his own views regarding Marx, property, and the impending collapse of the patriarchal family. To this, Wittels curtly snapped: "one cannot be a Freudian and a Social Democrat at the same time." Vienna Minutes, I:352–53.

175 **about to become an organization":** Vienna Minutes, I:301.

175 **could not always cite his source:** Ibid., I:299–300.

176 **any of his own remarks":** Ibid., I:301–02.

176 **forbidden to do so by the author:** Ibid., I:303.

176 must be acceptable to the others": Ibid., I:313–15.
176 he withdrew his proposal: Ibid., I:299–300, 317.
177 adhering to them or ignoring them": Ibid., I:121–23, 125.
178 transformed itself into a "Psychoanalytic Society.": Ibid., I:373.
178 *Methode der Psychoanalyse* in German: The German term "psychische Analyse"
     and "psychologisch-klinischen Analyse" first appeared in Freud, S. (1894), "Die
     Abwehr-Neuropsychosen," *G.W.* I:61, 73; also see "The Neuro-Psychoses of De-
     fence," *S.E.* 3:47, 59. The French—"la psychoanalyse"—was used in Freud, S. (1896),
     "L'hérédité et l'étiologie des névroses," *G.W.* I:419. Also see (1896), "Heredity and the
     Aetiology of the Neuroses," *S.E.* 3:151; The German—"Methode der Psychoanal-
     yse"—first found its way into print in Freud, S. (1896), "Weitere Bemerkungen über
     Die Abwehr-Neuropsychosen," *G.W.* I:379. In English, see "Further Remarks on the
     Neuro-Psychoses of Defence," *S.E.* 3:162.
178 postcathartic methods as psychoanalytic: Freud, S. (1904), "Freud's Psycho-Analytic
     Procedure," *S.E.* 7:249–56; Freud, S. (1905), "On Psychotherapy," *S.E.* 7:257–270;
     Freud, S. (1905), "Psychical (or Mental) Treatment," *S.E.* 7:283–304.

5. ZURICH

179 found their way to Switzerland: On Switzerland, see Craig, G. A. (1988), *The Tri-
     umph of Liberalism: Zurich in the Golden Age, 1830–1869*, New York: Collier Macmil-
     lan Publishers; and Bouvier, N., Craig, G., and L. Grossman, (1994), *Geneva, Zurich,
     Basel: History, Culture, and National Identity*, Princeton, N.J.: Princeton University
     Press, 1994.
180 had also been advocating change: Diethelm, O. (1975), "Switzerland," in *World His-
     tory of Psychiatry*, ed. J. G. Howells, New York: Brunner/Mazel, 250–51.
180 psychiatric teaching, research, and care: On the Zurich School, see Graf-Nold, A.
     (2001), "The Zurich School of Psychiatry in theory and practice. Sabina Spielrein's
     treatment at the Burghölzli Clinic," *Journal of Analytical Psychology* 46:73–104.
180 not always see what Meynert saw": Forel, A. (1937), *Out of My Life and Work*, 80, also
     see 53.
182 publishing widely on the subject: For example, Forel, A. (1889), *Der Hypnotismus:
     Seine Bedeutung und seine Handhabung*, Stuttgart: Ferdinand Enke.
182 in paranoid patients at the Burghölzli: I am indebted to Sonu Shamdasani, who
     pointed me to this claim in the seventh edition of Forel's book; see Forel, A. (1918),
     Der Hypnotismus oder die Suggestion und die Psychotherapie, Stuttgart: Ferdinand
     Enke, 233.
182 had been torn by mental illness: This motive was attributed to Bleuler by his son; see
     Bleuler, M. (1953), "Eugen Bleuler's Conception of Schizophrenia—An Historical
     Sketch," *Bulletin of the Isaac Ray Medical Library* 1:50–51.
182 account of being put under: Bleuler, E. (1887), "Der Hypnotismus," *Münchener Med-
     izinische Wochenschrift* 34:699–703, 714–17; and Bleuler, E. (1889), "Zur Psychologie
     der Hypnose," *Münchener Medizinische Wochenschrift* 36:76–77. Auguste Forel re-
     printed Bleuler's 1889 article, informing readers that Bleuler had conducted many
     hypnotic treatments and had "completely mastered the method"; see Forel, A. (1906),
     *Hypnotism or Suggestion and Psychotherapy: A Study of the Psychological, Psycho-
     Physiological and Therapeutic Aspects of Hypnotism*, trans. H. W. Armit, London and
     New York: Rebman Limited and Rebman Company, 360–66.

183 **Bleuler collapsed from exhaustion:** Forel, A. (1937), *Out of My Life and Work*, 164.

183 **formal dinner at the director's home:** Bleuler, M. (1953), "Eugen Bleuler's Conception of Schizophrenia—An Historical Sketch," 51–57.

184 **field of normal and pathological psychology":** See the nine book reviews composed by Bleuler in 1892 alone, covering authors such as Moll, Krafft-Ebing, Schrenck-Notzing, and Bernheim: Bleuler, E. (1892), *Münchener Medizinische Wochenschrift* 39:151, 169–70, 187–88, 259, 431, 450–51, 609, 676. The review of Breuer and Freud is in Freud Reviews, 74.

185 **578 in the same period:** Compare *Rechenschaftsbericht über die Zürcherische kantonale Irrenheilanstalt Burghölzli für das Jahr 1900*, Zürich: Buchdruckerei Berichthaus, 1901, with *Rechenschaftsbericht über die Zürcherische kantonale Irrenheilanstalt Burghölzli für das Jahr 1913*. Zürich: Buchdruckerei Berichthaus, 1914. These documents are to be found in the *Staatsarchiv* in Zurich.

185 **could even effect some cures:** Bleuler, E. (1898), *Die Allgemeine Behandlung der Geisteskrankheiten*, Zürich: Ed. Rascher, 44.

185 **split, double, or multiple consciousness:** Bleuler, E. (1894), "Versuch einer naturwissenschaftlichen Betrachtung der psychologischen Grundbegriffe," *Allgemeine Zeitschrift für Psychiatrie und psychish-gerichtliche Medicin* 50:133–68.

185 **counterreport that said he was wrong:** See Puenzieux, D., and B. Ruckstuhl (1994), *Medizin, Moral und Sexualität: Die Bekämpfung der Geschlechtskrankheiten Syphilis und Gonorrhöe in Zürich 1870–1920*, Zürich: Chronos.

186 **Forel published *The Sexual Question*:** Forel, A. (1905), *The Sexual Question: A Scientific, Psychological, Hygienic and Sociological Study*, trans. C. F. Marshall, New York: Physicians and Surgeons Book Company, 1925.

186 ***Neue Züricher Zeitung* and the *Züricher Post*:** Also, Forel, A. (1906), *Sexuelle Ethik: Ein Vortag gehalten am 23. März 1906 auf Veranlassung des "Neuen Vereins" in München*, München: Ernst Reinhardt, unnumbered page that comes after p. 55.

186 **the book as a moral aberration:** Puenzieux, D., and B. Ruckstuhl (1994), *Medizin, Moral und Sexualität*, 158.

187 **dragged into daylight and be recorded":** Galton, F. (1879), "Psychometric Experiments," *Brain*, 2:149.

187 **exert upon his conscious life":** Galton, F. (1879), "Psychometric Facts," *The Nineteenth Century*, March 5:425.

187 **before they faded from mind:** Ibid., p. 426.

187 **determining the laws of their relations:** Wundt, W. (1897), *Outlines of Psychology*, 28–275.

188 **future for scientific psychopathology:** Kraepelin, E. (1895), "Der psychologische Versuch in der Psychiatrie," in *Psychologische Arbeiten*, ed. E. Kraepelin, Leipzig: Wilhelm Engelmann, 1–91.

188 **foundation for the study of psychopathology:** Ziehen, T. (1894), *Psychiatrie für Ärzte und Studirende bearbeitet*, Berlin: Freidrich Wreden.

188 **based on sound (i.e., "bat-rat-cat"):** Aschaffenburg, G. (1895), "Experimentelle Studien über Associationen," in *Psychologische Arbeiten*, ed. E. Kraepelin, 209–99.

188 **emotionally charged idea-complex:** See Ziehen, T. (1898–1900), *Die Ideenassoziation des Kindes*, 2 vols., Berlin: Reuther and Reichard.

189 **a prominent Basel family, the Preiswerks:** See Ellenberger, H. (1991), "C. G. Jung and the Story of Helene Preiswerk: A Critical Study with New Documents," in

Beyond the Unconscious: Essays of Henri F. Ellenberger in the History of Psychiatry, ed. Mark S. Micale, Princeton, N.J.: Princeton University Press, 1993, 291–305.

189 **as mysterious as divinity itself:** Jung, C. G. (1989), *Memories, Dreams, Reflections*, ed. A. Jaffé, New York: Vintage Books, 68–69. On the difficulties with this text, see Shamdasani, S. (1995), "Memories, Dreams, Omissions," *Spring* 57:114–37. On recent Jung scholarship, see Taylor, E. (1996), "The New Jung Scholarship," *Psychoanalytic Review* 83:547–68.

189 **to understand what plagued him:** Jung, C. G. (1989), *Memories, Dreams, Reflections*, 94, also 101–03.

189 **university and its Zofingia Students Association:** Oeri, A. (1977), "Some Youthful Memories," in *C. G. Jung Speaking: Interviews and Encounters*, eds. W. McGuire and R. F. C. Hull, Princeton, N.J.: Princeton University Press, 7. Jung's nickname, "Walze," could also be rendered "the Roller," see Shamdasani, S. (2005), *Jung Stripped Bare by His Biographers, Even*, London and New York: Karnac, 87.

189 **domain for mystery and human subjectivity:** Jung, C. (1896), "The Border Zones of Exact Science," in *The Zofingia Lectures*, trans. Jan van Heurck, Princeton, N.J.: Princeton University Press, 1983, *C.W.* 20:3–19. In an 1897 lecture on psychology, Jung argued for a continued interest in metaphysics and the soul; see his "Some Thoughts on Psychology," in Ibid., 21–48.

190 **would parrot "Papa Du Bois-Reymond":** Jung, C. (1896), "The Border Zones of Exact Science," *C.W.* 20:6.

190 **in an otherwise mechanized world:** Jung, C. (1897), "Some Thoughts on Psychology," *C.W* 20:25.

190 **eventually Jung set it aside:** Jung, C. (1989), *Memories, Dreams, Reflections*, 107.

190 **assistant doctor at the Burghölzli:** *Rechenschaftsbericht über die Zürcherische kantonale Irrenheilanstalt Burghölzli für das Jahr 1900*. Zurich: Buchdruckerei Berichthaus, 1901, 17; Jung and others thereafter mistakenly put the date as Dec 10. see Jung, C. G. (1989), *Memories, Dreams, Reflections*, ed. A. Jaffé. New York: Vintage Books, 111.

190 **unconscious and psychotic mental phenomena:** In Jung's memoirs, Eugen Bleuler is conspicuous by his absence. Jung also spoke of the Burghölzli as a place barren of interest in psychological meaning, which is, I believe, untenable; Jung, C. (1989), *Memories, Dreams, Reflections*, 112.

190 **work on dreams by Sigmund Freud:** Jung, C. G. (1901), "Sigmund Freud: 'On Dreams' January 25, 1901," *C.W.* 18:361–68.

191 **nor did it allow for supernatural happenings:** Jung, C. (1902), "On the Psychology and Pathology of So-called Occult Phenomena," *C.W.* 1:3–88.

191 **a fellow Swiss, Théodore Flournoy:** See Shamdasani, S. (2003), *Jung and the Making of Modern Psychology: The Dream of a Science*, Cambridge, U.K.: Cambridge University Press.

191 **From India to the Planet Mars:** Flournoy, T. (1900), *Des Indes à la planète mars: Étude sur un cas de somnambulisme avec glossolalie*, Paris and Geneva: F. Alcan and Ch. Eggimann.

192 **lives himself into the suggestion":** Jung, C. (1902), "On the Psychology and Pathology of So-called Occult Phenomena," *C.W.* 1:87.

192 **allow lower mental states to emerge:** Janet's lectures are described in the (1902), *Annuaire du Collège de France*, Paris: E. Leroux, 27–28.

192 **Freud's studies of hysteria:** Jung, C. G. (1904), "On Hysterical Misreading," *C.W.* 1:92.

192 **Burghölzli researchers in 1906:** Bleuler, E. (1906), "Upon the Significance of Association Experiments," in *Studies in Word-Association, Experiments in the Diagnosis of Psychopathological Conditions Carried Out at the Psychiatric Clinic of the University of Zurich*, ed. C. G. Jung, trans. M.D. Eder, London: William Heinemann, 1918, 3.

192 **complete psychology of a man:** Ibid., 4.

193 **part of this new world, Bleuler announced:** Bleuler's introduction and the word-association studies were first published in a series of articles in the *Journal für Psychologie und Neurologie* in 1904. See that journal, vol. 3:49–54, 55–83, 145–164, 193–215, 283–308.

193 **baseline for normal associative patterns:** Bleuler, E. (1906), "Upon the Significance of Association Experiments," in *Studies in Word-Association*, 8.

193 **valuable stimulus in our investigations":** Jung, C., and Riklin, F. (1906), "The Associations of Normal Subjects," in Ibid., 8–172. The quotation can be found on p. 168.

193 **poisoning it when it awakes":** Meyer, A. (1905), "Psychological Literature: Normal and Abnormal Association," *The Psychological Bulletin* 2:253.

194 **whose family complex occurred in 54 percent:** Jung, C. G. (1906), "Reaction-Time in Association Experiments," in *Studies in Word-Association*, ed. C. G. Jung, 238.

194 **complex itself caused the illness:** Jung, C. G. (1906), "Psycho-analysis and Association Experiments," in Ibid., 299–300.

194 **Hermann von Helmholtz and Sigmund Freud:** Bleuler, E. (1906), "Consciousness and Association," in Ibid., 271, 276.

194 **secure framework for gathering data:** Jung, C. G. (1906), "Psychoanalysis and the Association Experiments," *C.W.* 2:290.

195 **emphasis on psychic etiologies and therapies:** Vogt, O. (1902–03), "Psychologie, Neurophysiologie und Neuroanatomie," *Journal für Psychologie und Neurologie* 1:1.

195 **"very important" association studies:** Flournoy, T. (1903), Review of S. Freud's "Die Traumdeutung," *Archives de Psychologie* 2:72–73; Flournoy, T. (1903), Review of C. G. Jung's "Zur Psychologie und Pathologie sogenannter occulter Phänomene," *Archives de Psychologie* 2:85–86; Claparède, E. (1906), Review of Freud's "Zur Psychopathologie des Alltagslebens," *Archives de Psychologie* 5:181; and Claparède, E. (1906), Review of Jung's "Diagnostische Assoziationsstudien," *Archives de Psychologie* 5:181–82

195 **experiments were generally warmly received:** See Decker, H. S. (1977), *Freud in Germany: Revolution and Reaction in Science, 1893–1907*, New York: International Universities Press, 111.

196 **printing six extensive, though not uncritical, reviews:** Lipmann (1905), "Review of "Diagnostische Assoziationsstudien," *Zeitschrift für Psychologie* 40:213–5; Ibid. (1906), 41:230–32; Ibid. (1906), 42:69–71; Ibid. (1906), 43:119–120; Ibid. (1907), 44:153; Ibid. (1907), 45:298–99.

196 **to psychopathology during the past year":** Meyer, A. (1905), "Psychological Literature: Normal and Abnormal Association," *The Psychological Bulletin* 2:242.

196 **same issue of the *Bulletin*:** Ibid., 242–59.

196 **experiment is an absolutely sure guide":** Carl Jung to Adolf Meyer, July 10, 1907, The Adolf Meyer Papers, The Alan Mason Chesney Medical Archives of The Johns Hopkins Medical Institutions.

196 **the outlines for conducting psychoanalysis:** Ibid. Also see the letter dated November 19, 1907.

197 **contemporary, experimentally based psychology:** See Meyer, A. and E.B Titchner (1990), *Defining American Psychology: The Correspondence Between Adolf Meyer and Edward Bradford Titchner*, eds. R. Leys and R.B Evans, Baltimore and London: Johns Hopkins University Press, 116.

197 **efforts of the moribund authorities":** Freud-Jung Letters, 18.

198 **to question Freud's judgment and logic:** Bleuler-Freud Letters, June, 9, 1905. Some excerpts of this important correspondence have been published in Alexander, F. (1965), "Freud-Bleuler Correspondence," *Archives of General Psychiatry* 12:1–9. This is the first extant letter from Bleuler, but it is clearly not the first between the two men. Unfortunately Freud's half of this correspondence has not been made available to scholars.

198 **revealed an amalgam of methods:** Riklin, F. (1905), "Analytische Untersuchungen der Symptome und Associationen eines Falles von Hysteria (Lina H.)," *Psychiatrisch-Neurologische Wochenschrift*, 46:449–52, 47:464–67, 48:469–75, 49:481–84, 50:493–97, 51:505–511, 52:521–25.

198 **interpreting my own dreams," he confessed:** Bleuler-Freud Letters, October 9, 1905.

199 **his wife's unconscious complexes, not his:** Ibid., October 14, 1905.

199 **He had intended to write "them":** Ibid.

199 **not an emotional resistance":** Ibid., October 17, 1905.

200 **hospital attendants and his wife's sister:** Ibid., November 5, 1905.

200 **colleagues were eagerly devouring:** Ibid., November 28, 1905.

200 **shall also gladly accept correction":** Freud-Jung Letters, 3.

200–01 **hysteria without a second thought:** Kraepelin, E. (1899), *Psychiatrie: Ein Lehrbuch für Studirende und Aerzte*, vol. 2, Leipzig: Johann A. Barth, 511–12.

201 **Freud had never heard of Aschaffenburg:** Freud-Jung Letters, 6.

201 **"Here we have two warring worlds":** Ibid.

201 **lately devoted to explaining Freud's theories:** Freud-Jung Letters, 4.

201 **Münchener Medizinische Wochenschrift:** Aschaffenburg, G. (1906), "Die Beziehungen des sexuellen Lebens zur Entstehung von Nerven- und Geisteskrankheiten," *Münchener Medizinische Wochenschrift* 53:1793–1798.

201 **tautological trap that offered nothing:** Ibid., 1796.

201 **baby with the bathwater":** Jung, C. G. (1906), "Freud's Theory of Hysteria: A Reply to Aschaffenburg," *C.W.* 4:3–9.

202 **position on sexuality over time:** Freud-Jung Letters, 4–6.

202 **check Freud's thought-processes experimentally":** Jung, C. G. (1906), "Freud's Theory of Hysteria," *C.W.* 4:3–4.

202 **no different than the association experiment":** Ibid., 7.

202 **Spielmeyer's critique of the Dora case:** Aschaffenburg, G. (1906), "Die Beziehungen des sexuellen Lebens zur Entstehung von Nerven- und Geisteskrankheiten," 1797.

202 **scientific research is imperiled":** Jung, C. G. (1906), "Freud's Theory of Hysteria," *C.W.* 4:9.

203 **in the fray by condemning Spielmeyer:** Bleuler, E. (1906), "Vermischtes," *Centralblatt für Nervenheilkunde und Psychiatrie* 29:460–61.

203 **storm of protest against you":** Freud-Jung Letters, 16.

203 **quite different from the Professor's:** Ibid., 14.

204 **link to Wundt's psychology:** Ibid., 15.

204 **treat patients in the same situation":** Ibid., 18.

204 **what difficulties defending you presents":** Bleuler-Freud Letters, February 18, 1907.

204 **reservations on that very subject:** Freud-Jung Letters, 32.

204 **desire to marry Freud's' daughter:** Binswanger, L. (1957), *Sigmund Freud: Reminiscences of a Friendship*, New York: Grune and Stratton, 2.

204 **Freud's commitment to sexual theory:** Jung, C. G. (1989), *Memories, Dreams, Reflections*, 149.

205 **black tide of mud . . . of occultism":** Ibid., 150.

205 **you have seen the gang":** Binswanger, L. (1957), *Sigmund Freud: Reminiscences of a Friendship*, 4.

205 **forty-nine "the large Jewish penis":** Vienna Minutes I, 144.

205 **who were still thinking a bit:** Bleuler-Freud Letters, March 21, 1907.

205 **swears by the master:** Freud-Jung Letters, 26.

206 **take a guess at the cause:** *Rechenschaftsbericht über die Zürcherische kantonale Irrenheilanstalt Burghölzli für das Jahr 1900.* Zurich: Buchdruckerei Berichthaus, 1901, 4.

207 **odd behaviors into this one illness:** Kraepelin, E. (1896), *Psychiatrie: Ein Lehrbuch für Studirende und Aerzte*, Leipzig: Johann Ambrosius Barth.

207 **states of negativity might be related:** Bleuler, E. (1904–05), "Die negative Suggestibilität, ein psychologisches Prototyp des Negativismus," *Psychiatrisch-neurologische Wochenschrift*, 249–69.

208 **found as pairs of opposites:** Freud, S. (1905), *Three Essays on the Theory of Sexuality*, S.E. 7:160.

208 **nervous and psychic areas":** Bleuler-Freud Letters, October 17, 1905.

208 **"play of contraries":** Jung, C. G. (1907), *The Psychology of Dementia Praecox, C.W.* 3:18.

208 **negative and positive affects:** Bleuler, E. (1906), *Affectivity, Suggestibility, Paranoia*, Utica, N.Y.: State Hospitals Press, 1912, 45–54.

209 **of libido on the analyst:** Freud-Jung Letters, 6, 8, 13.

209 **symptoms of dementia praecox:** Bleuler-Freud Letters, January 28, 1906.

209 **symbolized as a persecutory Other:** Bleuler, E. (1906), "Freud'sche Mechanismen in der Symptomatologie von Psychosen," *Psychiatrisch-neurologische Wochenschrift*, 316–18, 323–25, 338–40.

210 **I shall try to believe you":** Freud-Jung Letters, 13.

210 **by an unknown, internal poison:** Jung, C. G. (1907), *The Psychology of Dementia Praecox, C.W.* 3:1–152.

210 **"panegyric" to the Viennese doctor's ideas:** Bleuler-Freud Letters, March 21, 1907.

210 **published a biting critique:** Isserlin, M. (1907), "Ueber Jung's 'Psychologie der Dementia praecox' und die Anwendung Freud'scher Forschungsmaximen in der Psychopathologie," *Centralblatt für Nervenheilkunde und Psychiatrie*, 18: 329–43.

210 **started using the heavy artillery":** Freud-Jung Letters, 43.

210 **Gustav Aschaffenburg's textbook:** Ibid., 44.

211 **Wednesday nights in Vienna:** Ibid., 89–91, 101. The German name was "Freudsche

Gesellschaft von Ärzten," see Freud, S., and C. Jung (1974), *Briefwechsel*, eds. W. McGuire and W. Sauerländer, Frankfurt am Main: S. Fischer, 99.

211　**Budapest, Boston, and Switzerland:** Freud-Jung Letters, 68.

6. FREUDIANS INTERNATIONAL

213　**lines are converging on you":** Freud-Jung Letters, 67.

213　**reckon with their thick hides":** Ibid., 33.

214　**only Freudian in all of Amsterdam:** Ibid., 79–81.

214　**alcohol, asylum-related, and psychological societies:** Wayenburg, G. A., ed. (1908), *Compte rendu des travaux du 1er congrès international de psychiatrie, de neurologie, de psychologie et de l'assistance des aliénés.* Amsterdam: J. H. de Bussy.

214　**misguided in their emphasis on sexuality:** Ibid., 265–71.

215　**manifested itself through transferences:** Ibid., 273–84; compare with Jung, C. G. (1908), "The Freudian Theory of Hysteria," *C.W.* 4:10–24. The quotation is on p. 18.

215　**Jungian theory in his own paper:** Freud-Jung Letters, 84–85. Also see Wayenburg, G. A. ed. (1908), *Compte rendu des travaux du 1er congrès international de psychiatrie*, 293–302.

215　**French were supremely unimpressed:** Freud-Jung Letters, 94.

215　**penchant for grand philosophical speculation:** Wayenburg, G. A., ed. (1908), *Compte rendu des travaux du 1er congrès international de psychiatrie*, 301–02.

216　**panelists mention Freudian psychoanalysis:** Ibid., 855–69.

216　**opponents repeatedly assaulted Jung:** Jung reported that his *Psychology of Dementia Praecox* was harshly reviewed by another of Kraepelin's assistants, Wilhelm Weygandt, then a few weeks later attacked by Paul Näcke and Ernst Meyer. See Freud-Jung Letters, 51, 91, 94.

216　**over and done with":** Ibid., 98.

216　**"Many enemies, much honor":** Ibid., 178.

216　**much in a short visit:** Ibid., 99.

217　**"a focus of infection" to their hometowns:** Ibid., 98.

217　**congress of Freudian followers:** Ibid., 101–02.

217　**journal dedicated to Freudian studies:** Carl Jung to Ernest Jones, Dec. 7, 1907; also see Freud-Jung Letters, 59, 63, 88.

217　**dropped any reference to Freud's creation:** Freud-Jung Letters, 59, 63, 104.

217　**Freudian journal would be small:** Carl Jung to Ernest Jones, Dec. 7, 1907, Sigmund Freud Copyrights, Wivenhoe, Colchester, England.

217　**prospect Jung found troublesome:** Carl Jung to Ernest Jones, Jan. 21, 1908, Sigmund Freud Copyrights, Wivenhoe, Colchester, England.

218　**"proposed place of meeting is *Salzburg*":** Freud-Jung Letters, 110.

218　**in scientific work," Jones later recalled:** Jones, E. (1959), *Free Associations: Memories of a Psycho-Analyst*, New York: Basic Books, 165.

218　**"pretentious title of the circular":** Freud-Jung Letters, 114. In German, the flyer was sent out for the "I. Kongress für Freudsche Psychologie."

218　**mentioned in public at all:** Freud-Abraham Letters, 38.

218　**private, international meeting of Freudians:** There are discrepancies between Ernest Jones's tally (42) and the number of attendees listed on the official program (38); See Jones, E. (1953), *The Life and Work of Sigmund Freud*, vol 1, 41, and Vienna Minutes, 390–91.

218 **for the Viennese and the Swiss:** Jones, E. (1955), *The Life and Work of Sigmund Freud*, vol. 2, 40–41.

218 **detailed Freud's studies on hysteria:** Jones, E. (1959), *Free Associations: Memories of a Psycho-Analyst*, 125. On Jones, see Maddox, B. (2006), *Freud's Wizard, The Enigma of Ernest Jones*, London: John Murray.

219 **Jones made Jung's acquaintance:** Wayenburg, G.A., ed. (1908), *Compte rendu des travaux du 1er congrès international de psychiatrie*, 408–14.

219 **invited to visit the Burghölzli:** Freud-Jung Letters, 85–86. Carl Jung to Ernest Jones, Nov. 23, 1907, Sigmund Freud Copyrights, Wivenhoe, Colchester, England.

219 **establish a Freudian society there:** Carl Jung to Ernest Jones, Dec. 7, 1907, Sigmund Freud Copyrights, Wivenhoe, Colchester, England.

219 **manifestation of unconscious mental life:** Ferenczi, S. (1899), "Spiritizmus," *Gyógyászat*, 30:477–79. For the English, see Sagi, G. (1999), "Notes on Translating Ferenczi," *Bulletin of the Association for Psychoanalytic Medicine* 36: 40–50. On Ferenczi's early development see Mézáros, J. (1993), "Ferenczi's Pre-analytic Period Embedded in the Cultural Streams of the Fin de Siècle," in *The Legacy of Sándor Ferenczi*, ed. L. Aron and A. Harris, Hillsdale, N.J.: The Analytic Press, 41–51, and Lorin, C. (1993), *Sándor Ferenczi: De la medicine à la psychanalyse*, Paris: Presses Universitaires de France.

220 **equipment and became a convert:** Stanton, M. (1991), *Sandor Ferenczi: Reconsidering Active Intervention*, Northvale, N.J.: Jason Aronson Inc., 11.

220 **some way analyzed by Jung:** Falzeder, E. (1994), "The Threads of Psychoanalytic Filiations or Psychoanalysis Taking Effect," in *100 Years of Psychoanalysis*, eds. A. Haynal and E. Falzeder, Geneva: Cahiers Psychiatriques Genevois, 169–94.

220 **Budapest's physicians and literati:** On the Hungarian writers who gravitated to Ferenczi, see Moreau-Ricaud, M., ed. (1992), *Écrivains hongrois autour de Sándor Ferenczi*, Paris: Gallimard.

220 **were yet willing to consider:** Brill, A. A. (1908), "Psychological Factors in Dementia Praecox. An Analysis," *The Journal of Abnormal Psychology* 3:219–39.

221 **we could then build with confidence":** Lehrman, P. (1948), "A. A. Brill in American Psychiatry," *Psychoanalytic Quarterly* 17:157–58.

221 **promised much and delivered little:** Brill, A. A. (1943), "Max Eitingon," *Psychoanalytic Quarterly* 12:456. The 1858 novel, *Oblomov*, was written by the Russian author Ivan Goncharov.

221 **determined to work at the Burghölzli:** Jones, E. (1948), "Introductory Memoir," in Abraham, K., *Selected Papers of Karl Abraham M.D.* London: The Hogarth Press, p. 14.

221 **hardly anybody is willing to approach":** Freud-Abraham Letters, 1. Also see Abraham, K. (1907), "The Experiencing of Sexual Trauma as a Form of Sexual Activity," in *Selected Papers of Karl Abraham M.D.*, 1948, 47–63.

221 **Freud's support through patient referrals:** Freud-Abraham Letters, 8–9.

221 **win Berlin for the Freudian corps:** Ibid., 10, 34–35; Karl Abraham to Max Eitingon, cited in Abraham, H. C. (1974), "Karl Abraham: An Unfinished Biography," *I.R.P.* 1:38.

222 **perceptual states in an original way:** Gross, O. (1902), *Sekundärfunction*, Leipzig: Verlag von F. C. W. Vogel.

222 **predicated on an affect of helplessness:** Gross, O. (1903), "Beitrag zur Pathologie des Negativismus," *Psychiatrisch-neurologische Wochenschrift* 26:269–73.

222 **Freud's concept of unconscious repression:** Gross, O. (1904), "Zur Differentialdiagnostik negativistischer Phänomene," *Psychiatrisch-neurologische Wochenschrift* 37: 345–53.

222 **more general notion of psychic conflict:** Gross, O. (1907), *Das Freud'sche Ideogenitätsmoment und seine Bedeutung in manisch-depressivem Irresein Kraepelin's*, Leipzig: F. C. W. Vogel, 7–8. For an English translation of this and other early works, see Otto Gross, *Collected Works 1901–1907, The Graz Years*, ed. L. Madison, Hamilton, N.Y.: Mindpiece, 162–92.

223 **in front of the academics from Switzerland:** Freud-Jung Letters, 120.

223 **"degenerate and Bohemian crowd":** Jones, E. (1959), *Free Associations: Memories of a Psycho-Analyst*, 167.

223 **deeming most of them insignificant:** Abraham, H. C. (1974), "Karl Abraham: An Unfinished Biography," *I.R.P.* 1:35.

223 **Viennese followers were almost all Jewish:** The group was 88 percent Jewish. See Mühlleitner, E., and J. Reichmayr (1997), "Following Freud in Vienna: The Psychological Wednesday Society and the Viennese Psychoanalytical Society 1902–1938," *Int Forum Psychoanal* 6:73–102.

224 **thwarted by anti-Semitism:** See Oxaal, I. (1988), "The Jewish Origins of Psychoanalysis Reconsidered," in *Freud in Exile: Psychoanalysis and Its Vicissitudes*, eds. E. Timms, E. and N. Segal, New Haven and London: Yale University Press, 1988, 37–53.

224 **resulting in the repression of hate":** Jones, E. (1955), *The Life and Work of Sigmund Freud*, vol. 2, 42.

224 **Freud never mentions the Swiss doctor:** In 1923, Freud finally relented, adding a tepid footnote that still insisted on his own priority: "Bleuler subsequently (1910) introduced the appropriate term 'ambivalence' to describe this emotional constellation"; Freud, S. (1909), "Notes upon a Case of Obsessional Neurosis," *S.E.* 10:239. Bleuler, inspired by a passage in Freud's *Three Essays*, put forward a theory of ambivalence, without naming it as such, in a letter to Freud on Oct. 17, 1905.

225 **attributable to sexual trauma:** Abraham, K. (1907), "Ueber die Bedeutung sexueller Jugendträumen für die Symptomatologie der Dementia Praecox," *Centralblatt für Nervenheilkunde und Psychiatrie* 30:409–16.

225 **thinking about their psychosexual differences:** Freud-Abraham Letters, 2.

225 **deal with this subject very soon":** Abraham, H. C. (1974), "Karl Abraham: An Unfinished Biography," *I.R.P.* 1:35.

225 **"Psycho-sexual Differences between Hysteria and Dementia Praecox":** Freud-Abraham Letters, 24–45.

225 **considered an untrustworthy, competitive loner:** Freud-Jung Letters, 78.

226 **to abandon his plan":** Ibid., 149.

226 **characterized dementia praecox and negativism:** Abraham, K. (1908), "The Psychosexual Differences Between Hysteria and Dementia Praecox," in *Selected Papers of Karl Abraham M.D.*, 69.

226 **"recently discussed toxin theory":** Ibid., 78.

226 **it is our intention to remain":** Turner, J., C. Rumpf-Worthen, et al. (1990), "The Otto Gross-Frieda Weekley Correspondence: Transcribed, Translated, and Annotated," *The D. H. Lawrence Review*, 22:222. Also see Stekel, W. (1950), *The Autobiography of Wilhelm Stekel*, 122.

226 "**clearness of his utterances**": Wittels, F. (1924), *Sigmund Freud: His Personality, His Teaching, & His School*, trans. E. and C. Paul, New York, Dodd, Mead & Company Publishers, 136.

226 **domain of general science**": Ibid.

227 **Psychoanalytic and Psychopathologic Research**": See Freud-Jung Letters, 147–52.

227 **to his mind even I am reactionary**": Ibid., 130, 145.

227 **convinced by their own work**": Ibid., 195.

228 **extension of the libido concept**: Abraham, H. C. (1974), "Karl Abraham: An Unfinished Biography," *I.R.P.* 1:35.

228 **which is very gratifying for me**": Ibid.

228 "**Jewish national affair**": Freud-Abraham Letters, 38.

229 **impressions which this illness gives us**: Carl Jung–Ernest Jones Letters, February 25, 1909. Published by permission of Ulrich Hoerni, Erbengemeinschaft C. G. Jung.

229 **would not advance the movement**: Freud-Jung Letters, 144.

229 **when we are besieging Troy**": Ibid., 146.

229 **ought to be excluded amongst us**": Freud-Abraham Letters, 34.

229 **letter of resignation from Alfred Adler**: The letter dated May 31, 1908, is reprinted in Handlbauer, B. (1998), *The Freud-Adler Controversy*, Oxford, U.K.: Oneworld, 173.

229 **distinct form from the sexual drive**: Vienna Minutes I, 406.

229 **say that a long debate ensued**: Ibid., 410.

230 **with Else and Frieda von Richthofen**: Green, M. (1974), *The von Richthofen Sisters: The Triumphant and the Tragic Modes of Love*, New York: Basic Books.

230 **influence on the English writer D. H. Lawrence**: Turner, J., C. Rumpf-Worthen, et al. (1990), "The Otto Gross-Frieda Weekly Correspondence: Transcribed, Translated and Annotated," *The D. H. Lawrence Review* 22:167.

231 **associates you with Nietzsche**: Freud-Jung Letters, 90.

231 **no longer lies across my path**": Turner, J., C. Rumpf-Worthen, et al. (1990), "The Otto Gross-Frieda Weekly Correspondence: Transcribed, Translated and Annotated," *The D. H. Lawrence Review* 22:190.

231 **up to twelve hours a day**: Jones, E. (1959), *Free Associations: Memories of a Psycho-Analyst*, 174.

231 **people of that kind**": Freud-Abraham Letters, 72.

231 **do great harm to our cause**": Freud-Jung Letters, 162.

231 **but for the Dementia praecox**": Ibid., 156.

231–32 **idea of my polygamous components**": Ibid., 207.

232 **seeking counsel on this matter**: Minder, B. (1993), "Jung an Freud 1905: Ein Bericht über Sabina Spielrein," *Gesnerus* 50:113–20.

232 **suppress his feeling for me**": Letters from Spielrein to Freud are reprinted in Carotenuto, A. (1982), *A Secret Symmetry: Sabina Spielrein Between Jung and Freud*. New York: Pantheon Books, 107.

232 **not suffocate due to banal conventions**: Letters from Jung to Spielrein are reprinted in Carotenuto, A. (1986), *Tagebuch einer heimlichen Symmetrie: Sabina Spielrein zwischen Jung und Freud*, Freiburg: Kore, 189.

232 **covered in her own blood**: Carotenuto, A. (1982), *A Secret Symmetry: Sabina Spielrein between Jung and Freud*, 97, 100–4.

232 **a bit too much in my head**": Freud-Jung Letters, 228.

232 **needed to be controlled**: Ibid., 231.

232 **Bleuler and Gross," Freud wrote to Jung:** Freud-Jung Letters, 238.

233 **left London for the New World:** Ibid., 207.

233 **harem of women:** Jones, E. (1959), *Free Associations: Memories of a Psycho-Analyst*, 173–74. Freud-Jung Letters, 211.

233 **child molestation while in London:** Kuhn, P. (2002), "Romancing with a Wealth of Detail," *Studies in Gender and Sexuality*, 3:344–78; also see my critique, Makari, G. (2002), "On 'Romancing with a Wealth of Detail,' " *Studies in Gender and Sexuality*, 3:389–94.

233 **rein in his followers:** Timms, E. (1986), *Karl Kraus Apocalyptic Satirist; Culture and Catastrophe in Habsburg Vienna*, New Haven: Yale University Press, 105.

233 **one-time sidekick, Fritz Wittels:** Ibid., 96. Freud rejected Wittels's version of Freudian thought, critiquing the "assertion that suppression of sexuality is the root of all evil. But we go further, and say: we liberate sexuality through our treatment, but not in order that man may from now on be dominated by sexuality, but in order to make a suppression possible—a rejection of the instincts under the guidance of a higher agency," *Vienna Minutes*, II, 89.

233 **dedicated to Sigmund Freud:** Wittels, F. (1909), *Die sexuelle Not*, Wien and Leipzig: C.W. Stern. The title has also been translated as *Sexual Need*.

233 **vulgarization of his own ideas:** Wittels, F. (1995), *Freud and the Child Woman*, 99–100.

233 **portraying him as a deformed neurotic:** *Vienna Minutes*, II, 382–93.

233 **compensation for his own hideousness:** Wittels, F. (1995), *Freud and the Child Woman*, 93.

233 **if you publish this book":** Ibid., 98.

234 **psychology in the United States:** Letter from G. Stanley Hall to Sigmund Freud, in Rosenzweig, S. (1992), *The Historic Expedition to America (1909): Freud, Jung and Hall the King-maker*, St. Louis: Rana House, 339.

234 **they will drop us":** Freud-Jung Letters, 196.

235 **borderline field for our chief light":** Rosenzweig, S. (1992), *The Historic Expedition to America (1909): Freud, Jung and Hall the King-maker*, 361.

235 **answered his first day back:** Freud-Jung Letters, 249.

235 **all of the books is extraordinary":** Freud-Brill Letters, May 2, 1909.

236 **"King Edward VII of England?!":** See Freud-Jung Letters, 170, 127, 237, 195, 198, 214, 225, 262, 272, 239.

236 **percolating through to the public":** Ibid., 193.

236 **there's no stopping it now":** Ibid., 247.

236 **retained his leadership of the lab:** Freud-Abraham Letters, 51–52.

236 **"going unreservedly with us":** Freud-Ferenczi Letters, I, 17, also 50.

236 **"we're on top of the world":** Freud-Jung Letters, 268.

236 **dared not show their faces:** Ibid., 268.

236 **have the upper hand":** Freud-Brill Letters, Feb. 14, 1909.

236 **not use Freudian methods:** Abraham, H. C. (1974), "Karl Abraham: An Unfinished Biography," *I.R.P.* 1:36.

237 **with the women in another:** Freud-Abraham Letters, 35.

237 **understood Freudian sexual theory:** Ibid., 56, 64–65.

237 **into account as much as possible":** Freud-Jung Letters, 138.

238 **Freudians would not be welcome:** Jung, C. G. (1973), *Letters*, ed. G. Adler, and A. Jaffé, vol. 1, Princeton, N.J.: Princeton University Press, 11.

238 **sway members to join their ranks:** Freud-Jung Letters, 259.

238 **definition and defense of their cause:** Ibid., 282.

7. INTEGRATION/DISINTEGRATION

240 **adversaries were in retreat:** Freud-Ferenczi Letters, I, 152.

240 **"organization and propaganda":** Freud-Jung Letters, 292.

240 **standardization in the field:** Freud S. (1910), "The Future Prospects of Psychoanalytic Therapy," *S.E.* 11:139–52.

241 **this primer for two years:** Ibid.

241 *A General Exposition of the Psychoanalytic Method*: Freud-Ferenczi Letters, I, 27; Freud-Brill Letters, Nov. 8, 1908; Freud-Jung Letters, 175; Freud-Abraham Letters, 58.

241 **obstacles writing this work:** Freud-Jung Letters, 202.

241 **approximated a section of *A General Exposition*:** Ibid., 193.

241 **origin in repressed parental complexes:** Ferenczi, S. (1909), "Introjection and Transference," in *First Contributions to Psycho-Analysis*, 35–93. Also see Freud-Brill Letters, August 27, 1908.

242 **had to be surmounted, Freud warned:** Freud, S. (1910), "The Future Prospects of Psycho-analytic Therapy," *S.E.* 11:144–45.

243 **produce such a textbook:** Vienna Minutes, II:207–13.

243 **young German and Austrian doctors:** Hitschmann, E. (1911), *Freud's Neurosenlehre*. Leipzig und Wien, Franz Deuticke. The English translation is Hitschmann, E. (1911), *Freud's Theories of the Neuroses*, New York: The Journal of Nervous and Mental Disease Publishing Company, 1913. On Freud's view of this project, see Freud-Brill Letters, Oct. 17, 1910.

244 **dreams, myths, and folklore were constant and universal:** Freud, S. (1910), "The Future Prospects of Psycho-analytic Therapy," *S.E.* 11:142–43.

244 **constructing the various forms of neurosis":** Ibid., 144.

244 **inklings of a solution:** See Vienna Minutes, III, 30.

244 **part of normal psychology:** Freud-Fliess Letters, 272; Freud, S. (1900), *Interpretation of Dreams*, *S.E.* 4:261–63.

244 **account for the girl's neurosis:** Freud, S. (1905), *Fragment of an Analysis of a Case of Hysteria*, *S.E.* 7:56. See Makari, G. (1998), "Dora's Hysteria and the Maturation of Sigmund Freud's Transference Theory: A New Historical Interpretation," *Journal of the American Psychoanalytic Association* 45:1061–1096.

244 **beautiful mother and sleep with her":** Freud, S. (1909), "Analysis of a Phobia in a Five-Year-Old Boy," *S.E.* 10:111.

244 **love and hate for the Father:** Freud, S. (1909), "Notes upon a Case of Obsessional Neurosis," *S.E.* 10:155–249.

244 **core complex in psychosexual development:** Freud, S. (1910), "Five Lectures on Psychoanalysis," *S.E.* 11:47.

244 **coined the term: the Oedipus complex:** Freud, S. (1910), "A Special Type of Choice of Object Made by Men," *S.E.* 11:171.

245 **accused of endangering ideals," Freud warned:** Freud, S. (1910), "The Future Prospects of Psycho-analytic Therapy," *S.E.* 11:147.

245 **revolution from the couch:** Ibid., 149.

246 **redouble their efforts at repression:** Rank, O. (1910), "Bericht über die I. private psy-

choanalytische Vereinigung in Salzburg am 27. April 1908," *Zentalblatt für Psycho-analyse: Medizinische Monatsschrift für Seelenkunde* I:130.

246 **propaganda, but few serious followers:** Freud-Ferenczi Letters, I, 25, 34, 91, 103, 131.

246 **left him scientifically isolated:** Ibid., 83, 88.

246 **close relationships only with intelligent women:** Freud-Brill Letters, May 11, 1911.

246 ***propaganda* for our psychological movement":** Freud-Ferenczi Letters, I, 117.

246 **so as to exclude "undesirable elements":** Ibid., 119–20.

246 **prepared the Hungarian's Nuremberg speech:** Alexander, F., and S. T. Selesnick (1965), "Freud-Bleuler Correspondence," *Archives of General Psychiatry* 12:4.

247 **should maintain its hegemony":** Freud-Ferenczi Letters, I, 130.

247 **power to an enlightened few:** Ibid., 151.

247 **"International Psychoanalytical Association" (hereafter the "I.P.A."):** Ferenczi, S. (1911), "On the Organization of the Psycho-Analytic Movement," in *Final Contributions to the Problems and Methods of Psycho-analysis*, New York: Brunner/Mazel, Inc., 1980, 299–307. Also see Ferenczi, S. (1911) "Zur Organisation der psychoanalytischen Bewegung," in Ferenczi, S., *Bausteine zur Psychoanalyse*, vol. I, Leipzig: Internationaler Psychoanalytische Verlag, 274–89.

247 **kept in check by "mutual control":** Ibid., 303.

248 **gain respect in the outside world:** Ibid., 304.

248 **prevails on fundamental matters," Ferenczi reasoned:** Freud-Ferenczi Letters, I, 306.

248 **there was an outcry:** See, for instance, Freud-Brill Letters, Nov. 14, 1910.

249 **independent of the whims of Zurich!:** Stekel, W. (1950), *The Autobiography of Wilhelm Stekel*, 127–29. In these memoirs, Stekel mistakenly placed the congress in Weimar.

249 **acceptance in the world of science:** Stekel's account is corroborated in Wittels, F. (1924), *Sigmund Freud: His Personality, His Teaching, & His School*, New York: Dodd, Mead & Co., 140.

249 **will pass with me":** Alexander, F., and S. T. Selesnick (1965), "Freud-Bleuler Correspondence," 2. The letter is dated Sept. 28, 1910.

250 **what was not psychoanalysis:** Ibid.

250 **against rash fellow-workers":** Freud-Abraham Letters, 115.

250 **Jung was pleased too:** Freud-Brill Letters, Nov. 14, 1910. Freud-Jung Letters, 313.

250 **not an unconscious resistance:** Bleuler-Freud Letters, March 26, 1910.

250 **everyone in the group:** Freud-Jung Letters, 312–13.

250 **kick out the whole lot:** Ibid., 318.

251 **Nuremberg rules had been quite helpful:** Ibid., 320, 321.

251 **"stupid" situation to develop:** Ibid., 330–31.

251 **consolidation had proceeded too swiftly:** Ibid., 343, 345.

251 **would do just the opposite:** Alexander, F., and S. T. Selesnick (1965), "Freud-Bleuler Correspondence," 3.

251 **Isserlin accepted:** Bleuler-Freud Letters, Oct. 18/19, 1910.

251 **did not want such filth around:** Freud-Jung Letters, 299.

252 **bar him from attending:** Bleuler-Freud Letters, Oct. 13, 1910.

252 **render an opposition impossible," he reasoned:** Ibid.

252 **members who themselves accepted psychoanalysis:** Alexander, F., and S. T. Selesnick (1965), "Freud-Bleuler Correspondence," 4.

252 **getting his theory accepted:** Bleuler-Freud Letters, Oct. 18/19, 1910.

253 **looked into them," he observed:** Freud-Jung Letters, 366.

253 **"steadfastness for intolerance":** Alexander, F., and S. T. Selesnick (1965), "Freud-Bleuler Correspondence," 4.

253 **should psychoanalysts take in antipsychoanalysts?:** Freud-Jung Letters, 314.

253 **take on Freud's detractors:** Alexander, F., and S. T. Selesnick (1965), "Freud-Bleuler Correspondence," 4.

253 **be tested by science:** Bleuler E. (1910), "Die Psychanalyse Freuds," *Jahrbuch für psychoanalytische und psychopathologische Forschungen*, 2:623–70.

253 **against Kraepelin and Aschaffenburg:** Freud-Abraham Letters, 93–94.

254 **joined the Zurich society:** Vienna Minutes, III, 101–02.

254 **to be a towering achievement:** Bleuler, E. (1911), *Dementia Praecox or the Group of Schizophrenias*, trans. J. Zinkin, New York: International Universities Press, 1950.

254 **owed to the Viennese innovator:** Ibid., 1.

254 **his section on therapeutics:** Ibid., 461–89.

254 **"respect-filled horror":** Bleuler-Freud Letters, Oct. 6, 1911.

254 **mad jurist, Daniel Paul Schreber:** Freud, S. (1911), "Psychoanalytic Notes on an Autobiographical Account of a Case of Paranoia (Dementia Paranoides)," *S.E.* 12:3–84.

254 **Zuricher found quite dubious:** Bleuler-Freud Letters, Oct. 6, 1911.

255 **noxious to science," Bleuler wrote:** Ibid., Dec. 4, 1911. This letter is excerpted in Alexander, F., and S. T. Selesnick (1965), "Freud-Bleuler Correspondence," 5, where it is misdated as March 11, 1911.

255 **by the psychoanalysts themselves:** Bleuler-Freud Letters, Jan. 1, 1912. Also see Alexander, F., and S. T. Selesnick (1965), "Freud-Bleuler Correspondence," 7.

256 **resigned from the I.P.A.:** Freud-Brill Letters, Dec. 4, 1911.

256 **bring himself to believe it:** Ibid., Dec. 14, 1911.

256 **"What do you think about Freud?":** Freud-Abraham Letters, 118.

257 **deputy chair to Stekel:** Vienna Minutes, II, 463–64.

257 **Zurichers were of another "breed":** Ibid., 467.

258 **Zentralblatt für Psychoanalyse:** Ibid., 465–70.

258 **over one of Stekel's own submissions!:** Stekel, W. (1950), *The Autobiography of Wilhelm Stekel*, 131.

258 **any article he disliked as well:** Freud-Jung Letters, 367.

258 **greatest period of growth ever:** See Mühlleitner, E., and J. Reichmayr (1997), "Following Freud in Vienna: The Psychological Wednesday Society and the Viennese Psychoanalytical Society 1902–1938," *Int. Forum Psychoanal.* 6:73–102.

258 **feelings of inferiority and neurosis:** Vienna Minutes, II, 259–65.

259 **weak, inferior, and female:** Ibid., 425–46.

260 **between Alfred Adler and Sigmund Freud:** Vienna Minutes, III, 19–22.

260 **prevents young men from getting ahead":** Freud-Jung Letters, 373.

260 **radically different than ours":** Ibid., 376.

260 **organization were really psychoanalysts:** Freud, S. (1910), " 'Wild' Psychoanalysis," *S.E.* 11:221–27.

261 **increased by masculine protest:** Vienna Minutes, III, 102–03.

261 **"The Masculine Protest as the Central Problem of Neurosis":** Ibid., 140.

261 **another sign of masculine protest:** Ibid., 145.

261 **revenge of the offended goddess Libido":** Freud-Pfister Letters, 48.

261 **damage to the standing of psychoanalysis:** Vienna Minutes, III, 146–47.

261 **was to be found in Freud:** Ibid., 172.

261 **letting this go on so long:** Ibid., 175.

261 **his position in the society":** Ibid., 177.

262 **became deputy chair:** Freud-Jung Letters, 400.

262 **incompatible with the Society's:** Vienna Minutes, III, 179.

262 **at the recklessness of drivers":** Freud-Pfister Letters, 48.

262 **not rise above such pettiness:** Freud-Brill Letters, March 20, 1911.

262 **Society for Free Psychoanalytic Investigation:** Freud-Jung Letters, 422, 428. Adler's group met as the "Verein für Freie Psychoanalytische Forschung."

262 **The "Palace Revolution" was over:** Freud-Jung Letters, 403.

264 **position in the theoretical controversy:** Stekel, W. (1950), *The Autobiography of Wilhelm Stekel*, 141–42.

264 **criminal instincts motivated dreams:** Stekel, W. (1911), *Die Sprache des Traumes*, Wiesbaden, Bergmann. The first portion of this work is translated as Stekel, W. (1922), *Sex and Dreams: The Language of Dreams*, trans. J. S. Van Teslaar, Boston: The Gorham Press.

264 **expose deviations and discredit wild psychoanalysts:** Jung, C. G. (1911), "Annual Report (1910/11) by the President of the International Psychoanalytic Association," *C.W.* 20:424–25.

265 **create any physical illness:** Vienna Minutes, III, 336–38.

265 **open forum for all views:** Ibid., 339.

265 **their audience who may think otherwise":** Freud, S. (1912), "Contributions to a Discussion on Masturbation," *S.E.* 12:243.

266 **attack Jungian work in the *Jahrbuch*:** Freud-Ferenczi Letters, I, 409.

266 **join the *Zentralblatt* board, Stekel refused:** See Stekel, W. (1950), *The Autobiography of Wilhelm Stekel*, 142, Jones, E. (1955), *The Life and Work of Sigmund Freud*, vol. 2, 134–37.

266 **to get rid of him":** Freud-Ferenczi Letters, I, 414.

266 **"independent, and open it up to *everyone*":** Ibid., 41. Also see Alexander, F., and S. T. Selesnick (1965), "Freud-Bleuler Correspondance," 8.

266 **to take over the *Zentralblatt*:** Andreas-Salomé, L. (1964), *The Freud Journal*, London and New York: Quartet Encounters, 35.

266 **nothing short of treason:** Freud-Ferenczi Letters, I, 418.

267 **isolate the "Stekelblatt":** Freud-Jung Letters, 518–59.

267 **He is an intolerable person":** Freud-Ferenczi Letters, I, 424.

267 **but rather Stekel's unpleasant manner:** Freud-Abraham Letters, 12.

267 **Freud also now recognizes them":** Andreas-Salomé, L. (1964), *The Freud Journal*, 41.

267 **not remain a Freudian for long":** Stekel, W. (1950), *The Autobiography of Wilhelm Stekel*, 143.

268 **"Crown Prince" of the movement:** Freud-Jung Letters, 218.

268 **real father became God the Father:** Jung, C. G. (1909), "The Significance of the Father in the Destiny of the Individual," *C.W.* 4:321. Jung's argument was not free of

an anti-Semitic taint, for he noted that Judaism was an imperfect attempt at sublimation by a still too-barbarous people. Ibid., 320.

268 **come to terms with":** Freud-Jung Letters, 97.

268 **could only end in apostasy":** Ibid., 98.

268 **knowledge of the human psyche":** Ibid., 421. In this Jung was not alone. Ferenczi was quite fascinated by thought transference and mediums, and Freud too became interested in these phenomena. See Freud-Ferenczi Letters, II, 70, 216–18.

269 **dreams organized fairy tales:** Ricklin, F. (1907), *Wish Fulfillment and Symbolism in Fairy Tales*, trans. W. A. White, New York: Nervous and Mental Disease Pub., 1915. The author's name, usually rendered "Riklin," is spelled as above in this work.

269 **an infantile period of humankind:** Abraham, K. (1909), "Dreams and Myths: A Study in Folk-Psychology," in *Clinical Papers and Essays on Psycho-Analysis*, New York: Brunner/Mazel Publishers, 1955, 150–09.

269 **"core complex" of neurosis, the Oedipal complex:** Rank, O. (1909), *The Myth of the Birth of the Hero*, trans. F. Robbins and S. E. Jelliffe, New York: Brunner, 1952.

269 **it created a stir:** Honegger's paper entitled "Über paranoide Wahnbildung" was listed and abstracted with the other presentations at the 1910 Congress; see Rank, O. (1910), "Bericht über die I. private psychoanalytische Vereinigung in Salzburg am 27. April 1908," *Zentralblatt für Psychoanalyse* 1:130.

269 *Archives de Psychologie:* Miller, F. (1906), "Quelques faits d'imagination créatrice subconsciente," *Archives de Psychologie*, 5:36–51. On Miller, see Shamdasani, S. (1990), "A Woman Called Frank," *Spring* 50:26–56.

269 **who voiced his approval:** Freud-Jung Letters, 333–35.

270 **new aspect of human psychology:** Jung, C. G. (1912), "Wandlungen und Symbole der Libido," *Jahrbuch* 3: 120–227. In English, see Jung, C. G. (1912), *Psychology of the Unconscious: A Study of the Transformations and Symbolisms of the Libido*, ed. W. McGuire, Princeton, N.J.: Princeton University Press, *C.W.* Supplement B, 1991.

270 **two principles of mental functioning:** Freud, S. (1911), "Formulations on the Two Principles of Mental Functioning," *S.E.* 12:213–26.

271 **teemed in Miss Miller's unconscious:** Jung, C. G. (1912), *Psychology of the Unconscious*, *C.W.* Supplement B, 66–75.

271 **'Transformations and Symb. of the Li' ":** Freud-Jung Letters, 438.

272 **lay in the Oedipus complex:** Ibid., 441.

272 **"much inclined toward plagiarism":** Freud-Ferenczi Letters, I, 133.

272 **plunged Ferenczi into a self-analysis:** Ibid., 312. For Ferenczi's comedy of errors, see 304–07.

272 **Freud's opinion of it:** Freud-Jung Letters, 452, 455–56, 462.

272 **too happy to be tyrannized:** Ibid., 454.

273 **"You are a dangerous rival":** Ibid., 460–61.

273 **influence of someone else's ideas":** Ibid., 472.

273 **Riklin on universal myth:** Ellenberger, H. (1970), *The Discovery of the Unconscious*, 810.

273 **Talmudic-exegetic-theological interpretations":** Ibid., 810–14.

274 **psychological foundations of religion:** Abraham would suggest that this bitter public feud encouraged Jung to pivot against libido theory. Freud-Abraham Letters, 242.

274 **neglect his duties to the movement:** Freud-Jung Letters, 488.

274 **if one remains only a pupil":** Ibid., 491.

274 **nothing more than autobiography:** Ibid., 492–94, 502–07.
275 **matriarchal community predated patriarchy:** Freud-Jung Letters, 502–06. On matriarchal societies, Jung was likely influenced by Bachofen, B. (1967), *Myth, Religion, and Mother Right: Selected Writings of J. J. Bachofen*, trans. R. Manheim, Princeton, N.J.: Princeton University Press.
275 **"disastrous similarity" to Adler's:** Freud-Jung Letters, 507.
275 **rather a holistic life energy:** See Pfister, O. (1913), *Die Psychanalytische Methode: Eine erfahrungswissenschaftlich-systematische Darstellung*, Leipzig and Berlin: Julius Klinkhardt, 134–35, 140–41; and Pfister, O. (1917), *The Psychoanalytic Method*, New York, Moffat, Yard & Company, 156–57, 166–67.
276 **woman who was not his wife:** Freud-Binswanger Letters, 83–84. Gerhard Fichtner suggests either Toni Wolff or a former patient of Freud's was the woman in question, though in an earlier French edition of these letters, he suggested it was Maria Moltzer; Fichtner, G. ed. (1995), *Sigmund Freud, Ludwig Binswanger: Correspondance 1908–1938*, Paris: Calmann-Lévy, 151.
276 **disapproved of Jung's new theory:** Freud-Jung Letters, 509.
276 **are to be tolerated or not":** Ibid., 512.
276 **long-awaited conclusion of his studies:** Ibid., 514.
276 **could not be reduced to autoerotism:** See Jung, C. G. (1912), *Psychology of the Unconscious*, in *C.W.* Supplement B, 1991, 130–35.
276 **could be explained by race:** Freud-Ferenczi Letters, 399.
276 **overly concerned with sex:** Freud-Binswanger Letters, 92–93.
277 **lectures at Fordham University:** Jung, C. G. (1913), "The Theory of Psychoanalysis," *C.W.* 4:83–228.
277 **one-time father-confessor, Sigmund Freud:** Ibid., 192–93.
277 **new adherents to the field:** Freud-Jones Letters, 175.
278 **fool riddled with complexes":** Freud-Jung Letters, 516.
278 **"he considers psa. as his own":** Freud-Jones Letters, 182.
278 **his deviations were fine:** Freud-Jung Letters, 517.
278 **before their trip to America:** Carl Jung–Ernest Jones Letters, Dec. 19, 1959. Sigmund Freud Copyrights, Wivenhoe, Colchester, England.
278 **against him and had passed out:** Jones, E. (1959), *Free Associations: Memories of a Psycho-Analyst*, New York: Basic Books, 222.
278 **"rumors of my death are greatly exaggerated":** Freud-Binswanger Letters, 107.
278 **based on archaic complexes:** Freud-Jung Letters, 521–22, 524.
278 **A wretched theory!)":** Ibid., 526.
279 **especially against new ideas":** Ibid., 529.
279 **as one of *yours*":** Ibid., 533.
279 **either slavish sons or impudent puppies":** Ibid., 534.
279 **he is having an affair":** Freud-Ferenczi Letters, I, 446.
279 **break off all personal communications:** Freud-Jung Letters, 537–39.
280 **unite on psychoanalytic ground":** Freud-Rank Letters, Sept. 13, 1912.
280 **away from the rotten sex":** Freud-Brill Letters, undated letter from early 1914.
281 **have forced us to make":** Freud-Ferenczi Letters, I, 421.
282 **their own specific gravity":** Ibid., 335.
282 **thought-transference and psychic powers:** Freud-Jones Letters, 146.
283 **his existence and in his actions":** Ibid., 148.

283 **"kingdom and policy of their master":** Ibid., 149.

283 **reliability of Ferenczi and Rank:** Freud-Ferenczi Letters, I, 400; Freud-Jones Letters, 146.

283 **relied heavily on libido theory:** Freud-Jones Letters, 158.

283 **I am completely helpless:** Carl Jung–Ernest Jones Letters, Nov. 15, 1912. By permission of Ulrich Hoerni, Erbengemeinschaft C. G. Jung. The text is in Jung's English.

284 **Secret Committee came into being:** On the "Secret Committee," see Grosskurth, P. (1991), *The Secret Ring: Freud's Inner Circle and the Politics of Psychoanalysis*, Reading, Mass.: Addison-Wesley; and Wittenberger, G. (1995), *Das 'Geheime Komitee' Sigmund Freuds: Institutionalisierungsprozesse in der Psychoanalytischen Bewegung zwischen 1912 und 1927*, Tübingen: edition diskord.

284 **"I suppose you will stay loyal?":** Jones, E. (1959), *The Life and Work of Sigmund Freud*, Vol. III., 228.

284 **psychoanalysis and the cultural sciences.:** See his autobiography, Sachs, H. (1944), *Freud: Master and Friend*, Cambridge, Mass., Cambridge University Press. Sachs's first name appears in the literature as both "Hans" and Hanns."

284 **when the inevitable showdown came:** Freud-Ferenczi Letters, I, 411, 417, 437, 484.

284 **from all Aryan religiousness":** Freud-Ferenczi Letters, I, 486. Freud-Abraham Letters, 139.

285 **long roster of his enemies:** For accounts of this tense congress, see Andreas-Salomé, L. (1964), *The Freud Journal*, London, Quartet Encounters, 168–70; Jones, E. (1959), *Free Associations*, 224; and Freud-Brill Letters, Oct. 30, 1913.

285 **would have to compromise:** See Maeder, A. E. (1913), *The Dream Problem*, trans. F. M. Hallock and S. E. Jelliffe, New York: Nervous and Mental Disease Pub. Co., 1916.

285 **every intention of dissolving the I.P.A.:** Freud-Brill Letters, Oct. 30, 1913.

286 **"too good to be true," Jones wrote:** Freud-Jones Letters, 234.

286 **reconstitute it under his sole control:** Freud-Abraham Letters, 206.

286 **demand this action from Jung:** Ibid., 205–06.

286 **would be enough to secede:** Freud-Abraham Letters, 206.

286 **London Psychoanalytical Society with nine members:** Jung-Jones Letters, Oct. 31, 1913.

286 **might be said for the Americans:** Freud-Jones Letters, 234–35.

286 **go his own way:** Freud-Jones Letters, 238, 240.

287 **even in science," he grumbled:** Jung-Jones Letters, Dec. 2, 1913. Published by permission of Ulrich Hoerni, Erbengemeinschaft C. G. Jung.

287 **between 1912 and 1913:** Freud, S. (1913), *Totem and Taboo, S.E.* 13:1–164.

287 **of modern men and women:** Freud-Fliess Letters, 227.

287 **may have been writing from experience:** Freud, S. (1913), *Totem and Taboo, S.E.* 13:1–74.

287 **tribesmen, neurotics, and children did:** Ibid., 75–99.

288 **in a word, civilization:** Ibid., 100–161.

288 **let Jung go on his own:** Freud-Jones Letters, 354.

288 **becomes downright improbable to one":** Freud-Ferenczi Letters, I, 340.

288 **son killing the father:** Jones, E. (1955), *The Life and Work of Sigmund Freud*, vol. II, 354.

288 **no doubt somewhere in the feast:** Ibid., 355.

289 **could put me into retirement?":** Freud, S., and L. Binswanger (1992), *S. Freud and L. Binswanger Briefwechsel, 1908–1938*, ed. G. Fichtner, Frankfurt am Main: S. Fisher, 133, (my translation). In the English edition of the letters, this grand pronouncement is rendered into tepid English: "I often wish I were a pensioner and could leave the work to others. But what man alive will pension me off?" Freud-Binswanger Letters, 119.

289 **collapse of the I.P.A.:** Freud-Brill Letters, Jan. 22, 1914.

289 **changes in its editorship and format:** Freud, S. (1914), "On the History of the Psychoanalytic Movement," *S.E.* 14:7.

289 **upon which psychoanalysis was built:** See Makari, G. (1998), "The Seductions of History: Sexual Trauma in Freud's Theory and Historiography," *I.J.P.* 79:857–69.

290 **brought him nearly all his adherents:** Freud, S. (1914), "On the History of the Psychoanalytic Movement," *S.E.* 14:26–27.

290 **great works of culture:** Ibid., 61–62.

290 **was fired at Zurich:** Freud-Ferenczi Letters, II, 55.

290 **Ferenczi, Abraham, Eitingon, and Jones:** See the string of negative reviews of Jung's work by Abraham, Jones, and Ferenczi in the *Internationale Zeitschrift für Ärztliche Psychoanalyse*, 2:72–82, 83–86, 86–87, as well as the discussion by Eitingon, M. (1914), "Über das Ubw. bei Jung und seine Wendung ins Ethische," *Internationale Zeitschrift für Ärztliche Psychoanalyse* 2:99–104.

290 **his willingness to go quietly:** Jung's letter is dated April 20, 1914, and is reproduced in Freud-Jung Letters, 551. Also see Freud-Abraham Letters, 233, and Freud-Ferenczi Letters, 550.

291 **Maeder in Zurich—reluctantly agreed:** Freud-Abraham Letters, 234.

291 **give up his membership was Binswanger:** Ibid., 260. Freud-Jones Letters, 295. Also see Shamdasani, S. (1998), *Cult Fictions: C. G. Jung and the Founding of Analytical Psychology*, London and New York: Routledge, 18–19.

291 **outnumbered by Jung's supporters:** Freud-Jones Letters, 303.

292 **not rely on others:** Freud-Ferenczi Letters, 5.

292 **"Hurrah," he wrote Abraham:** Freud-Abraham Letters, 260.

PART THREE: MAKING PSYCHOANALYSIS

8. EVERYTHING MAY PERISH

295 **everything has sensed it may perish:** Valéry, P. (1919), "The Crisis of the Mind," in *Paul Valéry: An Anthology*, ed. J. R. Lawler, London: Routledge and Kegan Paul, 1977, 95.

296 **object of direct experience, behind consciousness:** Freud-Binswanger Letters, 233–34.

297 **authority of a charismatic leader:** On the nature of communities organized around charismatic leaders, see the classic work, Weber, M. (1947), *The Theory of Social and Economic Organization*. Oxford, U.K.: Oxford University Press. On the Jungian community, see Kirsch, T. B. (2000), *The Jungians: A Comparative and Historical Perspective*, London and Philadelphia: Routledge, 238. Also see Oskar Pfister's unpublished memoir, "Autobiographie," *Oskar Pfister Papers*, Zurich, Zentralbibliothek Zürich, which described his surprise at the demands placed on him by Jung. The Adlerians have received little scholarly attention. Despite Adler's seemingly rigid insistence

upon his notions in the Wednesday Psychological Society, a follower has written of the openness in that community; see Furtmüller, C. (1973), "A Biographical Essay," in *Alfred Adler: Superiority and Social Interest*, 311–91. Nonetheless, this group came to be rather strictly Adlerian in its definition.

298 **"unified world picture"**: Planck, M. (1908/9), "The Unity of the Physical World Picture," in *Ernst Mach—A Deeper Look*, ed. J. Blackmore, The Netherlands: Kluwer Academic Pubs., 1992, 130, 145–46. For Freud on Mach see, Freud-Ferenczi Letters, II, 91.

298 **Breuer, Adler, and Christian von Ehrenfels**: In the winter of 1911, the Society engaged in an extended debate on Mach's positivism as well as the nature of scientific laws. They even devoted one evening to "metapsychology," as seen by none other than Freud's former patient Hermann Swoboda. Blackmore, J., R. Itagaki, et al. (2001), "The University of Vienna Philosophical Society," in *Ernst Mach's Vienna 1895–1930*, eds. J. Blackmore, R. Itagaki, and S. Tanaka, The Netherlands: Kluwer Academic Publishers, 277–314.

298 **corresponding notions in psycho-analysis**: Freud, S. (1914), "On Narcissism: An Introduction," *S.E.* 14:77.

300 **collapse of their practices**: See, for example, Freud-Abraham Letters, 289.

300 **have to let run wild"**: Freud-Jones Letters, 309.

300 **forced to shut down**: Freud, S., and M. Eitingon (2004), *Sigmund Freud-Max Eitingon Briefwechsel (1906–1939)*, ed. M. Schröter, Tübingen: edition diskord, vol. I, 103–04. The date of the letter is January 17, 1915.

301 **treating only one patient**: Freud-Ferenczi Letters, II, 24.

301 **"On Narcissism: An Introduction"**: Freud, S. (1914), "On Narcissism: An Introduction," *S.E.* 14:67–102.

301 **capacity to love others**: Freud, S. (1905), *Three Essays on the Theory of Sexuality*, *S.E.* 7:222.

301 **fueled by sexual libido**: The idea of a separate "I" drive was first developed in Freud, S. (1910), "The Psycho-Analytic View of Psychogenic Disturbance of Vision," *S.E.* 11:209–18. In "On Narcissism," Strachey translated "Ichtriebe" as ego-instincts, rather than consistently rendering "Trieb" as drive. Compare Freud, S. (1914) "On Narcissism: An Introduction," *S.E.* 14:79, and Freud, S. (1914) "Zur Einführung des Narzissmus," *G.W.* 10:145.

302 **as a cause of depression**: Freud, S. (1917), "Mourning and Melancholia," *S.E.* 14: 243–59.

302 **Russian's neurosis miraculously came unglued**: Freud, S. (1918), "From the History of an Infantile Neurosis," *S.E.* 17:12.

303 **but rather dogs coupling**: Ibid., 56, 58, 97.

303 **interconnected meta-psychological papers**: Of twelve papers reportedly written by Freud, only five were published and the rest were deemed lost. However, one was recently discovered and published. See Freud, S. (1987), *A Phylogenetic Fantasy: Overview of the Transference Neuroses*, ed. I. Grubrich-Simitis, Cambridge, Mass.: Harvard University Press.

303 **psycho-analytic system could be founded"**: Freud, S. (1917), "A Metapsychological Supplement to the Theory of Dreams," *S.E.* 14:222.

303 **drives, repression, the unconscious, and dreams**: Freud, S. (1915), "Instincts and Their Vicissitudes," *S.E.* 14:117–40, Freud, S. (1915), "Repression," *S.E.* 14:146–58;

Freud, S. (1915), "The Unconscious," *S.E.* 14:166–215; Freud, S. (1917), "A Metapsychological Supplement to the Theory of Dreams," *S.E.* 14:219–35. Strachey's translation of *Triebe* as "Instinct" in the first of these meta-psychological papers has been controversial. I have adopted the alternate translation, "drive."

303–04   **The Psychoanalytic Method:** Pfister, O. (1913), *Die Psychanalytische Methode: Eine erfahrungswissenschaftlich-systematische Darstellung*, Leipzig and Berlin, Julius Klinkhardt.

304   **entry on his own work:** See Freud-Pfister Letters, 60.

304   **malicious refusal to be obedient:** This turn of events is described in Oskar Pfister, "Autobiographie," *Nachlass Oskar Pfister*, Zentralbibliothek Zürich, 61–62.

304   **essay on character types:** For sections cut from the first edition, see Pfister, O. (1913), *Die Psychanalytische Methode*, 140–42, 215–20.

304   **leaders of analysis," Pfister confessed:** Ibid., 371–72; also see Pfister, O. (1917), *The Psychoanalytic Method*, New York: Moffat, Yard & Company, 435.

304   **Fundamentals of Psychoanalysis appeared in 1914:** Kaplan, L. (1914), *Grundzüge der Psychoanalyse*, Leipzig and Vienna: Franz Deuticke.

304   **book attacking psychoanalytic rebels:** Freud-Abraham Letters, 245. Freud-Ferenczi Letters, II, 110, 123.

305   **originate from the French":** Régis, E., and A. Hesnard (1914), *La Psychanalyse: Des Névroses et des psychoses, ses applications médicales et extra-médicales*, Paris, Félix Alcan, 1929. This was not an empty complaint; see v–xv. Also see Freud-Ferenczi Letters, II, 8.

305   **introductory book in Dutch:** Freud-Ferenczi Letters, II, 65, 67.

305   **libido or sexuality or transference:** Freud, S. (1916–17), "Introductory Lectures on Psychoanalysis," *S.E.* 15:16. On psychoanalysis not being a closed system, see *S.E.* 16:244.

305–06   **in favour of this one portion":** Ibid., *S.E.* 16:346.

306   **past and the present:** Ibid., 350.

307   **shipped back to the front:** See Shephard, B. (2001), *A War of Nerves: Soldiers and Psychiatrists in the Twentieth Century*, Cambridge, Mass.: Harvard University Press, 97–98.

307   **"surprise shock attack":** Discussed in Ferenczi, S., Abraham, K., Simmel, E., and E. Jones (1919), *Zur Psychoanalyse der Kriegneurosen*, Leipzig and Wien: Internationaler Psychoanalytischer Verlag, 40–41. This was translated in Ferenczi, S., Abraham, K., Simmel, E., and E. Jones (1921), *Psychoanalysis and the War Neuroses*, London, Vienna, New York: International Psychoanalytical Press, 28–29.

308   **success with his hypnotic technique:** On Nonne, see Lerner, P. (2003), *Hysterical Men: War, Psychiatry, and the Politics of Trauma in Germany, 1890–1930*, Ithaca, N.Y.: Cornell University Press.

308   **psychoanalytic methods on the shell-shocked:** Freud, S., and M. Eitingon (2004), *Sigmund Freud—Max Eitingon Briefwechsel (1906–1939)*, vol. I, 109–10. The date of the letter is Jan. 1, 1916.

308   **Breuer and Freud's old cathartic strategy:** Ferenczi, S. (1916/17), "Two Types of War Neuroses," in *Further Contributions to the Theory and Technique of Psycho-analysis*. New York: Brunner/Mazel, 1980, 141.

308   **connection to disturbances of libido:** Ferenczi, S., Abraham, K., Simmel, E., and E.

Jones (1921), *Psychoanalysis and the War Neuroses*, 17. My thanks to Paul Lerner for pointing this out to me.

308 **in Posen for war neuroses:** Ernst Simmel C.V. and Bibliography, Ernst Simmel Papers, Los Angeles Psychoanalytic Society and Institute, Los Angeles, Cal.; also see Freud-Abraham Letters, 372.

308 **"painful electric currents, etc.":** Ferenczi, S., Abraham, K. Simmel, E., and E. Jones (1921), *Psychoanalysis and the War Neuroses*, 43.

308 **"dreams I do not know":** Ibid., 37.

309 **"German war medicine has taken the bait":** Freud-Ferenczi Letters, II, 265.

310 **number of members in attendance, nineteen:** The earlier Budapest Society had collapsed; on this rebirth see Freud-Ferenczi, II, 274.

310 **Hungarian Academy of Sciences:** Jones, E. (1955), *The Life and Work of Sigmund Freud*, vol. II, 197.

310 **failed attempts at catharsis:** Ferenczi, S., Abraham, K., Simmel, E., and E. Jones, (1921), *Psychoanalysis and the War Neuroses.* 37.

310–11 **done mythology an injustice:** Ibid., 7.

311 **to treat war neurotics:** Ibid., 29; also see Freud-Ferenczi Letters, II, 298.

311 **"is indeed not of this world":** Freud-Ferenczi Letters, II, 311.

312 **impervious to his influence:** Freud-Binswanger Letters, 6.

312 **transference with Herr J. had "failed":** Ibid., 35–56.

312 **"a distinctly erotic nature":** Vienna Minutes, III, 204.

312 **"The Dynamics of Transference":** Freud, S. (1912), "The Dynamics of Transference," *S.E.* 12:97–108.

312 **every attempt to aid her:** Freud, S., and M. Eitingon (2004), *Sigmund Freud-Max Eitingon Briefwechsel (1906–1939)*, vol. I., 70–71. The date of the letter is February 13, 1912.

312 **feelings could be repressed:** Freud, S. (1912), "The Dynamics of Transference," *S.E.* 12:106.

312 **Oedipal transferences were similarly split:** Riklin, F. (1910), "Aus der Analyse einer Zwangsneurose," *Jahrbuch für psychoanalytische und psychopathologischen Forschungen* 2:248–311.

313 **protest and rage:** Adler, A. (1910), "Beitrag zur Lehre vom Widerstand," *Zentralblatt für Psychoanalyse*, 1:214–19.

313 **which he called "Eros" and "Thanatos":** Vienna Minutes, I, 175. Stekel would later complain that Freud plagiarized him when he adopted these terms.

313 **not love but hatred:** Stekel, W. (1911), *Sex and Dreams: The Language of Dreams*, trans. James S. Van Teslaar, Boston: The Gorham Press, 1922. Discussed in Freud, S. (1913), "The Disposition to Obsessional Neurosis," *S.E.* 12:325.

313 **trial paper to the Vienna Society:** Vienna Minutes, III, 281.

313 **being and perishing, by Eros and Thanatos":** Ibid., 312.

313 **but she had not:** Carotenuto, A. (1982), *A Secret Symmetry: Sabina Spielrein Between Jung and Freud*, 20.

313 **contained in the sexual instinct itself":** Vienna Minutes, III, 316–17.

313 **"Destruction as a Cause of Coming into Being":** Spielrein, S. (1994), "Destruction as a Cause of Coming into Being," *Journal of Analytical Psychology* 39: 155–86.

314 **unimpressed by Reik's claims:** Vienna Minutes, III, 317–18.

314 **from genital sexual pleasure:** Freud, S. (1913), "The Disposition to Obsessional Neurosis," *S.E.* 12:317–26. As James Strachey pointed out, in 1915 Freud added this to the *Three Essays on the Theory of Sexuality*; see *S.E.* 7:197–99.

314 **primaeval man, a gang of murderers":** Freud, S. (1915), "Thoughts for the Times on War and Death," *S.E.* 14:296–97.

314 **any way that can improve them":** Freud, S. (1919), "Lines of Advance in Psycho-Analytic Therapy," *S.E.* 17:159.

315 **a lust for destruction:** Freud, S. (1921), "Introduction," in Ferenczi, S., Abraham, K., Simmel, E., and E. Jones (1921), *Psychoanalysis and the War Neuroses*, 1–4.

315 **paper on the genesis of masochism:** Freud-Ferenczi Letters, II, 329.

315 **less than revolutionary paper:** Young-Bruehl, E. (1988), *Anna Freud: A Biography*, New York: Summit Books, 1988, 104.

315 **"mysterious" title—"Beyond the Pleasure Principle":** Freud-Ferenczi Letters, II, 335.

315 **felt the essay remained confused:** Ibid., 341.

316 **address in an upcoming work:** Freud, S. (1920), "Beyond the Pleasure Principle," *S.E.* 18:4–5.

316 **real world did not help here:** Freud, S. (1911), "Formulations on the Two Principles of Mental Functioning," *S.E.* 12:218–26.

316 **choices between pleasure and pain:** Freud, S. (1920), "Beyond the Pleasure Principle," *S.E.* 18:8. See Fechner, G. (1873), *Einige Ideen zur Schöpfungs—und Entwickelungsgeschichte der Organismen*, Leipzig: Breitkopf und Härtel, 25–35.

317 **replaces him by a new one:** Freud, S. (1920), "Beyond the Pleasure Principle," *S.E.* 18:22.

317 **Wilhelm Stekel, Theodor Reik, and Sabina Spielrein:** See Praz, M. (1970), *The Romantic Agony*, Oxford, New York: Oxford University Press.

317 **human desire for death:** Freud, S. (1920), "Beyond the Pleasure Principle," *S.E.* 18:39.

318 **account himself sold to the devil":** Ibid., 7, 51, 59.

318 **"I am a wild analyst!":** Cited in Schacht, L. (1970), "Introduction," in Groddeck, G. (1970), *The Meaning of Illness*, London: Maresfield Library, 7.

318 **Freud stuck by his new friend:** Groddeck, G. (1921), *Der Seelensucher: Ein Psychoanalytischer Roman*, Leipzig: Internationaler Psychoanalytischer Verlag.

318 **"secret sources of knowledge":** Freud-Groddeck Letters, 54.

319 **in your eyes a little:** Freud-Groddeck Letters, 56.

319 **unconscious as an immutable truth:** See Wittels, F. (1924), *Sigmund Freud: His Personality, His Teaching, & His School*, New York: Dodd, Mead & Company Publishers, 142.

319 **its flank may be overturned tomorrow:** Hitschmann, E. (1956), "Freud Correspondence," *Psychoanalytic Quarterly*, 25:358.

9. SEARCHING FOR A NEW CENTER

324 **at least at all clearly":** Ernest Jones, Secret Committee RB, I, 92–96. The date of the letter is Oct. 19, 1920.

324 **Sigmund Freud was its prophet":** Isador Coriat to Ernest Jones, April 4, 1921, *Otto Rank Papers*, Rare Book and Manuscript Library, Columbia University, New York.

324 **poor at the Budapest Congress:** Freud, S. (1919), "Lines of Advance in Psycho-Analytic Therapy," *S.E.* 17:167–68.

324 **Melanie Klein, and István Hollós:** Freud-Ferenczi, III, 16.

324 **expelled from the local Medical Society:** Freud-Ferenczi Letters, II, 351, Freud-Ferenczi Letters, III, 22.

325 **in economically stable London:** (1920), "Reports of the International Psycho-Analytical Association," *International Journal of Psychoanalysis* 1:114–15.

325 **social problems of sexuality:** Abraham, K., Secret Committee RB, I, 65–66. The date of the letter is Oct. 6, 1920.

325 **four guineas a course":** Jones, E., Secret Committee RB, II, 67–71. The date is Feb. 11, 1921.

326 **make all publication impossible":** Jones, E., Secret Committee RB, II, 119–22. The date is March 21, 1921.

326 **"burnt by the common hangman":** Jones, E., Secret Committee RB, II, 141–42. The date is April 11, 1921.

326 **home for clinical work:** Abraham, K., Secret Committee RB, I, 170–74. The date is Nov. 17, 1920.

326 **doctor named James Glover:** Jones, E., Secret Committee RB, I, 140–44. The date is November 2, 1920.

326 **could get her hands on:** Abraham, K., Secret Committee RB, I, 170–74. The date is November 17, 1920.

327 **lectures or classes without prior approval:** Abraham, K., and Sachs, H., Secret Committee RB, I, 238–40. The date is Dec. 31, 1920.

327 **deal fell through:** Jones, E., Secret Committee RB, II, 223–25. The date is July 22, 1921.

327 **formed a coalition government:** For a first-person account of the atmosphere in Vienna, see Zweig, S. (1943), *The World of Yesterday*, 297.

327 **further stabilized the *Verlag*:** Freud-Ferenczi Letters, III, 20.

327 **profitable or at least prestigious:** Ibid., 35.

327 **efforts to open a free clinic:** See Danto, E. A. (2004), *Freud's Free Clinics: Psychoanalysis and Social Justice, 1918–1938*, New York: Columbia University Press.

327 **abandoned garrison hospital:** Hitschmann, E. (1932), "A Ten Years' Report of the Vienna Psycho-Analytical Clinic," *I.J.P.* 13:245.

327 **entrusted with such an operation:** Freud-Abraham Letters, 430.

327 **dream of a clinic appear unrealizable:** On the testimony of Freud, see Eissler, K. (1992), *Freud sur le front des névroses de guerre*, Paris: Presses Universitaires de France.

328 **cardiac unit in the Vienna General Hospital:** Hitschmann, E. (1932), "A Ten Years' Report of the Vienna Psycho-Analytical Clinic," *I.J.P.* 13:245.

328 **welcomed eight societies to the Association:** (1920), "Reports of the International Psycho-Analytical Association—Sixth Congress of the International Psychoanalytical Association," *I.J.P.* 1:208.

328 **proposal quietly died:** Abraham, K., and Sachs, H. Secret Committee RB, II, 254–56. The date is Oct. 21, 1921.

329 **short account of psychoanalytic technique:** Stekel, W. (1908), *Nervöse Angstzustände und ihre Behandlung*, Berlin and Vienna: Urban & Schwarzenberg.

329 **word-association method, and dream interpretation:** Jones, E. (1913), *Papers on Psychoanalysis*, New York: William Wood & Company, 182–205.

330 **"a pig finding truffles," Freud joked):** Freud-Jung Letters, 404.

330 **prehistory that could be translated:** Stekel, W. (1911), *Die Sprache des Traumes*, Wiesbaden: Bergmann.

330 **postulated in the unconscious:** Vienna Minutes III, 236.

330 **author's "perverse" unconscious:** Freud-Jung Letters, 404.

330 **"uncertain and superficial":** Ibid., 418. Also see Vienna Minutes, III, 236.

330 **the readers of the *Zentralblatt:*** Freud-Jung Letters, 458.

330 **never come out right":** Ibid., 475.

330 **conduct of the analysis throughout":** Freud, S. (1911), "The Handling of Dream-Interpretation in Psycho-Analysis," *S.E.* 12:89–96.

331 **engine of psychoanalytic cure:** Freud-Jung Letters, 12–13.

331 **case the Welshman had just published:** Carl Jung–Ernest Jones Letters, May 21, 1908, Sigmund Freud Copyrights, Wivenhoe, Colchester, England.

331 **in which transference was central:** My thanks to Ulrike May who informed me of this little known work. Sadger, J. (1908), "Fragment der Psychoanalyse eines Homosexuellen," *Jahrbuch für sexuelle Zwischenstufen* 9:339–424.

331 **key to the case:** Binswanger, L. (1909), "Versuch einer Hysterieanalyse," *Jahrbuch für psychoanalytische und psychopathologischen Forschungen* 1:234–352.

331 **Freud's "most significant discoveries":** Ferenczi, S. (1909), "Introjection and Transference," in *First Contributions to Psycho-Analysis*, 35.

331 **free patients from these forces:** Ibid., 93.

331 **transference was essential for success:** Stekel, W. (1912), *Nervöse Angstzustände und ihre Behandlung*. Berlin and Vienna: Urban & Schwarzenberg, 27; also see Stekel, W. (1912), *Conditions of Nervous Anxiety and Their Treatment*, trans. Rosalie Gabler, London: Kegan Paul, 1923, 409.

331 **analyst's office or apartment:** Stekel, W. (1912), *Nervöse Angstzustände und ihre Behandlung*, 27–30; Stekel, W. (1912), *Conditions of Nervous Anxiety and Their Treatment*, 411–13.

332 **to the Dora case in 1905:** Freud, S. (1912), "The Dynamics of Transference," *S.E.* 12:97–108.

332 **neuroses were due to psychosexuality:** Freud-Jung Letters, 8.

333 **curing and knowing, medicine and science:** Freud, S. (1912), "The Dynamics of Transference," *S.E.* 12:108.

333 **suggestions of the analyst:** Ibid., 105–06.

333 **dismiss analysis as suggestion:** Freud-Ferenczi Letters, I, 342–43.

333 **than with anything else:** Vienna Minutes, III, 204.

333 **translation was prepared for study:** Freud-Binswanger Letters, 113; Freud-Ferenczi Letters, I, 342; Freud-Jones Letters, 266.

333 **free himself from these feelings:** Freud-Binswanger Letters, 112.

333 **transference problems with Jung and Pfister:** Falzeder, E. (1994), "My Grand-Patient, My Chief Tormentor: A Hitherto Unnoticed Case of Freud's and the Consequences," *Psychoanalytic Quarterly* 63:297–331.

333 **countertransference was sorely needed:** Freud-Jung Letters, 476.

334 **what will you do for me?":** Ibid., 478–79.

334 **falsify what he may perceive":** Freud, S. (1912), "Recommendations to Physicians Practising Psycho-Analysis," *S.E.* 12:112.

335 **what was inside the patient:** Ibid.

335 **could be perceived without mediation:** On Freud's interest in telepathy see Freud, S.

(1921), "Psychoanalysis and Telepathy," *S.E.* 18:177–94; and Freud, S. (1922), "Dreams and Telepathy," *S.E.* 18:195–220.

335 **an unconsious "compulsion to repeat":** Freud, S. (1914), "Remembering, Repeating and Working-Through (Further Recommendations on the Technique of Psycho-Analysis II)," *S.E.* 12:151.

335 **untangling of unconscious repetitions:** Ibid., 155.

335 **female patient, Toni Wolff:** On Jung and Spielrein, see Kerr, J. (1993), *A Most Dangerous Method: The Story of Jung, Freud, and Sabina Spielrein*, New York: Alfred A. Knopf. On Wolff and Jung, see Kirsch, T. B. (2000), *The Jungians: A Comparative and Historical Perspective*, London and Philadelphia: Routledge. On Jung and Moltzer, see Freud-Ferenczi Letters, I, 446.

336 **both of whom he had analyzed:** On Ferenczi and mother and daughter Pálos, see Axel Hoffer, "Introduction" in Freud-Ferenczi Letters, II, xix.

336 **conducting affairs with patients:** See Stanton, M. (1992), "The Case of Otto Gross: Jung, Stekel and the Pathologization of Protest," in *Psychoanalysis in Its Cultural Context*, eds. Edward Timms and Ritchie Robertson, Edinburgh: Edinburgh University Press, 49–56. Stekel openly wrote about having an affair with a patient; Stekel, W. (1950), *The Autobiography of Wilhelm Stekel*, 176–79.

336 **falling in love with a male doctor:** Freud, S. (1915), "Observations on Transference-Love (Further Recommendations on the Technique of Psycho-Analysis III)," *S.E.* 12:157–71.

336 **sway of countertransference:** The German word used was *Indifferenz*, and this was incorrectly translated as "neutrality" rather than indifference. Ibid., 164. On indifference and "neutrality," see Hoffer, A. (1985), "Toward a Definition of Psychoanalytic Neutrality," *Journal of the American Psychoanalytic Association* 33:771–95.

336 **remember and work through the past:** Freud, S. (1915), "Observations on Transference-Love," *S.E.* 12:165.

337 **"who does not belong to it":** Freud-Groddeck Letters, 32.

337 **member of the wild army":** Ibid., 36.

337 **against every tradition," Stefan Zweig recalled:** Zweig, S. (1943), *The World of Yesterday*, 299.

337 **theosophy, occultism, and spiritualism, sprang up:** Ibid., 301.

337 **came with no adult control:** Mosse, G. L. (1964), *The Crisis of German Ideology: Intellectual Origins of the Third Reich*, 171–89.

338 **disregard for conventional modes of love:** Zweig, S. (1943), *The World of Yesterday*, 299.

338 **Wilhelmine prudery and hypocrisy:** Interview of Mitzi Mills by Myron Sharaf, July 5, 1972, Myron Sharaf Papers, Courtesy of Giselle Sharaf.

338 **newer developments should stay afterward":** Interview of Grete Bibring with Myron Sharaf, May 30, 1971. Myron Sharaf Papers, Courtesy of Giselle Sharaf. Also see Reich, W. (1973), *The Function of the Orgasm: Sex-Economic Problems of Biological Energy*. New York, Simon & Schuster, 21.

338 **vanguard of the psychoanalytic movement:** Mühlleitner, E. (1992), *Biographisches Lexikon Der Psychoanalyse*, 93.

338 **thought the work was ludicrous:** Interview of Grete Bibring with Myron Sharaf, May 30, 1971, Myron Sharaf Papers, Courtesy of Giselle Sharaf.

338 **saw sexuality everywhere:** Reich, W. (1988), *Passion of Youth: An Autobiography,*

1897–1922, ed. M. B. Higgins and C. Raphael, New York: Farrar, Straus & Giroux, 80.

338 **spiritual life of the individual, revolves":** Ibid.

339 **young Reich was left devastated:** Reich, W. (1920), "A Case of Pubertal Breaching of the Incest Taboo," in *Early Writings*, trans. P. Schmitz. New York: Farrar, Straus and Giroux, 1975, 65–72.

339 **Reich took over the seminar:** Reich, W. (1973), *The Function of the Orgasm: Sex-Economic Problems of Biological Energy*, trans. V. Carfagno, New York: Simon & Schuster, 22, 106, 127.

339 **bring more order into their studies:** Ibid., 30.

339 **giving him books and reprints:** Ibid., 35.

339 **from Forel to Freud and Jung:** Reich, W. (1922), "Drive and Libido Concepts from Forel to Jung," in *Early Writings*, 86–124.

339 **sexological seminars, Isidor Sadger:** Sharaf, M. (1983), *Fury on Earth: A Biography of Wilhelm Reich*, New York: St. Martin's Press, 64.

339 **blessings of Freud:** Reich, W. (1988), *Passion of Youth: An Autobiography, 1897–1922*, 98.

340 *The Impulsive Character:* Reich, W. (1925), "The Impulsive Character: A Psychoanalytic Study in Ego Psychology," in *Early Writings*, 237–52.

340 **but rather her analyst:** Reich, W. (1988), *Passion of Youth: An Autobiography, 1897–1922*, 29, 124.

340 **psychoanalysis and then committed suicide:** Reich, W. (1988), *Passion of Youth: An Autobiography, 1897–1922*, 124–44, and Sharaf, M. (1983), *Fury on Earth: A Biography of Wilhelm Reich*, New York: St. Martin's Press/Marek, 60.

340 **analyze, analyze, ana-ana-anal!":** Ibid., 145, 155.

341 **sent to me by Freud himself":** Reich, W. (1988), *Passion of Youth: An Autobiography, 1897–1922*, 147.

341 **isn't all love a transference?:** Ibid., 152.

341 **"resolve *my* transference to *her*":** Ibid., 156, 166, 167.

341 **struggled with his countertransference:** Ibid., 173–74.

341 **Later, their affair began:** Ibid., 176.

342 **for analysis with Freud:** Freud-Jones Letters, 364.

342 **two other Anglo-Americans in treatment:** Freud-Ferenczi Letters, III, 12.

342 **analysis at the Brunswick Square Clinic:** James Glover to James Strachey, Feb. 3, 1920, James Strachey Papers, The British Library.

342 **enchanted by Freud's method:** James Strachey to Lytton Strachey, Oct. 15, 1920, James Strachey Papers, The British Library.

343 **shows you out the door:** James Strachey to Lytton Strachey, Nov. 6, 1920. James Strachey Papers, The British Library.

343 **paying attention to the therapeutic problems":** Kardiner, A. (1977), *My Analysis with Freud: Reminiscences*, New York: W. W. Norton and Company Inc., 17, 68–69, 82. Also see Weiss, E. (1970), *Sigmund Freud as a Consultant: Recollections of a Pioneer in Psychoanalysis*, New York: Intercontinental Medical Book Corporation. Freud-Jones Letters, 446.

343 **"He never says a word.":** Kardiner, A. (1977), *My Analysis with Freud: Reminiscences*, 77–8.

344 **cells, embryology, and organic illnesses:** (1923), "Report of the International Psycho-

Analytical Congress in Berlin September 25–27, 1922," *Bul. Int. Psychoanal. Assn.* 4:358–81.

344 **was at work inside us:** Ibid., 372.

344 **"to do with Groddeck":** Freud-Ferenczi Letters, III, 84.

344 *The "I" and the "It":* (1923), "Report of the International Psycho-Analytical Congress in Berlin September 25–27, 1922," *Bul. Int. Psychoanal. Assn.* 4:366.

344 **understanding of inner experience:** As we shall see, when these common words—*Es* and *Ich*—were translated into English, they were not given their literal equivalents but rather put into Latin as the "Id" and the "Ego." Following Bruno Bettelheim's trenchant critique, I render these German words as "It" and "I" until the primary language of psychoanalysis changes from German to English after 1938. See Bettelheim, B. (1983), *Freud and Man's Soul*, New York: Alfred A. Knopf.

344 **before eventually coming around:** Weiss, E. (1970), *Sigmund Freud as a Consultant*, 13, 18. Sterba, R. F. (1982), *Reminiscences of a Viennese Psychoanalyst*, Detroit: Wayne State University Press, 75–77.

345 **had been devoted to technique:** See (1920), "Reports of the International Psycho-Analytical Association—Sixth Congress of the International Psycho-Analytical Association," *I.J.P.* 1:208–21; and (1923), "Report of the International Psycho-Analytical Congress in Berlin September 25–27, 1922," *Bul. Int. Psychoanal. Assn.* 4:358–81.

345 **at the present time":** (1922), "Prize Essay," *I.J.P.* 3:521.

345 **aggressive psychotherapeutic measures:** Freud-Ferenczi Letters, II, 125.

345 **wish for a full recovery:** Freud, S. (1919), "Lines of Advance in Psycho-Analytic Therapy," *S.E.* 17:163.

345 **"premature substitutive satisfactions":** Ibid., 164.

345 **see that it is good":** Ibid.

345 **never moved forward in the analysis:** Ferenczi, S. (1919), "Technical Difficulties in the Analysis of a Case of Hysteria," in *Further Contributions to the Theory and Technique of Psycho-Analysis*, 189–97.

346 **crossing her legs during sessions:** Ibid., 191.

346 **analyst's neutral disengagement useless:** Freud-Ferenczi Letters, II, 281.

346 **back to their sources:** Ferenczi, S. (1919), "On the Technique of Psycho-Analysis," in *Further Contributions to the Theory and Technique of Psycho-Analysis*, 184.

346 **adhere to "psa. Abstinence":** Freud-Ferenczi Letters, III, 100, also 95.

346 **failure to maintain abstinence:** Ibid., 97.

346 **instruct him to have intercourse:** Ferenczi, S. (1919), "On Influencing of the Patient in Psycho-Analysis," in *Further Contributions to the Theory and Technique of Psycho-Analysis*, 235–37.

347 **against the direction of pleasure":** Ferenczi, S. (1920), "Further Extension of the Active Technique in Psycho-Analysis," Abstracts of the Sixth Inter. Psa. Congress, *I.J.P.* 1:354; and Ferenczi, S. (1920), "The Further Development of an Active Therapy in Psycho-Analysis," in *Further Contributions to the Theory and Technique of Psycho-Analysis*, 198–216.

347 **"a basically disgusting person" and "quite crazy":** Freud-Ferenczi Letters, III, 29.

347 **calling him a sadist:** Ibid., 24, 49.

348 **to Gizella for him:** Freud-Ferenczi Letters, II, 188–89.

348 **a decision" in an analysis:** Hoffer suggests Ferenczi's active technique was linked to his own analysis with Freud, see Ibid., xvii–xliv. Ferenczi, S. (1919), "On the Tech-

nique of Psycho-Analysis," in *Further Contributions to the Theory and Technique of Psycho-Analysis*, 184.

348 **with improvement in technique":** Freud-Ferenczi Letters, III, 73.

348 **their paper deserved the prize:** Ibid., 120.

348 **new publishing house Rank directed:** See, for instance, Ernest Jones–Otto Rank Letters, especially Nov. 16, 1920, *Otto Rank Papers*. Rare Book and Manuscript Library, Columbia University, New York. Also Ernest Jones, Secret Committee RB, III, 193–94, in which he accuses Rank of treating colleagues like "puppets." The date of the letter is August 22, 1922.

349 **obvious choice as Freud's successor:** Freud-Rank Letters, August 4, 1922.

349 **Rank from the Professor:** Freud-Rank Letters, July 20, 1922, and August 7, 1922.

349 **"obstacle" of his presence:** Freud, S., Secret Committee RB, III, 231–35. The date is November 26, 1922.

350 **three-hour radical surgery:** Ferenczi-Groddeck Letters, 93.

350 **Freud wrote to Groddeck:** Freud-Groddeck Letters, 84.

350 **The "I" and the "It," was published:** Freud, S. (1923), *The Ego and the Id, S.E.* 19: 12–59.

350 **"It" is the right name:** Freud-Groddeck Letters, 58.

351 **The Book of the It: Psychoanalytic Letters to a Friend:** Groddeck, G. (1923), *Das Buch von Es: Psychoanalytische Briefe an eine Freundin*, Vienna: Internationaler Psychoanalytischer Verlag.

351 **dealing with similar issues:** Freud-Groddeck Letters, 77.

351 **taken from Stekel and Spielrein":** Quoted in Groddeck, G. (1970), *The Meaning of Illness*, London: Maresfield Library, 13.

351 **attachments that settled in the "I":** Freud, S. (1923), *The Ego and the Id, S.E.* 19:29.

351 **it came from a child's parents:** Ibid., 31.

352 **The Development of Psychoanalysis:** Ferenczi, S., and O. Rank (1924), *The Development of Psychoanalysis*, Madison, Ct.: International University Press, Inc., 1986.

352 **"antiquated," and required modification:** Ibid., 2.

352 **overall analytic situation:** Ibid., 28–30.

352 **patient's personality was addressed:** Ibid., 30–32.

352 **true task of analysis:** Ibid., 34.

353 **effectiveness for treating patients:** Ibid., 36–44.

353 **acquisition of new knowledge:** Freud-Rank Letters, July 10, 1922.

353 **in those reactions lay infantile repetitions:** Ferenczi, S., and Rank, O. (1924), *The Development of Psychoanalysis*, 25.

354 **sought to reform the discipline:** Ferenczi-Groddeck Letters, 91.

354 **trauma of their birth:** Freud-Ferenczi Letters, III, 125.

354 **fixation on the mother, a maternal transference:** Rank, O. (1924), *The Trauma of Birth*, New York: Harcourt, Brace and Company, 1929, 4.

354 **weren't given larger significance:** Ferenczi-Groddeck Letters, 77.

355 **would be similarly constructed:** Rank, O. (1924), *The Trauma of Birth*, 6, 20, 30–74, 209.

355 **method created by Freud":** Ibid., xi, xiii.

355 **waiting for this to emerge:** Ibid., 213.

356 **"Not all of me shall die":** Freud-Rank Letters, Dec. 1, 1923.

356   publication in the *International Journal*: Jones, E., Secret Committee RB, IV, 59–60. The date is March 1, 1923.

356   accused Jones of plagiarizing his work: Jones, E. Secret Committee RB, IV, 76–7, 127–130, and 144. The dates are April 5, November 12, and December 15, 1923. Also Ferenczi, S., Secret Committee RB, IV, 121–22. The date is November 2, 1923.

356   reading him their draft in 1922: Freud-Rank Letters, Sept. 8, 1922.

356   only 33% or 66% is true": Sándor Ferenczi to Otto Rank, March 30, 1924, *Otto Rank Papers*, Rare Book and Manuscript Library, Columbia University, New York.

356   founded his own school: Ibid.

357   sexuality through the ages: Ferenczi-Groddeck Letters, 88.

357   out of water into air: Ferenczi, S. (1924), *Thalassa: A Theory of Genitality*, London: Karnac and Maresfield Library, 1989.

357   follow his way freely: Freud, S., Secret Committee RB, IV, 169–72. The date is Feb. 15, 1924. This letter is reproduced in Freud-Abraham Letters, 480, and I have employed Ernst Falzeder's translation save for the final phrase which he renders "to follow his way freely."

358   their own unconscious lives: Ibid., 480–82.

358   those outside our circle": Ibid.

358   concerning vital matters of Ψ": Ibid., 483.

358   other had underrated it": Ibid., 487.

358   Abraham's warning to Rank: Jones, E. (1957), *The Life and Work of Sigmund Freud*, III, 65–66.

359   appear of their own accord: Freud-Rank Letters, Feb. 15, 1924.

359   "boundless ambition and jealousy": Freud-Ferenczi Letters, III, 127.

359   without trying to test them: Ibid., 131–32.

359   had not worked it out well: Ibid., 135.

359   were on the committee: Ferenczi, S., Secret Committee RB, IV, 184–85. The date is March 6, 1924.

359   "lead[s] away from Ψ": Freud-Abraham Letters, 489.

359   test out their new techniques: Freud-Ferenczi Letters, III, 133.

359   Secret Committee was dead: Otto Rank to Sándor Ferenczi, March 20, 1924, *Otto Rank Papers*, Rare Book and Manuscript Library, Columbia University, New York.

359   Freud wrote of his inner circle: Freud-Rank Letters, March 20, 1924; also see Freud-Ferenczi Letters, III, 128–29.

360   he lamented to Ferenczi: Otto Rank to Sándor Ferenczi, March 20, 1924, *Otto Rank Papers*, Rare Book and Manuscript Library, Columbia University, New York.

360   declarations of the Professor: Sándor Ferenczi to Otto Rank, March 30, 1924, *Otto Rank Papers*, Rare Book and Manuscript Library, Columbia University, New York.

360   two men came to some agreement: Freud-Abraham Letters, 521.

360   seeing Abraham elected president: Ibid., 564.

361   not criticized too sharply," Freud advised: Freud-Abraham Letters, 502.

361   "merely an innovation in technique": Trigant Burrow's letter to Freud is reprinted in Burrow, T. (1958), *A Search for Man's Sanity: The Selected Letters of Trigant Burrow*, New York: Oxford University Press, 76–89; Freud's unpublished response is quoted in Freud-Ferenczi Letters, III, 162.

361   retreat from his new theory: Freud-Rank Letters, July 23, 1924.

361 **could not and would not:** Ibid., August 9, 1924.

361 **psychology of his own neurosis:** Ibid., August 27, 1924.

361 **only later grew dubious:** Freud-Ferenczi Letters, III, 166, 174.

361 **editor of the *Zeitschrift*:** Freud-Ferenczi Letters, III, 166.

361 **"supposed to be kicking the bucket":** Strachey, J. and A. Strachey (1985), *Bloomsbury/ Freud: The Letters of James and Alix Strachey, 1924–1925*, eds. P. Meisel and W. Kendrick, New York: Basic Books, 112, 115.

361 **"boiling with rage":** Freud-Ferenczi Letters, III, 168–78.

361 **had come to an end:** Freud-Jones Letters, 559. Also Freud-Ferenczi Letters, III, 186.

362 **Rank was again his old self:** Freud-Ferenczi Letters, III, 196; Ferenczi-Groddeck Letters, 96.

362 **would now be corrected:** Rank, O., Secret Committee RB. IV, 210–12. The date is December 20, 1924. Eitingon, M., Abraham, K., and Sachs, H., Secret Committee RB, IV, 213–14. The date is December 25, 1924. Max Eitingon to Otto Rank, Otto Rank Archives, December 26, 1924.

362 **Rank's book on birth trauma:** Glover, E. (1924), " 'Active Therapy' and Psycho-Analysis—A Critical Review," *I.J.P.* 5:269–311. Sachs, H. (1925), "Review of 'Das Trauma der Geburt und seine Bedeutung für die Psychoanalyse' by Otto Rank," *I.J.P.* 6:499–508. Sachs is referred to as both "Hanns" and "Hans."

362 **Ferenczi and Rank's book on technique:** Alexander, F. (1925), "Review of 'Entwicklungsziele der Psychoanalyse,' " *I.J.P.* 6:484–96.

363 **directly to birth trauma:** Freud-Brill Letters, Nov. 19, 1924; Freud-Ferenczi Letters, III, 188.

363 **questioned about his analyst's methods:** Freud-Ferenczi Letters, III, 175.

363 **led to these notions:** Sachs, H. (1925), "Review of 'Das Trauma der Geburt,' " *I.J.P.* 6:498.

363 **could open a separate school:** Freud-Ferenczi Letters, III, 178.

363 **mass application of quick therapies:** Rank, O. (1927), *The Practical Bearing of Psychoanalysis*. New York, The National Committee for Mental Hygiene, Inc., 20, 41.

363 **fought Freud on theoretical grounds:** Freud-Rank Letters, August 9, 1924.

364 **my own winding paths":** Freud, S., Secret Committee RB, IV, 169–72. The date is Feb. 15, 1924. I have used Falzeder's translation in the Freud-Abraham Letters, 480.

364 **before Rank left Vienna for good:** Freud, S. (1926), "Inhibitions, Symptoms and Anxiety," *S.E.* 20:87–178.

364 **fear of the "Over-I's" punishment:** Ibid., 141–42.

365 **moment of birth:** See Ferenczi, S. (1927), "Review of 'Technik der Psychoanalyse: I. Die Analytische Situation' by Otto Rank," *I.J.P.*, 8:93.

365 **blame Rank for that, too:** Ferenczi, S. (1925), "Contra-Indications to the 'Active' Psycho-Analytical Technique," in *Further Contributions to the Theory and Technique of Psycho-Analysis*, 217–29.

365 **his former friend's proposals:** Ferenczi, S. (1927), "Review of 'Technik der Psychoanalyse,' " *I.J.P.* 8:92.

10. A NEW PSYCHOANALYSIS

368 **when it joined the I.P.A.:** Abraham, H. C. (1974), "Karl Abraham: An Unfinished Biography," *I.R.P.* 1:22.

368 **took hold in Weimar Berlin:** Craig, G. A. (1978), *Germany: 1866–1945*, New York

and Oxford, U.K.: Oxford University Press, 470. For a first-person account of these changes, see Grosz, G. (1998), *George Grosz: An Autobiography*, 65, 88, 90, 113.

368 **especially in Germany, dried out here":** Canetti, E. (1982), *The Torch in My Ear*, trans. J. Neugroschel, New York, Farrar Straus & Giroux, 299.

368 **"mad, corrupt, and fantastic Berlin":** Grosz, G. (1998), *George Grosz: An Autobiography*, 119, 140.

369 **like the Adonis all operated freely:** Gordon, M. (2000), *Voluptuous Panic: The Erotic World of Wiemar Berlin*, Venice, Cal.: Feral House, 232.

369 **"Berlin is clamouring for psychoanalysis":** Freud-Abraham Letters, 405.

369 **193 patients came for consultations:** Fenichel, O. (1930), "Statistischer Bericht über die therapeutische Tätigkeit, 1920–1930," in *Zehn Jahre Berliner Psychoanalytisches Institut*, Wien: Internationaler Psychoanalytischer Verlag, 16.

370 **were improved or cured:** Eitingon, M. (1923), "Report of the Berlin Psycho-Analytical Policlinic," *Bul. Int. Psychoanal. Assn.* 4:254–69.

370 **professorship in psychoanalysis for Karl Abraham:** Freud-Abraham Letters, 412–26.

370 **Abraham told Freud in 1920:** Ibid., 418.

370 **attracted over eighty people:** Ibid., 470.

370 **"headquarters" of international psychoanalysis:** Ibid., 433.

370 **wanted to study at the clinic:** Ibid., 429.

370 **most of the members bred trouble:** Abraham, K., and M. Eitingon, Secret Committee RB, I, 122–25, and Abraham, K. Ibid., 170–74. The dates are Oct. 27 and Nov. 17, 1920.

371 **proved himself an unwavering Freudian partisan:** Sachs, H. (1944), *Freud: Master and Friend*, Cambridge, Mass., Cambridge University Press.

371 **full member of the Berlin Society:** Abraham, K., and M. Eitingon, Secret Committee RB, II, 232–34. The date is August 14, 1921.

371 **protect clinic patients from harm:** Müller-Braunschweig, C. (1930), "Historische Übersicht über das Lehrwesen, seine Organisation und Verwaltung," in *Zehn Jahre Berliner Psychoanalytisches Institut*, Wien: Internationaler Psychoanalytischer Verlag, 31. Also Eitingon, M. (1923), "Report of the Berlin Psycho-Analytical Policlinic," *Bul. Int. Psychoanal. Assn.* 4:266.

371 **enrolled in the Poliklinik's training program:** Eitingon, M. (1923), "Report of the Berlin Psycho-Analytical Policlinic," *Bul. Int. Psychoanal. Assn.* 4:256.

372 **had entered analytic training:** See Abraham, K., Sachs, H., and M. Eitingon, Secret Committee RB IV, 44–47. The date is February 17, 1923.

372 **of those, 91 were Berliners:** (1923), "Report of the International Psycho-Analytical Congress in Berlin September 25–27, 1922," *Bul. Int. Psychoanal. Assn.* 4:358–81.

372 **with clear requirements:** Abraham, K., Sachs, H., and M. Eitingon, Secret Committee RB, IV, 44–47. The date is February 17, 1923.

372 **first psychoanalytic training institute:** Horney, K. (1930), "Die Einrichtungen der Lehranstalt," in *Zehn Jahre Berliner Psychoanalytisches Institut*, 48.

373 **similar guidelines at other local societies:** Schrötter, M. (2002), "Max Eitingon and the Struggle to Establish an International Standard for Psychoanalytic Training (1925–1929)," *I.J.P.* 83:875–93.

373 **parricidal wishes of his intellectual heirs:** Alexander, F. (1940), "Recollections of Berggasse 19," *Psychoanalytic Quarterly* 9:199.

373 **infantile sexuality, and human ambivalence:** Sachs, H. (1930), "Die Lehranalyse," in

Zehn Jahre Berliner Psychoanalytisches Institut, Wien: Internationaler Psychoanalytischer Verlag, 53.

374 **his analyst-as-mirror style:** While Sachs's influence in analyzing this fecund generation of Berlin analysts might seem large, his brief analyses did not seem to win him followers or much admiration. Radó voiced contempt for Sachs and recalled that Alexander spoke derisively of this analyst's "coffeehouse hokum"; see Roazen, P., and B. Swerdloff (1995), *Heresy: Sandor Rado and the Psychoanalytic Movement*, Northvale, N.J.: Jason Aronson Inc., 102. Also see Bannach, H. (1971), "Die wissenschaftliche Bedeutung des alten Berliner Psychoanalytischen Instituts," *Psyche*, 14:242–54.

374 **as well as agoraphobia:** Abraham, K. (1911), "Notes on the Psycho-Analytical Investigation and Treatment of Manic-Depressive Insanity and Allied Conditions," in *Selected Papers of Karl Abraham M.D.* London: The Hogarth Press, 1948, 137–56; and Abraham, K. (1913), "On the Psychogenesis of Agoraphobia in Childhood," In *Clinical Papers and Essays on Psycho-Analysis*. New York: Brunner/Mazel, Publishers, 1955, 42–43.

374 **experience-near clinical phenomena:** See Roazen, P., and B. Swerdloff (1995), *Heresy: Sandor Rado and the Psychoanalytic Movement*, Northvale, N.J.: Jason Aronson Inc., 80–88.

375 **ineptness of the Berliners (especially Abraham)":** Freud-Ferenczi Letters, III, 170.

375 **specimen for his examination:** Horney, K. (1980), *The Adolescent Diaries of Karen Horney*, New York: Basic Books, Inc., 241.

375 **"wildest wishes twine around" her doctor:** This entry is not included in the published account of Horney's diaries. See Karen Horney, "5. Tagebuch," April 10, 1910, Karen Horney Papers, Manuscripts and Archives, Yale University Library, New Haven, Conn. Quoted by permission of Dr. Marianne Eckhardt.

375 **were insoluble in the analysis:** Horney, K. (1980), *The Adolescent Diaries of Karen Horney*, 267. Quinn has commented on the seemingly poor fit between Horney's emotional state and Abraham's interpretations. See Quinn, S. (1987), *A Mind of Her Own: The Life of Karen Horney*, New York: Summit Books, 159–64.

375 **structure of their neurosis:** Abraham, K., Secret Committee RB, I, 80–84. The letter is dated Oct. 13, 1920.

375 **clinically superior to the Professor:** Strachey, J. and A. Strachey (1985), *Bloomsbury/Freud: The Letters of James and Alix Strachey, 1924–1925*, eds. P. Meisel and W. Kendrick, 198, 252; Radó on the other hand, while admiring Abraham's clinical acumen and humane touch, saw his analysis between 1922 and 1924 as a quest for theoretic confirmation; Roazen, P., and B. Swerdloff (1995), *Heresy: Sandor Rado and the Psychoanalytic Movement*, 77.

375 **to his junior colleagues:** Abraham, K. (1919), "A Particular Form of Neurotic Resistance Against the Psycho-Analytic Method," in *Selected Papers of Karl Abraham M.D.*, 303–11.

375 **stages of sexual development:** See, for instance, Freud-Abraham Letters, 303–05. Here Abraham developed the idea that the melancholic fantasizes that he has devoured the loved object, a theory that would become central to the work of Abraham's student Melanie Klein.

376 **incorporation of a psychic object:** Abraham, K. (1916), "The First Pregenital Stage of the Libido," in *Selected Papers of Karl Abraham M.D.*, 248–79.

376 **result of intense repressed sadism:** Abraham, K. (1917), "Ejaculatio Praecox," in Ibid., 280–302.

376 **unconsciously desired to attack men:** Abraham, K. (1920), "Manifestations of the Female Castration Complex," in Ibid., 338–69.

376 **six kinds of love relationships:** Abraham, K. (1924), "A Short Study of the Development of the Libido, Viewed in the Light of Mental Disorders," in Ibid., 418–501.

377 **enlarged that theory:** Fenichel, O. (1924), "Review of '*Psychoanalytische Studien zur Charakterbildung*' by Karl Abraham," *I.J.P.* 6:496–99.

377 **based on differing body types:** Kretschmer, E. (1921), *Körperbau und Charakter: Untersuchungen zum Konstitutionsproblem und zur Lehre von den Temperamenten*, Berlin: Springer.

377 **to develop his typology:** Jung, C. G. (1921), *Psychologische Typen*, Zurich: Rascher Verlag.

377 **character types encountered in analysis:** Freud, S. (1908), "Character and Anal Erotism," *S.E.* 9:169–75; and Freud, S. (1916), "Some Character-Types Met with in Psychoanalytic Work," *S.E.* 14:309–36.

377 **bad character was beyond redemption:** See, for example, Freud, S. (1908), "Character and Anal Erotism," *S.E.* 9:175; also Weiss, E. (1970), *Sigmund Freud as a Consultant: Recollections of a Pioneer in Psychoanalysis*, New York: Intercontinental Medical Book Corporation, 27.

378 **character types were convincingly human:** Abraham, K. (1921), "Contribution to the Theory of the Anal Character," in *Selected Papers of Karl Abraham M.D.*, 370–92.

378 **demanding, aggressive, and devouring:** Abraham, K. (1924), "The Influence of Oral Erotism on Character-Formation," in Ibid., 393–406.

378 **but also for character analysis:** Abraham, K. (1925), *Psychoanalytische Studien zur Charakterbildung*, Leipzig: Internationaler Psychoanalytischer Verlag.

378 **essential story of a patient":** Roazen, P., and B. Swerdloff (1995), *Heresy: Sandor Rado and the Psychoanalytic Movement*, 88.

379 **worst blow psychoanalysis had yet faced:** Jones, E. (1926), "Karl Abraham 1877–1925," *I.J.P.* 7:155–81.

379 **"had made his own":** Glover, E. (1928), "Review of *Selected Papers of Karl Abraham*," *I.J.P.* 9:123.

379 **maintaining the center themselves:** Freud-Ferenczi Letters, III, 241–43, 246.

380 **also to biological science:** Horney, K. (1924), "On the Genesis of the Castration Complex in Women," *I.J.P.* 5:49–50.

381 **develop castration terrors:** Ibid., 62. This paper was delivered at the Salzburg Congress in 1922. One of Horney's biographers suggests that Freud adopted Horney's idea without attribution when he would publish on the differences between the sexes a year later; see Quinn, S. (1988), *A Mind of Her Own: The Life of Karen Horney*, New York: Summit Books, 213–14.

381 **papers on masculine women:** Horney, K. (1926), "The Flight from Womanhood: The Masculinity-Complex in Women, as Viewed by Men and by Women," *I.J.P.* 7:324–39.

381 **new physics of relativity:** Radó, S. (1922), "The Paths of Natural Science in the Light of Psychoanalysis," in *Psychoanalysis of Behavior*, New York and London: Grune & Stratton, 1956, 3–15.

381 **might account for drug addiction:** Radó S. (1926), "The Psychic Effect of Intoxicants:

An Attempt to Evolve a Psychoanalytical Theory of Morbid Craving," in Ibid., 25–39.

381 **dangers she was saving him from:** Radó, S. (1927), "An Anxious Mother: A Contribution to the Analysis of the Ego," in Ibid., 40–46.

381 **bad mother in causing psychopathology:** Radó, S. (1927), "The Problem of Melancholia," in Ibid., 47–63.

382 **go to Berlin and study psychoanalysis:** Alexander, F. (1940), "A Jury Trial of Psychoanalysis," *J. Abn. and Soc. Psychology*, 35:306–13.

382 **that old fear: castration anxiety:** Alexander, F. (1923), "The Castration Complex in the Formation of Character," in *The Scope of Psychoanalysis, 1921–1961: Selected Papers of Franz Alexander*, New York: Basic Books, Inc., 1961, 3–30.

383 **seek to repair that structure:** Alexander, F. (1925), "A Metapsychological Description of the Process of Cure," *I.J.P.* 6:.13–34.

383 **all future psycho-analytic therapy":** Ibid., 32.

384 **before the ink dried:** Alexander, F. (1925), Review of "Entwicklungsziele der Psychoanalyse," *I.J.P.* 6:484.

384 **"I liked Alexander's critique very much":** Freud-Ferenczi Letters, III, 213.

384 *Psychoanalysis of the Total Personality:* Alexander, F. (1927), *The Psychoanalysis of the Total Personality: The Application of Freud's Theory of the Ego to the Neuroses*, New York: Nervous and Mental Disease Publishing Co., 1930.

384 **" 'Total Director' of the 'Total Press' ":** Bluma Swerdloff's oral history with Sándor Radó was published as Roazen, P., and B. Swerdloff (1995), *Heresy: Sandor Rado and the Psychoanalytic Movement*, 78.

384 **"a new era in psychoanalysis":** Alexander, F. (1927), *The Psychoanalysis of the Total Personality*, xx, 2, 55.

384 **whole psychic apparatus":** Ibid., 3.

384 **"I," the long-suffering citizen:** Ibid., 50, 55.

385 **"justification for commiting it":** Ibid., 94.

385 **study of the mind:** (1953), *Psychoanalysis as Seen by Analyzed Psychologists*, Washington, D.C.: American Psychological Association, Inc., 314.

386 **adopted a different approach:** Freud-Ferenczi Letters, III, 239.

387 **they knew everything:** Jack Rubins interview with Abram Kardiner, Karen Horney Papers, Manuscripts and Archives, Yale University Library, New Haven, Conn.

387 **best Analytic Institute in the World":** Ives Hendrick's correspondence cited in Gifford, S. "Ives Hendrick Abroad," unpublished ms., 104–09. My thanks to Dr. Sanford Gifford for making this manuscript available to me.

387 **Abraham's contributions to "I" psychology:** (1930), *Zehn Jahre Berliner Psychoanalytisches Institut*, Wien: Internationaler Psychoanalytischer Verlag, 41.

387 **shaping of the "I" by identifications:** Fenichel, O. (1923), "Psychoanalysis and Metaphysics: A Critical Inquiry," in *The Collected Papers of Otto Fenichel*, eds. H. Fenichel and D. Rapaport, vol. 1, New York: W. W. Norton, 1953, 8–26; Fenichel, O. (1925), "Introjection and the Castration Complex," in Ibid., 39–70; Fenichel, O. (1925), "The Clinical Aspect of the Need for Punishment," in Ibid., 71–96; and Fenichel, O. (1926), "Identification," in Ibid., 97–112.

387 **did not need a didactic analysis:** Ekstein, R. (1966), "Siegfried Bernfeld, 1892–1953," in *Psychoanalytic Pioneers*, eds. Alexander, F., S. Eisenstein, M. Grotjahn, New York: Basic Books, 418.

388   **some medical illnesses:** On Schloss Tegel, see Simmel, E. (1929), "Psycho-Analytic Treatment in a Sanatorium," *I.J.P.* 10: 70–89.

388   **conducted in the clinic:** Radó, S. (1929), "German Psycho-Analytical Society," *Bul. Int. Psychoanal. Assn.* 10: 532–36.

388   **mix of Marxism and psychoanalysis:** Brecht, K., V. Friedrich, et al., eds. (no date), *"Here Life Goes on in a Most Peculiar Way . . .": Psychoanalysis Before and After 1933*, trans. C. Trollope, Hamburg: Kellner, 56–57. Also see (1928), *Bul. Int. Psychoanal. Assn.* 9:390.

388   **criminals and delinquents:** The result of a 1929 lecture series was Alexander, F. and H. Staub (1931), *The Criminal, the Judge, and the Public: A Psychological Analysis*, New York: Macmillan. On Staub, see Alexander, F. (1943), "Hugo Staub—1886–1942," *Psychoanalytic Quarterly* 12:100–05.

388   **Poliklinik's, 1,955 consultations and 721 analyses:** The diagnoses of these patients were revealing. While psychoneuroses and melancholia were common diagnoses, the second most common diagnosis in Berlin was a problem with the "I" or neurotic inhibition, a diagnosis the Viennese didn't use. Furthermore, in Berlin it was common to receive a diagnosis of character "disruption" (*Charakterstörungen*), while the Viennese diagnosed nothing of that sort and very rarely considered their patient's primary problem to have anything to do with their character. (1930), *Zehn Jahre Berliner Psychoanalytisches Institut*, Wien: Internationaler Psychoanalytischer Verlag, 16, compare with (1932), *Zehn Jahre Wiener Psychoanalytisches Ambulatorium (1922–1932)*, Wien: Internationaler Psychoanalytischer Verlag, 10.

388   **courses offered to teachers and educators:** (1930), *Zehn Jahre Berliner Psychoanalytisches Institut*, 7–74.

389   **membership levels only in 1926:** Mühlleitner, E., and J. Reichmayr (1997), "Following Freud in Vienna: The Psychological Wednesday Society and the Viennese Psychoanalytical Society 1902–1938," *Int Forum Psychoanal* 6:75.

390   **Vienna Technical Seminars were born:** Reich, W. (1973), *The Function of the Orgasm: Sex-Economic Problems of Biological Energy*, trans. V. Carfagno, New York: Simon and Schuster, 59–60. Also see Bibring-Lehner, G. (1932), "Seminar for the Discussion of Therapeutic Technique," *I.J.P.* 13: 257–59.

390   **maze of metapsychological speculations":** Reich, W. (1973), *The Function of the Orgasm*, 60.

390   **after only one semester:** Lobner, H. (1978), "Discussions on Therapeutic Technique in the Vienna Psycho-Analytic Society (1923–1924)," *Sigmund Freud House Bulletin* 2:31.

390   **analysis in the world:** Reich, W. (1924), "On Genitality: From the Standpoint of Psychoanalytic Prognosis and Therapy," in *Early Writings*, trans. P. Schmitz, New York: Farrar, Straus & Giroux, 1975, 158–79.

390   **not in the Berlin lexicon:** Compare the statistics listed in (1930), *Zehn Jahre Berliner Psychoanalytisches Institut*. Wien: Internationaler Psychoanalytischer Verlag, 16, with (1932), *Zehn Jahre Wiener Psychoanalytisches Ambulatorium, (1922–1932)*, Wien: Internationaler Psychoanalytischer Verlag, 10.

390   **result of repressed sexuality:** Reich, W. (1926), "The Sources of Neurotic Anxiety: A Contribution to the Theory of Psycho-Analytic Therapy," *I.J.P.* 7: 380.

391   **more discussion would be profitable:** Lobner, H. (1978), "Discussions on Therapeutic Technique in the Vienna Psycho-Analytic Society (1923–1924)," *Sigmund Freud House Bulletin* 2:20–21.

391 **their own clinical debacles:** Reich, W. (1927), "Bericht über das 'Seminar für psycho-analytische Therapie' am Psychoanalytischen Ambulatorium in Wien (1925/26)," *I.Z.P.* 13: 241–44.

391 **focus on transference and resistance:** Reich, W. (1925), "A Hysterical Psychosis in Statu Nascendi," in *Early Writings*, 1975, 235–36.

392 **succeed in modifying neurotic character":** Ibid., 238. Also Reich, W. (1967), *Reich Speaks of Freud*, eds. M. Higgins and C. Raphael, New York: Farrar, Straus & Giroux, 146.

392 **imperious rule of Paul Federn:** On Federn's behavior see, for example, the unflatter-ing portrait in Menaker, E. (1989), *Appointment in Vienna: An American Psychoanalyst Recalls Her Student Days in Pre-War Austria*, New York: St. Martin's Press, 65–66.

392 **focused intently on resistances:** Reich, W. (1927), "Bericht über das 'Seminar für psy-choanalytische Therapie' am Psychoanalytischen Ambulatorium in Wien (1925/26)," *I.Z.P.* 13:243.

393 **unconscious contents that changed nothing:** Reich, W. (1973), *The Function of the Orgasm*, 120–121.

393 ***one's own associations, one's own self":*** Landauer, K. (1924), " 'Passive' Technique: On the Analysis of Narcissistic Illnesses," in the superb compendium, Bergmann, M. S. and F. R. Hartman, eds. (1976), *The Evolution of Psychoanalytic Technique*, New York: Columbia University Press, 175.

393 **latent hatred would destroy treatments:** Reich, W. (1973), *The Function of the Orgasm*, 118, 244.

393 **Richard Sterba, mined this thought:** Sterba, R. F. (1927), "Über latente negative Übertragung; Aus dem 'Seminar für psychoanalytische Therapie' in Wien," *I.Z.P.* 13:160–65.

394 **these hidden hatreds:** Reich, W. (1927), "Zur Technik der Deutung und der Wider-standsanalyse," *I.Z.P.* 13:141–59; Reich, W. (1928), "Über Charakteranalyse," *I.Z.P.* 14:180–96; Reich, W. (1929), "Der genitale und der neurotische Charakter," *I.Z.P.* 15:435–55; Reich, W. (1930), "Über kindliche Phobie und Charakterbildung," *I.Z.P.* 16:285–300.

394 **focus on the patient's resistance:** Reich, W. (1933), *Charakteranalyse: Technik und Grundlagen für Studierende und praktizierende Analytiker*, Wien: Im Selbstverlage des Verfassers, 36–55.

395 **it was "character armor":** His phrase was "Charakterliche Panzerung"; Ibid., 57.

395 **attack on Franz Alexander:** Reich, W. (1928), "Criticism of Recent Theories of the Problem of Neurosis," *I.J.P.* 9:227–40.

395 **He left it at that:** Alexander, F. (1928), "Reply to Reich's Criticism," *I.J.P.* 9:243.

395 **death drive was just a hypothesis:** Reich, W. (1973), *The Function of the Orgasm*, 128–29.

396 **his battle was not with Freud:** Reich, W. (1928), "Criticism of Recent Theories of the Problem of Neurosis," *I.J.P.* 9:228.

396 **landmark study, *Character Analysis*:** Reich, W. (1933), *Charakteranalyse: Technik und Grundlagen für Studierende und praktizierende Analytiker*, Wien: Im Selbstverlage des Verfassers.

396 **"the founder of modern technique":** Reich, W. (1967), *Reich Speaks of Freud*, 66.

396 **so much as disorders of character:** Angel, K., E. Jones, et al., eds. (2003), *European Psychiatry on the Eve of War: Aubrey Lewis, the Maudsley Hospital, and the Rockefeller Foundation in the 1930s*, London: The Wellcome Trust, 110.

397 **manner with which he applied it:** See, for example, Sterba, R. F. (1982), *Reminiscences of a Viennese Psychoanalyst*, 82–87, and Deutsch, H. (1973), *Confrontations with Myself: An Epilogue*, New York: W. W. Norton, 157.

397 **opposed by Federn behind the scenes:** Letter from Wilhelm Reich to Paul Federn, February 12, 1926, in Reich, W. (1973), *The Function of the Orgasm*, 148–52.

397 **Reich did not:** See Nunberg, H. (1926), "The Sense of Guilt and the Need for Punishment," in *Practice and Theory of Psychoanalysis*, New York: Nervous and Mental Disease Monographs, 1948, 89–101; also see Federn, P. (1927), "Narcissism and the Structures of the Ego," in *Ego Psychology and the Psychoses*, New York: Basic Books, 1952, 38–59, and Federn, P. (1931), "Die Wirklichkeit des Todestriebes: Zu Freud's 'Unbehagen in der Kultur,' " *Almanach der Psychoanalyse*, 68–97. On this topic, see Sterba, R. F. (1982), *Reminiscences of a Viennese Psychoanalyst*, 75–77.

397 **one-dimensional models of psychoanalytic theory:** Nunberg, H. (1928), "Problems of Therapy," in *Practice and Theory of Psychoanalysis*, New York: Nervous and Mental Disease Monographs, 1948, 105.

397 **intellectual life of the association":** Letter from Sigmund Freud to Paul Federn, November 22, 1928, Sigmund Freud Archives, Library of Congress. Reprinted by permission of Sigmund Freud Copyrights. Also see Reich, W. (1976), *People in Trouble*, New York: Farrar, Straus and Giroux, 75.

398 **bring psychoanalysis to the downtrodden:** Reich discusses joining the *Internationalen Arbeiterhilfe* in Reich, W. (1976), *People in Trouble*, 31. On Reich's radicalization, see Rabinbach, A. (1973), "The Politicization of Wilhelm Reich: An Introduction to 'The Sexual Misery of the Working Masses and the Difficulty of Sexual Reform,' " *New German Critique*, 1:90–110.

398 **most belligerent temperament":** Theodor Reik Interview with Bluma Swerdloff (1966), Columbia University Oral History Research Office Collection, 87.

399 **sexual, mental, and nervous hygiene:** Federn, P. and H. Meng (1926), *Das Psychoanalytische Volksbuch*, Stuttgart-Berlin: Hippokrates-Verlag.

399 **taking place in the Soviet Union:** On psychoanalysis in the Soviet Union, see Etkind, A. (1997), *Eros of the Impossible: The History of Psychoanalysis in Russia*, Boulder, Colo.: Westview Press; and Miller, M. (1998), *Freud and the Bolsheviks: Psychoanalysis in Imperial Russia and the Soviet Union*, New Haven, Conn.: Yale University Press.

399 *The Future of an Illusion:* Freud, S. (1927), "The Future of an Illusion," *S.E.* 21: 5–56.

399 **into a "hornet's nest":** Reich, W. (1976), *People in Trouble*, 74.

399 **seen some seven hundred cases:** See Sharaf, M. (1988), *Fury on Earth: A Biography of Wilhelm Reich*, 129–33. Also see Gardner, S., and G. Stevens (1992), *Red Vienna and the Golden Age of Psychology, 1918–1938*, New York: Praeger.

399 **younger analyst had to say":** Reich, W. (1967), *Reich Speaks of Freud*, 153.

400 **curative power of orgasms:** Sterba, R. F. (1982), *Reminiscences of a Viennese Psychoanalyst*, 87.

400 *Civilization and Its Discontents:* Freud, S. (1930), *Civilization and Its Discontents*, *S.E.* 21:64–145.

400 **be right at any price":** Sterba, R. (1982), *Reminiscences of a Viennese Psychoanalyst*, 112.

400 **began to consider emigration:** Large, D. C. (2000), *Berlin*, New York: Basic Books, 234–38.

401 **criminality and the penal system:** See Gregory Zilboorg's introduction to Alexander,

F., and H. Staub (1931), *The Criminal, the Judge, and the Public: A Psychological Analysis*, New York: The Macmillan Company, v.

401 **bring sexual revolution to the people:** Reich, W. (1976), *People in Trouble*, 135–57.

401 **New York Psychoanalytic Institute:** Roazen, P., and B. Swerdloff (1995), *Heresy: Sandor Rado and the Psychoanalytic Movement*, 84. Sterba, R. (1982), *Reminiscences of a Viennese Psychoanalyst*, 88.

402 **could beget different forms of masochism:** Reich, W. (1932), "Der masochistische Charakter: Eine sexualökonomische Widerlegung des Todestriebes und des Wiederholungszwanges," *I.Z.P.* 18:303–51.

402 **activity of the capitalist system":** Freud-Ferenczi Letters, III, 426.

402 **psychoanalytic thinking to their political beliefs:** See correspondence reprinted in Reich, W. (1967), *Reich Speaks of Freud*, 155; also Freud-Ferenczi Letters, III, 426.

402 **Freud had no right to interfere:** Freud-Ferenczi Letters, III, p. 426.

402 **supposed fellow traveler, Siegfried Bernfeld:** Bernfeld, S. (1932), "Die kommunistiche Diskussion um die Psychoanalyse und Reich's 'Widerlegung der Todesbtriebhypothese,' " *I.Z.P.* 18:352–85.

403 **members of the Berlin Communist Party:** Reich, W. (1976), *People in Trouble*, 193.

403 **rights of the citizenry:** Craig, G. A. (1978), *Germany: 1866–1945*, 576.

403 ***The Mass Psychology of Fascism:*** Reich, W. (1933), *Charakteranalyse: Technik und Grundlagen für Studierende und praktizierende Analytiker*. Wien: Im Selbstverlage des Verfassers. The English translation of this 1933 work comes from the revised third edition; Reich, W. (1949), *Character-Analysis*, trans. T. Wolfe, New York: Farrar, Straus & Giroux. Reich, W. (1933), *Massenpsychologie des Faschismus*, Kopenhagen: Verlag für Sexualpolitik. This English translation also comes from a revised third edition; Reich, W. (1946), *The Mass Psychology of Fascism*, trans. V. Carfagno: New York: Simon & Schuster, 1970.

404 **"has forced psychoanalysis to become political":** Cited in Steiner, R. (1989), " 'It is a New Kind of Diaspora' " *I.R.P.* 16:59.

404 **curtail another analyst's political activities:** Ibid., 60.

11. THE PSYCHO-POLITICS OF FREEDOM

407 **over lap each other":** Brierley, M. (1934), "Present Tendencies in Psychoanalysis," *British J. of Medical Psychology* 14:226.

407 **abstaining from the satisfaction of breathing:** Ferenczi's Clinical Diary, 2.

407 **"new oppositional analysis" was in the works:** Freud-Ferenczi Letters, III, 372–73.

407 **heal them by a "neo-catharsis":** Ibid., 376, 396.

407 **empathy and the analyst's "elasticity":** Ferenczi, S. (1928), "The Elasticity of Psycho-Analytic Technique," in *Final Contributions to the Problems and Methods of Psychoanalysis*, 87–101.

407 **speak more openly and heal:** Ferenczi, S. (1930), "The Principle of Relaxation and Neocatharsis," in Ibid., 108–25.

407 **back to Papa Freud:** Freud-Ferenczi Letters, III, 422.

408 **resistances were clearly in play:** Ferenczi's Clinical Diary, 71, 85. On Elizabeth Severn, see Fortune, C. (1993), "The Case of 'R.N.' Sándor Ferenczi's Radical Experiment in Psychoanalysis," in *The Legacy of Sándor Ferenczi*, eds. L. Aron and A. Harris, Hillsdale, N.J.: Analytic Press, 101–21.

408 **her own howling inner world:** Ferenczi's Clinical Diary, 14, 194.

409   **catharsis as a therapy:** Ferenczi, S. (1930), "The Principle of Relaxation and Neoca-
      tharsis," and his (1933) "Confusion of Tongues Between Adults and Children," in
      Final Contributions to the Problems and Methods of Psycho-analysis, 1980, 108–25, 156–
      67. Also see Freud-Ferenczi Letters, III, 443–44.

409   **consolidating what already exists":** Freud-Ferenczi Letters, III, 441.

409   **Ferenczi to defect like Rank:** Ibid., 441, 445.

409   **would mar its reception:** Ferenczi's Clinical Diary, 219.

409   **would not be published for fifty-two years:** After his death from pernicious anemia,
      Ferenczi also suffered the indignity of being called psychotic by his analysand and
      rival, Ernest Jones. See Haynal, A. E. (2002), *Disappearing and Reviving: Sándor Fe-
      renczi in the History of Psychoanalysis*, London and New York: Karnac.

409   **contests that would define psychoanalysis:** On the Budapest School, see Moreau-
      Ricaud, M. (1996), "The Founding of the Budapest School," and Vikár, G. (1996),
      "The Budapest School in Psychoanalysis," in *Ferenczi's Turn in Psychoanalysis*, eds. P.
      Rudnytsky, A. Bókay, and P. Giamperio-Deutsch, New York: New York University
      Press, 41–59, 60–76.

409   **except the Jewish Hospital:** Large, D. C. (2000), *Berlin*, New York: Basic Books, 291.

410   **represent the institute:** (1934), "Reports of Proceedings of Societies," *Bul. Int. Psycho-
      anal. Assn.* 15: 525–34.

410   **would not be psychoanalysts by profession":** Fenichel Rundbriefe, I, 35.

410   **go in hopes of appeasement?:** Ibid.

411   **International Training Commission (I.T.C.):** Schrötter, M. (2002), "Max Eitingon
      and the Struggle to Establish an International Standard for Psychoanalytic Training
      (1925–1929)," *I.J.P.* 83: 875–93.

411   **psychoanalyist had been answered:** Max Eitingon's I.T.C. report is contained in
      Freud, A. (1929), "Report of the Eleventh International Psycho-Analytical Congress,
      Bull. of the I.J.P. 10:505

412   **where lay members could be trained:** Hitschmann, E. (1932), "A Ten Years' Report
      of the Vienna Psycho-Analytical Clinic," *I.J.P.* 13:245, 254.

412   **Karl Abraham in Berlin:** Theodor Reik Interview with Bluma Swerdloff, 1966, Co-
      lumbia University Oral History Research Office Collection, 2.

412   **"The Question of Lay Analysis":** Freud, S. (1926), "The Question of Lay Analysis,"
      S.E. 20, 183–258. On the history of the debates about lay analysis, see Wallerstein,
      R. S. (1998), *Lay Analysis: Life Inside the Controversy*, Hillsdale, N.J. and London, The
      Analytic Press.

413   **purely medical specialty":** Freud-Ferenczi Letters, III, 339.

413   **but a gigantic mistake":** Cited in Gay, P. (1988), *Freud: A Life for Our Time*, London
      and New York: W. W. Norton, 563.

413   **published their positions:** (1927), "Discussion: Lay Analysis," *I.J.P.* 8:174–283.

414   **assistants supervised by physicians:** Jones, E. (1927), "Discussion: Lay Analysis," *I.J.P.*
      8:173–76, 181, 195–96.

414   **treatment of an illness:** Reik, T. (1927), "Discussion: Lay Analysis," *I.J.P.* 8:241–44.
      Overall, thirteen of the personal commentaries and the statement put out by the New
      York Society went against lay analysis, while eleven individuals and the Hungarian
      contingent voiced their approval.

414   **"definite position on the topic":** Freud, A. (1929), "Report on the Eleventh Interna-
      tional Psycho-Analytical Congress, *Bull. Int. Psa. Assoc.*, 10:506.

414 **committee was at an impasse:** Ibid., 503–19.

414 **professional group, the I.P.A.:** (1929), "Report of Psycho-Analysis Committee," *British Medical Journal*, Appendix II, June 29, 1929, 266, 277.

414–15 **discern true analysts from wild ones:** Ibid., 262–70.

415 **new rules solved nothing:** (1935), "Standing Rules of the I.T.C. in Relation to Training Institutes and Training Centres (Lucerne Standing Rules)," *Bull. Int. Psa. Assoc.* 16:245–46.

415 **American secession hung in the air:** Glover, E. (1937), "I. Conclusion of Report of the Fourteenth International Psycho-Analytical Congress," Ibid., 18:346–58; also see Wallerstein, R. S. (1998), *Lay Analysis: Life Inside the Controversy*, 27–49.

416 **Deep in the Germanic psyche:** Jung, C. G. (1934), "The State of Psychotherapy Today," *C.W.* 10, 166.

417 **Adolf Hitler's *Mein Kampf*:** (1934), "Notes," *Psychoanalytic Quarterly* 3:150–51.

417 **same party, the Fatherland Front:** Gardner, S., and G. Stevens (1992), *Red Vienna and the Golden Age of Psychology, 1918–1938*, New York; Westport, Ct.; London: Praeger, 67–72. The group's name was the "Vaterländische Front."

417 **any better than oil and water":** Glover, E. (1934), "Report of the Thirteenth International Psycho-Analytical Congress," *Bull. Int. Psa. Assoc.* 15:487.

418 **Aryanization of the Berlin Institute:** Ibid., 513.

418 **establish our political nonpartisanship":** Freud-Ferenczi Letters, III, 425.

418 **done little to advance the field:** Reich W. (1927), "Discussion: Lay Analysis," *I.J.P.* 8:252.

419 **way of corroboration and amplification":** Obendorf, C. (1927), "Discussion: Lay Analysis," *I.J.P.* 8:206.

419 ***Moses and Monotheism:*** Freud, S. (1939), "Moses and Monotheism," *S.E.* 23:98–101.

419 **"both unproven and biologically improbable":** Jones, E. (1928), "Review of 'Sex and Repression in Savage Society,' " *I.J.P.* 9:370.

420 **cognitive, linguistic, and ethical development:** Hug-Hellmuth, H. (1913), *A Study of the Mental Life of the Child*, trans. J. Putnam and M. Stevens, Washington: Nervous and Mental Disease Publishing Company, 1919, ix. She also appears in the literature as Hermine von Hug-Hellmuth.

421 **letter from Freud as its introduction:** Anonymous (1919), *Tagebuch eines halbwüchsigen Mädchens*, Leipzig: Internationaler Psychoanalytischer Verlag. For Freud's letter to Hug-Hellmuth, see Freud, S. (1919), "Letter to Dr. Hermine von Hug-Hellmuth," *S.E.* 14:341.

421 **subject of psychic development":** Hug-Hellmuth, H. (1920), "Child Psychology and Education," *I.J.P.* 1:320.

421 **editor of the volume:** Hug-Hellmuth, H., ed. (1922), *Tagebuch eines halbwüchsigen Mädchens*, Leipzig: Internationaler Psychoanalytischer Verlag.

421 **believe she did, to support Freud:** Geissman, C., and Geissman, P. (1992), *A History of Child Psychoanalysis*, 59.

421 **too strict, too lenient, or inconsistent:** Hug-Hellmuth, H. (1920), "Child Psychology and Education," *I.J.P.* 1:320. Also see (1920), "Abstracts from the Proceedings of the Sixth International Psycho-Analytical Congress—Held at the Hague, September 8th to 12th, 1920," *I.J.P.* 1:361–62.

422 **remedial work with troubled children:** For a personal reminiscence of this movement, see Willi Hoffer Interview by Bluma Swerdloff, 1968, Columbia University Oral History Research Office Collection.

422   **neurotics on whom its studies began":** Freud, S. (1925), "Preface to Aichorn's *Wayward Youth*," *S.E.* 19:273.

423   **open to nonmedical workers:** Freud, S. (1926), "Dr. Reik and the Problem of Quackery: A Letter to the Neue Freie Presse," *S.E.* 21:247.

423   **number of her papers:** Graf-Nold, A. (1988), *Der Fall Hermine Hug-Hellmuth: Eine Geschichte der frühen Kinder-Psychoanalyse*, München and Wien: Verlag Internationale Psychoanalyse.

423   **Hug-Hellmuth's work at the Ambulatorium:** Hitschmann, E. (1932), "A Ten Years' Report of the Vienna Psycho-Analytical Clinic," *I.J.P.* 13:245–55.

423   **elected to the Society in 1922:** According to her biographer, this was a disguised piece of autobiography, since Anna had yet to see patients; see Young-Bruehl, E. (1988), *Anna Freud: A Biography*, 103.

424   **she had done beautifully:** Freud, S. (1926), "Dr. Reik and the Problem of Quackery: A Letter to the Neue Freie Presse," *S.E.* 21:247.

424   **field of child psychoanalysis herself:** Menaker, E. (1989), *Appointment in Vienna*, 146–47.

424   **precisely to exclude the unwed Anna:** Deutsch, H. (1973), *Confrontations with Myself: An Epilogue*, 167–68.

425   **"supplementary conceptions," she declared:** The paper was published two years later: Freud, A. (1929), "On the Theory of Analysis of Children," *I.J.P.* 10:29–38.

425   **value of the new "I" psychology:** Young-Bruehl, E. (1988), *Anna Freud: A Biography*, 161–62.

426   **pursue child psychoanalytic work:** On Klein's life, see Grosskurth, P. (1987), *Melanie Klein: Her World and Her Work*, Cambridge, Mass.: Harvard University Press. On her intellectual debts to Ferenczi, Abraham, and Jones, see Aguayo, J. (1997), "Historicising the Origins of Kleinian Psychoanalysis," *I.J.P.* 78:1165–1182.

426   **performing child analysis in Budapest:** Abraham, K., Secret Committee RB, I, 65–66. The date is Oct. 6, 1920.

426   **insight into infantile instinctual life":** Freud-Abraham Letters, 471.

426   **anxiety and the development of personality:** Klein, M. (1922), "Eine Kinderentwicklung. Die infantile Angst und ihre Bedeutung für die Entwicklung der Persönlichkeit," lecture delivered in Berlin, Feb., 1922; Melanie Klein Papers, Wellcome Institute for the History of Medicine, London.

426   **Abraham came to her defense:** Strachey, J. and A. Strachey (1985), *Bloomsbury/Freud: The Letters of James and Alix Strachey, 1924–1925*, eds. P. Meisel and W. Kendrick, 145.

426   **proof she was an "imbecile":** Ibid., 263.

426   **open or secret sentimentalist":** Ibid., 146.

427   **"an extraordinarily deep impression":** Freud-Jones Letters, 577.

427   **won over to Klein's views:** Ibid., 617. Also see Freud-Ferenczi Letters, III, 313.

427   **women analysts as teachers:** Klein, M. (1921), "The Development of a Child," in *Love, Guilt and Reparation and Other Works, 1921–1945*, New York: The Free Press, 1984, 22, 53.

427–28 **didn't seem to engender more inhibition:** Klein, M. (1923), "The Role of the School in the Libidinal Development of the Child," in Ibid., 75–76.

428   **theory of the child's inner life:** See for instance, Klein, M. (1939), Unpublished Lecture on Technique, No. 2, p. 4, Melanie Klein Papers, Wellcome Institute for the History of Medicine.

428 **around her to take notice:** Klein, M. (1926), "Infant Analysis," *I.J.P.* 7:34–36.

428 **be in place by that time:** Ibid.

428 **had no opinion on the matter:** Freud-Jones Letters, 579.

429 **he could prove "in detail":** Ibid., 617–18.

429 **advancements in empirical knowledge:** Ibid., 619.

429 **attacks on Anna's little book:** Klein, M., J. Riviere, M. N. Searl, E. Sharpe, E. Glover, E. Jones (1927), "Symposium on Child Analysis," *I.J.P.* 8:339–91.

430 **not analyzing the unconscious:** Ibid., 340–44, 363, 370.

430 **with a long denial:** Freud-Jones Letters, 623–24, 631.

430 **Max Eitingon said to Melanie Klein:** Cited in King, P., and R. Steiner, eds. (1991), *The Freud-Klein Controversies, 1941–45*, New Library of Psychoanalysis, London and New York: Tavistock/Routledge, 90.

431 **world of competing, triangular relationships:** Klein, M. (1928), "Early Stages of the Oedipus Complex," in *Love, Guilt and Reparation and Other Works, 1921–1945*, New York: The Free Press, 1984, 186–98.

431 **projecting their aggression onto others:** Klein, M. (1929), "Personification in the Play of Children," in *Ibid.*, 199–209. By using projection and introjection as the building blocks of the infant psyche, Klein followed Ferenczi's early work; her emphasis on the power of early oral aggression wove in the insights of Karl Abraham.

431 **more a transitional document:** Klein, M. (1932), *The Psycho-Analysis of Children*, trans. A. Strachey and H. A Thorner, New York: The Free Press, 1975.

431 **book on the analytic situation:** Klein, M. (1939), Unpublished lectures on technique, Melanie Klein Archives, Wellcome Institute for the History of Medicine.

431 **Daddy and Mummy's thingummy":** Klein, M. (1932), *The Psycho-Analysis of Children*, 17.

432 **inhabited only by enemies:** Klein, M. (1929), "Personification in the Play of Children," in *Love, Guilt and Reparation and Other Works, 1921–1945*, 208. Also see her (1930), "The Importance of Symbol-Formation in the Development of the Ego," in Ibid., 219–32.

432 **nothing more than damaged little boys:** Jones, E. (1927), "The Early Development of Female Sexuality," *I.J.P.* 8:459–72.

432 **he stood "defenceless":** Freud, S. (1933), "Femininity," *S.E.* 22:126, 132.

432 **girl to her mother:** Freud, S. (1931), "Female Sexuality," *S.E.* 21, 242. Also see Freud, S. (1933), "Femininity," *S.E.* 22:130–31.

432 **to say the least, highly ornate:** Klein, M. (1932), *The Psycho-Analysis of Children*, 194–239.

433 **clumsy and illogical inferences:** Alexander, F. (1933), "Review of *Die Psychoanalyse Des Kindes* by M. Klein," *Psychoanalytic Quarterly* 2:141.

433 **equal to Freud's classic contributions:** Glover, E. (1933), "Review of *The Psycho-Analysis of Children* by Melanie Klein," *I.J.P.* 14:119.

433 **book which it contains":** Strachey, J. (1924), "Review of *A Critical Examination of Psycho-Analysis* by A. Wohlgemuth," *I.J.P.* 5:222.

434 **transference interpretations made therapy effective:** Strachey, J. (1934), "The Nature of the Therapeutic Action of Psycho-Analysis," *I.J.P.* 15:126–59.

436 **guilt for one's own destructiveness:** Klein, M. (1935), "A Contribution to the Psychogenesis of Manic-Depressive States," in *Love, Guilt and Reparation and Other Works, 1921–1945*, 262–89.

436 **fourth and fifth month of life:** Ibid., 285.

436 **assigned to the preverbal child's mind:** It is possible that Klein was led astray by a continued adherence to Lamarckian thinking. In one of her earliest unpublished works, she argued for a similarity between the thinking of young children and primitives, an argument that would lead one to conclude that young children would have the capacity for highly developed fantasy like primitives; see Klein, M. (1921), "Eine Parallele zwischen kindlichen und primitiven Vorstellungen," unpublished ms., Melanie Klein Archives, Wellcome Institute for the History of Medicine.

436 **found this new model untenable:** Roazen, P. (2000), *Oedipus in Britain: Edward Glover and the Struggle over Klein*, New York: Other Press, 75–76.

436 **"no longer the Goddess" of London:** Willi Hoffer interview with Bluma Swerdloff, 1968, Columbia University Oral History Research Office Collection.

437 **said about Melanie Klein:** Angel, K., E. Jones, et al., eds. (2003), *European Psychiatry on the Eve of War*, 94, 110.

437 **at the institute in 1926:** Freud, A. (1927), "Four Lectures on Child Analysis," in *Introduction to Psychoanalysis: Lectures for Child Analysts and Teachers 1922–1935*, New York: International Universities Press, 1974, 3–72. Also Freud, A. (1966), "A Short History of Child Analysis," *Psychoanalytic Studies of the Child* 21:8.

437 **an artist, Erik Homburger Erikson:** See Freud, A. (1992), *Anna Freud's Letters to Eva Rosenfeld*, ed. P. Heller, trans. M. Weigand, Madison, Connecticut: International Universities Press, Inc.; and Blos, P. (no date), "A Contemplative Tale: Autobiographical Notes on How I Became a Psychoanalyst," Peter Blos Archives, Oskar Diethelm Library, Weill Medical College of Cornell University.

437 **began to lecture on pedagogy:** (1929/30) "Lehrausschuss der Wiener Psychoanalytischen Vereinigung," Sigmund Freud Museum, London.

438 **child psychoanalysis continued to grow:** (1932/1933) "Lehrkurse," Wiener Psychoanalytischen Vereinigung, Sigmund Freud Museum, London.

438 **one of Aichhorn's protégés, Kurt Eissler:** Hitschmann, E. (1932), "A Ten Years' Report of the Vienna Psycho-Analytical Clinic," *I.J.P.* 13:254–55.

438 **Hitler's shadow loomed over Austria:** Jeanne Lampl-de Groot, M.D., interview with David Milrod, M.D., March 11, 1973, A. A Brill Library, Archive and Special Collections, New York Psychoanalytic Society and Institute.

438 **Edith Buxbaum, all Viennese:** See Freud, A. (1935), "Introductory Notes," *Psychoanalytic Quarterly*, 4:1.

438 **Glover considered "tame and uninspired":** Glover, E. (1931), "Review of 'Introduction to Psycho-Analysis for Teachers,' by Anna Freud," *I.J.P.* 12:369–70.

438 **altered the psychoanalytic community:** Freud, A. (1936), *Das Ich und die Abwehrmechanismen*, Wien: Internationaler Psychoanalytischer Verlag. The book was first published in English in 1937; in discussing this work, I will refer to Freud, A. (1936), *The Ego and the Mechanisms of Defense*, trans. Cecil Baynes, New York: International Universities Press, 1966.

439 **accounts on the very same issue:** Popper, K. (1962), *Conjectures and Refutations*, New York: Basic Books, 34–36

440 **study of the "I," she wrote:** Freud, A. (1936), *The Ego and the Mechanisms of Defense*, 4.

440 **"health and disease, virtue or vice":** Ibid.

440 **all cases with the "I":** Ibid., 5–6.

441 **"I," and the "Over-I":** Ibid., 28.

442 **solution to all of these problems:** Wälder, R. (1936), "The Principle of Multiple Function: Observations on Over-Determination," *Psychoanalytic Quarterly* 5:49. First published as Wälder, R. (1930), "Das Prinzip der mehrfachen Funktion," *I.Z.P.* 16:285–300. Sometime after his immigration to America, the spelling of Wälder's name became "Waelder."

442 **different aspects of the same whole:** Wälder, R. (1936), "The Principle of Multiple Function," 60.

442 **definitely ascertainable knowledge":** Jones, E. (1938), "Review of *The Ego and the Mechanisms of Defence* by Anna Freud," *I.J.P.* 19:115.

443 **can hope to achieve success:** Ibid.

443 **clarity into the prevailing confusion":** Ibid., 116–17.

443 **psychoanalysts and academic psychologists:** Kris, E. (1938), "Review of *The Ego and the Mechanisms of Defence* by Anna Freud," *I.J.P.* 19:146.

443 **study at the analytic society:** Dora Hartmann, M.D., interview with Stephen Firestein, M.D., Jan. 31, 1973, A. A Brill Library, Archive and Special Collections, New York Psychoanalytic Society and Institute.

444 **had been saying all along:** Marianne Kris, M.D., interview with Robert Grayson, M.D., Nov. 15, 1972, A. A Brill Library, Archive and Special Collections, New York Psychoanalytic Society and Institute.

444 **outside world," she reasoned:** Freud, A. (1936), *The Ego and the Mechanisms of Defense*, 57.

445 **stoked the populace's anti-Semitism:** Craig, G. A. (1978), *Germany: 1866–1945*, 612.

445 **"Auto-da-fé of the Mind":** Roth, J. (2003), *What I Saw: Reports from Berlin, 1920–1933*, trans. M. Hofmann, New York: W. W. Norton, 207. The Nazi institute was called the "Deutsches Institut für Psychologische Forschung und Psychotherapie," and is also referred to in the literature as the Göring Institute.

445 **will step in, as people think":** Freud, S. and A. Zweig (1970), *The Letters of Sigmund Freud and Arnold Zweig*, ed. E. Freud, New York: Harcourt Brace Jovanovich, 64.

445 **under Bolshevism and National Socialism":** Ibid., 108.

445–46 **by the Austrian authorities:** Ibid., 91–92.

446 **Freud wrote wistfully to Zweig:** Ibid., 21.

446 **in order to be punished:** Freud, S. (1916), "Some Character-Types Met with in Psycho-Analytic Work," *S.E.* 14:333.

446 **masochistic need for punishment:** Alexander, F., and H. Staub (1929), *The Criminal, the Judge and the Public*, New York: Macmillan, 1931.

446 **temper their innate aggressiveness:** Rickman, J. (1957), *Selected Contributions to Psycho-Analysis*, London: The Hogarth Press and the Institute of Psycho-Analysis, 45.

446 **matters such as degrading surroundings:** Klein, M. (1934), "On Criminality," British Journal of Medical Psychology, 14:312–315

446 **aggression, delinquency, and crime:** Simmel, E. (1932), "National Socialism and Public Health," *Los Angeles Psychoanalytic Bulletin*, 1989, 25.

447 **nation, church, or army:** This was rendered "Group Psychology and the Analysis of the Ego" by Strachey, but I have taken the liberty of translating the German "Massenpsychologie" more literally as "Mass Psychology"; see Freud, S. (1921), "Group Psychology and the Analysis of the Ego," *S.E.* 18:67–144.

447 **began to practice psychoanalytic sociology:** Storfer, A. J., ed. (1931), "Psychoanalyse der Politik," *Psychoanalytische Bewegung* 3:385–478. This issue of the journal included six articles on the subject by Erich Fromm, Fritz Wittels, Hugo Staub, and others.

447 **lose himself in a mass:** Brecht, K., V. Friedrich, et al., eds. (no date), "*Here Life Goes on in a Most Peculiar Way . . .": Psychoanalysis Before and After 1933*, 60.

447 **lectures in London by Edward Glover:** Freud, S., and A. Einstein (1933), "Why War?" *S.E.* 22:199–215.

448 **their own unconscious self-destructiveness:** Glover, E. (1933), *War, Sadism and Pacifism, Three Essays*, London: George Allen & Unwin Ltd.

448 **land of Goethe and Schiller:** Ibid., 46.

448 **irrationality that had gripped Europe:** Guttman, S. A. (1969), "Robert Waelder 1900–1967," *I.J.P.* 50:269–73; Guttman, S. A. (1986), "Robert Waelder and the Application of Psychoanalytic Principles to Social and Political Phenomena," *Journal of the American Psychoanalytic Association* 34:835–62.

448 **emphasis to biology and fantasy:** Wälder, R. (1937), "The Problem of the Genesis of Psychical Conflict in Earliest Infancy—Remarks on a Paper by Joan Riviere," *I.J.P.* 18:406–73.

449 **decided to become a psychoanalyst:** Heinz Hartmann interview with Bluma Swerdloff, 1963, Columbia University Oral History Research Office Collection, 2.

449 **neurology, psychiatry, and psychoanalysis:** On Schilder, see Ziferstein, I. (1966), "Paul Ferdinand Schilder, 1886–1940," in *Psychoanalytic Pioneers*, eds. F. Alexander, S. Eisenstein, M. Grotjahn, New York: Basic Books, 457–68.

449 *The Foundation of Psychoanalysis:* Hartmann, H. (1927), *Die Grundlagen Der Psychoanalyse*, Leipzig: Georg Thieme.

449 **on the other hand, offered explanations:** In German, understanding is *Verstehen*, while explanation is *Erklären*. The distinction between these two terms was critical to, among others, Hartmann's teacher, Max Weber.

449 **Willi and Hedwig Hoffer:** Deutsch, H. (1973), *Confrontations with Myself*, 166–67.

450 **if he stayed in Austria:** Dora Hartmann, M.D., interview with Stephen Firestein, M.D., January 31, 1973, A. A Brill Library, Archive and Special Collections, New York Psychoanalytic Society and Institute; Heinz Hartmann interview with Bluma Swerdloff, 1963, Columbia University Oral History Research Office Collection, 72.

450 **offer he could not refuse:** Heinz Hartmann interview with Bluma Swerdloff, 1963, Columbia University Oral History Research Office Collection, 57.

450 **sided with "Reich's Bolshevism":** Freud-Ferenczi Letters, III, 433, 439.

450 **values inherent in irrationality":** Wälder, R. (1929), "The Influence of Psychoanalysis and the Outlook on Life of Modern Man," in *Psychoanalysis: Observation, Theory, Application*, ed. S. Guttman, New York: International Universities Press, 1976, 387.

451 **religious and Marxist worldviews:** Then rather self-consciously, the old liberal opined that perhaps the Bolshevik experiment—despite its repressive nature and "disagreeable details"—might hold "the message of a better future"; see Freud, S. (1933) "The Question of a Weltanschauung," in *New Introductory Lectures on Psycho-Analysis, S.E.* 22:181.

451 **clearly state their assumptions:** Hartmann, H. (1933), "Psychoanalyse und Weltan-schauung," *Psychoanalytische Bewegung* 5:416–29.

451 **more clearly comprehend their world:** Mannheim, K. (1929), *Ideology and Utopia: An Introduction to the Sociology of Knowledge*, trans. L. Wirth and E. Shils, New York: Harcourt, Brace and World, 1936.

452 **profoundly influenced by Mannheim:** See Hartmann, H. (1939), *Ego Psychology and the Problem of Adaptation*, trans. D. Rapaport, New York: International Universities Press, 1958, 70–71. This influence was discussed by Friedman, L. (1989), "Hart-mann's 'Ego Psychology' and the Problem of Adaptation," *Psychoanalytic Quarterly* 58:526–50.

452 **psychiatric debates on heredity:** Hartmann, H. (1934–35), "Psychiatric Studies of Twins," in *Essays on Ego Psychology: Selected Problems in Psychoanalytic Theory*, New York: International Universities Press, 1964, 420.

452 **academic psychology with psychoanalysts:** Heinz Hartmann interview with Bluma Swerdloff, 1963, Columbia University Oral History Research Office Collection, 58–59.

452 **"*I*" Psychology and the Problem of Adaptation:** This 1937 lecture was reported on by Wälder, R. (1939), "Vienna Psycho-Analytical Society," *Bull. Int. Psa. Assoc.*, 20:136. Hartmann, H. (1939), "Ich-Psychologie und Anpassungsproblem," *Inter Zeit. f. Psa. und Imago*, 24:62–135. I will refer to the English translation: Hartmann, H. (1939), *Ego Psychology and the Problem of Adaptation*, New York: International Universities Press, 1958.

452 **but rather individual adaptation:** Hartmann, H. (1939), *Ego Psychology and the Problem of Adaptation*, 24.

452 **impact on their character:** Hartmann, H. (1934–35), "Psychiatric Studies of Twins," 444.

453 **uncompromised by drives and defenses:** Hartmann, H. (1939), *Ego Psychology and the Problem of Adaptation*, 8.

454 **capacity for creativity and adaptation:** Ibid., 12.

454 **in terms of human freedom:** Wälder presented this work at the Lucerne Congress; see Glover, E. (1934), "Report of the Thirteenth International Psycho-Analytical Congress," *Bul. Int. Psychoanal. Assn.* 15:485–524. It was published two years later; Wälder, R. (1936), "The Problem of Freedom in Psycho-Analysis and the Problem of Reality-Testing," *I.J.P.* 17:89–108.

454 **might mean something different in another:** Freud, S. (1933), *New Introductory Lectures on Psychoanalysis, S.E.* 22:80. Strachey translated "Wo Es war, soll Ich warden" into "Where id was, there ego shall be." I use Jonathan Lear's retranslation. For his discussion of this phrase, see Lear, J. (1990), *Love and Its Place in Nature: A Philosophical Interpretation of Freudian Psychoanalysis*, New York: Farrar, Straus & Giroux, 156–82.

455 **aggressive and political complexion":** Hartmann, H. (1939), *Ego Psychology and the Problem of Adaptation*, 9.

455 **made scientific work possible:** Glover, E. (1937), "Report of the Fourteenth International Psycho-Analytic Congress," *Bull. Int. Pa. Assoc.* 18:72–107, 72.

455 **editor of the *Zeitschrift*:** Freud-Ferenczi Letters, III, 433.

456 **in the United States was abysmal:** On the fate of the Marxist analysts, see Jacoby, R.

(1983), *The Repression of Psychoanalysis: Otto Fenichel and the Political Freudians*, New York: Basic Books; and Harris, B., and A. Brock (1991), "Otto Fenichel and the Left Opposition in Psychoanalysis," *Journal of the History of the Behavioral Sciences* 27: 157–65.

457   **preparation for the Flood:** Fenichel, O. (1934), *Outline of Clinical Psychoanalysis*, New York: The Psychoanalytic Quarterly Press and W. W. Norton.

457   **they had fallen into depressions:** Fenichel Rundbriefe, I, 205.

457   **in the I.P.A., like Fenichel:** Ibid., 199–200. On the conflict between Fenichel and Reich, see Harris, B., and A. Brock (1992), "Freudian Psychopolitics: The Rivalry of Wilhelm Reich and Otto Fenichel, 1930–1935," *Bulletin of the History of Medicine* 66:577–611.

457   **preserve psychoanalysis for the future:** Fenichel Rundbriefe, I, 181–82, 208, 225.

457   **on weekends at the Society:** Ibid., 441.

458   **actual technique in practice:** Glover, E., and M. Brierley, eds. (1940), *An Investigation of the Technique of Psycho-Analysis*, Baltimore: The Williams & Wilkins Company, v.

458   **method of shattering character defenses:** Fenichel, O. (1936), *Problems of Psychoanalytic Technique*, New York: The Psychoanalytic Quarterly, 1941, 9, 34, 44, 47, 86, 103–05. The German text has been reprinted with commentary in Fenichel, O. (2001), *Otto Fenichel: Probleme der Psychoanalytischen Technik*, eds. Giefer, M. and E. Mühlleitner, Giessen: Psychosozial-Verlag. Also see Fenichel Rundbriefe, I, 207.

458   **not always able to do so:** Fenichel, O. (1936), *Problems of Psychoanalytic Technique*, 70, 73–74.

458   **judge reality and act accordingly:** Ibid., 90.

458   **"the social structure of the particular society":** Fenichel, O. (1940), "Review of 'Ich-Psychologie und Anpassungsproblem' by Heinz Hartmann," *Psychoanalytic Quarterly* 9:445.

459   **of Glover and Ella Sharpe:** Fenichel, O. (1936), *Problems of Psychoanalytic Technique*, 107.

459   **called death by silence:** Fenichel, O. *Rundbriefe*, I, 567.

459   **over the next fifty years:** The book appeared in installations; see Fenichel, O., (1938), "Problems of Psychoanalytic Technique," *Psychoanalytic Quarterly* 7:421–42, 8:57–87, 164–85, 303–24, 438–470.

460   **"Anna-Antigone" as he once called her:** Freud-Zweig Letters, 106.

460   **leaving behind a holocaust:** Craig, G. (1978), *Germany 1866–1945*, 635–37.

460   **no more in Vienna:** Sterba, R. F. (1982), *Reminiscences of a Viennese Psychoanalyst*, 156–66. The extent to which any semblance of psychoanalysis continued under Göring has been a source of controversy among scholars. See Cocks, G., (1997), *Psychotherapy in the Third Reich: The Göring Institute*, New Brunswick: Transaction Pub.; and Goggin, J., and E. Goggin (2001), *Death of a "Jewish Science": Psychoanalysis in the Third Reich*, W. Lafayette, Ind.: Purdue U. Press.

462   **looked broken and depressed:** Walter Briehl interview with Myron Scharaf, March 1972, Myron Scharaf Papers. Courtesy of Giselle Scharaf.

463   **intelligent but not always learned:** Freud, S. (1926), *The Question of Lay Analysis*, S.E. 20:195.

463   **call "das Es" "the Id":** Strachey, J. and A. Strachey (1985), *Bloomsbury/Freud: The Letters of James and Alix Strachey, 1924–1925*, eds. P. Meisel and W. Kendrick, 83.

464 *The I and the It* might be better": Ibid., 307.

464 appeared in 1927: Freud, S. (1923), *The Ego and the Id*, trans. J. Riviere, London: Hogarth Press, 1927.

464 *Ich* as *le moi*, or "the me": Roudinesco, E. (1986), *LaBataille de cent ans: Histoire de la psychanalyse en France.1, 1885–1939*, Paris: Seuil. 376–78.

465 this was a tremendous feat: Glover, E. (1939), "Fifteenth International Psycho-Analytical Congress," *Bul. Int. Psychoanal. Assn*. 20: 116–127, 124.

465 Ernest Jones and the I.P.A.: See Steiner, R. (1989), " 'It Is a New Kind of Diaspora,' " *I.R.P.* 16:35–72.

465 for scientific purposes only": Glover, E. (1939), "Fifteenth International Psycho-Analytical Congress," *Bul. Int. Psychoanal. Assn*. 20:121.

466 future of the I.P.A looked dismal: Ibid., 121–22, 124.

466 lethal dose of morphine: Freud's assisted suicide was first determined by Peter Gay; see Gay, P. (1988), *Freud: A Life for Our Time*, London and New York: W. W. Norton, 648–51.

466 all fall under Nazi rule: Craig, G. A. (1978), *Germany: 1866–1945*, 743.

EPILOGUE

467 Ernest Jones and Eduard Hitschmann remained: Freud, A. (1949), "Bulletin of the International Psycho-Analytical Association," *Bull. of I.P.A* 30:178–208.

468 center of the psychoanalytical movement": Freud Letters, 458.

469 Mrs. K's ideas are totally subversive: Cited in King, P., and R. Steiner, eds. (1991), *The Freud-Klein Controversies, 1941–45*. London: Tavistock/Routledge, 32–33.

470 *International Journal for Psychoanalysis*: Ibid., 40–41.

470 as "psychoanalytic science": Ibid., 59.

470 science of psychoanalysis, founded by Freud": Ibid., 63.

470 ensure against such exploitation: Ibid., 53.

471 "draw the line somewhere": Ibid., 84–6.

471 "science founded by Freud": Ibid., 89.

471 who wants to live in it": Ibid., 77, 100.

472 differences over theory without fostering anarchy: Ibid., 602–09.

472 not other major European psychoanalysts: Ibid., 631–32.

472 her followers a home: Ibid., 907.

473 academic psychiatry, and medicine: See Burnham, J. C. (1958), "Psychoanalysis in American Civilization Before 1918," Ann Arbor, Mich.: University Microfilm; Hale, N. (1995), *Freud and the Americans: The Beginning of Psychoanalysis in the United States, 1876–1917*, New York and Oxford: Oxford University Press; and Hale, N. (1995), *The Rise and Crisis of Psychoanalysis in the United States: Freud and the Americans, 1917–1985*, New York: Oxford University Press.

473 assume a "more scientific character": Alexander, F. (1938), "Psychoanalysis Comes of Age," *Psychoanalytic Quarterly* 7:299–306.

474 seed analysts throughout the country: American Psychoanalytic Association Archives, Oskar Diethelm Library, Weill Medical College of Cornell University, File Box 9–10.

474 relocated somewhere in the United States: Ibid.

474 there had been little original work: Oberndorf, C. P. (1953), *A History of Psychoanalysis in America*, New York: Grune and Stratton, 179.

475 "concentrated in America for many years": Sándor Radó to David Levy, May 27, 1937, David Levy Papers.

475 in the school which he left": Radó, S. (1938), "Scientific Aspects of Training in Psychoanalysis," in *Psychoanalysis of Behavior*. New York and London: Grune & Stratton, 1956, 126.

476 work of Walter B. Cannon: Radó, S. (1939), "Psychoanalytic Conception and Treatment of the Neuroses," *Archives of Neurology and Psychiatry* 42:1195–1198.

477 range of scientific method: See the transcript of the proceedings along with a letter from the secretary of the New York Neurological Society, Clarence Hare to Lawrence Kubie, March 20, 1940, David Levy Papers.

477 he must be opposed: Ibid.

478 so thoroughly dialectic: David Levy to Lawrence Kubie, Nov. 14, 1939, David Levy Papers.

478 superseded by something better: Lawrence Kubie to David Levy, Nov. 17, 1939, David Levy Papers.

478 had become a "little Hitler": Adolph Stern to Lawrence Kubie, undated, circa fall of 1939, David Levy Papers.

478 accusing Levy of "vilifying Federn": Lawrence Kubie to David Levy, May 2, 1940, David Levy Papers.

478 to the Viennese analyst Bertha Bornstein: David Levy to Lawrence Kubie, Oct. 13, 1939, David Levy Papers.

478 would not endorse lay analysis: Author's interview with Henry Nunberg, M.D., July 27, 2006.

478 New York Society meetings than English: Millet, J. A., (1966) "Psychoanalysis in the United States," in *Psychoanalytic Pioneers*, eds. F. Alexander, S. Eisenstein, M. Grotjahn, New York: Basic Books, 557.

479 on certain theoretical issues": Lawrence Kubie to Samuel Atkin, Dec. 11, 1940, David Levy Papers.

479 what I present is psychoanalysis: Horney, K. (1937), *The Neurotic Personality of Our Time*, New York: W. W. Norton & Company Inc., ix.

479 critique of psychoanalytic fundamentals: Horney, K. (1939), *New Ways in Psychoanalysis*, New York: W. W. Norton & Company Inc.

480 scientific society and not a club": Abram Kardiner to Lawrence Kubie, Oct. 24, 1939, David Levy Papers.

480 he asked, what was left?: Fenichel, O. (1940), Review of 'New Ways in Psychoanalysis' by K. Horney," *Psychoanalytic Quarterly* 9:114–121.

480 not truths but different errors: Alexander, F. (1940), "Psychoanalysis Revised" *Psychoanalytic Quarterly* 9:1–36.

480 should recant or resign: Fritz Wittels to Lawrence Kubie, March 13, 1940, David Levy Papers.

481 grossly impeded by the committee: Gregory Zilboorg to David Levy, Dec. 6, 1940, David Levy Papers. The undated students' petition entitled "Resolutions Submitted to the New York Psychoanalytic Society" was enclosed in a letter dated March 25, 1941, from Bernard S. Robbins to David Levy; David Levy Papers.

481 name of scientific freedom, of course: David Levy to Franz Alexander, March 31, 1942; Franz Alexander to David Levy, April 16, 1942, David Levy Papers. On the

founding of the Columbia Center, see Tomlinson, C. (1996), "Sandor Rado and Adolph Meyer: A Nodal Point in American Psychiatry and Psychoanalysis," *I.J.P.* 77:963–82.

482 **gets questions about Freud":** Oral Communication from Theodore Shapiro, 2005.

482 **siding with the orthodox for now:** Fenichel Rundbriefe, II, 1613.

483 **in American psychoanalysis had begun:** See Bergmann, M. S. (2000), "The Hartmann Era and Its Contribution to Psychoanalytic Technique," *The Hartmann Era*, ed. M. S. Bergmann, New York: The Other Press, 1–78.

483 **so often opposed Freud, Ernest Jones:** For an extensively documented account of the censoring of Freud's biography, letters, and papers between 1945 and 1955, see Borch-Jacobsen, M., and Shamdasani, S. (2006), *Le Dossier Freud: Enquête sur l'histoire de la psychoanalyse*, Paris: Les Empêcheurs de penser en rond, 333–432.

483 **suffered from grave mental illness:** Jones claimed that Ferenczi, Fliess, and Rank were psychotic, and that Stekel suffered from moral insanity. Numerous other opponents were dismissed as mentally flawed as well; see Ibid., 394–95. Also see Bonomi, C., (1999), "Flight into Sanity: Jones's Allegation of Ferenczi's Mental Deterioration Reconsidered," *I.J.P.* 80:507–42.

Permissions

1. Quotations from Sigmund Freud's published and unpublished writings are reprinted with the kind permission of Sigmund Freud Copyrights, Paterson Marsh, Ltd.
2. Quotations from unpublished correspondence of Jean-Martin Charcot to Sigmund Freud are reprinted with the permission of the Johns Hopkins University Press.
3. Excerpts from the unpublished letters and diaries of Otto Rank, part of the Otto Rank Papers held by the Rare Book and Manuscript Library of Columbia University, are reprinted with the permission of Ruhama Veltfort and The Trustees of Columbia University.
4. Excerpts from the unpublished correspondence of Eugen Bleuler are reprinted with the gracious permission of Mrs. Tina Joos-Bleuler.
5. Quotes from the published and unpublished writings of C. G. Jung are reproduced here with the kind permission of Mr. Andreas Jung and the Familienarchiv C. G. Jung.
6. Quotes taken from the unpublished correspondence of Adolph Meyer are reprinted with the permission of The Alan Mason Chesney Medical Archives of the Johns Hopkins Medical Institutions.
7. Quotes from the unpublished writing of Karen Horney, held in Manuscripts and Archives, Yale University Library, are published by permission of Dr. Marianne Eckardt and Yale University.
8. Quotes taken from the unpublished letters of James Strachey are printed by permission of The British Library, and The Society of Authors in its capacity as agent for The Strachey Trust.
9. Quotes from Bluma Swerdloff's interviews of Theodor Reik, Willi Hoffer, and Heinz Hartmann are reprinted with the permission of the Columbia University Oral History Research Office, Butler Library, Columbia University.
10. Excerpts from interviews with Dr. Jeanne Lampl-de Groot, Dr. Dora Hartmann, and Dr. Marianne Kris are reprinted with the permission of A.A. Brill Library, Archives and Special Collections, New York Psychoanalytic Society and Institute.
11. I thank the publishers of the *Bulletin of the History of Medicine, History of Psychiatry*, the *International Journal of Psychoanalysis, International Review of Psychoanalysis*, and the *Journal of the American Psychoanalytic Association*, for permission to use material from these prior publications of mine: (1998) "Between Seduction and Libido: Sigmund Freud's Masturbation Hypotheses and the Realignment of His Etiologic Thinking, 1897–1905," *Bulletin of the History of Medicine*, 72:638–662; (1997) "Towards defining the Freudian unconscious: seduction, sexology and the negative of perversion (1896–1905)," *History of Psychiatry*, 8:459–486; (1998) "The Seductions of

History: Sexual Trauma in Freud's Theory and Historiography," *I.J.P.*, 79:857–869; (1992) "A History of Freud's First Concept of Transference," *I.R.P.*, 19:415–432; (1994) "In the Eye of the Beholder: Helmholtzian Perception and the Origins of Freud's 1900 Theory of Transference," *Journal of the American Psychoanalytic Association*, 42:549–80; and (1998) "Dora's Hysteria and the Maturation of Sigmund Freud's Transference Theory: A Historical Interpretation," *Journal of the American Psychoanalytic Association*, 45:1061–1096.

Illustration Credits

1. "Somnambulisme Provoqué, Sommeil." Reprinted from Borneville and Regnard, *Iconographie Photographique de la Salpêtrière*, Paris, Bureaux du Progrès Medical, V. A. Delahaye & Lecrosnier, 1879–1880, Plate 7.
2. Sigmund Freud, 1885. Reprinted courtesy of the United States Library of Congress.
3. A Jewish "neuropathic" family. Reprinted from J. M. Charcot, *Clinique des Maladies du Système Nerveux*, Paris: Felix Alcan, 1893, p. 318.
4. Hermann Ludwig Ferdinand von Helmholtz, 1894. Reprinted frontispiece of John Gray McKendrick, *Hermann Ludwig Ferdinand von Helmholtz*, New York: Longmans, Green & Co., 1899.
5. A Diagram of Conscious Movement as the Result of Cerebral Reflexes by Theodor Meynert, Vienna, 1885. Reprinted from Theodor Meynert *Psychiatry*, New York and London: G. P. Putnam & Sons, 1885, p. 158.
6. Gustav Theodor Fechner. Photogravure from an earlier painting or photograph, 1910. Courtesy of Science Museum/Science and Society Picture Library, United Kingdom.
7. Postcard of male transvestite from the collection of Richard von Krafft-Ebing. Reproduced courtesy of The Wellcome Trust Medical Photographic Library.
8. Diagram showing the origins of Freudian Theory; by the author.
9. Medical students in the dissecting room at the University of Vienna. Reprinted by permission of the Medizinische Universität Wien, Institut für Geschichte der Medizin, Bildarchiv.
10. Karl Kraus, photo by d'Ora, 1908. © IMAGNO/Austrian Archives
11. "The Morning Report" at the Burghölzli, c. 1890. Reprinted from August Forel *Out of My Life and Work* trans. Bernard Miall, New York: W. W. Norton, 1937, Plate 4.
12. Eugen Bleuler, 1902. Reprinted courtesy of the Oskar Diethelm Library of Weill Medical College of Cornell University.
13. Word Associations from the Case of Catterina H. Taken from Franz Riklin "Cases Illustrating the Phenomena of Association in Hysteria" in C. G. Jung, ed., *Studies in Word-Association: Experiments in the Diagnosis of Psychopathological Conditions Carried Out at the Psychiatric Clinic at the University of Zurich Under the Direction of C. G. Jung*, M. D. Eder trans., London, Heinemann, 1918, p. 326.
14. Otto Gross. International Otto Gross Society (*www.ottogross.org*)/Otto Gross Archive, London.
15. Congress of the International Psychoanalytical Association, Weimar, 1911. Courtesy of the United States Library of Congress.

16. Alfred Adler. Reprinted frontispiece from Ansbacher, H. and R. Ansbacher, eds., *The Individual Psychology of Alfred Adler: A Systematic Presentation in Selections from His Writings*, New York: Basic Books, 1956.

17. Wilhelm Stekel. Reprinted courtesy of the Oskar Diethelm Library of Weill Medical College of Cornell University.

18. Carl Gustav Jung and Emma Jung, 1903. Reprinted courtesy of Mr. Andreas Jung and Familienarchiv C. G. Jung.

19. The Battle of Verdun, February 1916. © Hulton-Deutsch Collection/CORBIS

20. Curvatures and Paralyses due to War Neurosis. Taken from G. Roussy and J. L'Hermitte, *The Psychoneurosis of War*, W. B. Christopherson, trans., London: University of London, 1918, Plate 4.

21. Caricatures of figures attending the Psychoanalytic Congress in Salzburg, 1924 by Robert Berény and Olga Székely-Kovacs. Reproduced in *Almanach der Psychoanalyse 1927*, Wien: Internationaler Psychoanalytischer Verlag, 1927, p. 181.

22. Sándor Ferenczi. Reprinted courtesy of the Oskar Diethelm Library of Weill Medical College of Cornell University.

23. Karl Abraham. Reproduced from Theodor Reik, "Gedenkrede über Karl Abraham," *Almanach der Psychoanalyse 1927*, Wien: Internationaler Psychoanalytischer Verlag, 1927, after page 80.

24. Staff of the Vienna Ambulatorium. Courtesy of The Freud Museum, London.

25. Wilhelm and Annie Reich at the beach, 1928. Reproduced courtesy of The New York Psychoanalytic Society and Institute Archive.

26. Erwin Stengel, Grete Bibring, Rudolph Lowenstein, and Wilhelm Reich, Lucerne, 1934. Photograph by Edward Bibring. Courtesy of Boston Psychoanalytic Society and Institute Archives. Also see Gifford, S. D. Jacobs, V. Goldman, eds., (2005), *Edward Bibring Photographs the Psychoanalysts of His Time, 1932–1938*, Giessen: Psychosozial-Verlag.

27. Ernest Jones, Lucerne, 1934. Photograph by Edward Bibring. Courtesy of Boston Psychoanalytic Society and Institute Archives.

28. Anna Freud, Budapest, 1937. Photograph by Edward Bibring. Courtesy of Boston Psychoanalytic Society and Institute Archives.

29. Princess Marie Bonaparte, Melanie Klein, Anna Freud and Ernest Jones, Paris, 1938. Photograph by Edward Bibring. Courtesy of Boston Psychoanalytic Society and Institute Archives.

30. Bertha Bornstein, Otto Fenichel, and an unidentified woman, Marienbad, 1936. Photograph by Edward Bibring. Courtesy of Boston Psychoanalytic Society and Institute Archives.

31. Sigmund and Anna Freud arrive in Paris, 1938. © Hulton-Deutsch Collection/CORBIS.

Index

Note: Page numbers in italic indicate illustrations.